Automotive Gasoline Direct-Injection Engines

Other SAE books on this topic

**Direct Injection Systems for Spark-Ignition
and Compression-Ignition Engines**
Edited by Cornel Stan
(Order No. R-289)

For more information or to order this book, contact SAE at 400 Commonwealth Drive, Warrendale, PA
15096-0001; (724) 776-4970; fax (724) 776-0790; e-mail: publications@sae.org;
web site: www.sae.org/BOOKSTORE.

Automotive Gasoline Direct-Injection Engines

Fuquan (Frank) Zhao
DaimlerChrysler Corporation

David L. Harrington
General Motors Corporation

Ming-Chia Lai
Wayne State University

SAE INTERNATIONAL®

Society of Automotive Engineers, Inc.
Warrendale, Pa.

Library of Congress Cataloging-in-Publication Data

Zhao, Fuquan
    Automotive gasoline direct-injection engines / Fuquan (Frank)
Zhao, David L. Harrington, Ming-Chia Lai.
      p. cm.
    Includes bibliographical references and index.
    ISBN 0-7680-0882-4
    1. Automobiles—Motors—Fuel injection systems. I. Harrington,
D.L. (David L.) II. Lai, Ming-Chai. III. Title.

TL214.F78 Z49 2002
629.25'3—dc21

                    2002017747

Copyright © 2002   Society of Automotive Engineers, Inc.
                           400 Commonwealth Drive
                           Warrendale, PA 15096-0001 U.S.A
                           Phone: (724) 776-4841
                           Fax: (724) 776-5760
                           E-mail: publications@sae.org
                           http://www.sae.org

ISBN 0-7680-0882-4

TL214.F78 Z49 2002
0134109657780
Zhao, Fuquan.

Automotive gasoline
  direct-injection engines
    c2002.

2007 04 16

SAE Order No. R-315

Contents

Foreword

A profound increase in the level of research and development of gasoline direct-injection engines has occurred over the last decade. This is directly attributable to the significant potential of this technology for enhancing engine fuel economy. Numerous innovative strategies for mixture preparation, combustion control and emissions reduction have been proposed and developed by automotive companies, component suppliers and research institutions, with a significant number successfully incorporated into production vehicles in both the Japanese and European markets. Accompanying this work is the continuing generation of a very large volume of technical information, with the growing need for systematic organization, description of fundamental processes, and identification of key trends and insights on technical issues. This book was developed to meet this essential need.

The material in this book constitutes a very comprehensive examination of the fundamental processes that comprise combustion and emissions formation in gasoline direct-injection engines. The authors have done an outstanding job of incorporating all of the experimental and analytical work that has been reported worldwide, and in developing a highly organized treatise on the subject. This provides a very broad knowledge base from which many important observations, trends and guidelines are developed. Additionally, both historical and current trends related to nearly every aspect of the field are analyzed, and all prototype and production gasoline direct-injection engines are discussed in depth and fully documented.

This book will become a primary reference manual that both addresses and documents all aspects of combustion system development, operation and optimization for gasoline direct-injection engines. In addition to the main issues of mixture preparation, combustion, emissions and fuel economy, the related key areas of deposit formation and the fuel system are considered in significant detail. Very extensive consideration of the requirements and the performance characteristics of the fuel spray and injection system is provided, perhaps the most detailed consideration that is available in any publication, with the key metrics clearly identified. The crucial topics of aftertreatment strategies and injector and combustion chamber deposits are addressed in depth, with guidelines that reflect the current state of the art presented and discussed. Many recommendations regarding application guidelines, interpretations of data and parameter measurement considerations have been incorporated, which in my view is an invaluable aid to project engineers in achieving continuing improvements in engine performance, fuel economy and durability.

As the first major book to provide organized and authoritative discussions of both fundamental issues and practical considerations in gasoline direct-injection engine development, there is little doubt that this monumental work will prove to be of significant value in many related fields of endeavor worldwide, including advanced engine development, education, strategic planning, engineering training and academic research. I highly recommend that engineers and researchers first check the related sections of this book before finalizing any project proposal or technical paper in the domain of gasoline direct injection. It is my firm belief that this book will become the primary reference handbook of gasoline direct injection, and one of the most valuable information resources of project engineers, academic and government researchers, engineering managers and authors of technical documents in this discipline. It is a volume that should be on the reference shelf of all professionals who are engaged in projects related to this field.

Hiromitsu Ando, Ph.D.
Executive Officer
Deputy Corporate General Manager
Car Research & Development Center
Mitsubishi Motors Corporation

Preface

The development of four-stroke, spark-ignition engines that are designed to inject gasoline directly into the cylinder is an important worldwide initiative of the automotive industry. The thermodynamic potential of such engines for significantly enhanced fuel economy has led to a significant number of research and development projects that have the goal of understanding, developing and optimizing gasoline direct-injection (G-DI) systems. The processes of fuel injection, spray atomization and vaporization, mixture preparation, combustion, deposit formation and emissions reduction represent key areas of optimization, and all are being actively investigated. New, enabling technologies such as the high-pressure, common-rail, gasoline injection system and real time computer control of process parameters have enabled the current new examination of an old objective: the direct-injection, stratified-charge (DISC) gasoline engine. A very substantial database of technical information has been generated on this subject over the past decade.

This treatise addresses the spectrum of critical developmental issues in the rapidly progressing area of direct-injection, spark-ignited gasoline engines, and examines the contribution of each process and subsystem to the efficiency of the overall system. The material reflects the very latest global technical initiatives that are being incorporated or investigated within the automotive and research communities, and includes discussions, data and figures from many technical papers and proceedings that are not available in the English language. The focus is on both the fundamental processes of mixture preparation, combustion and emissions formation and the practical issues related to engine system development, such as cold start, deposits, oil dilution and aftertreatment systems. The intricate phenomena that are associated with direct fuel injection, spray formation, entrainment and mixing, combustion and emissions are presented in a clear and organized manner, and are placed within an organized framework of current state-of-the-art hardware and processes. All published work on this topic is considered and discussed in detail.

The goal is to enable a rapid learning curve on this complex topic, as well as to provide an invaluable desk reference for any G-DI subject or direct-injection subsystem that is being developed worldwide. A further objective is to provide a basic understanding of, and appreciation for, the parameters that must be carefully considered in developing any new direct-injection combustion system. Detailed comparisons of the G-DI engine with port fuel injection and diesel engines are provided where appropriate. Nearly all topics that the reader in the field may wish to comprehend, including terminology, the review of worldwide research and development, and explanations of physical processes, are presented and documented. It would be of significant benefit to read the material in this book before drafting a proposal or a technical paper, conducting an experiment or making a management decision on any aspect of gasoline direct-injection systems.

The material is logically organized and presented in ten chapters which contain more than 330 figures, 70 tables and 500 technical references. The data on fuel economy and emissions for actual engine configurations are of significant importance to engine researchers and developers, hence these data have been obtained and assembled for all of the available G-DI configurations, and are reviewed and discussed in detail. The types of G-DI engines are arranged in four classifications of decreasing complexity, and the advantages and disadvantages of each class are noted and explained, with emphasis placed on important trends and consensus conclusions. This enables engine developers, researchers and engineering managers to be informed on both trends and consensus areas where conflicting data and opinions exist. Thus, the reader is informed, for example, as to the degree to which engine volumetric efficiency and compression ratio can be increased under optimized conditions, and as to the extent to which HC, NOx and particulate emissions can be minimized for specific control strategies.

In Chapter 1 the fundamental processes and classifications that are discussed in detail in later chapters are introduced. The mixture preparation processes that are inherent in port-fuel-injection and direct-injection engines are discussed and the resultant benefits of G-DI technology are outlined. The generic classification of

G-DI engines and the specifications of all production, prototype and research G-DI gasoline engines are summarized.

In Chapter 2 the combustion system configurations and control strategies for creating a stratified charge are described in detail. The advantages and challenges of spray-guided, wall-guided and air-guided combustion systems are extensively compared in terms of system requirements, combustion stability, system flexibility, engine performance and emissions.

The requirements of the fuel injector and the overall fuel system are outlined in Chapter 3, with the operating principles and unique characteristics of all production and prototype injectors for G-DI applications described. A comprehensive description of injector actuation and fuel delivery dynamics is provided, along with a detailed discussion of the capabilities for multiple injections per cycle.

A very comprehensive discussion of G-DI fuel spray dynamics and characteristics is provided in Chapter 4— perhaps the most detailed consideration of this area that is available. Very extensive coverage is provided for the areas of spray atomization requirements, sac spray characteristics, spray cone angle, spray penetration and after-injection dynamics. The characteristics of the offset spray and the sprays resulting from split injection are reviewed, and the essential characteristics of sprays from various production and prototype injectors are described in detail. Included here are sprays from the swirl (both inwardly opening and outwardly opening), the multihole, the slit and the shaped-spray type injectors, and the fuel sprays from the pulse-pressurized, air-assisted injector design. A very important and underestimated area, the influence of ambient conditions on the structure of sprays from all categories of G-DI injectors, is also detailed in Chapter 4. The best-practice performance of the entire group of contemporary G-DI injectors is tabulated, and the critical issues related to G-DI spray measurement and characterization are provided.

In Chapter 5 the special requirements for the processes of fuel-air mixing, as well as the associated constraints on the air flow field of a G-DI engine and spray-wall interactions are discussed in detail. Extensive discussions have been devoted to unintended spray-wall impingement, which may lead to increased HC emissions and oil dilution. The mixture preparation process during a cold crank and start is discussed at length, and the effects of various engine operating parameters on cold start performance is reviewed.

Chapter 6 incorporates a detailed prescription of the combustion characteristics of G-DI engines under various engine operating modes. A spectrum of combustion and emissions control strategies are explained in detail, and the associated influences on the combustion process are discussed.

The critical area of G-DI deposits and the associated effect on combustion, performance and emissions degradation is reviewed, and important system guidelines for minimizing deposition rates and deposit effects are presented in Chapter 7.

The emissions formation mechanisms for HC, NOx and particulates are described in detail in Chapter 8. The capabilities and limitations of contemporary emissions control techniques and aftertreatment hardware are reviewed in depth, and areas of consensus on attaining future emissions standards are compiled and discussed extensively. The influence of lean-NOx catalysis on the development of late-injection, stratified-charge G-DI engines is reviewed, and the relative merits of lean-burn, homogeneous, direct-injection engines as an option requiring less control complexity are analyzed.

All aspects of the potential of G-DI technology for improving fuel economy are considered in Chapter 9. The range of sources that comprise the total discrepancy between the thermodynamically predicted fuel economy and current best-practice fuel economy are itemized and examined, and the possible short-term and long-term solutions to narrow this gap are discussed. The options for combining G-DI with other emerging vehicle technologies are analyzed with the objective of realizing more of the potential of G-DI, while alleviating some of the deficiencies of other technologies.

The prior work on DISC engines that remains relevant to current G-DI engine development is reviewed and discussed in the introductory portion of Chapter 10. The main body of Chapter 10 contains a comprehensive assembly of all known production, prototype and research G-DI engines worldwide. These engines and the associated combustion and aftertreatment systems are reviewed as to performance, emissions and fuel economy advantages, and areas requiring further development are discussed. The engine schematics, system control diagrams and specifications are compiled, and the emissions control strategies are illustrated and discussed.

The inclusion of extensive advanced content related to all aspects of gasoline direct-injection engines means that this book can be used as a textbook for engineering graduate students or as an advanced reference text for an undergraduate course in internal combustion engines. The compiled information in this volume would also constitute an excellent base for an industrial training program, engineering seminar or workshop in the areas of gasoline fuel systems, gasoline spray technology, mixture preparation, combustion system development, deposit issues or aftertreatment technology for lean-burn engines. In each of Chapters 2 through 10, a specialized topic is addressed in significant detail, thus the text, figures and tables can be used independently for any of the special purposes listed above. All researchers and engineers who are employed by vehicle manufacturers, suppliers of engine components and aftertreatment systems, additive and oil companies, research institutions and engineering schools will benefit from being cognizant of the information that is compiled here. Engineers and researchers in this field will find this volume to be invaluable in developing optimum systems for direct-injection gasoline engines, and in understanding and interpreting test data. Corporate managers who are responsible for product decisions in this area may benefit significantly from the clear statements of the advantages and disadvantages of a spectrum of sub-systems, and from the discussions of current global best practice.

Fuquan (Frank) Zhao
David L. Harrington
Ming-Chia Lai

Acknowledgements

The authors would like to acknowledge the following reviewers for their constructive comments and suggestions: Dr. R.W. Anderson of Ford Motor Company, Dr. H. Ando of Mitsubishi Motors Corporation, Dr. A.A. Aradi of Ethyl Corporation, Dr. T.W. Asmus of DaimlerChrysler Corporation, Prof. D.N. Assanis of University of Michigan, Prof. W.K. Cheng of Massachusetts Institute of Technology, Dr. G.K. Fraidl of AVL, Prof. J.B. Heywood of Massachusetts Institute of Technology, Mr. B. Imoehl of Siemens Automotive, Dr. T.E. Kenney of Ford Motor Company, Prof. R.R. Maly of DaimlerChrysler AG, Mr. P.M. Najt of General Motors Corporation, Dr. R.B. Rask of General Motors Corporation, Prof. R.W. Reitz of University of Wisconsin at Madison, Dr. R.J. Tabaczynski of Ford Motor Company, and Dr. Y. Takagi of Nissan Motors Corporation (currently Musashi Institute of Technology).

Acknowledgement and appreciation are expressed by the authors to Elsevier Science Ltd. for the permission to publish a portion of their earlier review article, "Automotive Spark Ignited Direct Injection Gasoline Engines." This was published in 1999 in the international review journal *Progress in Energy and Combustion Science* (Vol. 25, 1999, pp. 437–562).

The authors also want to express their gratitude to Mr. A.S. Horton, SAE Editorial Consultant, for his initiation of the work, and Ms. L.I. Moses, SAE Product Manager, for her continuous support in preparing the book.

Finally, the authors would like to express their sincere acknowledgements to their families, especially their spouses, Dannie, Glenna and Wen-Haw, for their encouragement, understanding and patience through the lengthy process of writing this book.

Nomenclature

AF	air/fuel ratio
A/F	air/fuel ratio
AT	automatic transmission
AFR	air/fuel ratio
ATDC	after top dead center
BDC	bottom dead center
BMEP	brake mean effective pressure
BSFC	brake specific fuel consumption
BSHC	brake specific HC
BSNOx	brake specific NOx
BSU	Bosch smoke unit
BTDC	before top dead center
CA	crank angle
CAD	crank angle degree
CC	close-coupled
CCD	combustion chamber deposit
CFD	computational fluid dynamics
CIDI	compression-ignition direct-injection
CMC	charge-motion-control
COV	coefficient of variation
CR	compression ratio
CVCC	compound vortex combustion chamber
CVT	continuously variable transmission
D32	Sauter mean diameter of a fuel spray
DI	direct-injection
DISC	direct-injection stratified-charge
DMI	direct-mixture-injection
DOHC	double overhead cam
DV10	spray droplet diameter for which 10% of the fuel volume is in smaller droplets
DV50	spray droplet diameter for which 50% of the fuel volume is in smaller droplets
DV80	spray droplet diameter for which 80% of the fuel volume is in smaller droplets
DV90	spray droplet diameter for which 90% of the fuel volume is in smaller droplets
EC	exhaust valve closing
EDM	electric discharge machining
EFI	electronic fuel injection
EGT	exhaust gas temperature
EGR	exhaust gas recirculation
ELP	end of the logic pulse
EMS	engine management system
EO	exhaust valve opening

EOA	end of air injection
EOB	end of burn
EOI	end of injection
EPMA	electron probe microanalysis
ETC	electronic throttle control
EURO III	European Stage III emissions standards
EURO IV	European Stage IV emissions standards
EVC	exhaust valve closing
EVO	exhaust valve opening
FID	flame ionization detector
FPW	fuel pulse width
FTIR	Fourier transform infrared
FTP	federal test procedure
G-DI	gasoline direct-injection
GMEP	gross mean effective pressure
HC	hydrocarbon
HCCI	homogeneous-charge, compression-ignition
HCEI	hydrocarbon emission index
HEV	hybrid electric vehicle
HSDI	high speed direct injection
IC	intake valve closing
ICP	inductively coupled plasma
IDI	indirect injection
IMEP	indicated mean effective pressure
IO	intake valve opening
IPTV	incidents per thousand vehicles
ISCO	indicated specific CO
ISFC	indicated specific fuel consumption
ISHC	indicated specific HC
ISNOx	indicated specific NOx
IVD	intake valve deposit
IVO	intake valve opening
kPa	kilopascal
LDV	laser Doppler velocimetry
LEV	low-emission-vehicle emission standard
LIF	laser-induced fluorescence
MAP	manifold absolute pressure
MBT	minimum spark advance for best torque
MFB	mass fraction burned
MMR	main-spray, maximum-penetration rate
MPa	megapascal
MPI	multi-point port injection
MT	manual transmission
MVEG	motor vehicle emissions group

NCP	new combustion process
NEDC	New European Driving Cycle
NMEP	net mean effective pressure
NSCO	net specific CO
NSHC	net specific HC
NSNOx	net specific NOx
OBD	on-board diagnostics
OEM	original equipment manufacturer
Oh	Ohnesorge number
P_{amb}	ambient back pressure
PCV	positive crankcase ventilation
P_{cyl}	cylinder pressure
PDA	phase-Doppler anemometry
PDI	phase-Doppler interferometry
PFI	port fuel injection
PI	port injection
P_{inj}	injection pressure
PIV	particle imaging velocimetry
PLIF	planar laser-induced fluorescence
PM	particulate matter
PPAA	pulse-pressurized, air-assisted
PROCO	Ford programmed combustion control system
RB	rapid burning
R.H.R.	rate of heat release
RON	research octane number
RVP	Reid vapor pressure
SC	stratified charge
SCR	selective-catalysis-reduction
SCRC	stratified-charge rotary combustion
SCV	swirl control valve
SD	standard deviation
SEM	scanning electron microscopy
SFC	specific fuel consumption0
SFT50	sac-spray flight time to a location 50 mm from the injector tip
SI	spark ignition
SLP	start of the logic pulse
SMD	Sauter mean diameter of a fuel spray
SMR	sac-spray maximum penetration rate
SOA	start of air injection
SOB	start of burn

SOHC	single overhead cam
SOI	start of injection
SPI	spark plug injector
SULEV	super ULEV
TBI	throttle body injection
TC	top dead center
TCCS	Texaco controlled combustion system
TDC	top dead center
THC	total hydrocarbon
TTL	transistor-to-transistor logic
TWC	three-way catalyst
UBHC	unburned hydrocarbons
ULEV	ultra-low-emission-vehicle emission standard
VCO	valve covered orifice
VGT	variable geometry turbine
VO	value opening
VOF	volume of fluid
VVT	variable valve timing
We	Weber number
WFR	working flow range
WOT	wide open throttle

Various names for gasoline direct-injection technology that are encountered in the literature

D-4	Toyota DI engine
DGI	direct gasoline injection
DIG	direct injection gasoline
DI-G	direct injection gasoline
DISI	direct-injection spark-ignited
ECOTEC DIRECT	Adam Opel DI engine
FSI	Volkswagen DI engine
GDI	Mitsubishi DI engine
G-DI	gasoline direct injection
HPi	PSA DI engine
IDE	Renault DI engine
NEODi	Nissan DI engine
SCC	Saab DI engine
SIDI	spark-ignited direct-injection

Chapter 1

Introduction

1.1 Overview

With the increasing emphasis on achieving substantial improvements in automotive fuel economy, automotive engineers are striving to develop engines that provide both significantly reduced brake-specific fuel consumption and compliance with future stringent emissions requirements. In comparing gasoline spark-ignition engines to the compression-ignition diesel engines that are in production in automotive applications today, it is evident that the reduced brake-specific fuel consumption, and hence the associated vehicle fuel economy of the compression-ignition, direct-injection diesel engine is superior to that of the port-fuel-injected, spark-ignition engine. This is mainly due to the use of a higher compression ratio, coupled with unthrottled operation. The diesel engine, however, generally exhibits a slightly higher noise level, a more limited speed range, diminished startability and higher particulate and NOx emissions than is exhibited by gasoline spark-ignition engines.

Over the past three decades, a research goal has been to develop an internal combustion engine for automotive applications which combines the best features of the gasoline and the diesel engines [164, 169, 421, 428], with the specific objective being to combine the specific power of the gasoline engine with the efficiency of the diesel engine at part load [224, 253, 411]. Such an engine would exhibit a brake-specific fuel consumption (BSFC) approaching that of the diesel engine, while maintaining the operating characteristics and specific power output of the gasoline engine. Significant technical work over this time has verified that this goal may be approached using a direct-injection, four-stroke, spark-ignition engine that does not throttle the inlet mixture to control the load. In this engine, a fuel spray plume is injected directly into the cylinder, generating a fuel-air mixture with an ignitable composition at the spark gap at the time of ignition. This overall class of engine is designated as a direct-injection, stratified-charge (DISC) engine, and it has been verified that this engine type generally exhibits an improved tolerance for fuels of lower octane number and poorer driveability index. In fact, a significant segment of the early work on prototype DISC engines was focused on the inherent multi-fuel capability [239, 287, 311].

In a manner similar to that of the diesel, the power output of the DISC engine is controlled by varying the amount of fuel that is injected into the cylinder. The induction air is not significantly throttled, thus the negative work that is associated with the pumping loop of the cycle is significantly reduced. By using a spark source to ignite the fuel-air mixture in the cylinder, the engine is provided with direct ignition, thus avoiding many of the stringent ignition quality specifications required of fuels for the diesel engine. Furthermore, by means of the relative alignment of the spark plug and the fuel injector, overall ultra-lean operation may be achieved, thus yielding a further reduced BSFC [41, 87, 99, 179, 485].

From a historical perspective, the very early work on the gasoline DISC engine did result in a number of important technical developments. Several detailed combustion strategies were proposed and implemented, including the Texaco Controlled Combustion System (TCCS) [8], MAN-FM of Maschinenfabrik Auguburg-Nurnberg [161, 307, 453], and the Ford programmed combustion control (PROCO) system [401, 410]. Most of these earlier systems were based upon engines having two valves per cylinder, and incorporating a bowl-in-piston combustion chamber. Late-injection operation was achieved by utilizing mechanical pump-line-nozzle fuel injection systems from diesel engine applications. In these systems that are representative of the early developmental work, unthrottled operation was obtained throughout most of the load range, and BSFC values were achieved which were competitive with the indirect-injection diesel engine of that era. A major drawback was that late injection timing was maintained even at full load due to the limitations of the mechanical (non-electronic) fuel injection system. This resulted in smoke-limited combustion for air/fuel ratios richer than approximately 20:1. The necessity of using diesel fuel injection equipment, coupled with the need for a turbocharger to provide adequate power output, culminated in an engine that exhibited performance characteristics that were similar to those of a diesel engine, but which had poor, part-load, unburned hydrocarbon (HC) emissions. Even with the PROCO combustion system, which utilized a centrally located injector to provide a hollow-cone spray with injection occurring early in the compression stroke, the control of HC emissions was extremely difficult for light-load operation. The combination of relatively poor air utilization and the use of mechanical fuel injection equipment that was limited in speed range meant that the engine-specific power output was quite low. A brief discussion of the geometric configuration of these pioneering DISC systems is provided in Section 10.1.

Many of the basic limitations encountered in the earlier work on DISC engines can now be circumvented due to advances in electronics and computer control. This is particularly true of the significant control limitations that existed for the direct-injection injectors and pump controls of two decades ago. New technologies and computer-control strategies are currently being invoked by a number of automotive companies to reexamine the extent to which the potential benefits of the gasoline direct-injection (G-DI) engine can be realized in production engines [2, 13–17, 53, 55, 56, 65, 84, 104, 117, 134, 156, 208, 209, 211, 219, 259, 387, 398, 402, 414–420, 438–442]. These engines, their mixture preparation and combustion control strategies,

emissions characteristics, fuel economy potential, deposit formation mechanism and other critical related issues are discussed in detail in this book.

1.2 Gasoline Direct Injection versus Port Fuel Injection

Figure 1.2-1 shows the layout of a typical G-DI engine system [261]. The major difference between the gasoline port-fuel-injected (PFI) engine and the G-DI engine lies in the mixture preparation strategies, which are illustrated schematically in Fig. 1.2-2 [521]. In the PFI engine, gasoline is injected into the intake port of each cylinder, and there is an associated time lag between the injection event and the induction of the fuel and air into the cylinder. The vast majority of current automotive PFI engines worldwide utilize timed fuel injection onto the back of the intake valve when the intake valve is closed. The PFI injector may be mounted either in the cylinder head upstream of the intake valve (20% of applications) or in the intake manifold near the cylinder head (80% of applications). During cranking and cold starting, a transient film, or puddle, of liquid fuel forms in the intake valve area of the port. Some portion of this steady, oscillatory film is drawn into the cylinder during each induction event [348]. The fuel film in the

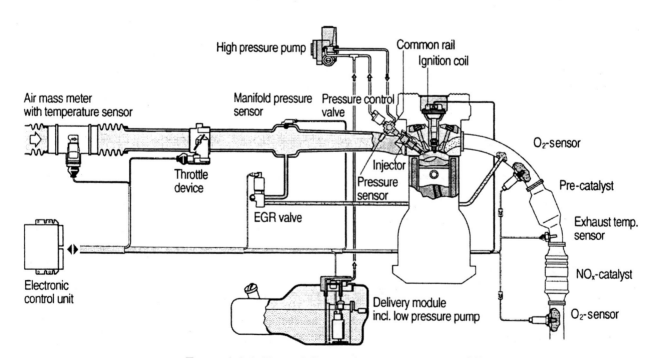

Figure 1.2-1 Typical G-DI engine system layout [261].

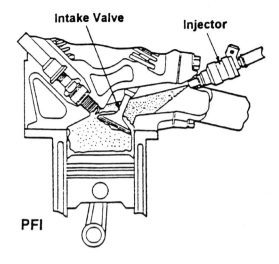

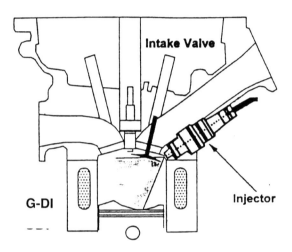

Figure 1.2-2 Comparison of G-DI and PFI mixture preparation systems [521].

Alternatively, injecting fuel directly into the engine cylinder avoids the problems associated with fuel wall wetting in the port, while providing enhanced control of the metered fuel for an individual combustion event, as well as a reduction in the fuel transport time. The actual mass of fuel entering the cylinder on a given cycle can be more accurately controlled with G-DI than with PFI, thus providing the potential for leaner combustion, less cylinder-to-cylinder and cycle-by-cycle variations in the air/fuel ratio and reduced BSFC. It is well established that the direct injection of gasoline with little or no cold enrichment can provide starts on the first cranking cycle [9], and can exhibit significant reductions in hydrocarbon spikes during load transients. Hence, the G-DI engine requires much less fuel to start the engine, and this difference in the minimum fuel requirement becomes greater as the ambient temperature decreases [341]. However, the injection of gasoline directly into the cylinder significantly reduces the time available for evaporation and mixing, thus the requirements for fuel spray atomization are an order of magnitude more stringent. A substantially higher operating fuel pressure is generally utilized in the G-DI system, thus the fuel entering the cylinder is much better atomized than that of the PFI system, in fact an order of magnitude better, to provide the required higher rates of fuel vaporization. It is important to note, however, that injection of fuel directly into the cylinder is not a guarantee that fuel film problems are not present. The wetting of piston crowns or other combustion chamber surfaces, whether intentional or unintentional, does introduce the important variable of transient wall film formation and evaporation inside the cylinder.

Another limitation of the PFI engine is the requirement of throttling for basic load control. Even though throttling is a well-established and reliable mechanism of load control in the PFI engine, the thermodynamic loss associated with throttling is substantial. Figure 1.2-3 shows a comparison of indicated specific fuel consumption (ISFC) between throttled and unthrottled operations as a function of engine load. It is worth noting that in this analysis, the ISFC is based on the full 720 degrees of the cycle. Clearly any system that utilizes throttling to control or adjust load levels will experience the thermodynamic loss that is associated with the negative pumping loop, and will exhibit thermal efficiency degradation, particularly at low levels of engine load. In the G-DI stratified-charge mode, the engine load is controlled by varying the amount of fuel injected into the cylinder. This

intake port of a PFI engine acts as an integrating capacitor, and the engine actually operates on fuel inaccurately metered from the pool in the film, *not* from the current fuel being accurately metered by the injector [516]. This causes a fuel delivery delay and an associated inherent metering error due to partial vaporization. This, coupled with the relatively poor evaporation from a cold puddle of fuel, makes it necessary to supply amounts of fuel for cold start that significantly exceed that required for the ideal stoichiometric ratio, even though the catalyst temperature may be below the light-off threshold. This puddling and time lag may cause the engine to experience an unstable burn on the first 4 to 10 cycles of a cold start, with an associated significant increase in the engine-out emissions of unburned hydrocarbons [66].

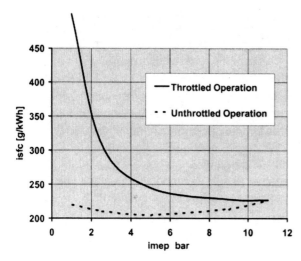

Figure 1.2-3 Comparison of ISFC between throttled and unthrottled operation as a function of engine load [108].

circumvents the limitation of the PFI engine with respect to load control, and can significantly enhance the engine thermodynamic efficiency. This is discussed in detail in Chapter 9.

Current advanced PFI engines still utilize throttling for basic load control. They also have, and will continue to have, an operating film of liquid fuel in the intake port. These two basic PFI operating requirements represent major impediments to achieving significant breakthroughs in PFI fuel economy or emissions. Continuous incremental improvements in the older PFI technology will be made, but it is unlikely that substantial improvements in the areas of fuel economy and emissions can be simultaneously achieved. The G-DI engine, in theory, does not have these two significant limitations, or the performance boundaries that are associated with them.

In addition to the capabilities of eliminating the fuel wall film in the intake port and reducing throttling loss by controlling the engine load through fueling rate adjustment, the G-DI concept also offers a number of other theoretical advantages over the contemporary PFI engine, as summarized in Table 1.2-1. The significantly higher injection pressures used in the majority of common-rail G-DI injection systems increase both the degree of fuel atomization and the fuel vaporization rate. This permits stable combustion from the first or second injection cycle *without* supplying excess fuel. Therefore,

G-DI engines have the potential to achieve cold-start HC emissions that can approach the level observed for steady operating conditions [438]. Another potential advantage of the G-DI engine is the option of invoking fuel cut-off on deceleration. If implemented successfully, fuel cut-off can provide additional incremental improvements in both fuel economy and engine-out HC emissions, as is discussed in Chapters 8 and 9, respectively. For the PFI engine, which operates from an established film of fuel in the intake port, fuel cut-off during vehicle deceleration is not a viable option, as it causes a reduction or elimination of the liquid fuel film in the port. The reestablishment of the stable pool of fuel in the port is a transient event that requires a few engine cycles, and this process can generate very lean mixtures in the combustion chamber, which may lead to an engine misfire or backfire. Another significant potential advantage of direct, in-cylinder injection is the cooling of the inducted charge of air. The evaporation of the finely atomized fuel can substantially cool the air, particularly if the latent heat of vaporization is supplied mainly from the air, rather than from the metal wall surfaces. The cooling effect will be present for both early and late injection, but the volumetric efficiency can also be enhanced if injection occurs during the induction event. Detailed discussions of the benefits related to autoignition suppression and peak power, as well as the guidelines for the utilization of the process, are presented in Chapter 5.

In spite of the important potential advantages of G-DI engines noted above, PFI engines do have some limited advantages due to the fact that the intake system acts as a prevaporizing chamber. When fuel is injected directly into the engine cylinder, the time available for mixture preparation is reduced significantly. As a result, the atomization level of the fuel spray must be fine enough to permit fuel evaporation in the limited time available between injection and ignition. Fuel droplets that do not vaporize are very likely to participate in diffusion burning, and may exit the engine as HC emissions. Also, injecting fuel directly into the engine cylinder can result in unintended fuel impingement on the piston crown and/ or on the cylinder wall. These factors, if present in a specific engine design, can contribute to increased levels of HC and particulate emissions, and to cylinder bore wear rates that can easily exceed that of an optimized PFI engine. Some other advantages of PFI engines such as a low-pressure fuel system, the feasibility of using three-way catalysis and higher exhaust temperatures for

Table 1.2-1
Theoretical advantages of G-DI engines over PFI engines

Fuel Economy	• Improved fuel economy (up to 25% potential improvement, depending on test cycle) ❑ Substantially reduced pumping loss (unthrottled, stratified-charge mode) ❑ Less heat losses (unthrottled, stratified mode) ❑ Higher compression ratio may be employed (due to charge cooling with injection during induction) ❑ Lower octane requirement (due to charge cooling with injection during induction) ❑ Increased volumetric efficiency (due to charge cooling with injection during induction) ❑ Fuel cut-off during vehicle deceleration may be employed ❑ Less acceleration-enrichment required
Driveability	• Improved transient response • Improved cold startability
Air/Fuel Ratio Controllability	• More precise air/fuel ratio control ❑ More rapid starting and combustion stabilization ❑ Less cold enrichment required during cold start ❑ Less acceleration-enrichment required
Combustion Stability	• Extended EGR tolerance limit (to minimize the use of throttling and to reduce NOx emissions)
Emissions	• Selective emissions advantages ❑ Reduced cold-start HC emissions ❑ Reduced HC spikes on engine transients ❑ Reduced CO_2 emissions
System Optimization	• Enhanced potential for system optimization

Note: improvements are relative to a comparable, optimized PFI engine.

improved catalyst efficiency present an evolving challenge to the G-DI engine.

The displacement of the PFI engine by the G-DI engine as the primary production automotive powerplant has been, and is still, constrained by the areas of concern listed in Table 1.2-2. These concerns must be addressed and alleviated in any specific design if the G-DI engine is to supplant the current PFI engine in specific global markets. If future emissions regulations can be achieved using PFI engines without the requirement of complex new hardware, the market penetration rate for G-DI engines will be reduced, as there will most assuredly be a G-DI requirement for sophisticated fuel injection hardware, a high-pressure fuel pump and a more complex engine control system. An important constraint on G-DI engine designs has been the relatively high HC and NOx emissions, and the fact that a three-way catalyst (TWC) could not be effectively utilized.

Operating the G-DI engine under overall lean conditions does reduce the engine-out NOx emissions, but this generally cannot achieve the minimum 90% reduction level that can be attained using a TWC. Much work is under way worldwide to develop lean-NOx catalysts, but at the moment the attainable conversion efficiency over a wide range of engine operating conditions and fuels is still significantly lower than that of a TWC. The excessive HC emissions at light load also represent a significant research and development task to be solved.

In spite of these concerns and difficulties, the G-DI engine offers a horizon for future applications that expands beyond that of the well-developed PFI engine.

Finally, it is worth noting that design engineers, managers and researchers who must evaluate and prioritize the published information on the advantages of G-DI engines over PFI engines should be aware of one area of data comparison and reporting that is disconcerting. In many reports and research papers G-DI performance is compared to PFI baselines that are not

Table 1.2-2
Major obstacles that hinder G-DI development and application

Emissions Challenge	• High local NOx production under part-load, stratified-charge operation. As a result, three-way catalysts cannot be utilized to full advantage. • Relatively high light-load HC emissions • Relatively high high-load NOx emissions • Increased particulate emissions
Combustion Stability and Control Challenge	• Difficulty in controlling stratified-charge combustion over the required operating range • Complexity of the control and injection technologies required for seamless load changes • High EGR tolerance required for NOx emissions reduction • Relatively high rate of formation of injector deposits and/or ignition fouling
Fuel Economy Challenge	• Elevated fuel system pressure and fuel pump parasitic loss • Possible fuel consumption increase to enable technologies for fast catalyst light-off and catalyst regeneration • Increased electrical power and voltage requirements of the injectors and drivers
Performance and Durability Challenge	• Relatively high rate of formation of injector deposits and/or ignition fouling • Increased fuel system component wear due to the combination of high pressure and low fuel lubricity • Increased rate of cylinder bore wear • Other deposits such as intake valve and combustion chamber
System Complexity Challenge	• More complex emissions control system and control strategies required • Sophisticated fuel system and combustion system designs required to compromise the needs for various loads from cold start to WOT operations • Significantly increased number of calibration variables for system optimization

well defined, thus making it very difficult for the reader to make a direct engineering comparison between G-DI and PFI performance. One extreme example is the comparison of G-DI and PFI fuel economy data obtained using two different vehicles with two different inertial weights. An example of a more subtle difference is the evaluation of the BSFC reduction resulting from the complete elimination of throttling in a G-DI engine, but not noting or subtracting the parasitic loss of a vacuum pump that would have to be added for braking and other functions. A number of published comparisons lie between these two extremes. The readers are cautioned to review all claims of comparative G-DI/PFI data carefully as to the precise test conditions for each, and the degree to which the systems were tested under different conditions or constraints.

1.3 Classification of Gasoline Direct-Injection Engines

Many new combustion systems have been proposed and developed for automotive, four-stroke, G-DI engines in the last decade, and a detailed description of these prototype and production systems and control strategies is presented in Chapter 10. Each combustion system has unique features that reflect specific strategies of mixture preparation, combustion control and emissions reduction. All have in common the goal of achieving substantial fuel economy improvement while simultaneously realizing large reductions in engine-out and tailpipe emissions.

Based upon the associated combustion systems and control strategies, G-DI engines may be classified into eight logical categories. Table 1.3-1 summarizes both the categories used and the resulting G-DI engine classifications. The first five categories in the table are based upon the combustion system layouts, whereas the final three categories are related to the control strategy employed. Table 1.3-2 is a very imformative compilation that summarizes the key features of all production, prototype and research G-DI engines that have been reported in the literature.

1.4 Summary

The potential advantages of the gasoline direct-injection concept are too significant to receive other than priority status. The concept offers many opportunities for achieving significant improvement in engine fuel consumption and emissions reductions. The current high-technology PFI engine, although highly evolved, has nearly reached the limit of the potential of a system that is based upon operation using throttling and a port fuel film; however, the technical challenge of competing with, and displacing, a proven, evolved performer such as the PFI engine is not to be underestimated. Since the late 1970s, when a significant portion of the DISC engine work was conducted, the spark ignition (SI) engine has continued to evolve monotonically as an ever-improving baseline. The fuel system has also evolved continuously from carburetion to throttle-body injection, then to simultaneous-fire PFI, and more recently to phased, sequential-fire PFI. Enhanced systems such as variable valve timing and variable (demand) displacement continue to be incorporated. The result is that today the spark-ignited PFI engine remains the benchmark standard for automobile powerplants. In order to displace this standard, future G-DI engines have to meet the practical development targets that are outlined in Table 1.4-1, and are derived from current best-practice PFI engines. In addition, reliable and efficient hardware and control strategies for G-DI technology will have to be developed and verified under field conditions. For the same reasons that port fuel injection gradually replaced the carburetor and the throttle-body injector, a G-DI combustion configuration that will be an enhancement of one of the concepts outlined in this book will emerge as the predominant engine system, and will gradually displace the sequential-fire PFI applications.

Table 1.3-1
Master chart of G-DI engine classifications

Category	Classification	Description
Distance between Injector Tip and Spark Plug Gap	Narrow-Spacing	The spark plug gap is located close to the injector tip in order to ignite the spray periphery directly. It is also referred to as a spray-guided combustion system.
	Wide-Spacing	A relatively large distance exists between the injector tip and the spark plug gap. Stratification is created by air flow or spray/wall impingement. These are also referred to as air-guided and wall-guided systems, respectively.
Approach to Create Stratified Charge	Spray-Guided	Stratification results from fuel spray penetration and mixing. The spark plug gap is located close to the injector tip to ignite the spray periphery. It is a narrow-spacing concept.
	Wall-Guided	The fuel spray is directed toward a shaped cavity in the piston crown. Stratification is created through spray/wall interaction. It is a wide-spacing concept and is currently the most commonly used approach in production applications.
	Air-Guided	Stratification is created through the interaction of the fuel spray with the in-cylinder charge motion. It is normally a wide-spacing concept.
Charge Motion	Tumble-Based	Tumble is used to create or assist in charge stratification.
	Swirl-Based	Swirl is used to create or assist in charge stratification.
Injector Location	Centrally-Mounted	Injector is located at the center of the chamber. The spark plug gap is usually in close proximity to the injector tip.
	Side-Mounted	Injector is located near the periphery of the chamber, on the intake side. The spark plug is usually located near the center of the chamber.
Injector Type	Single-Fluid	Single-liquid fuel is used; normally at high pressure. It is currently the most commonly utilized injector.
	Pulse-Pressurized, Air-Assisted	A mixture of air and fuel is injected into the cylinder at the time of fuel injection into the cylinder. It employs moderate fuel and air pressures.
Fuel Distribution	Homogeneous	A homogeneous mixture is formed in the cylinder.
	Stratified	A stratified mixture is formed in the cylinder.
Injection Timing	Early Injection	Fuel is injected during the intake stroke to form a homogeneous mixture. A certain degree of stratification may still occur.
	Late Injection	Fuel is injected during the compression stroke to form a stratified charge.
Air/Fuel Ratio	Richer Than Stoichiometric	Engine is operated at an overall air/fuel ratio richer than stoichiometric. Depending on the fuel injection timing, the charge can be homogeneous or stratified.
	Stoichiometric	Engine is operated at an overall air/fuel ratio of stoichiometric. Depending on the fuel injection timing, the charge can be homogeneous or stratified.
	Lean	The overall air/fuel ratio is leaner than stoichiometric. Depending on the fuel injection timing, the mixture distribution can be either homogeneous-lean or stratified (locally richer than the overall air/fuel ratio).

TABLE 1.3-2
Key features of production, prototype and research G-DI engines

Manufacturer and institution	Displacement (cm³)	Bore × stroke (mm)	Cylinder Configuration	Valves / cylinder	Compression ratio	Stratification approach	Piston geometry	Charge motion	Injector location	Spark plug location	Fuel pressure (MPa)	Comments
ADAM OPEL	2200	86 × 94.6	I-4	4	11.5	Wall-guided	Bowl	Swirl	Intake side	Central	8	Prototype
AUDI	1196		I-3	5		Wall-guided	Bowl	Tumble	Intake side	Central	10	Prototype
AVL	2000		I-4	4		Wall-guided	Bowl	Swirl	Intake side	Central		Prototype
FEV				4		Air-guided	Bowl	Tumble	Intake side	Central		Prototype
FIAT	1995	90.2 × 90.0	I-4	4	12.0		Flat	Tumble, swirl	Central	Central		Prototype; homogeneous only
FORD	575		1	4	11.5		Bowl	Tumble	Central	Central	5	Homogeneous only
FORD	1125	79 × 76.5	I-3	4	11.5	Wall-guided	Bowl	Swirl	Intake side	Central	12	Prototype
HONDA	1000		I-3	4		Spray-guided	Bowl	Swirl	Central	Central		Prototype for hybrid application
ISUZU	528	93.4 × 77.0	1	4	10.7		Flat	Tumble, swirl	Central, intake side	Central	5	Homogeneous only
ISUZU	3168	93.4 × 77.0	V-6	4	10.7		Flat	Tumble, swirl	Central, intake side	Central	5	Prototype; homogeneous only
MAZDA	1992	83 × 92	I-4	4	11.0	Wall-guided	Bowl	Tumble, swirl	Intake side	Central	7	Prototype
MERCEDES	538.5	89 × 86.6	1	4	10.5	Spray-guided	Bowl		Central			
MITSUBISHI	1864	81 × 89	I-4	4	12.0	Wall-guided	Bowl	Reverse tumble	Intake side	Central	5	Production in Japan
MITSUBISHI	1468	75.5 × 82	I-4	4	11.0	Wall-guided	Bowl	Reverse tumble	Intake side	Central	5	Production in Japan
MITSUBISHI	3496	93 × 85.8	V-6	4	10.4	Wall-guided	Bowl	Reverse tumble	Intake side	Central	5	Production in Japan
MITSUBISHI	1864	81 × 89	I-4	4	12.5	Wall-guided	Bowl	Reverse tumble	Intake side	Central	5	Production in Europe
MITSUBISHI	4500	66 × 80	V-8	4		Wall-guided	Bowl	Reverse tumble	Intake side	Central		Production in Japan
MITSUBISHI	1094		I-4	4		Wall-guided	Bowl	Reverse tumble	Intake side	Central		Production in Japan; combined with idle stop system
NISSAN	1838	82.5 × 86	I-4	4	10.5	Wall-guided	Bowl	Swirl	Intake side	Central	10	Prototype
NISSAN	1769	80 × 88	I-4	4	10.5	Wall-guided	Bowl	Swirl	Intake side	Central	7	Production in Japan; combined with CVT
NISSAN	2987	93 × 73.3	V-6	4	11.0	Wall-guided	Bowl	Swirl	Intake side	Central	Variable: 7–9	Production in Japan
NISSAN	2500		V-6			Wall-guided	Bowl	Swirl	Intake side	Central		Prototype for hybrid application
ORBITAL	1796	80.6 × 88	I-4		10.4	Spray-guided	Bowl		Central	Central	Fuel: 0.72; air: 0.65	Prototype: air-assisted
PSA	1998		I-4	4	11.4	Wall-guided	Bowl	Reverse tumble	Intake side	Central	3–10	Production in Europe
RICARDO	325	74 × 75.5	1	4	12.7	Wall-guided	Bowl	Reverse tumble	Intake side	Central		
RENAULT	2000		I-4	4	11.5	Spray-guided	Bowl	Tumble	Central	Central	4–10	Production in Europe; homogeneous only
SAAB			I-4	4			Bowl		Central			Prototype: air-assisted injector; spark-plug-injector-integrated; homogeneous only
SUBARU	554	97 × 75	1	4	9.7	Spray-guided	Bowl	Tumble	Central	Intake side	7	
TOYOTA	1998	86 × 86	I-4	4	10.0	Wall-guided	Bowl	Swirl	Intake side	Central	8–13	Production in Japan (1st generation D-4)
TOYOTA	1998	86 × 86	I-4	4	9.8	Wall-guided	Bowl	Tumble	Intake side	Central		Production in Europe; fan-spray combustion system; homogeneous only (2nd generation D-4)
TOYOTA	2997	86 × 86	I-6	4	11.3	Wall-guided	Bowl	Tumble	Intake side	Central	~13	Production in Japan; fan-spray combustion system (2nd generation D-4)
VOLKSWAGEN	1390		I-4	4	12.1	Wall-guided	Bowl	Tumble	Intake side	Central		Production in Europe; use downstream NOx sensor

Table 1.4-1
Practical development targets for future G-DI engines

Fuel Economy	Up to 15% reduction in BSFC over an integrated cycle
Emissions	Compliant with stringent future emissions regulations
Specific Power Output	Comparable to PFI engines
Driveability	Comparable to or superior to PFI engines for cold starts, warm-up and load transients
Durability	Comparable to PFI engines

Chapter 2

Combustion System Configurations

2.1 Introduction

A well-designed, full-function G-DI combustion system provides for operation in both the homogeneous-charge and stratified-charge modes, and allows for a seamless transition between the two. This requires not only an optimal combustion system configuration, but also an associated, matched control system. Achieving superior performance in the stratified-charge operating mode requires much more development than does the corresponding homogeneous-charge option, mainly because of the difficulty in preparing and maintaining an ignitable mixture at the spark gap at the time of ignition over a wide range of engine operation. As illustrated in Fig. 2.1-1, the location and mixture strength of the combustible air-fuel mixture cloud can vary significantly from cycle to cycle during stratified-charge operation.

As summarized in Fig. 2.1-2, many and varied approaches to G-DI combustion systems have been proposed and developed over the years, with these systems incorporating a wide range of combinations of in-cylinder charge motion (swirl, tumble, squish), combustion chamber shape, piston geometry, and spark plug and injector locations [108]. The systems that are illustrated in Fig. 2.1-2a use a swirl-based, in-cylinder flow field to stabilize the mixture stratification. Ignition stability is maintained in the majority of the depicted systems by positioning the spark gap at the periphery of the fuel spray. Additional designs that incorporate a central spark plug and a non-central injector position are shown schematically in Fig. 2.1-2b. The concept of an off-axis piston bowl, fuel injection onto the floor of the piston bowl, and a central spark gap in the combustion chamber is depicted in Fig. 2.1-2b-a, with an adaptation of the flow-collision concept to the G-DI engine represented in Fig. 2.1-2b-b. This latter concept invokes a flow-to-flow collision at the center of the combustion chamber where ignition subsequently occurs. An open chamber designed to generate two quasi-divided regions near top dead center (TDC) is illustrated in Fig. 2.1-2b-c. Conceptually, this quasi-divided chamber limits and controls the air quantity that mixes with the late-injected fuel. Further, three examples of G-DI combustion systems using a tumble motion of

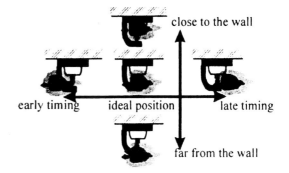

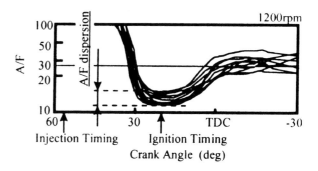

Figure 2.1-1 Mixture characteristics in the vicinity of the spark gap during stratified-charge operation [155].

the in-cylinder air flow field are shown in Fig. 2.1-2c, with a direct injection (DI) system that utilizes squish flow illustrated schematically in Fig. 2.1-2d.

Each G-DI combustion system that has been designed for stratified-charge operation can be assigned to one of the following major classifications: spray-guided, wall-guided and air-guided, on the basis of the strategy used for realizing stratified-charge operation during part load. The specific classification depends upon whether the spray dynamics, the spray impingement on the piston surface or the mixture flow field is primarily utilized to achieve stratification. A number of these systems will be described in this chapter, and their advantages and limitations will be compared. Guidelines will be provided for the development of future combustion systems.

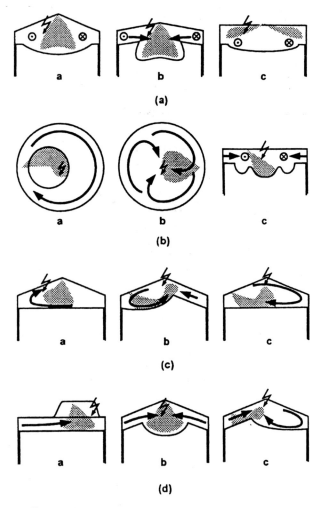

Figure 2.1-2 Typical G-DI combustion system configurations: (a) swirl-based systems with centrally mounted fuel injector; (b) swirl-based systems with centrally mounted spark plug; (c) tumble-based systems; (d) squish-based systems [108].

2.2 Relative Position of Fuel Injector and Spark Plug

The location and orientation of the fuel injector relative to the ignition source are critical geometric parameters in the design and optimization of a G-DI combustion system. The spark plug gap must always be positioned in a zone of ignitable and combustible mixture when the spark occurs, and this zone is affected by the swirl and tumble ratios of the in-cylinder air, as well as spray cone angle, mean droplet size, and the injection and spark timings [14, 108]. During high-load operation of the engine the selected injector spray axis orientation and spray cone angle must promote rapid and complete mixing of fuel

with the induction air in order to maximize the air utilization. For late injection, the spark plug and injector locations should ideally provide an ignitable mixture cloud at the spark gap at an ignition timing that will yield the maximum work from the cycle for a reasonable range of engine speeds. In general, no single set of positions is optimal for all speed-load combinations, thus the positioning of the injector and spark plug is nearly always a compromise. Hardware packaging constraints will also be an important consideration.

In extending the proven guidelines for designing PFI combustion chambers to the design of G-DI combustion chambers, a number of additional requirements must be satisfied. To minimize the flame travel distance, and to increase the knock-limited power for a specified octane requirement, a single spark plug is generally positioned in a near-central location. As with PFI systems, this location usually provides the lowest heat losses during combustion. The eccentricity of the spark gap position should be less than 12% of the bore diameter if a low octane requirement is to be obtained [202]. For a majority of the concepts the only factor that is constant seems to be the near-central position of the spark plug in the cylinder head [425]. Thus the main advantages inherent in the use of a near-centrally mounted spark plug are to obtain a symmetric flame propagation, to maximize the burn rate and specific power, and to decrease the heat losses and autoignition tendency. Although it is an additional complexity, the use of two spark plugs could increase the probability of effective ignition. Two ignition sources may enhance the robustness of the combustion process, but will contribute to the already difficult packaging problem. It should also be noted that the use of two spark plugs could require the injected fuel to be divided into two plumes and directed to each individual ignition source, which could significantly reduce the lean capability of the system.

Once a near-central location has been specified for the spark plug, numerous additional factors must be taken into account in positioning and orienting the fuel injector. The location selection process can be aided by laser diagnostics [102, 107, 119, 124, 125, 471] and computational fluid dynamics (CFD) analyses [101, 152, 153, 210, 289, 310, 403] that include spray and combustion models. However, the use of laser-based optical diagnostics generally requires a modification of the engine combustion chamber, as will be discussed in Chapters 4 and 5 on fuel sprays and spray-wall interactions. Spray models and wall-film sub-models for direct injection (DI) engines are considered to still be in the development stage, with many evolutionary improvements being incorporated each year. This means that, although CFD can be used to screen some possible locations, much of the final verification work on injector spray-axis positioning will consist of prototype hardware evaluation and proof-of-concept testing on an engine dynamometer [466, 467]. The basic considerations are that the injector position should provide a stable, stratified charge at light load, a homogeneous mixture with good air utilization at high load, and avoid the unintended impingement of fuel on the cylinder wall and on the piston crown outside of any piston cavity. Other important factors include the operating temperature of the injector tip, the fouling tendencies of the spark plug and injector, the compromise between intake valve size and injector location, and the design constraints of injector access and service. The effective area available for the intake valves is noted because most proposed injector locations negatively impact the space available for the engine valves, forcing smaller valve flow areas to be used. The important parameters influencing the final selected location of the spark plug and injector are summarized in Table 2.2-1.

As outlined in Table 1.3-1, gasoline direct-injection combustion systems may be differentiated in various ways depending on the categories that are used. In general, all G-DI combustion systems can be categorized as narrow-spacing and wide-spacing systems, based on the relative positions of the fuel injector and the spark plug, as depicted schematically in Fig. 2.2-1. Because the spark plug is usually positioned near the center of the chamber, the narrow-spacing concept has the fuel injector located with only moderate eccentricity (<12%) from the center of the chamber. Stratification of the mixture is promoted simply by the juxtaposition of the fuel injector and the spark gap. Therefore, the stratification capability of the narrow-spacing design is significantly higher than that of the wide-spacing design. In contrast, the wide-spacing design generally has the injector placed at or near the periphery of the combustion chamber. With the wide-spacing concept the fuel spray must be transported through a greater distance, and over a longer time period, from the injector tip to the spark gap. This transport is realized by means of the combined influences of spray momentum, in-cylinder air motion and piston crown geometry. In general, the G-DI injector should

Table 2.2-1
Important parameters for consideration in the selection of spark plug and injector locations

Design Issues	• Head, port and valve packaging constraints • Piston crown and bowl geometry, including bowl shape, volume, depth, and exit lip angle • Spark plug design and allowable extent of electrode extension into the chamber • Combustion chamber geometry • Injector access and serviceability
Injector-Related Issues	• Spray characteristics, including cone angle, penetration, sac volume, and mean drop size • Injector body and tip temperature limits • Injector access and serviceability
Charge Motion	• Structure and strength of the in-cylinder flow field • Velocity history at the proposed injector tip position (injector cooling)

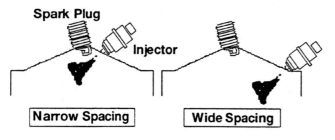

Figure 2.2-1 Comparison of narrow-spacing and wide-spacing concepts [196].

not be located on the exhaust side of the chamber, as injector tip temperatures of more than 175°C may result, exacerbating durability problems and promoting deposit formation, as discussed in detail in Chapter 7. The advantages and shortcomings of various injector locations are compared and summarized in Table 2.2-2.

The maximum intake valve size that can be utilized in conjunction with a head-mounted injector in a four-valve engine is depicted in Fig. 2.2-2 [108], where

Table 2.2-2
Comparisons of various injector positions and orientations

Centrally Mounted Position	• Higher ignition stability over the engine operating map • High degree of mixture stratification possible, but in a narrow time window • Good at distributing fuel uniformly throughout the chamber • Advantageous for homogeneous-charge operating mode • Mixture formation relatively independent of piston crown geometry • More difficult installation and removal for service • Reduced valve size • Possible spark plug fouling • Higher tip temperature and injector deposit tendency • Sensitive to variations in fuel spray characteristics • Higher probability of spray impingement on piston crown • Special spark plug with extended electrodes may be required
Intake-Side-Mounted Position	• Larger valve size permitted • Increased mixture preparation time available • Facilitated injector installation and removal • Less effect of variations in fuel spray characteristics • Injector tip more readily cooled by intake air • Lower tip temperature and deposit tendency • Less fuel impingement on spark plug electrodes • Standard spark plug possible • More constraints on fuel-rail positioning • Low degree of stratification and large scale fluctuations • Higher probability of spray impingement on cylinder wall • Increased probability of oil dilution
Exhaust-Side-Mounted Position	• Should be avoided

the two large open circles indicate the positions of two intake valves, and the two small open circles indicate the two exhaust valves. The two configurations on the left-hand side of the figure show the possible layouts for a narrow-spacing system having the injector and spark gap in close proximity. The combustion system schematic on the right shows several possible configurations for a system in which the spark gap and injector are widely separated [330]. The significant design constraints on valve size for a narrow-spacing system are evident, and it is usually necessary to invoke auxiliary methods for improving the volumetric efficiency, such

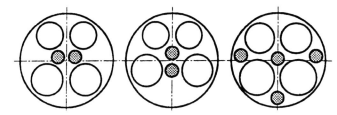

Figure 2.2-2 Valve size limitation for optional injector and spark plug locations [108].

as optimizing the intake port design for achieving a high flow coefficient.

Comparison of the high-load, homogeneous-charge performance between a side-mounted DI injector configuration and a centrally mounted DI injector configuration have revealed that charge homogeneity is normally superior for a centrally mounted injector, resulting in decreased CO and smoke emissions, as well as a higher attainable engine torque. At a comparable mid-range engine speed, the side-mounted injector is found to exhibit a 4.5% BMEP (brake mean effective pressure) advantage, while the centrally mounted injector provides a 6% BMEP advantage, over the PFI baseline. The side-mounted injector is found to yield the highest volumetric efficiency, as this configuration permits larger intake valve dimensions. Results for part load show a slight advantage in HC emissions for the central injector configuration. An overall comparison of the two systems indicates that the performance differences between the two systems are incremental, with only a very slight advantage exhibited for the central injector geometry. The choice of the configuration

for production is therefore governed mainly by manufacturability considerations rather than the performance increment [20]. It should be taken into consideration that nearly all such studies comparing DI configurations use combustion chamber geometries that are not fully optimized.

Finally, it should be noted that the spark plug electrodes must have the proper protrusion into the combustion chamber, as measured relative to the chamber wall surface, and must be positioned to the greatest extent possible to shield the gap from being affected by the bulk flow. The importance of the protrusion of the spark gap is not to be underestimated. Depending on the details of the combustion system design, a minimum protrusion (properly shielded) is required to achieve a stable stratified-charge combustion [31]. A platinum electrode with a protrusion of 7 mm is utilized in the Mitsubishi GDI ignition system. Other promising improvements include very small electrode diameters, individual ignition coils for each cylinder and the use of a larger discharge current and a longer discharge time [221].

2.3 Approaches to Achieving a Stratified Charge

On the basis of strategies for realizing stratified-charge operation during part load, the DI systems that are designed to operate in the stratified-charge mode are assigned to one of three major classifications: spray-guided, wall-guided and air-guided, illustrated schematically in Fig. 2.3-1. The specific classification for a particular system depends upon whether the spray dynamics, the spray impingement on a piston surface or

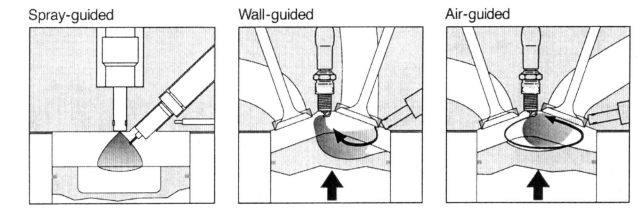

Figure 2.3-1 Schematic representations of spray-guided, wall-guided and air-guided combustion systems [369].

the mixture flow field is primarily utilized to achieve stratification. It should be emphasized that regardless of the classification, the mixture stratification of real systems is generally achieved *by some combination of these three mechanisms* in which the relative contribution of each individual process varies. Based on the separation distance between the spark gap and the tip of the fuel injector as discussed in Section 2.2, a spray-guided combustion system can be further defined as a narrow-spacing concept, whereas the wall-guided and air-guided systems generally utilize a wide-spacing concept.

2.3.1 Spray-Guided Combustion Systems

The concept of a spray-guided combustion system has its origin in the early days of DISC engine development. For example, the Ford PROCO system invoked this concept to achieve lean combustion, with mixture stratification resulting from the proximity between the spray periphery and the spark gap. The stratification capability of this design is significantly higher than that of a wide-spacing system and stratified operation with a lambda value exceeding 8.0 has been demonstrated [166]. In addition, stratification can be achieved without a strong dependence on charge motion control or piston cavity design. In this system concept, the in-cylinder charge motions and the associated turbulent fluctuations exert only a minor influence on the mixture transport. Compared to the other two classifications, a spray-guided combustion system configuration requires the fewest alterations from a conventional PFI design [275]. This is a consideration if an existing PFI engine design is being converted to G-DI operation.

Many shortcomings, however, limit the wide application of this approach. The proximity of the injection nozzle and the spark gap, and the very brief interval between fuel injection and spark discharge, exacerbate problems such as spark plug fouling and soot generation. This results from the presence of many fuel droplets near the spark gap [128, 197, 250]. The combustion characteristics exhibited by this system for stratified-mode operation are known to be quite sensitive to variations in the fuel spray characteristics. For example, variations in spray symmetry, skew or cone angle due to production variations or injector deposits can result in excessive values of the coefficient of variation (COV) of the indicated mean effective pressure (IMEP), and misfires and partial burns can occur for significant distortions of the expected baseline spray geometry. Given the fact that G-DI fuel spray characteristics are affected by so many variables,

accommodating the constantly changing environment that is associated with various engine operating conditions is a significant challenge in designing a spray-guided combustion system. The engine performance of such a system is known to be sensitive to both the injection and ignition timings, and is affected by the piston geometry [227]. The requirement for very close spacing of the injector and spark plug in some of these designs results in an additional reduction of intake valve sizes, and also further increases the sensitivity to variations in spray characteristics. The proximity of the injector and spark gap also introduces considerations of ignition fouling resulting from the impingement of the fuel droplets from the spray periphery onto the electrodes. Although the use of higher ignition energy can indeed decrease the susceptibility to fouling, this approach can also have a negative impact on the durability of the spark plug. As the formation of injector deposits is always a concern, positioning the injector tip closer to the ignition source could aggravate any deposit tendencies that are already present.

The spray atomization quality is quite critical to the success of spray-guided combustion systems due to the limited time available for fuel vaporization. That is why the air-assisted fuel system is considered to have a relatively high potential in spray-guided G-DI combustion systems. In addition to the improved mixture quality, the air-assisted fuel injection system uses pneumatic atomization to achieve mixture ignitability without depending on the bulk air motion. In fact, a reduction in the bulk flow velocity is effective in improving the stratification process, resulting in improved fuel economy for part-load stratified operation [63]. In order to take full advantage of the spray and fuel preparation qualities of the air-assisted DI fuel injector, the preferred injector location is aligned with the cylinder axis, with the spark plug also located near the center of the chamber. This arrangement has the advantage of minimizing wall wetting, reducing sensitivity to in-cylinder flow conditions and achieving a favorable alignment of the spray to the piston bowl. In this regard, a spark plug injector (SPI) module that integrates a spark plug and an air-assisted fuel injector into one unit has been explored to facilitate packaging in a modern 4-valve combustion chamber [353]. In contrast, the side-mounted injector with a central spark plug requires a very careful refinement of the piston bowl and combustion chamber geometries in order to achieve similar combustion and emissions capabilities [426]. It should also be noted that the insensitivity to in-cylinder flow conditions also makes the air-assisted fuel injection system relatively well suited to engine families with different valve configurations, without the need for active or passive flow control systems such as the swirl control valve.

A spray-guided combustion system that has been extensively investigated is one that positions the fuel injector near the center of the chamber and has the spark plug located near the periphery of a conical spray [9, 10, 95, 290, 407, 408, 477, 494]. Figure 2.3-2 shows a schematic representation of such a combustion system. This basic configuration is preferable in many chamber designs, as it virtually assures the presence of a rich mixture near the spark gap at the time of ignition. In addition, the vertical, centrally mounted injector location has the advantage of distributing the atomized fuel symmetrically in the cylindrical geometry, thus promoting good air utilization. For late injection, however, the centrally mounted, vertical location directs the fuel spray at the piston and is likely to yield higher HC emissions and smoke than an optimally inclined orientation. This unintended spray impingement on the piston crown can be minimized by extending the distance between the injector tip and piston surface by means of a deep bowl in the piston or a dome in the combustion chamber [128]. In some studies using a combustion chamber layout with a centrally mounted injector and spark plug, it was recommended that the injector be located on the exhaust side of the chamber, with the spark plug on the intake side. This allows conventional tumble in the correct direction to carry the fuel cloud to the spark plug during the compression stroke [275, 473]. This is illustrated schematically in Fig. 2.3-3. This configuration avoids spark plug wetting during early injection by taking advantage of the intake charge motion; however, the higher thermal loading of the injector on the exhaust side could increase the rate of injector deposit formation, and should not be implemented without a careful evaluation.

In general, it is more difficult to apply G-DI technology to engines having very small bore diameters, as the additional packaging of a fuel injector into a multivalve, small-bore engine is a significant challenge [64]. For the small-bore G-DI engine with a side-mounted injector, it is difficult to avoid the impingement of fuel onto the opposite cylinder wall, but this indeed must be avoided from the perspectives of HC emissions and dilution of lubricating oil. Due to the increased probability of fuel wetting on the combustion chamber surfaces, it is also likely that the effective increase in volumetric efficiency and the net cooling effect resulting from

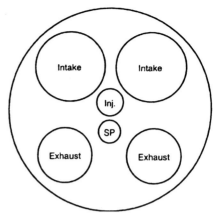

Figure 2.3-2 Schematic of spray-guided combustion system with centrally located fuel injector (Inj.)and spark plug (SP)[10].

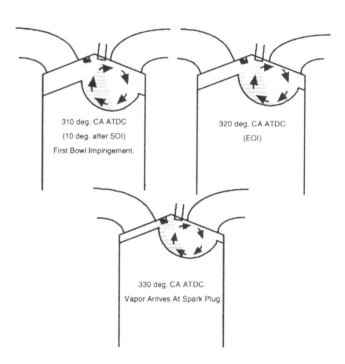

Figure 2.3-3 Schematic of spray-guided combustion system with fuel injector located on the exhaust side [275].

in-cylinder fuel vaporization will be somewhat less for a small-bore engine. Moreover, a small-bore engine tends to operate at a higher speed than a large-bore engine. Achieving the capability of operating in the stratified-charge mode at a moderate load without significant smoke emissions is a development challenge.

The narrow-spaced, spray-guided concept is regarded as a viable design for G-DI engines, particularly when the packaging constraints of a small-bore engine are considered. With this type of engine, valve diameters have to be kept large enough to ensure good breathing, and the injector position has to be optimized in order to maximize the positive effect of charge cooling on the volumetric efficiency. For G-DI engines with a bore diameter of less than 75 mm, a 3-valves-per-cylinder configuration with the spark plug close to the injector, with a piston cavity that confines the fuel spray to the vicinity of the spark gap, has been proposed [279]. Figure 2.3-4 shows a schematic of this proposed combustion chamber. To allow for a large intake valve diameter, the intake valve stems are vertical, and the exhaust valve is located in the pentroof of the combustion chamber. This single exhaust valve is designed to provide a more rapid catalyst light-off than is achievable with two exhaust valves. An eccentric location of the spark plug is utilized for reasons of packaging. Due to the smaller bore size, such a location does not result in a large degradation of combustion efficiency resulting from heat loss and autoignition. Both the spark plug and injector are displaced from the zone of highest temperature, and are reasonably accessible for service. The confinement of the fuel spray to the vicinity of the spark gap is generally achieved by an optimized bowl in the piston. The bowl has an inclined axis to restrict the fuel-air mixture to the vicinity of the ignition position and to guide the charge to the spark gap. The inclination angle of the injector and the spray characteristics of cone angle and penetration are considered to be the key parameters that must be selected carefully to minimize any possibly unintended fuel impingement on the cylinder wall and piston crown.

An additional spray-guided combustion system configuration that incorporates a dome in the cylinder head to confine the fuel close to the spark plug is illustrated in Fig. 2.3-5 [128, 370]. This combustion system utilizes a central spray plug and a vertically mounted, swirl-type injector that provides a relatively narrow fuel spray. For this spray-guided concept the best ignition condition is obtained in the tail of the spray. This system also exhibits an optimum ignition timing that is highly dependent on the dynamics of the particular spray being utilized. Because the combustion chamber must incorporate a recess for locating both the injector and the spark plug in the center of the chamber, it is difficult to incorporate this concept into a multi-valve small-bore engine.

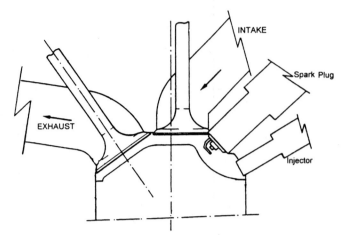

Figure 2.3-4 Schematic of a narrow-spacing, three-valve, G-DI combustion chamber for a small-bore engine [279].

One of the first vehicle manufacturers to use a spray-guided DI combustion system in production was Renault. A cut-away of the Renault DI system is illustrated in Fig. 2.3-6. It is important to note that this engine is fully designed to operate only in the

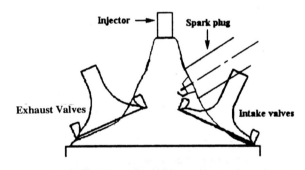

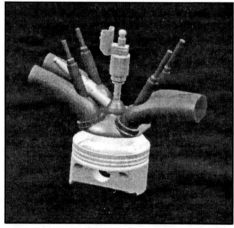

Figure 2.3-5 Schematic of a spray-guided combustion chamber incorporating a dome in the cylinder head [128, 370].

homogeneous, stoichiometric mode in order to take full advantage of the three-way catalyst aftertreatment system. A deep bowl is incorporated into the piston crown to minimize fuel dispersion toward the cylinder wall [138].

Figure 2.3-6 Cut-away view of the Renault spray-guided combustion system [138].

2.3.2 Wall-Guided Combustion Systems

System Layout. In concept, two important ways to move toward obtaining a stable stratified mixture are to incrementally decrease the time interval between injection and ignition and to decrease the distance between the injector tip and the spark gap [446]. However, a decrease in the distance between the injector and the spark gap, which is the near-limiting case for a spray-guided combustion system, also decreases the time available for mixture preparation. This generally has a negative effect on HC emissions and soot formation. Improved fuel-air mixture formation can be obtained with the possible loss of some combustion stability by increasing the separation distance between the spark gap and the injector tip, or by increasing the time delay between fuel

injection and ignition. This is the fundamental reasoning behind the use of a wide-spacing concept.

For open chamber designs in which the stratification is supported mainly by charge motion, a stable stratification can be obtained with directed spray impingement on a shaped piston crown or a bowl cavity, with subsequent transport of the fuel vapor toward the spark plug by the charge motion and the momentum of the spray-induced air flow field. This wall-guided concept is illustrated schematically in Fig. 2.3-7 [196]. With the use of a specially contoured piston and an optimized balance of tumble and swirl, the transport of the fuel that impinges on the piston can be controlled to obtain stratified combustion. As will be discussed in Section 5.4.1, only a minority fraction of the liquid droplets actually impinge on the wall and form a fuel film for G-DI fuel sprays. The majority of the liquid droplets, and the bulk of the liquid mass injected, follows the wall contour while entrained within a highly transient air flow field that is established by the injection event. This majority of droplets never actually impinge on the bowl floor, nor does the film that is formed move along the bowl floor. The overall effect is certainly wall guidance of the charge by the wall contour, but the actual phenomenon is more complex than is normally depicted in describing the mechanics of the wall-guided system.

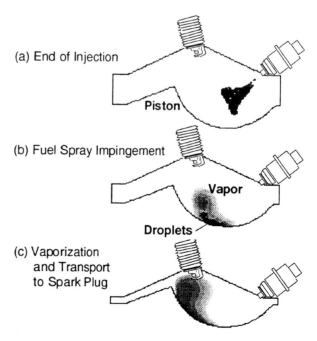

Figure 2.3-7 Fuel transport process in the wall-guided DI combustion system [196].

19

For a four-valve gasoline engine with a centrally located spark plug, the possible injector locations for increasing the separation between the spark gap and the injector tip are illustrated in Fig. 2.3-8. One is to locate the injector on the intake valve side of the combustion chamber, with the other location being at the cylinder periphery between the intake and exhaust valves. An analytical study of the mixture formation process associated with the latter location indicates that when the spray is directed toward the center of the combustion chamber at an angle of 70° down from the horizontal plane, this injector location and orientation produces a poor mixture distribution near the spark gap when used with a flat piston [310]. This may be improved substantially by modifying the piston crown geometry [274]. The option of placing the injector at the periphery of the cylinder between the intake and exhaust valves does exist, but this must be weighed against the likely higher thermal loading for the injector tip. Almost all of the current production and prototype wall-guided G-DI systems have the fuel injector located underneath the intake port between the intake valves. This injector location, along with positioning the spark plug at the center of the cylinder, has been extensively investigated [182, 183, 259, 273, 387, 438–440]. This geometric configuration is considered to be a key element of a compact design for a multi-cylinder G-DI engine, particularly for those having smaller bore diameters [201]. It provides an improved entrainment of the induction air into the fuel spray for early injection, as well as an enhanced cooling of the injector tip [273]. Figures 2.3-9, 2.3-10, 2.3-11, and 2.3-12 show the schematics of four production, wall-guided, DI combustion systems, all of which are described in detail in Chapter 10.

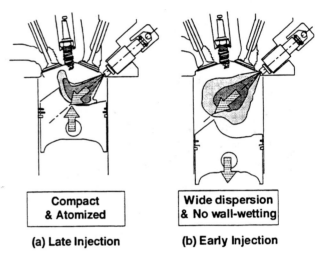

Compact & Atomized | **Wide dispersion & No wall-wetting**

(a) Late Injection | **(b) Early Injection**

Figure 2.3-9 Schematic of the Mitsubishi GDI combustion system [259].

As opposed to the configuration using a centrally mounted injector, systems using spray impingement on piston surfaces to create a stratified mixture generally exhibit a reduced sensitivity to variations in spray characteristics. Hence, these systems tend to be more robust regarding the effects of spray degradation from deposits, pulse-to-pulse variability or part-to-part production variances in the fuel spray envelope. However, locating the injector at the periphery of the chamber may lead to an increase in unintended fuel impingement on chamber surfaces. This is likely to reduce the benefit of in-cylinder charge cooling for the early-injection mode. For example, locating an injector at the periphery of the cylinder on an Isuzu single-cylinder G-DI engine produced higher HC emissions and fuel consumption than was obtained with the same injector located at the center of the combustion chamber [407]. This was theorized to be due to fuel penetrating to the cylinder wall on the exhaust valve side, with an increased absorption of

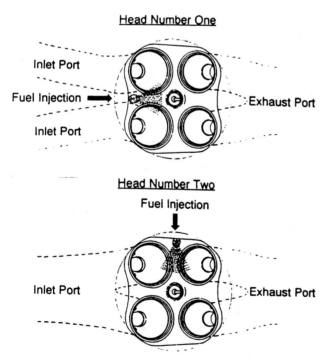

Figure 2.3-8 Possible injector locations for increased separation between injector tip and centrally located spark plug [273].

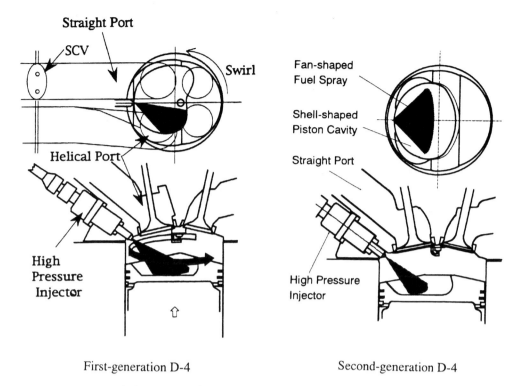

First-generation D-4   Second-generation D-4

Figure 2.3-10 Schematics of the Toyota first-generation and second-generation D-4 combustion systems (SCV: Swirl Control Valve) [156, 223].

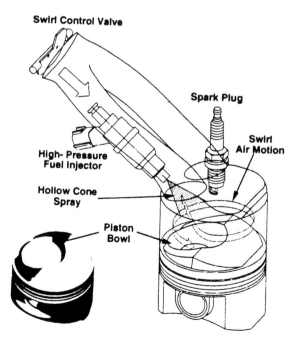

Figure 2.3-11 Schematic of the Nissan NEODi combustion system [440].

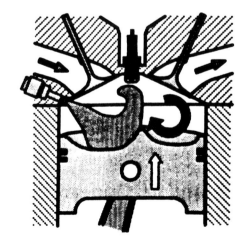

Figure 2.3-12 Schematic of the Volkswagen FSI combustion system [49].

liquid fuel and desorption of fuel vapor by the lubricating oil film [521]. The direct impingement of fuel on the cylinder wall or the piston crown forms a film of liquid fuel on the surface, which can result in some pool-burning and an increase in HC emissions, smoke, or both [281, 423]. The use of a side-mounted injector should be carefully evaluated for any increased oil dilution and more rapid combustion chamber deposit (CCD) formation, especially for high-load operation [436].

Piston Bowl Design. It is well established that the specific geometry of the piston bowl is a very important factor in the performance of a wall-guided G-DI combustion system, and bowl optimization and spray/bowl matching are two of most critical phases of combustion system development. Figures 2.3-13, 2.3-14 and 2.3-15 show photos of the piston bowls that are utilized in three production G-DI engines. Obviously piston bowl design varies significantly, depending on specific applications, and the bowl depth is known to be a key parameter. Generally the mixture at the time of ignition tends to be too rich with a small bowl in the piston at light and medium load. A deeper bowl is effective in extending stratified-charge operation over a wide range of operating conditions, but will negatively impact full-load combustion [503]. As shown in Fig. 2.3-16, both the ISFC and the HC emissions are reduced by utilizing an open piston-bowl configuration (bowl-shape 2 in the figure), but this also results in an increase in the NOx emissions [227]. The use of a reentrant bowl is considered to degrade both the spray dispersion and fuel-air mixing significantly, which may yield an increase in soot emissions.

Figure 2.3-14 Photograph of the piston used in the Toyota first-generation D-4 engine [492]. (Courtesy of Toyota Motor Sales, USA, Inc.)

Figure 2.3-15 Image of the piston used in the Nissan NEODi engine [351].

Figure 2.3-13 Photograph of the piston used in the Mitsubishi GDI engine [259].

An important consideration in developing a wall-guided, spray-impingement system is the necessary compromise that must be made in shaping the piston crown to obtain part-load stratification. This compromise will most likely be somewhat detrimental to air utilization at high-load conditions for which a homogeneous mixture is required. As compared to a flat, horizontal piston crown, the introduction of a piston crown profile will increase the heat losses from the combustion gases and will also likely increase the manufacturing complexity. The effect of the piston crown profile and piston bowl geometry on the full-load performance of the engine must also be well evaluated. The engine power output for WOT operation is generally reduced markedly when a bowl is incorporated into the piston crown to accommodate stratified-charge combustion. As illustrated in Fig. 2.3-17, with

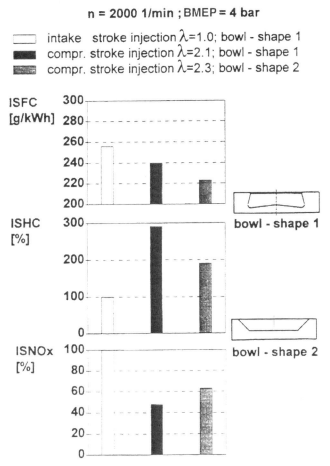

Figure 2.3-16 *Effect of piston bowl geometry on ISFC, ISHC and ISNOx* [227].

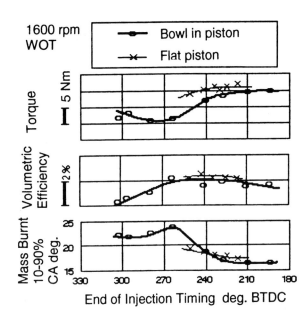

Figure 2.3-17 *Effect of piston bowl design on DI gasoline engine WOT performance* [183].

early-injection, homogeneous-charge operation, the piston bowl designed for promoting stratified-charge operation has a pronounced negative impact on the mixture formation process as compared to a flat piston. An extended torque range is obtained with a flat piston because the use of a flat piston promotes a more uniform mixture for homogeneous operation [183, 339]. The penalty in peak engine torque nearly always increases with increased piston bowl depth, or more specifically, with increased bowl volume, and this very important compromise must be evaluated carefully.

In-Cylinder Air Flow. For a wall-guided G-DI system the in-cylinder air flow field is important in transporting the fuel plume to the spark gap. It contributes to the entrainment of air into the spray plume and to transporting the vapor cloud along the proper path to the spark gap. An effective scavenging flow is also necessary to enhance fuel film vaporization from the piston cavity surfaces. This topic is discussed in detail

in Section 5.4. The majority of G-DI combustion systems that have been developed and reported, including the early DISC engine concepts, utilize in-cylinder air swirl as the primary in-cylinder air motion. A swirl-dominated, in-cylinder flow field is generally combined with either a simple, open combustion chamber or a bowl in the piston as a framework for a wall-guided combustion system. As illustrated in Figs. 2.3-10 and 2.3-11, the Toyota first-generation D-4 and Nissan NEODi are both classical, swirl-based, wall-guided DI systems.

With regard to the application of tumble-dominated, in-cylinder air flow fields, engineers at Mitsubishi [259] and Ricardo [199, 200, 273] first proposed the reverse tumble concept in conjunction with a specially profiled piston cavity to create a stratified charge near the spark gap. As illustrated in Fig. 2.3-9, the cavity is designed to control the spray impingement and subsequent flame propagation by enhancing the reverse tumble flow throughout the compression stroke, with the subsequent squish flow from the exhaust to the intake side of the chamber increasing the tumble motion and flame speed inside the piston cavity. Reverse tumble as the dominant in-cylinder air motion may indeed be effective for designs in which the spark plug is centrally located and the injector is positioned below the intake valve. In such designs a reverse tumble can be effective in transporting the vapor and liquid fuel droplets toward the spark gap after spray impingement/redirection on the walls of the piston cavity. The reverse tumble of the Mitsubishi GDI engine is achieved by a straight, vertical intake port, which also provides additional space in the cylinder head to accommodate the injector. As illustrated in Fig. 2.3-12, engineers at Volkswagen applied a conventional tumble flow field to a wall-guided DI combustion system having a side-mounted injector. The DI systems that utilize squish flow were illustrated earlier in Fig. 2.1-2d.

Packaging constraints often have significant weight in the selection of a charge motion for a particular G-DI system. A key design step in this selection is the establishment of the basic combustion chamber shape, which is significantly influenced by the valve angle. The pentroof combustion chamber favors both the creation and preservation of the large-scale tumble motion that is required for stratified combustion. Because the combustion chamber height required for such a G-DI system can be obtained to a large extent within the pentroof, the increase in the compression height of the piston is only marginal. Such systems can operate using both conventional and reverse tumble with a valve angle larger than 45°. However, if smaller valve angles are considered, the preservation of an intake-generated swirl motion is more efficient, mainly due to reduced momentum dissipation and less transformation into secondary flows. The height of the combustion chamber can thus be reduced, leading to a reduced piston crown height and a piston of less mass. Therefore, for a small to medium valve angle, swirl-based systems offer benefits that may be realized with less modification to existing PFI engines. This is part of the reason why the majority of G-DI combustion systems that have been developed to date are swirl-based [482]. The influence of the valve angle on the requirements of the combustion chamber, piston bowl and flow field is illustrated in Fig. 2.3-18 [481]. The tumble-flow concept is recommended for a 4-valve engine having a valve angle that exceeds 40°, whereas the swirl-flow concept is recommended for 4-valve and 3-valve engines having valve angles of less than 30°. With valve angles from 30° to 40°, either charge-motion strategy can be applied effectively.

Injector Mounting and Orientation. For implementing a side-mounted injector the alignment of the fuel injector relative to the intake port requires design compromises. The use of an intake port configuration that is largely identical to a PFI port leads to a very shallow injector elevation angle above the horizontal plane [109]. Such a configuration does provide an intake port

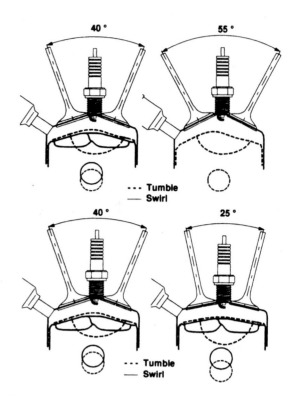

Figure 2.3-18 Impact of valve angle on combustion chamber, piston design and flow field requirement [481].

with known, good flow characteristics, but the spray may impact the intake valves for early injection, possibly causing deposit formation, and may penetrate across the cylinder to cause bore wetting and top-land crevice fuel loading, as illustrated in Fig. 2.3-19. This can be partially circumvented by using either an offset spray or a larger injector inclination angle. As will be discussed in Chapter 4, an offset swirl spray is usually of slightly lower atomization quality than one that is not offset. In addition, some injector nozzle types are not available in an offset spray configuration. Therefore, the feasibility of increasing the injector inclination angle should be the first consideration. With a larger injector inclination angle the interaction between the spray and the target piston cavity is less dependent on the engine crank angle, which geometrically leads to a wider speed range for stratified operation. Based on this fact, Mitsubishi engineers [15, 342] modified the configuration of the Mitsubishi GDI combustion system for the European market by installing the injector more vertically. This does indeed extend the range of stratified operation to higher engine speeds, as shown in Fig. 2.3-20. This also improves the combustion stability

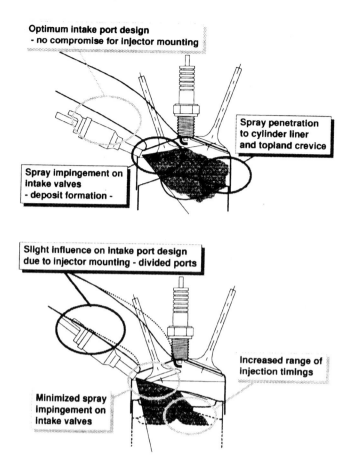

Figure 2.3-19 Effect of injector inclination on spray development inside the cylinder [481].

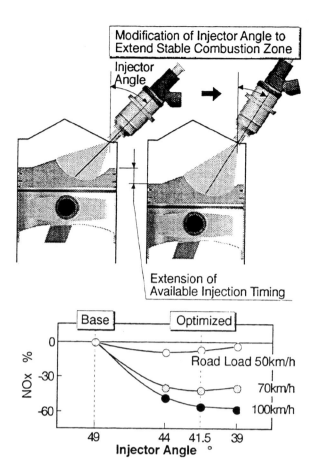

Figure 2.3-20 Comparison of spray targeting for different injector inclination angles in the Mitsubishi GDI engine [342].

under high exhaust gas recirculation (EGR) operation. As a result, a significant engine-out NOx reduction is achieved.

A compromise exists regarding the overall effects of injector inclination angle on fuel wall wetting, engine performance and exhaust emissions [31, 182, 183, 339]. Figure 2.3-21 shows the predicted fuel amounts, as computed by CFD, on various surfaces of the Nissan DI combustion system for three injector inclination angles relative to the horizontal plane. The injection timing was fixed at 270° before top dead center (BTDC) on compression for all three cases. By increasing the injector inclination angle, namely mounting the injector more vertically, the amount of fuel impinging on the cylinder wall is predicted to decrease only slightly. The amount of fuel wetting the head surface is reduced significantly, but the amount of fuel impinging on the piston crown exhibits a marked increase. With a more vertical injector mounting, the amount of fuel wetting the total internal surfaces decreases during early injection, implying that a

more vertical injector mounting is advantageous for reducing wall wetting. However, a more vertical spray axis degrades the engine WOT performance, as is illustrated in Fig. 2.3-22, due to a lower degree of mixture homogeneity. As a result, increasing the injector inclination angle degrades both the smoke emissions and the engine BSFC. There is indeed a consensus in the literature that the injector inclination angle is an important parameter that is to be considered carefully in developing and optimizing a G-DI combustion system.

It should be noted that, in addition to the injector mounting orientation, the injector protrusion into the combustion chamber must also be such that fuel spray impact on the intake valves is eliminated or at least minimized. As illustrated in Fig. 2.3-23, a flush mounting of the injector tip in the combustion chamber can significantly reduce the fuel wetting of the valves, thus assisting in avoiding deposit formation.

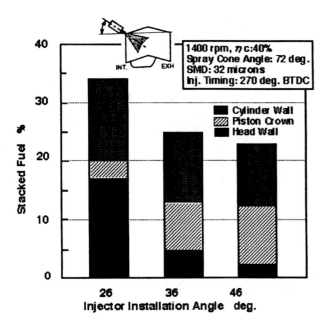

Figure 2.3-21 *Effect of injector inclination angle on the amounts of fuel impinging on the cylinder head, piston crown and cylinder wall* [440].

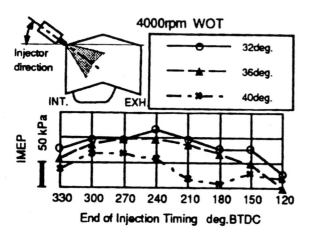

Figure 2.3-22 *Effect of injector inclination angle on the WOT performance; engine speed: 4000 rpm* [339].

2.3.3 Air-Guided Combustion Systems

For an air-guided combustion system, charge stratification is controlled by the interaction of the fuel spray and the in-cylinder, bulk air motion. Figure 2.3-24 illustrates a tumble-based, air-guided DI combustion system, in which a strong tumble flow is generated by a flow control valve, which, in turn, is utilized to establish a stratified charge. This type of system generally has the

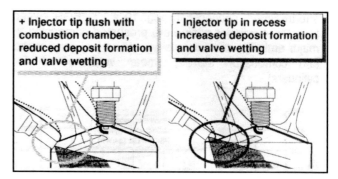

Figure 2.3-23 *Effect of injector tip recess on spray behavior inside the cylinder* [481].

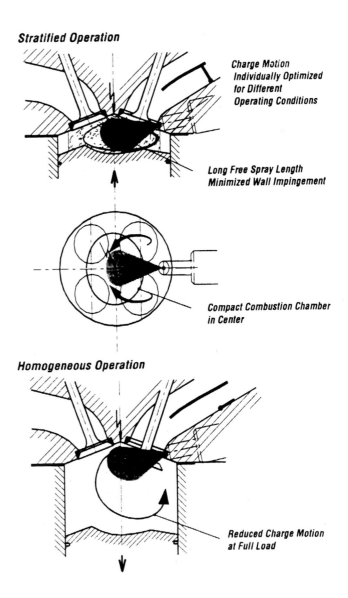

Figure 2.3-24 *Schematic representation of an air-guided DI combustion system* [127].

injector tip and spark gap separated by a large distance, thus exhibiting the advantages of the wide-spacing concept. Because charge stratification is conceptually achieved without relying on spray-wall impingement, some of the difficulties associated with wall-guided combustion systems such as wall-film-related HC emissions can be avoided. However, any factors that cause fluctuations in the in-cylinder flow field may lead to combustion instabilities. At lower engine speeds with an air-guided system, the combustion stability degrades slightly because the overall strength of the charge motion is reduced. For this reason a careful matching of the charge motion and spray characteristics to the combustion chamber geometry and spark gap position is required over the engine speed range [43, 250].

Theoretically, an air-guided combustion system does not require special piston geometry for stratified mixture preparation; however, specially designed piston geometry is generally necessary to assist in the generation of a desired charge motion. Although the piston shape for the air-guided combustion system is considered to be favorable for homogeneous operation, the accommodating requirements for a strong charge motion will generally penalize full-load performance.

2.4 Summary

Many different approaches have been proposed to generate a controlled stratified charge during light-load operation. Although the three basic system classifications of spray-guided, wall-guided and air-guided are distinct, charge stratification in an operating engine is generally achieved by some combination of the three different mechanisms. In order to obtain the best engine combustion characteristics with any system type, the spark plug should be located near the center of the combustion chamber. For a spray-guided DI system, mixture preparation is controlled primarily by the fuel spray dynamics, as opposed to domination by the spray-wall interaction at the piston cavity surface in a wall-guided system and a strong dependence on the flow field in an air-guided system. For a spray-guided system ignition occurs in a region with large gradients in the local mixture fraction, and is therefore sensitive to variations in the spray geometry. Any direct impingement of the spray on the spark plug electrodes can lead to plug wetting, and difficulties with cold starting and spark plug durability may be encountered.

In the wall-guided concept, injected fuel is directed toward the spark gap by an interaction with a combustion chamber surface, generally a well-defined piston cavity. The in-cylinder flow field is important, but is not dominant. In comparison to the spray-guided concept, the wall-guided system exhibits a reduced sensitivity to variations in spray characteristics, but a diminished degree of charge stratification and increased HC emissions are the principal limitations. In the air-guided concept, the vaporized fuel is directed toward the spark gap using a well-defined in-cylinder flow field. The interactions between the injected spray and the in-cylinder flow field must be coordinated over a large region of the engine operating map, and any factors that lead to variations in the engine flow field may result in combustion stability degradation. Configurations that utilize wide spacing between the injector and the spark gap have fewer inherent geometric and thermal constraints for the design of the combustion chamber. The increased time scale for mixture transport from the injector to the ignition location can enhance mixture preparation, but fluctuations in turbulence and related cyclic variability become more influential. Both the narrow-spacing and wide-spacing configurations can be combined with the charge-motion concepts of swirl and tumble to achieve viable G-DI combustion systems that can operate in the stratified mode. A disadvantage of the narrow-spacing concept is the associated restriction on valve sizes in multi-valve engines. In the spray-guided concept, the appropriate air/fuel ratio at different engine loads must be achieved by the selection of the required cone angle and penetration characteristics of the injected spray, and is therefore mainly determined by the spray characteristics. Due to the short distance between the injector and the spark gap, only a brief mixture formation time window is available prior to ignition, thus the engine requires a late injection. Moreover, liquid fuel may impinge onto the spark plug electrodes, increasing the propensity for spark plug fouling, which can lead to misfiring. Due to the relatively small zone of ignitable mixture at the periphery of the spray, the spray-guided system is strongly sensitive to spray geometry variations and injector installation tolerances, and requires fine atomization. Nevertheless, the narrow-spacing concept does provide the potential for ultra-lean combustion if the listed obstacles can be circumvented.

Chapter 3

Fuel Injection System

3.1 Introduction

Unthrottled operation with the load controlled by the fuel quantity has been shown to be a very efficient operating mode for the internal combustion engine, as the volumetric efficiency is increased and the loss associated with pumping work is significantly reduced. This control method is very successful in the diesel engine, as ignition occurs spontaneously at points within the combustion chamber where the mixture is well prepared for autoignition. The fixed location of the ignition source in the spark ignition (SI) engine, however, makes it quite difficult to operate in the unthrottled mode for other than high loads. A critical additional requirement is imposed on the mixture formation process of this type of engine in that the mixture cloud that results from fuel vaporization and mixing must be controlled both spatially and temporally in order to obtain stable combustion. This implies that the load must be reduced and controlled by invoking mixture leanness, mixture stratification, or both. Preparing and locating a desired mixture inside the combustion chamber over the full range of engine operating conditions is quite difficult, as the fuel-air mixing process is influenced by many time-dependent variables. The development of a successful unthrottled SI combustion system depends upon the optimized design of the fuel injection system and the proper matching of the system components to control the in-cylinder flow field and burn rate history.

The fuel injectors of early DISC engines were derived from the basic diesel fuel system. For example, the Texaco TCCS engine [8] utilized a diesel-type injector that produced a spray with relatively poor atomization and fuel-air mixing quality, and with high penetration rates relative to sprays from current pressure-swirl atomizers. The Ford PROCO engine [401] used an outwardly opening pintle atomizer with vibration to enhance fuel atomization; however, the poppet opening pressure was on the order of 2.0 MPa (gauge pressure), which is relatively low. The early work on gasoline injection using single-hole and multihole, narrow-angle sprays from converted diesel injection systems resulted in a high penetration velocity and in substantial HC emissions. In order to circumvent this problem, alternative combustion systems using direct wall impingement and fuel film formation, such as the MAN-FM [453], were developed. These systems were less sensitive to the degree of atomization achieved, and were optimized for controlled evaporation from a liquid film. For early DISC engine experiments using multihole injectors, the singularity of the ignition source proved to be a significant constraint, as compared to a diesel combustion system where multiple ignition sites occur simultaneously. Another constraint of the early diesel-based DISC injection systems was the lack of variability of injection characteristics between part load and full load. For full-power DISC engine operation, injection of maximum fuel quantities early in the intake stroke generally resulted in significant wall impingement of the fuel, mainly due to the high penetration rate of the fuel spray.

In recent years, significant progress has been made in the development of advanced, computer-controlled, fuel injection systems, which has had much to do with the expansion of research and development activities related to G-DI engines. In this chapter the key components that are utilized in the fuel system and the associated contributions to the processes of mixture preparation and combustion are described in detail.

3.2 Fuel System Requirements

The injector in a G-DI system is a key component that must be carefully matched with the specific in-cylinder flow field to provide the desired mixture cloud over the entire operating range of the engine. G-DI combustion systems, including wall-impingement systems, generally require a well-atomized fuel spray for all operating conditions, with a "well-atomized" G-DI spray generally considered to be one in which the Sauter mean diameter (SMD) is less than 20 μm. It is necessary for the fuel system to provide for at least two, and possibly three or more, distinct engine operating modes [259]. Fuel injection systems for full-feature G-DI engines must have the capability of providing both late injection for stratified-charge combustion at part load, as well as injection during the intake stroke for homogeneous-charge combustion at full load. At full load, a well-atomized and dispersed fuel spray is desirable to ensure a homogeneous charge for even the largest fuel quantities. This is generally achieved by early injection at low cylinder pressure, similar to the mode of open-valve fuel injection in a PFI engine. For unthrottled, part-load operation, a well-atomized compact spray is desirable to achieve rapid mixture formation and controlled stratification. The injection system should provide the capability for rapid injection moderately late in the compression stroke into an ambient pressure of up to 0.9 MPa, which requires a relatively elevated fuel injection pressure. In total, the G-DI fuel system requirements are more comprehensive than those of the PFI injection systems, as illustrated in Table 3.2-1; however, common-rail injection systems with electromagnetically activated injectors have been identified to meet these requirements [55, 144, 292, 305, 361, 366, 378, 432]. In principle and basic function, this common-rail fuel system is comparable to those utilized in current PFI and advanced diesel engines, with the G-DI application having an intermediate fuel pressure level.

The enhanced capabilities of G-DI combustion systems do offer some advantages in terms of effective use of well-atomized fuel sprays. It could perhaps be logically argued that the fuel system requirements of a G-DI combustion system are not an advantage, but a disadvantage. Such a logic position might be that G-DI combustion systems require fuel sprays that have mean diameters of under 20 μm, whereas PFI combustion systems can operate quite well using fuel sprays of 120–200 μm SMD, or even 20–200 μm SMD. To achieve such stringent atomization requirements the majority of G-DI fuel systems must utilize levels of fuel pressure that are 15–40 times higher than those of production PFI fuel systems. Even air-assisted G-DI injectors utilize a fuel pressure that is nearly double that of PFI, although some piezoelectric and multi-stage injectors do not require such high primary pump pressures. Other G-DI injectors, however, require a more complex and substantial pump, rail, injector, regulator and control system. Thus, there may be some merit in noting that it is not entirely fair to compare the atomization level and evaporation rate of a G-DI fuel spray with that of a PFI fuel spray, as the PFI engine can operate on both fuel systems, whereas the G-DI engine cannot. In this regard the relative advantage or disadvantage is in the

Table 3.2-1

Comparison of PFI and G-DI fuel systems

Fuel System	PFI	G-DI
Advantages	• Relatively simple • Robust in terms of spray parameter variations such as mean drop size • Low pressure operation <0.5 MPa • Cooler injector operation	• Enhanced fuel atomization and drop vaporization rate • More rapid rate-of-injection • Faster opening and closing of injector
Disadvantages	• 6 to 10 ms spray flight time to valve target • 85 to 250 μm mean drop diameter • Most of the injected fuel is not vaporized prior to puddle impact • Slow injector opening and closing	• Increased system complexity • Requirements on pumps, rail, regulator and injector are more substantial • More stringent spray requirements (combustion is less robust regarding spray parameter variations) • Requires mean drop diameter of 20 μm or less • Injector generally runs hotter • Increased pump load

eye of the beholder. With regard to fuel atomization levels, what *is* a true, and technically more correct, advantage of the G-DI engine is that G-DI combustion systems exhibit an enhanced capability for converting fuel atomization improvements into BSFC, COV of IMEP and emissions enhancements than do PFI combustion systems. In short, G-DI combustion systems can take greater advantage of the more complex fuel systems. As shown in Table 3.2-1, PFI combustion systems are more robust in terms of operating with a wide range of delivered fuel spray atomization levels, but the corresponding disadvantage is that very significant improvements in PFI operating efficiency and emissions do not occur as the mean drop size is reduced continuously from 120 μm to 20 μm. Gradual improvements in PFI operation are observed, and certain options such as open-intake-valve injection for rapid starting become viable, but one very important fact becomes evident: Once a fuel system has been developed to provide a sub-20-μm spray to a PFI engine, the same fuel system can be used to obtain even greater benefits if used in the direct, in-cylinder-injection mode. This is a true advantage of G-DI over PFI, not one that results from comparing a 12.0 MPa injection system to one operating at 0.3 MPa.

3.3 Fuel Injector Requirements and Considerations

The fuel injector is indeed a key component in a G-DI combustion system; in fact, it is arguably the most influential component, as it is the main determinant of the spatial and temporal location of the fuel. The phasing of the injection, along with the spray cone boundary and the rate of injection, will significantly influence the fuel-distribution history within the combustion chamber. Secondary effects such as the presence of spray asymmetries, a sac spray, after-injections or pulse-to-pulse variability in any of the above six parameters will also alter the resultant fuel-distribution history. In turn, this will alter the distribution of mixture stoichiometry within the chamber at the time of ignition and during the early stages of heat release.

The fuel injector should have the general attributes listed in Table 3.3-1, with many of the required characteristics of the G-DI injector being equivalent to those of the PFI injector. In certain critical areas the specification tolerances of a G-DI injector design are more stringent than those of the port fuel injector, and there are certain requirements that are not present for PFI applications.

Table 3.3-1
Requirements of G-DI injectors

General Requirements for Gasoline Injectors (PFI and GDI)	Specific Requirements for G-DI Injectors
• Accurate fuel metering (generally a ±2% band over the flow range) • Desirable fuel mass distribution pattern for the application • Minimal spray skew for both sac and main sprays • Good spray axisymmetry over the operating range • Zero drippage and fuel leakage, particularly for cold operation • Small sac volume • Good low-end linearity between the dynamic flow and the fuel pulse width • Small pulse-to-pulse variation in fuel quantity and spray characteristics • Minimal variation in the above parameters from unit to unit	• Significantly enhanced atomization level; a reduced value of spray mean drop size; application dependent, but certainly under 20 μm SMD • Ability to deliver desired spray shape and penetration at different ambient back pressure conditions • Expanded dynamic range • Combustion sealing capability • Avoidance of pintle bounce that creates unwanted secondary injections • Reduced bandwidth tolerance for static flow and flow linearity specifications • More emphasis on spray penetration control • More emphasis on the control of the sac volume spray • Enhanced resistance to deposit formation • Smaller flow variability under larger thermal gradients • Ability to operate at higher injector body and tip temperatures • Leakage resistance at elevated fuel and ambient pressures • Minimized leakage and improved sealing capability at –40°C and 10–20 MPa fuel pressure • More emphasis on enhancements in packaging options • Flexibility in producing off-axis sprays in various inclined axes to meet different combustion system requirements • Ability to conduct multiple injections within one engine cycle

The transient injection of gasoline directly into the combustion chamber is typically accomplished using an electronically actuated injector that is connected to a high-pressure, common-rail fuel system. The event occurs in a matter of milliseconds, and involves the injection of only milligrams of fuel. The typical duration of injection may range from 0.9 to 6.0 ms, with the amount of fuel delivered ranging from 5 to 60 mg, all dependent on the engine type and load, as well as on the static flow capacity of the particular injector. This flow capacity is typically stated by injector manufacturers in units of volumetric flow rate of fuel at the rated fuel rail pressure with the injector locked open. Nearly all production and prototype G-DI injectors for automotive application have static flow capacities that fall within the range of 9 to 24 cm^3/s. The fuel is metered to the engine in the same manner as with port fuel injection or common-rail diesel injection; that is, by precisely controlling the duration of a square-wave voltage pulse that is commonly designated as the fuel pulse width (FPW). The required FPW is computed for each injection event from calibration algorithms, and is based upon the output of sensors for the engine speed, load-demand and aftertreatment hardware operation. This computer-controlled pulse is sent to an injector driver module, where it is converted to a shaped current pulse that activates the opening solenoid. Each injector design has a linear working flow range within which the delivered mass and volume of fuel is incrementally proportional to the increment of fuel pulse width. This range of minimum and maximum flow rates and fuel pulse widths define the working flow range and dynamic range of the injector.

The G-DI injector should be designed to deliver a precisely metered fuel quantity with very repeatable spray geometry, and must provide a highly atomized fuel spray having an SMD of generally less than 25 μm, and with a DV90 not exceeding 45 μm [82, 83]. Smaller values than these are even more beneficial provided sufficient spray penetration is maintained for good air utilization. The SMD is sometimes denoted more formally in the spray community as D32. The DV90 statistic is a quantitative statistical measure that is a metric for the largest droplets in the total distribution of all droplet sizes. They will be described in detail in Sections 4.2 and 4.16.4. The fuel pressure required is at least 4.0 MPa for a single-fluid injector, with 5.0 to 7.0 MPa being more desirable if the late-injection, stratified-charge mode is to be invoked. Even if successful levels of atomization could be achieved at fuel pressures lower than 4.0 MPa, significant metering errors could result from the variation of metering pressure differential ($P_{inj} - P_{cyl}$) with cylinder pressure, although this could be corrected by the engine control system if cylinder pressure were monitored. The sac volume within the injector tip is basically the volume of fuel, resulting from the previous injection, which is not at the fuel line pressure. It always includes fuel that is in residence downstream from the sealing surface, but for swirl-type injectors it also includes the fuel mass that does not experience the full increase in angular momentum prior to exiting the injector. This is approximated by the volume of fuel that resides in the lower portion of the swirl channels. This fuel mass retards the acceleration of the injected fuel and generally degrades both the fuel atomization and the resulting combustion. There is also a general consensus that the volume and geometry of the sac spray can influence the air entrainment process and the resulting spray cone angle for a swirl injector, particularly at elevated ambient densities. In general, the smaller the sac volume, the fewer large peripheral drops will be generated when the injector opens. Needle bounce on closure is generally to be avoided, as a secondary injection generally results in uncontrolled atomization consisting of larger droplets of lower velocity or possible unatomized fuel ligaments. This contributes to a small degree to increases in the HC and particulate emissions. Needle bounce on closure also degrades the fuel metering accuracy. Needle bounce on opening is not nearly as important in this regard as that on closure, but should be controlled. The result of a needle bounce on opening is a small modulation in the injection rate and spray cone angle.

The ability to provide a high rate of injection, which is equivalent to delivering the required fuel with a short fuel pulse, is much more important for the G-DI engine than for the PFI engine, particularly for light-load, stratified-charge operation. Therefore, significant importance is attributed to the low-pulse-width region of the G-DI injector, effectively increasing the importance of the injector dynamic range requirement. Standard measures of G-DI injector capabilities are the dynamic (linear) working range, the shortest operating pulse width on the linear flow curve and the shortest stable fuel pulse width. Fuel pulse widths that are shorter than the linear range limit can optionally be used in the calibration by means of a lookup table of pulse width if the fuel mass delivery is stable at that operating point. The optimal design of the injector to resist deposit formation (coking) is also one of the important requirements of the G-DI injector, discussed in detail in Chapter 7. Zero leakage under full rail pressure, from –40°C to +140°C operating temperature, must be achieved for a G-DI injector. Drippage, which differs from leakage in that it is liquid fuel that gradually accumulates on the injector tip from either poor closure or a recirculation of fuel droplets around the injector tip. With combustion occurring each cycle, liquid fuel does not accumulate until it drips, as is relatively common with PFI injectors. Liquid fuel accumulated in the injector tip, however, does have an important impact on injector deposit formation. Sometimes overlooked are the voltage and power requirements of the injector solenoids and drivers. A number of prototype G-DI injectors have power requirements which would be considered unacceptable for a production application. It is also worth noting that it is advantageous to injector and rail packaging to have the body as small as possible. This provides more flexibility in optimizing the injector location and in sizing and locating the engine ports and valves.

In some applications having significant packaging constraints, the G-DI injector is required to deliver a spray that is offset from that of the injector axis [156, 351]. This requirement makes the G-DI injector design somewhat more complex, as some of the techniques for producing a symmetric and well-atomized spray may not be compatible with spray-offset methods. For example, swirling flow, as an effective technique to enhance the atomization of a symmetric spray, is generally less effective in an injector design having an offset spray. As a result, a higher injection pressure may be required to achieve the same degree of spray atomization of an on-axis spray design. Also, achieving an offset spray fundamentally may not be possible with some injector designs such as an outwardly opening pintle.

3.4 Fuel Pressure Requirements

The fuel pressure within the injector body has been determined to be very important for obtaining effective spray atomization and the required level of spray penetration. This pressure is very closely related to the regulated pressure within the fuel rail. A higher injection pressure is effective in reducing the mean diameter of the spray approximately as the inverse square root of the pressure differential ($P_{inj} - P_{cyl}$), whereas the use of a lower pressure generally reduces the fuel pump parasitic load, system priming time and injector noise, and generally increases the operating life of the fuel pump

system. Even with the current level of fuel pressure, the noise emitted from G-DI injectors and fuel pumps is of concern, and noise abatement research is continuing. The use of a very elevated fuel rail pressure, such as 20 MPa, will enhance the atomization of the fuel but may yield an over-penetrating spray that wets the combustion system boundaries, especially when a sac volume is present. Excessive penetration is not assured for high fuel pressures because there are counteracting effects. A higher fuel pressure also produces smaller droplets that attain terminal velocity sooner, and may yield lower penetration for some droplet size distributions. The fuel pressures that have been selected for most of the current prototype and production G-DI engines range from 4 to 13 MPa—quite low when compared with diesel injection system pressures of 50 to 160 MPa, but relatively high in comparison with typical PFI injection pressures of 0.25 to 0.45 MPa.

Although direct-injection gasoline fuel systems typically are designed for fuel rail pressures in the range of 4.0 MPa to 13.0 MPa, the trend over recent years in both production and prototype fuel systems is to increase the levels of fuel pressure. Although for reasons of fuel pump loading, life and priming time there is continuing interest in achieving adequately atomized G-DI fuel sprays using reduced fuel rail pressures of 1.0 to 2.0 MPa, the trend has nonetheless been in the opposite direction. Fuel pressures of 5.0 MPa to 7.5 MPa were common in the mid-1990s, but operating pressures of 10.0 MPa to 13.0 MPa are more typical only seven years later. Such pressure levels, coupled with a typical road-load fuel pulse width on the order of 1.5 ms, make the injection event a highly transient process that incorporates a large momentum exchange between the injected droplets and the ambient air. The momentum exchange also occurs rapidly, typically becoming 90% completed within 6.0 ms for a 1.5 ms fuel pulse. A flow field of entrained air may be quickly established, depending upon the specific geometry of the spray. Certainly, for a swirl-type injector with an initially hollow-cone spray near the injector tip, the resultant pressure field creates a highly transient flow of entrained ambient air that interacts with the penetrating liquid drops to alter the dynamics of spray cone formation. This is why spray cones from swirl-type G-DI injectors can "collapse" to smaller effective cone angles than might be indicated by the initial exit angle, and why this transition in spray cone is affected by the ambient air density.

Vehicle operation in the field incorporates many modes of engine and fuel system operation, including cranking and starting, restarting after hot soak, very cold start and warm-up, variable fuel rail pressure operation and selected limp-home capabilities that will partially accommodate equipment operating problems. One common operating mode is the crank and start, which, in the absence of auxiliary, parallel fuel systems, must rely on obtaining adequate fuel pressure to achieve stable combustion without excessive emissions. The mechanical, high-pressure G-DI fuel pump requires a certain time interval, or engine revolutions, to achieve a reasonable fraction of the design pressure. For the rapid start times that are expected with current automotive engines, standard high pressure G-DI fuel pumps do not achieve a very significant fraction of their design pressure. Until very rapid priming pumps are developed and proven to be durable, high pressure G-DI fuel systems must be designed to accommodate the crank-and-start using significantly reduced levels of fuel pressure. Alternatively, auxiliary pressure storage or parallel fuel systems must be added. All current production G-DI engines operate on the low-pressure, in-tank fuel pump for starting until the high-pressure pump achieves a certain level of design pressure. If the G-DI injectors that are designed for use at 7 to 10 MPa are to be used to provide fuel at pressures of less than 0.5 MPa, then the spray should be characterized for that condition. Some G-DI injectors will perform better than others under these off-design conditions, with the low-pressure, air-assisted injector being the least degraded. The air-assisted system, however, requires a supply of compressed air, which may also have to be pumped to design pressure for a crank and start, thus some degradation of the air-assisted fuel spray may also be encountered. Obviously the calibration of the high-pressure injector must accurately account for the greatly reduced fuel pressure, and significantly extended fuel pulse widths must be utilized. The fuel spray will be very degraded in terms of atomization level, even attaining 100 μm SMD, and the optimum cycle time to inject the fuel must be determined. The transition to high-pressure operation must be controlled by the engine management system (EMS) using the fuel pressure signal from the regulator. This is also true if an auxiliary fuel delivery system such as a throttle-body-injection (TBI) unit is utilized.

Spray visualization at the cranking fuel pressure of 0.3 MPa revealed that the swirl-type injector produces a hollow-cone spray; however, the atomization is noticeably degraded [405]. For such a low fuel pressure the injection duration required for a stoichiometric mixture is approximately four times that required when operating from the high-pressure main fuel pump, as the injection duration required is approximately proportional to the square root of the fuel injection pressure. If the fuel pressure is too low, or if the amount of the fuel required is too large under the conditions of low-temperature cold start and transient acceleration, the G-DI injector may not be able to deliver a sufficient amount of fuel within the allowable time window. For example, a low-pressure fuel injector in an auxiliary throttle body was used in the first-generation Toyota D-4 engine in order to meet the fueling demand for enhanced cold startability [37]. It is evident that being able to quickly deliver a high rail pressure during cold start is crucial for the G-DI engine if the potential of G-DI technology is to be realized during cold start. This potential includes a more rapid cold start, lower cold-start HC emissions and less fuel enrichment.

To circumvent the priming-time problem, the Mitsubishi GDI engine uses an in-tank fuel feed pump that is similar to that used in PFI engines. A bypass valve is used to allow the fuel to bypass the high-pressure regulator during engine crank and start. As a result, the electric feed pump supplies the fuel directly to the fuel rail. As the engine speed and fuel rail pressure increase, the bypass valve is closed and the high-pressure regulator begins to regulate the fuel pressure to the design pressure of 5.0 MPa. With this strategy, the engine can be started within 1.5 seconds under both cold and hot restart conditions [196, 197].

A constant fuel-rail pressure is utilized in the common-rail configuration for most current G-DI applications; however, a strategy using a variable fuel injection pressure does offer an alternative method of obtaining the required fuel flow rate range while reducing the linear-dynamic-range requirements of the injector [364]. It is also an option for meeting different fuel-spray requirements over a range of engine loads [302]. This is an important consideration that must be evaluated for each G-DI application.

The working flow range (WFR) of an injector may be effectively extended by the use of a variable fuel rail pressure. The dynamic range of an injector that uses a fixed fuel pressure is determined by the minimum and maximum flow rate limits for the linear flow curve. These flow rates are directly related to the minimum and maximum fuel pulse widths that can be specified while still maintaining a linear flow relationship between flow rate and FPW. If, however, a second, lower fuel rail pressure is utilized, then there is a second complete flow curve for the alternate pressure. The engine control system can use these two flow curves to an advantage, thus extending the usable FPW to values lower than the original limits. For example, the required fuel for idle operation might be injected using an FPW of 0.88 ms for a 10 MPa fuel rail pressure, or alternatively injected using an FPW of 1.38 ms at 7 MPa. The longer FPW may be in the linear range of the low-pressure flow curve, whereas the shorter pulse may not be in the linear flow range of the high-pressure flow curve. In effect, the dynamic range of the injector is extended significantly if a much lower fuel rail pressure can be used as an option for some portion of the engine operating map. Two practical limitations on this concept are that the fuel atomization and rate of injection will be degraded as the fuel rail pressure is significantly reduced; and that a strategy using dual fuel pressures or a fully variable rail pressure is obviously more complex.

In principle, the required fuel to be injected is easily computed from the existing signals plus the fuel rail pressure signal and a lookup table of the particular flow calibration. Thus the required fuel mass may, in principle, be provided using an infinite number of fuel rail pressures. However, both the rate of injection and the atomization level achieved will be degraded as approximately the square root of the pressure ratio. The alternative use of a 5 MPa fuel pressure in a nominal 10 MPa system may be expected to yield an approximately 40% reduction in the rate of injection and a 40% increase in the SMD of the spray, from perhaps 16 to 22 μm. The spray quality may prove to be the limiting criterion for many injector designs, as substantial reductions in fuel pressure would have to be used to extend the injector dynamic range by even 20%. Before any injector is used as part of the strategy of variable-fuel-pressure or dual-fuel-pressure, the design should be evaluated to ensure that a spray of reasonable quality can be provided at the low pressure limit desired.

Finally, it is worth noting that a wide range of gasolines, having variable alcohol content, sulfur content and driveability indices, are available in the field. This quality range makes it a necessity that a robust fuel system be developed and proven for G-DI production engines. To avoid dilution of the engine oil, most G-DI fuel pumps are lubricated with gasoline [432]. However, gasoline has a lower lubricity, a lower viscosity and a higher volatility than diesel fuel, generally resulting in higher friction and a greater potential for wear and leakage as compared with diesel fuel pumps. It should be noted, however, that hydrodynamic lubrication may be used at high fuel pressures to compensate for the low viscosity.

3.5 Fuel Injector Classification

The area of G-DI injector design is currently characterized by rapid change. The developments in this area may be classified as those involving basic atomization type, those involving injector packaging and those involving continuous improvement of delivered injector performance (the area in which the most extensive change is occurring). G-DI injectors are currently supplied for automotive production or combustion system development by many manufacturers. Such injectors may be first broadly categorized by actuation type, and may be additionally classified by atomization mechanism, nozzle configuration or spray geometry, as listed in Table 3.5-1. In principle, any combinations of these classifications are possible for different types of injector designs. For example, an air-assisted, swirl-type injector actuated by piezoelectric actuators can theoretically be designed. G-DI injectors are available among the various types in many spray configurations, including hollow-cone, solid-cone, fan (line), multi-plume, offset or custom-shaped designs. All of these configurations relate to the geometric distribution of fuel mass within the spray plume, which is a very important consideration in the matching of the fuel spray to the combustion chamber geometry. Within many of the classifications is a range of available spray cone angles from 30° to 90°, as well as a range of injector static flow rates, or flow capacities.

Although the term "piezoelectric injector" is commonly used in the vernacular of G-DI fuel sprays, it would be more precise to use the designation "piezoelectric actuator." For example, the terms "piezoelectric" and "multihole" are not necessarily mutually exclusive, as a multihole injector with a piezoelectric actuator could certainly be constructed. Although the "piezoelectric injector" may be considered as perhaps a special case, it is generally true that the resultant G-DI fuel spray is very much determined by the particular nozzle tip design, and only to a lesser degree by the method and speed of actuation. Hence the classification of injectors and fuel sprays is most meaningfully accomplished by using the categories listed in Table 3.5-1.

Table 3.5-1
Classification categories for G-DI injectors

Actuation Mechanism	• Single-solenoid • Dual-solenoid • Piezoelectric • Hydraulic • Cam
Fluid State	• Single fluid • Air-assist (two phase)
Primary Atomization Method	• Sheet (swirl-plate) • Pressure (hole-type) • Pressure (slit-type) • Turbulence (compound plate) • Pneumatic (air-assist) • Cavitation • Impingement
Nozzle Configuration	• Swirl • Slit • Multihole • Cavity
Pintle Opening Direction	• Inwardly opening • Outwardly opening
Spray Configuration	• Hollow-cone • Solid-cone • Fan • Offset • Multi-plume • Shaped

Table 3.5-2 summarizes the principal production and prototype injectors that have been developed for G-DI applications. The high-pressure, single-fluid, swirl-channel injector using an inwardly opening pintle has been the most common G-DI injector type. However, additional types of inwardly opening, high-pressure, single-fluid G-DI injectors are now being developed. These include the multihole injector and the slit injector. The multihole injector offers the flexibility of adjusting the spray cross-section and spatial distribution of fuel within the cross-section by varying the hole pattern, number of holes, individual hole diameter and hole inclinations. This is analogous to programming the holes in PFI injector director plates. Pulse-pressurized, air-assist (PPAA) injectors are also available that utilize a significantly reduced fuel pres-

sure of less than 1.0 MPa. This injector design incorporates two separate solenoids and requires a supply of compressed air. The PPAA design generally employs an outwardly opening poppet, although this is not required. The outwardly opening designs, whether air-assist or single-fluid, have no sac-volume spray in the classic sense, which tends to yield an improved overall mean drop size. However, it is generally agreed that it is more difficult to develop a portfolio of off-axis, angled sprays with this constraint. The development and optimization of off-axis sprays is a key factor in the packaging options for injection systems. Component packaging in G-DI design is indeed a primary consideration, and the continued evolution from a 12 mm to a 10 mm to an 8 mm injector mounting diameter is being driven by the need to enhance packaging options. The availability of injectors that deliver fuel sprays being offset 10°, 20° or 30° from the injector-mounting axis can offer significant flexibility in the original design configuration of a G-DI combustion chamber.

3.5.1 Inwardly Opening, Single-Fluid Swirl Injector

Currently, the most widely utilized G-DI injector is the single-fluid, swirl-type unit that utilizes an inwardly opening pintle, a single exit orifice and a fuel pressure in the range of 7.0 to 10 MPa. A schematic of this type of injector is illustrated in Fig. 3.5-1. This swirl-spray injector is designed to impart to the fuel in the injector nozzle a strong rotational momentum that adds vectorially to the axial momentum. In a number of nozzle designs, liquid flows through a series of tangential holes or slots into a swirl chamber. The liquid emerges from the single discharge orifice as an annular sheet that spreads radially outward to form an initially hollow-cone spray. The initial spray cone angle may range from a design minimum of 25° to nearly 150°, depending on the requirements of the application, with a delivered spray SMD ranging from 14 μm to 23 μm. In the swirl-type injector the pressure energy is effectively transformed into rotational momentum, which enhances atomization [365, 487, 488]. The swirl nozzle generally produces a spray having a narrower distribution of drop sizes (DV90-DV10) than is obtained with the standard hole-type nozzle, with the best atomization occurring at high delivery pressures and wide spray angles. However, increased surface roughness of the orifice wall in swirl injectors tends to exacerbate the formation of streams or fingers of fuel in the fuel sheet exiting the nozzle, resulting in the formation of pockets

Table 3.5-2
Principal fuel injectors developed for G-DI application

Injectors	Comments
Inwardly-Opening, Single-Fluid, High-Pressure, Swirl Injector	• Hollow-cone spray • Solid-cone spray • Symmetric spray: spray is symmetrically distributed across the spray axis. The spray axis can be either aligned with the injector axis or offset from the injector axis (also denoted as a "bent spray"). • Asymmetric (shaped) spray: mass distribution across the spray axis is not symmetric. The spray axis may or may not be on the injector axis.
Outwardly-Opening, Single-Fluid, High-Pressure, Swirl Injector	• In general, it produces a hollow-cone spray without a sac spray. It is difficult to produce an offset spray from this type of injector.
Pulse-Pressurized, Air-Assisted Injector	• This type of injector uses a significantly reduced fuel pressure, but requires two separate solenoids and a source of compressed air for improved spray atomization and dispersion.
Hole-Type Nozzle	• Single-hole • Multihole pattern: spray structure varies significantly with the number and arrangement of the holes.
Slit-Type Nozzle	• The geometry of the nozzle is in the form of a slit, which delivers a fuel spray in a fan shape.

of locally rich air-fuel mixture. In order to minimize such mixture inhomogeneity, precise control of the swirl channel surface finish and nozzle tip quality is required [446].

3.5.2 Shaped-Spray Injector

With an axi-symmetric fuel distribution such as is delivered by the conventional cone-shaped spray, the piston surface in a wall-guided system tends to be wetted non-uniformly when the injector is installed with the injector axis inclined to the piston surface. Due to pakaging constraints, this is quite common in many G-DI applications. Consequently, the region of the piston closer to the injector tends to have a larger fraction of impinging fuel, as shown in Fig. 3.5-2a. This may result in a substantially thicker film of fuel being formed at that location, causing increased HC and particulate emissions. To avoid this problem a shaped-spray injector, also called a "casting net injector" in the literature, has been developed to provide an angled spray with a uniform distribution of fuel on the piston crown, as depicted in Fig. 3.5-2b [351, 445]. As illustrated in Fig. 3.5-2c, the offset angle of the spray axis from that of the injector axis, β, referred to as the "deflected angle," and the spray short-side and long-side lengths L2 and L1 are key parameters that mustbe optimized. Both the ratio L1/L2 and the deflected angle β have marked impacts on engine combustion stability, torque and smoke emissions, as illustrated in Fig. 3.5-3. As shown in Fig. 3.5-4, the injection-timing window is about 45 crank angle degrees wider with the shaped-spray injector than with the conventional injector. With early injection timing, liquid film formation on the piston crown is diminished by weakening the vertical component of spray velocity with this injector. With late injection timing, it is claimed that improved mixture homogeneity can be achieved due to

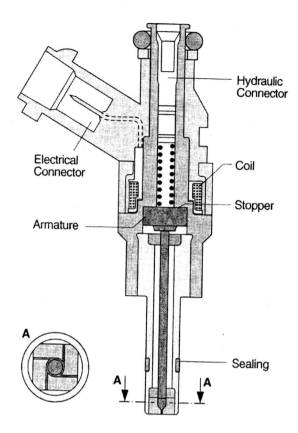

Figure 3.5-1 Schematic of inwardly opening, single-fluid, swirl-type DI injector [261].

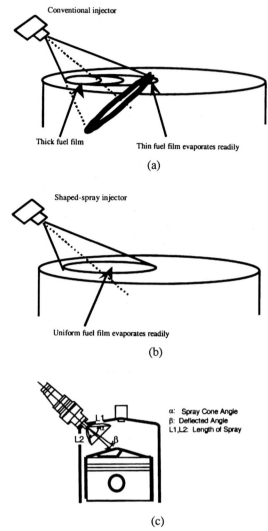

Figure 3.5-2 Comparison of impinged spray characteristics for the conventional and shaped-spray DI injectors: (a) conventional G-DI injector; (b) shaped-spray injector; (c) key parameters affecting the characteristics of the shaped spray [351].

an enhanced horizontal velocity component. In both cases, a reduction in smoke is achieved.

Figure 3.5-5 shows two nozzle designs (L-cut and taper-cut) for the shaped-spray injectors [312]. The spray pattern of the shaped-spray injector can be tailored to produce an inclined hollow-cone spray by varying the symmetric shape at the tip of the orifice [238, 312]. The advantage of this class of spray nozzle is that customization may be achieved by modifying the nozzle tip, while other components are the same as in a conventional swirl-type G-DI injector.

3.5.3 Slit-Type Nozzle

Figure 3.5-6 shows a schematic of the slit-type nozzle. The slit-type injector nozzle generally has a single rectangular orifice, although in principle multiple or tapered slits could be incorporated. The slit may be positioned to direct the fan-shaped spray either on-axis or off-axis. The spray from a slit-type injector has a cross-section that reflects the slit geometry, and, for a single rectangular orifice, expands downstream to form a spray in

the shape of a fan. There is generally a small sac volume downstream of the valve seat. Depending upon the effective L/W_M of the slit, where W_M is the major slit dimension and L is the length of the slit cavity along the flow direction, a range of nominal fan included angles may be generated. The slit nozzle may be produced by a high-resolution electric discharge machining (EDM) process [443].

3.5.4 Multihole Nozzle

A number of injector manufacturers have designed and developed a portfolio of multihole injectors for use in

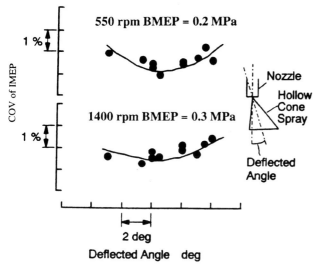

Figure 3.5-3 *Effect of spray deflected angle on BMEP and engine combustion stability* [445].

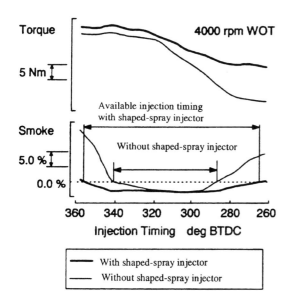

Figure 3.5-4 *Comparison of engine torque and smoke emissions between the conventional and shaped-spray injectors* [445].

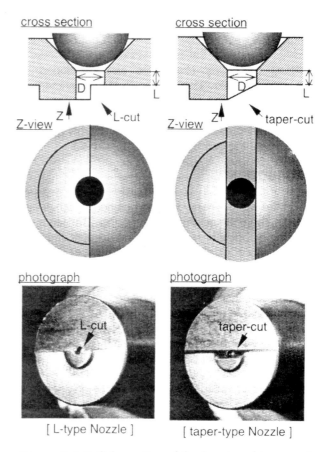

Figure 3.5-5 *Schematics of the L-cut and taper-cut shaped-spray nozzles* [312].

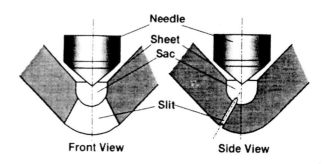

Figure 3.5-6 *Schematic of the slit-type DI nozzle* [443].

G-DI combustion systems. These injectors utilize a nozzle not unlike that of a diesel valve covered orifice (VCO) tip, and having 4 to 10 holes for fuel delivery. Whereas the diesel nozzle tip injects with a very wide conical pattern (130° to 170°), the gasoline multihole injector typically injects in a much more compact pattern, such as 30° to 90°. The specific variables that constitute the portfolio of multihole injectors includes the number of holes, the included angle or angles of the spray pattern, the offset of the centroid of the pattern from the injector axis and the pattern itself. The hole pattern, in fact, does not have to be circular, but may, in principle, be any shape.

39

Multihole injectors may be designed to provide a wetted footprint in the shape of a circle, an ellipse, a line, a crescent, a half-circle or a triangle. As with PFI injectors having a multihole director plate, this type of injector offers a number of distinct advantages for both manufacturability and combustion system optimization. In principle, an infinite number of spray patterns may be generated by the injector manufacturer by programming the hole distribution and sizes, thus creating many potential applications from a base injector of 4–10 holes using complex algorithms that relate the final spray pattern (single, dual, wide and narrow) to the hole pattern, size and orientation.

The advantage to G-DI combustion system optimization is that, in principle, nearly any spatial distribution of fuel mass can be obtained with a multihole nozzle. Having eight to ten holes in a nozzle, each with its own position, diameter and orientation relative to the injector axis, provides great flexibility in placing the fuel where it is needed. Unfortunately, during combustion system development it is not known precisely where and how the fuel should be targeted; only the engine knows this, although CFD modeling could be of significant assistance. Therefore, some trial-and-error testing is inevitably required, even for an injector design that provides a multiplicity of available spray patterns. This is not to imply that all combinations of multihole nozzle tips are currently available; they are not. The contemporary, electronically actuated, multihole G-DI injector is a recent development, and only a moderate number of tip options are available. A number of options just discussed are currently unavailable. For example, an eight-hole tip using four, or even two, different hole diameters is not available. The discussion cites the flexibility of the design concept, which does provide for tailoring of individual holes, *if necessary*. If it is later shown to be particularly advantageous for combustion-system optimization, it can be incorporated.

An additional advantage of the multihole nozzle is that the spray can be offset from the main injector axis with little or no penalty in terms of atomization. The penalty that is associated with offsetting a swirl-nozzle spray by an angle of 20° is such that a multihole injector that is offset by 20° will generally provide an equal or better-atomized spray than that of a swirl spray that is offset by an equivalent angle. It must be realized, however, that basic, non-enhanced, pressure atomization is used to obtain the fuel droplets, hence relatively high fuel rail pressure (9.5 to 12.0 MPa) must be employed. The atomization quality, particularly the DV90 value, degrades rather rapidly as the fuel rail pressure is lowered.

3.5.5 Outwardly Opening, Single-Fluid Swirl Injector

Even though most current G-DI injectors are inwardly opening, the outwardly opening pintle design has some advantages that should be considered [75, 460, 461, 487]. For example, the outwardly opening pintle injector is able to avoid the initial sac spray generated by the sac volume of most inwardly opening G-DI injectors. Moreover, the initial liquid sheet thickness is directly controlled by the pintle stroke rather than by the angular velocity of the swirling fuel in an inwardly opening nozzle. As a result, the outwardly opening injector has a design flexibility that allows the spray angle, penetration and droplet size to be controlled with less coupling. Swirling flow may also be used in the outwardly opening pintle design for reducing the spray penetration and increasing the spray cone angle. In addition, the sheet thickness provided by the outwardly opening pintle is smaller than that of the inwardly opening pintle during the valve opening and closing events, mainly due to the reduced pressure drop at the swirler. As a result, the atomization level obtained during opening and closing could be better than is obtained during the main spray portion, which is the opposite of what is generally obtained using inwardly opening pintles. As no nozzle holes are directly exposed to the combustion chamber environment, the outwardly opening design may prove to be more robust to combustion product deposition. More development and evaluation experience is required on this issue. With regard to the relative advantages of inwardly opening versus outwardly opening needles, the inwardly opening needle generally provides better pulse-to-pulse repeatability of spray cone geometry, especially when a flow-guide bushing is present at the needle tip. The outwardly opening geometry is, however, acknowledged to have enhanced leakage resistance [364, 388] because the combustion gas pressure assists in sealing the injector positively.

The design and operational concerns include the requirement for a very precise flow surface on the interior side of the pintle. It is well established that this surface must have a superior microfinish if excellent symmetry of the spray geometry and atomization is to be attained. Without such a microfinish the resultant spray will exhibit striations, or even voids, on portions of the conical sheet, and individual fingers of spray will form, with one finger for each swirl channel. The requirement for a superior surface finish leads to the operational concern, which is the sensitivity to injector deposits. Deposits are not any more likely to form on the outwardly opening pintle than on one that opens inwardly. However, the outwardly opening design is generally more sensitive to small amounts of surface deposits, as such deposits modify the microfinish of the original part. Thus, what can occur with cumulative engine operation is that spray fingers are formed when the original microfinish is degraded by deposits. In addition, it is generally agreed that it is more difficult to develop a portfolio of off-axis, angled sprays with the outwardly opening nozzle.

3.5.6 Piezoelectrically Actuated Injector

Piezoelectric actuation utilizes the very rapid incremental change in the lattice dimensions of certain crystals when a voltage is applied. If such crystals are properly configured in series, forming what is called a piezo stack, then the total dimensional shift is the sum of the changes for the individual crystals. This total dimensional change for the stack can be employed as a means to move the needle of a diesel injector, or to actuate the pintle of a G-DI injector, thus providing an injector opening time that is more than an order of magnitude faster than typical solenoid systems. The more rapid opening translates into less time spent in the period of low pintle lift and large pressure losses across the pintle seat curtain, which in turn provides improved atomization levels during injector opening. This also provides a very significant extension of the minimum operating pulse width to shorter values. The ability to use much shorter pulse widths with repeatable actuation dynamics and fuel delivery yields a substantial improvement in a very important injector performance parameter—that is, the dynamic range. The variation in the characteristics of the opening process from actuation to actuation is also superior for the piezo stack, which contributes to an expansion of the WFR of an injector that utilizes this principle. Therefore the piezoelectric actuator stack is ideal for controlled multiple pulses, as the lift, opening time and fuel delivery are very accurate and well controlled compared to the typical G-DI unit. An associated advantage of piezoelectric actuation is a reduction in injector power consumption, although the tradeoff is in the need for increased system component precision. An additional advantage over inwardly opening solenoid actuation is the ability to open against much higher levels of fuel pressure (15 to 25 MPa), which would

require very large coils and very high current levels with a solenoid. A piezo stack system is no more complex than a solenoid-actuated system; however, the required level of precision of the individual components is higher for the piezoelectric actuator. It should be noted that the term "piezoelectric" technically refers to the method of actuation rather than to an injector tip design. In fact, this type of actuation could be used with any of the nozzle tip designs listed in Table 3.5-1.

The inherent extended dynamic range and a wider injector WFR are significant enablers for multiple-injection strategies. For typical piezoelectrically actuated injectors, three, four, and even five injections per cycle can be utilized in either G-DI or diesel combustion systems if the need for such a strategy is clearly demonstrated. An important caveat (discussed in Section 4.8) is that the spray from the specific nozzle tip that is used may limit the number of injections per cycle, even though the actuator does not. If the nozzle has a sac volume, then the very short pulses that are associated with multiple injections can yield sprays that are only slightly better than a sac spray in terms of the level of atomization. Figure 3.5-7 shows a schematic of a piezoelectrically actuated DI injector [387].

3.5.7 Pulse-Pressurized, Air-Assisted (PPAA) Injector

Most air-assisted G-DI injectors utilize outwardly opening pintles, whereas most single-fluid swirl injectors are inwardly opening. Pulse-pressurized, air-assisted (PPAA) injectors also use two solenoids per injector, although it

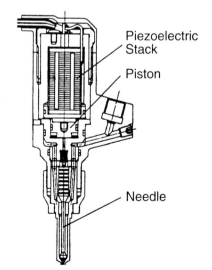

Figure 3.5-7 Schematic of a piezoelectrically actuated DI injector [387].

should be noted that there is some use of two solenoids on single-fluid swirl injectors to enhance the opening and closing characteristics. A further point of information is that injector designs that rely on a poppet cracking pressure tend to exhibit some injection rate variations due to lift oscillations, and tend to exhibit poppet bounce on closure. Each injector design should be evaluated for these tendencies. Air-assisted fuel injection systems provide an interesting alternative for future low emission concepts and this option is receiving increased attention among four-stroke G-DI developers [162, 426].

Figure 3.5-8 shows a schematic of a pulse-pressurized, air-assisted G-DI injector. The injection sequences for both fuel and air are illustrated in Fig. 3.5-9. The air injector used in the fuel injection system is a solenoid-activated, outwardly opening, poppet designed

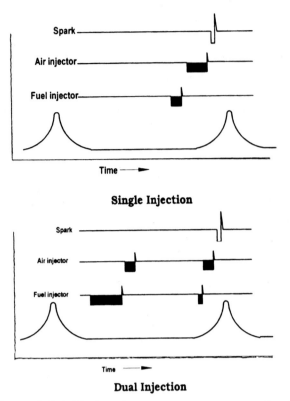

Figure 3.5-9 Fuel and air injection sequences for the pulse-pressurized, air-assisted injector [177].

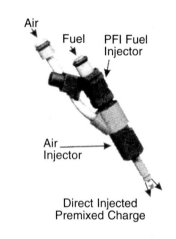

Axial Type

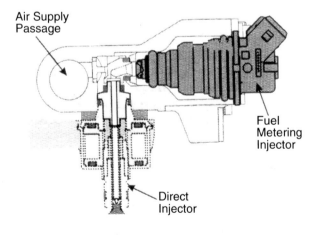

Lateral Type

Figure 3.5-8 Schematics of pulse-pressurized, air-assisted injector [177].

to inject precise amounts of fuel and air directly into the cylinder in the form of a finely atomized air-fuel cloud. Liquid fuel is metered and delivered into the top of this air injector by a separate PFI fuel injector calibrated for operation at a lower pressure differential of 0.07 MPa as compared to 0.35 MPa for a typical PFI application. The fuel normally remains in this holding cavity for a certain dwell period until the air solenoid opens, allowing the compressed air to flow through and purge the fuel. With an air rail pressure of 0.65 MPa and the differential pressure across the fuel injectors of 0.07–0.35 MPa, the required fuel pump pressure is in the range of 0.72–1.0 MPa. The same in-tank pump module is used for the air-assisted fuel injection system as is currently used in the conventional PFI engine, and a second fuel pump is installed in series to achieve the required fuel rail pressure at the engine. An in-tank, two-stage pump capable of achieving the specified fuel rail pressure could also be used if necessary. A camshaft or belt-driven compressor generates the required air rail pressure, which is regulated by a fixed mechanical pressure regulator. The fuel rail pressure is referenced to the air rail pressure for a constant pressure difference across the fuel metering injectors [255, 426]. The vacuum generated by the air compression for the air-assisted injection system can also be used to purge the canister. One of the advantages of purging the vapor through the compressor is that the fuel vapor is delivered to the direct injector. This makes it possible to maintain a high degree of charge stratification, thus the engine can maintain the same combustion mode while the canister is being purged [255, 426]. The main functions of the air injector portion of a PPAA injector are summarized in Table 3.5-3.

Variation in the ratio of injected air mass to injected fuel mass is plotted in Fig. 3.5-10 [177]. For injection corresponding to low load the ratio of injected

air mass to injected fuel mass is in the range of 1:1 to 2:1, and exhibits a gradual decrease to 0.2:1 at full load. Further optimization of the quantity of injected air can be made independent of the metered fuel quantity by adjusting either the opening duration or the supply air pressure. The absolute injected air mass is relatively constant over the engine operating map [426]. The amount of air injected by a PPAA system is not inconsequential, and as the engine air flow rate is decreased the percentage of the total air mass injected by the PPAA injector will increase. Typically the percentages will vary from about 1.5% for full-load operation to about 15% for idle operation. Detailed measurements and analyses of

Table 3.5-3
Main functions of the air injector for pulse-pressurized, air-assisted injector

---

- Control the timing of each injection event.
- Isolates fuel system from combustion chamber between injection events. As a result, fuel metering is not influenced by deposits in the air injector, with the time available for fuel metering being independent of the air injection event.
- During injection, fuel can be effectively atomized into fine droplets through the air injector using pneumatic atomization.
- The spray structure is directly controlled by the air injector nozzle design.
- During engine start, reverse flow from the engine to the air injector can be used as a "compressor" to rapidly elevate the air rail pressure to the normal operating level.

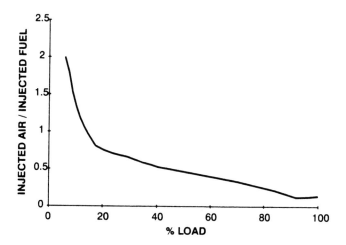

Figure 3.5-10 Ratio of injected air mass to injected fuel mass for the pulse-pressurized, air-assisted injector as a function of engine load [177].

the droplet flow field near the nozzle exit of the PPAA injector reveal that the fuel mass flow rate across the plane of the pintle seat is not a constant, therefore the ratio of air to fuel exiting the injector varies throughout the injection event [475].

The crank-to-run time, which is defined as the time delay between "key-on" and the attaining of an engine speed of 700 rpm, is less than 0.75 seconds for a four-cylinder, four-stroke DI engine, which is comparable to that of equivalent PFI engines. The first fuelled cylinder fires with an overall air-rail pressure of only 0.2 MPa (gauge), as compared to 0.65 MPa under steady state operation. Reliable combustion with this lower air-rail pressure can be achieved in the crank-to-run mode because the air-fuel mixture is injected early and the combustion process during this period of operation is homogeneous. Also during the crank-to-run, the air-rail pressure can be achieved rather quickly by taking the compressed air directly from the engine cylinders as well as from the compressor by optimal sequencing of the air and fuel injectors [426].

In contrast to the single-fluid, high-pressure swirl injector, in which fuel metering and spray structure may both be affected significantly by deposits, only the spray characteristics of the PPAA injector are affected by deposits, as the fuel is metered separately by another low-pressure injector. However, because the air-assisted injector is more often used for spray-guided combustion, even a minor change in spray geometry caused by injector deposits can affect the combustion characteristics dramatically. For this reason, some PPAA injectors incorporate a cleaning routine in which the surface temperature of the injector nozzle is temporally increased to oxidize the soot that has accumulated within the nozzle by modifying the timing of the direct injection event. This cleaning routine is activated on a periodic basis during normal operation in order to prevent excessive deposit formation on the injector delivery nozzle. During the cleaning routine the air injector is held open as the piston passes the compression TDC. As a result of the pressure differential, hot combustion gases are forced into the air injector nozzle. Each injector is in turn cycled through this cleaning routine. This cleaning process is made transparent to the driver by maintaining the output torque by means of stored calibration tables. Special care is required to prevent combustion inside the injector or air rail [426].

3.6 G-DI Injector Actuation and Dynamics

An understanding of the transient spray-development process must be preceded by a knowledge of the interrelationships among the electronic logic pulse, the pintle or armature lift and the fuel delivery rate. As depicted in Fig. 3.6-1, these interrelationships can be somewhat complex, involving a number of specific injector and driver performance parameters and characteristics. The initial input to the system occurs with the start of the fuel logic pulse, which is normally a standard transistor-to-transistor logic (TTL) square-wave pulse. Some systems use a true-low pulse, which is a grounding of the signal during the pulse duration, and some are designed for a true-high pulse; there is currently no industry standard in this area. The required pulse duration computed by the EMS is supplied to an injector driver, which shapes the pulse in terms of time, voltage and current. Again, there is not yet an industry standard, as there is with a saturated-switch PFI driver, and a wide range of delivered time delays, peak voltages

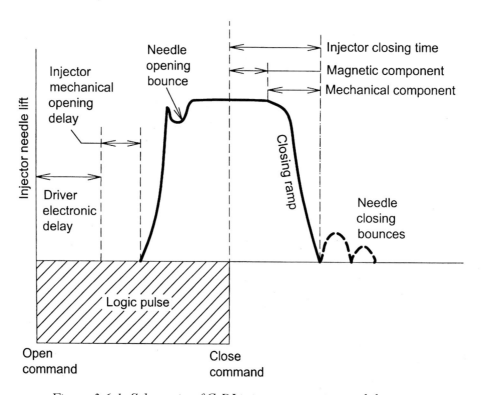

Figure 3.6-1 Schematic of G-DI injector actuation and dynamics.

and electrical current waveforms may be provided to the electrical solenoid on the injector. The peak voltage delivered by the driver may vary from 12 volts to more than 100 volts, and optional built-in time delays are sometimes incorporated to charge a capacitor on systems that deliver more than 36 volts to the injector. This enhances the injection-to-injection repeatability for a wide range of dwell periods between injection events. One example of this might be cranking and starting. The very long period between pulses might allow the capacitive charge of the driver to partially decay, particularly for higher voltage systems, which could degrade the subsequent injection event. By designing for a fixed, brief capacitive recharging time, the COV of peak voltage delivered to the injector solenoid is reduced. Typical capacitive time delays incorporated into high-voltage drivers are 0.25, 0.50 and 1.0 ms. It is very important to be cognizant of this time delay, as it represents an offset in both time and engine crank angle during which nothing occurs relative to injection activity.

Following the driver electronic (capacitive) delay, a voltage waveform is delivered to the electrical terminals of the injector solenoid. This voltage waveform, which is normally specific to the manufacturer of the driver and injector, is coupled with the electrical resistance and inductance of the solenoid windings to produce a current waveform that "pulls in" the armature and subsequently holds it in the fully open position. As illustrated in Fig. 3.6-1, every G-DI injector has a mechanical opening delay, even if there is no electronic driver delay. In fact, the mechanical opening delay, which ranges from 160 μs to 450 μs among all solenoid-actuated G-DI injectors, is independent of any driver capacitive delay, and is simply added algebraically to it to obtain the time that fuel first appears. For an injector with a 0.50 ms capacitive delay and a 0.20 ms mechanical opening delay, no fuel is observed to exit the injector tip until at least 0.70 ms elapses after the injector is initially commanded to open. This cumulative phase delay must be taken into account when evaluating spray interactions with the bowl cavity of a moving piston.

Following the injector and driver system delays, the armature, seat and pintle assembly moves from the fully closed to the fully opened position. This opening ramp normally occurs in just tens of microseconds, and is accompanied by a very rapid increase in the mass flow rate of fuel exiting the injector. The fuel within the injector and near the sealing surface has zero velocity and zero momentum immediately prior to the lifting of the needle. Also, depending upon the injector design, there may be a very small amount of residual fuel within the tip cavity downstream from the sealing surface. The residual fuel from the previous injection is often referred to as the "fuel hang-up," and can yield both metering errors and a degradation of the subsequent injection event. For swirl-type injectors there is always some fuel residing at or near the swirl-channel exit plane that does not experience an increase in angular momentum as the pintle opens, but is simply ejected along the injector axis. This fuel, which may range from 0.2 to 1.2 mg is the first to be ejected from the injector orifice, and is generally deficient in atomization level achieved. This sac fuel may not be subjected to the full fuel pressure or the full angular momentum transfer interval, and may not have enough time to achieve the levels of velocity and angular momentum that are achieved by subsequent fuel packets.

The armature velocity associated with rapid opening is such that there is normally a bounce that occurs when it strikes the internal surface that limits the opening lift, but this bounce has a relatively minor effect on the delivered flow rate of fuel and spray development. At this point in time the armature and seat assembly is in the fully opened position, and the delivered, instantaneous, fuel mass flow rate has attained the rated static flow of the injector. This is the steady flow rate that would be obtained if the injector were to be operated at the design fuel pressure and locked in the open position.

All of the above discussion on flow and spray development is independent of the length of the fuel logic pulse command, as the injector cannot know during the early delay and opening periods exactly when the closing command (end of logic pulse) will occur. That is why, except for very short pulse widths of less than a millisecond, the initial spray development and atomization characteristics are independent of the fuel pulse width utilized. The delivered fuel rate continues at values approximating the static flow rate for the duration of the fuel logic pulse with the exception of flow effects resulting from time variations in fuel rail pressure, as illustrated in Fig. 3.6-2. These variations generally occur because of wave dynamics in the fuel rail, a result of transient flow rates from multiple injectors in a fixed-volume rail. Nearly all gasoline direct-injection fuel rails have internal pressure dampers that are optimized to reduce the magnitude of such pressure variations, but such variations may still be present under certain operating conditions.

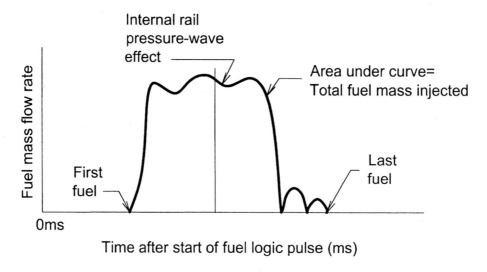

Figure 3.6-2 Typical fuel mass flow rate for G-DI injectors.

It is worth noting that the very first fuel to exit the nozzle tip on nearly any type of fuel injector is generally sub-standard in terms of atomization. In addition to the residual downstream fuel from the sac volume, which must be given momentum and moved out of the flow path, the opening process itself creates sub-standard atomization for many injector types. For swirl and multihole injectors the first micrometers of pintle lift result in a large fractional pressure loss across the needle seat area. Thus the effective fuel pressure is substantially lower than the operating rail pressure. Although this fuel is not considered to be part of a sac volume, it will generally have a larger mean drop diameter than fuel that exits a few hundreds of microseconds later, when the pintle is fully lifted and any initial blocking fuel has been ejected.

Finally, it is important to note that even after the command is given by the EMS to the driver to close the injector, fuel continues to exit the injector delivery orifice for a very significant time period, typically 0.32 ms to 0.65 ms for solenoid-actuated injectors. This time interval from the closing command (end of fuel logic pulse) until the needle is fully closed and fuel flow is initially reduced to zero is known as the injector closing time. There is an electrical component of closing in which the magnetic field collapses, and there is a mechanical component in which the differential between the spring return force and the magnetic force increases. When the spring return force exceeds the magnetic force the armature-seat assembly begins to accelerate toward the closed position. This, in turn, begins to reduce the instantaneous fuel mass flow rate being delivered, and begins to degrade the spray atomization that is achieved as the fuel velocity and/or angular momentum are reduced. The armature continues to accelerate, and, after the mechanical closing delay, strikes the sealing seat. In many injector designs the momentum and elasticity of the assembly are sufficient to result in one or more rebounds. This increases the fuel flow area slightly, yielding one or more after-injections of fuel, as schematically illustrated in Fig. 3.6-2. The atomization level of this injected fuel is always inferior to that of the main pulse, because most of the fuel pressure is dissipated across the small seat opening. Closing times normally exceed opening times by a significant margin—0.36 ms versus 0.20 ms, for example—because the closing is by spring force and the opening is powered by the magnetic force associated with a high peak current through the solenoid windings. Some relatively sophisticated direct-injection drivers control the armature impact velocity and eliminate or reduce after-injection by supplying a short, carefully timed, opening electrical pulse just prior to the armature closing impact. This reduces the armature closing velocity such that the always-present hydraulic damping due to the extruding of liquid fuel from the sealing area is sufficient to prevent a rebound. Through the interpretation of timed laser sheet photographs of fuel spray development and delay in G-DI systems, Table 3.6-1 summarizes the key features of G-DI injector actuation and dynamic fuel injection process.

Table 3.6-1
Key features of G-DI injector actuation and dynamic fuel delivery

- Actual fuel delivery is always displaced a non-trivial amount of time and engine crank angle from the fuel logic pulse.
- Many gasoline direct-injection drivers utilize capacitor-charging delays, which further displace the fuel delivery time from the initiation of the fuel logic pulse time.
- Actual fuel delivery, even without after-injection, occurs for a time interval that exceeds the fuel logic pulse. It occurs for the duration of the fuel logic pulse, plus the closing time minus the opening time.
- The initial spray development, the initial penetration rate and the initial level of atomization are independent of the operating fuel logic pulse width.

3.7 Requirements Regarding Multiple Injections

The capabilities of current control systems permit complex strategies for mixture formation and control. For example, in the Toyota first-generation D-4 system [302], a two-stage injection strategy is utilized to improve the transition between part-load and full-load operation. Similar two-stage, or split-injection strategies have also been proposed to avoid engine knock, thus increasing engine torque, by injecting the fuel partially during the intake stroke and partially during the compression stroke [15, 506]. In addition, a late injection strategy during the expansion stroke has been employed in the Mitsubishi GDI system to increase the exhaust gas temperature for quicker catalyst light-off during cold starts [16, 266]. For such a split-injection strategy, the split fraction is very critical. Detailed descriptions of the potential applications for split-injection strategy in mixture preparation, combustion and emissions control will be given in Section 6.3.

The electronics and the response times of modern G-DI injectors are certainly capable of providing multiple injection pulses, even during a single stroke such as compression, and certainly during two different strokes, such as one injection during intake and one injection during compression. The piezoelectric drive stack is certainly the best in this regard; however, the main concern in evoking such a strategy is not the electronics, but the low pulse width quality and stability of the injection event. This is not directly a multiple-pulse characteristic, but for small fuel delivery it can be associated with a range of single-pulse operation that is normally avoided. Using two or more injection pulses per cycle forces the use of smaller fuel pulse widths, thus moving toward non-linear and less stable fuel delivery and degraded spray quality. As will be discussed in significant detail in Chapter 4, for G-DI injectors other than piezoelectric, as the fuel pulse width is decreased the first constraint that is encountered is the lower limit of linearity of fuel delivery. For pulse widths that are less than this value, which defines the lower limit of the dynamic range, the delivered fuel per pulse is not linearly related to the FPW within the acceptable bandwidth, normally 3%. The spray repeatability acceptance limit will be the second, and most limiting constraint. Injection-to-injection variability in many spray parameters are encountered for nearly all injectors as the fuel pulse width is significantly reduced beyond the linear limit, with mass delivered, penetration, cone angle, spray skew, SMD and DV90 all experiencing degradation in both absolute value and COV. As the fuel pulse width is reduced to these very small values, any sac spray also becomes more predominant, with the main spray finally disappearing entirely. Before this point is reached the injection-to-injection variability in the mass delivered will normally become too large to be usable. The shortest fuel pulse width that can be used in a split-injection strategy must be determined for each injector model and operating condition, but this can be ascertained for single injection operation.

3.8 Summary

The fuel injection system is one of the key elements to be considered in the development of a G-DI combustion system. A G-DI fuel injection system must be able to accommodate at least two, and possibly three or more, distinct operating modes. A fuel system that is able to quickly deliver a high rail pressure during cold start is crucial for the G-DI engine to realize the potential of G-DI technology during cold crank and start. There are intense programs by injector manufacturers to continue to not only improve injector and spray performance, but to reduce unit-to-unit variability in spray characteristics. Non-spray performance parameters of the injector include the opening time, closing time, pintle bounce, durability, dynamic range, working flow range, noise level, power consumption, leakage and operating pressure range. Spray performance parameters include the mean diameters of the delivered main spray and the sac spray, as well as the associated statistical parameters that result from the drop size distribution, such as DV10, D32, DV50, DV80, and DV90. Other key spray parameters include the spray cone angles (both the initial angle and the final collapsed spray angle), the spray-deviation (skew) angles of the main and sac-volume sprays, the spray tip penetration rates and maximum velocities, the drippage, the after-injections or ligament formation upon injector closure and the fuel mass distribution within the spray. Additional key measures of performance, both spray and non-spray, are related to injection-to-injection and unit-to-unit variability in all of the above parameters. Some of the key requirements for the next generation of G-DI fuel systems are listed in Table 3.8-1.

Injection system hardware for G-DI engines has been evolving rapidly in recent years. In the last twelve years, G-DI combustion systems have progressed to the exclusive use of electronically controlled, common-rail, fuel system hardware. This class of fuel injection system permits full monitoring and computer control of injection timing and fuel pulse width, as well as providing the capability for adding fuel makeup pulses or multiple injections per cycle. Median fuel rail operating pressures have doubled over the last six years from 5 to 10 MPa, with a number of applications specifying 11 MPa, or even 12 MPa. There is not yet sufficient field data to evaluate the impact of such an increase on the wear parameters and mean life of the fuel pump, but such input could eventually moderate the trend line for fuel pressure increases.

As an individual component of the fuel system, the G-DI fuel injector is also improving rapidly as more combustion systems are implemented in production applications. Significant enhancements in performance,

functionality, durability, and physical size have been made in the past six years, and this trend is continuing. There is little doubt that as more G-DI production applications are introduced worldwide, additional enhancements in the areas of deposit resistance, injector noise and power consumption will be developed and incorporated. Although retaining the electronic pulse-width-modulation method of fuel metering, the methods of fuel atomization and spray geometry control have broadened. The simple poppet nozzle with a spring-controlled cracking pressure has now evolved into an available spectrum of electronic solenoid and piezoelectric G-DI injectors that includes swirl-plate, pulse-pressurized air-assist, multihole, slit and pressure boost designs. The swirl-plate injector could arguably be designated as the most common type of G-DI injector today, with variants available from nearly all injector manufacturers. This single-solenoid design uses a swirl-plate in proximity to the nozzle exit orifice to introduce angular momentum into the fuel stream by means of a number of tangential (or proprietary design) swirl channels. This forms a thin sheet of fuel on the micro-smooth surface of the delivery orifice, which becomes thinner as it moves downstream from the nozzle at a cone angle that is determined by the exit orifice angle. This thin sheet then becomes unstable and rapidly breaks up into very small droplets.

The slit and multihole types of injectors are becoming much more prevalent, with the portfolio of available modes expanding each year. Multihole injectors typically have four to eight holes, which may be arranged in a number of geometric patterns including circular, elliptical, half-circle and straight line. The slit, multihole and swirl injectors are all available with sprays that are offset by up to 20° from the axis of the injector body. The pulse-pressurized, air-assisted G-DI injector has been available for the past decade, and offers comparable fuel atomization levels to those of swirl and multihole injectors at a very much reduced fuel rail pressure. But this type of injector does require the use of two separate injector solenoids having individual drivers and pulse widths, and does require a supply of external compressed gas, usually air. This class of injector utilizes one solenoid to meter fuel to a mixing chamber, and a second solenoid to introduce a timed amount of compressed air to both atomize the fuel and purge the chamber through the exit orifice.

Table 3.8-1
Key requirements for next-generation G-DI fuel system

Items	Requirements
Fuel Injection Characteristics	• Minimum pulse width for stable injection <0.70 ms • Opening time <0.22 ms • Closing Time <0.35 ms • Linearity range suitable for highly boosted engines
Design Fuel Pressure	• Up to 20 MPa
Spray Quality	• SMD for main spray <15 μm • SMD for sac spray <20 μm • DV90 for main spray <28 μm • DV90 for sac spray <32 μm • Sac volume <5% of idle fuel delivery • Low sensitivity to injector deposits • Spray symmetry: sufficient to permit injector rotation or rail mounting tolerance without combustion degradation • Clean injector closure; no drops over 50 μm; no ligaments • Spray footprint: narrower (more cone collapse) for injection into higher ambient density • Fair spray quality (< 40 μm SMD) during cold start at 15% of design pressure
Others	• High resistance to deposit formation • Capability for two or more injections per cycle • No after-injections • Reduced priming time for pump • Reduced power consumption • Reduced noise level • Zero leakage at fuel rail pressure and temperatures from −40°C to +140°C

Chapter 4

Fuel Spray Characteristics

4.1 Introduction

The detailed characteristics of the gasoline spray that is injected directly into the combustion chamber are of paramount importance to the combustion efficiency and the resultant engine-out emissions. As summarized in Fig. 4.1-1, spray characteristics such as the spray cone angle, mean drop size, spray penetration rate and fuel delivery rate are known to be critical to the processes of vaporization, mixing, charge stratification and combustion stability, and are affected by many design and operating parameters. The optimum matching of these parameters to the in-cylinder flow field, chamber geometry and spark location usually constitutes the essence of a G-DI combustion system development project. In contrast, the fuel atomization characteristics of a port fuel injector generally have much less influence on the subsequent combustion event, mainly due to the integrating effects of the fuel residence time on the backface of the intake valve, and due to the secondary atomization of the liquid fuel film that occurs as the induction air subsequently flows through the valve opening curtain. For direct injection in both G-DI and diesel engines, however, the mixture preparation time is significantly less than is available for port fuel injection, and there is much more dependence on the primary spray characteristics to prepare and distribute the fuel to the optimum locations. It is well established that a port-injected gasoline engine can operate quite acceptably

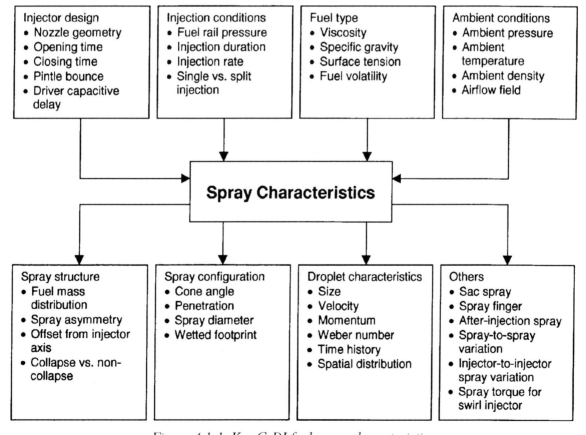

Figure 4.1-1 Key G-DI fuel spray characteristics.

51

using a spray having a 200-μm SMD, whereas both the G-DI and diesel engines require at least an order of magnitude finer atomization. Most G-DI applications will require a fuel spray having an SMD of less than 25 μm, and may require an SMD as low as 15 μm, if operating in the late-injection, stratified-charge mode, in order to achieve acceptable levels of both HC emissions and COV of IMEP. The diesel engine requires a fuel spray having an SMD that is less than 8 μm, as the available time for mixture preparation is even shorter, and the fuel is less volatile than gasoline. This is achieved in the contemporary common-rail diesel by using a designed fuel rail pressure that exceeds 100 MPa.

The characterization of the transient spray from a direct-injection gasoline fuel injector should be considered as a critical enabler in the development of a G-DI engine combustion system. It is required at some level of sophistication regardless of whether the development is totally experimental or is based upon a combination of CFD analyses and engine dynamometer testing. A knowledge of quantified parameters for the fuel spray is required at nearly all stages of combustion system development, including the initial selection of the range of test injectors, the supplying of input data for the tuning of CFD models, the interpretation of engine combustion data and the finalization of procurement specifications for the G-DI fuel system. At each stage some level of spray characterization data is very advantageous, almost to the point of being a requirement, with the level of advantage generally increasing as the system development project progresses. Such parameters as the spray cone angle, the spray-tip penetration and the mean drop size at either a single location or along a line through the spray are bare minimum requirements for initiating the design of the G-DI combustion system, whereas having the values for many more spray metrics may prove to be highly advantageous in interpreting combustion data and improving the design. Strictly speaking, the data are not just for the resultant spray, but may include injector performance parameters such as the mechanical opening and closing times, the maximum operating pressure for which the injector will open reliably, the pintle bounce characteristics (after-injections) and the effect of injector operating temperature on the delivered spray. Detailed spray metrics should be considered almost mandatory for initiating and tuning sophisticated CFD spray models, particularly if spray-wall sub-models are being used. These may include

drop-size data at multiple distances from the injector tip, drop-size data from one edge of the spray to the opposite edge, and time-resolved droplet arrival data (drop diameter and velocity versus time) at multiple locations in the spray. For spray-wall interactions, detailed spray measurements may be necessary at locations upstream and downstream from an impact location. These data may be required for operating conditions corresponding to those that exist in an engine combustion chamber at the time of injection; that is, at elevated ambient pressures and temperatures, and with elevated fuel and injector body temperatures. Such data must be obtained either on an operating optical engine or in an optical spray chamber that has the capability of using real fuels with a heated injector and fuel rail.

Even though a relatively complete correlation database has been established for diesel sprays [170], the bulk of these correlations unfortunately cannot be applied to predict the characteristics of DI gasoline sprays. This is the result of significant differences in fuel properties, injection pressure levels, droplet velocities and size ranges, ambient pressure and temperature levels and droplet drag regimes. It may thus be seen that the correlation and predictive characterization of the fuel sprays from G-DI injectors represents a new and important research area. Until such time as a comprehensive and proven correlation database is available, the spray parameters for individual injector designs will have to be measured in order to provide data for CFD model initiation and design comparisons.

4.2 Spray Atomization Requirements

An important operating criterion of a well-designed G-DI engine is that the fuel must be highly atomized, and subsequently vaporized, before the spark event occurs; otherwise combustion will be significantly degraded. The atomization requirement is directly related to the required rate of evaporation of the liquid droplets. A prime factor that controls the time required to evaporate the injected mass of liquid fuel is the surface area to volume ratio of the liquid. For example, if light-load operation of a G-DI engine is to be sustained, approximately 10 mg of gasoline must be injected into each cylinder per cycle. As illustrated in Fig. 4.2-1, this amount of fuel can conceptually have an infinite number of surface areas, depending on the level of atomization. In the limiting case, an unatomized, single 2.98 mm (2980 μm) drop can be injected, having a surface area of 28 mm^2. If that same amount of liquid fuel

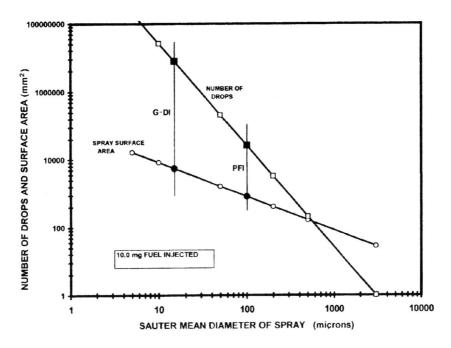

Figure 4.2-1 Number of drops and spray surface area versus SMD.

is atomized to the level of that achieved by a contemporary PFI injector, which is about 100 μm SMD, then there will be about 26,000 drops, with a total surface area of 30 times that of a single drop. It is well established that a late-injection G-DI engine will operate only marginally if the fuel is introduced with an SMD of 35 μm or greater. This threshold value has over 600,000 drops and a surface area of 85 times that of a single drop, but it is still inadequate to achieve fuel vaporization in the brief time interval available, which is less than 8 ms. The G-DI line that is drawn at 15 μm SMD in Fig. 4.2-1 is about what is required for efficient G-DI combustion. This typical G-DI fuel spray atomizes 10 mg of gasoline into approximately 8 million drops, with a total surface area that is 200 times that of a single drop. It is known experimentally that this surface area is sufficient for vaporizing the fuel in the 5 to 8 ms that are available between injection and ignition. Although of primary benefit for engine combustion, this number of fuel droplets does cause some drop size measurement problems in both laser-diffraction and phase-Doppler systems. This is due to the high number density and the associated obscuration and scattering, as well as laser-beam steering, that may result from the presence of fuel vapor. The difficulty that is associated with moving to the next level of G-DI spray atomization, which is an SMD of 10 μm or less, is clearly conveyed in Fig. 4.2-1. It is

readily evident that the 10 mg of fuel would have to be atomized into 26 million drops instead of 8 million, which could require on the order of 20 MPa of fuel pressure in a conventional common-rail system.

The rapid vaporization of very small droplets helps to make the concept of direct gasoline injection feasible [9, 10]; therefore, many techniques have been proposed for enhancing the level of spray atomization of G-DI injectors. The most common technique for G-DI combustion systems over the past decade is to use an elevated fuel pressure in combination with a swirl nozzle [259, 285, 302, 356, 392, 488]. The required fuel rail pressure level is generally at least 5 MPa, and in some cases up to 13 MPa [156], in order to atomize the fuel to the acceptable range of 15 μm to 25 μm SMD. The pulse-pressurized, air-assist injector has also been applied to G-DI engines [176, 177, 255, 315, 426], and certainly provides a spray with an SMD of less than 18 μm. However, numerous considerations such as fuel retention in the mixing cavity (fuel hang-up), the requirement of a medium-pressure, secondary air compressor and the use of two solenoids and timing events per cylinder have thus far limited its wide application to G-DI combustion systems [108].

The required spray characteristics and parameter threshold change significantly with the G-DI engine operating conditions. In the case of fuel injection during the induction event (early injection), a widely dispersed fuel spray is generally required in order to achieve good air utilization for the homogeneous mixture; however, impingement of the fuel spray on the cylinder wall should be avoided. For injection that occurs during the compression stroke (late injection), a more compact spray having a penetration rate within specified limits is generally preferred in order to achieve a stratified mixture distribution. As a concurrent requirement, the spray must be very well atomized, as the fuel must vaporize in a very short time [259]. An important point to consider is that the increased droplet drag that is associated with finer atomization reduces the spray penetration rate and maximum spray penetration, which can degrade air utilization. Even though fuel that impinges and wets the bowl surface of a hot piston may vaporize relatively rapidly, a wall film generally will not vaporize as quickly as the individual droplets that formed the film. It may be surmised that a suitable control of spray cone angle and penetration over the engine operating map could be advantageous; however, this is difficult to achieve with available hardware.

Based upon an analysis of the in-cylinder droplet evaporation process using a spray model, it has been demonstrated that a mean droplet size of 15 μm SMD or smaller should be utilized for G-DI combustion systems [82,83], as illustrated in Figs. 4.2-2, 4.2-3 and 4.2-4. Based on this calculation, a differential fuel pressure of at least 4.9 MPa is found to be required for a pressure-swirl atomizer to achieve this required degree of fuel atomization. Calculations indicate that the additional time available with early injection does not significantly advance the crank angle positions at which complete droplet vaporization is achieved. This is because the high compression temperatures are very influential in vaporizing the droplets, and these temperatures occur during the latter stages of the compression stroke. For a wall-guided system the atomization level associated with a 4.9 MPa fuel pressure may not be sufficient to avoid the excessive HC emissions associated with fuel impingement and the reduced film evaporation rate from a solid surface. Even with a well-atomized spray the HC emissions may or may not be reduced significantly, depending upon the in-cylinder turbulence level, due to small pockets of very lean fuel-air mixtures [180, 514, 515]. A strong turbulence level in the combustion chamber is required to enhance the fuel-air mixing process by eliminating these pockets.

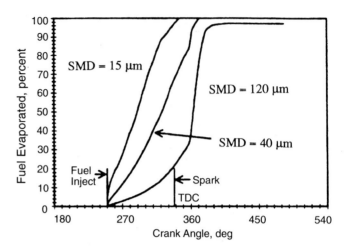

(a) Early-cycle direct-injection

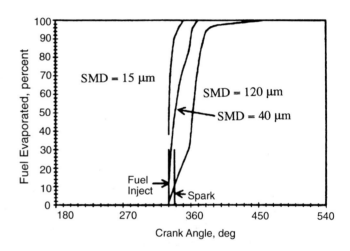

(b) Late-cycle direct-injection

Figure 4.2-2 Predicted fuel evaporation rate during low-temperature cold cranking; 120 rpm cranking speed;−29°C temperature; three levels of atomization [83].

The SMD, which is the most widely used metric for describing the level of spray quality, may not, in fact, be the single best indicator of the spray quality required for the G-DI engine. This is because a very small percentage of large droplets is sufficient to degrade engine HC emissions, even though the SMD may be quite small. Each 50-μm fuel droplet in a spray size distribution having an SMD of 25 μm not only has eight times the fuel mass of the mean droplet, but also will remain as liquid long after the 25-μm drop has evaporated. In fact, it is very informative to consider

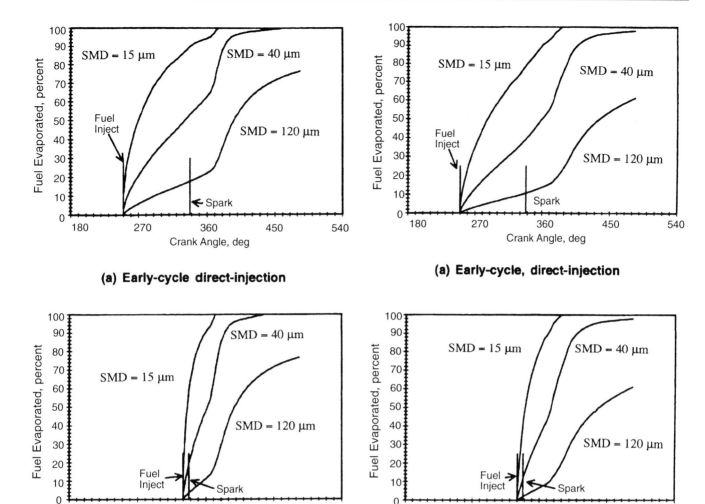

(a) Early-cycle direct-injection

(a) Early-cycle, direct-injection

(b) Late-cycle direct-injection

(b) Late-cycle direct-injection

Figure 4.2-3 Predicted fuel evaporation rate during warmed-up operation at 2000 rpm and moderate load; 45 kPa MAP; three levels of atomization [83].

Figure 4.2-4 Predicted fuel evaporation rate during warmed-up operation at 6000 rpm and WOT; 100 kPa MAP; three levels of atomization [83].

that when all of the 25-μm droplets are evaporated the original 50-μm droplets will still have a diameter of about 47 μm. An injector that delivers a well-atomized spray, but which has a wide spread in the drop-size distribution, may require an even smaller SMD than quoted above to operate satisfactorily in a G-DI engine combustion system. This spread may be quantified by the parameter of (DV90-DV10), or by (DV80-DV10), which will be discussed in detail later in Section 4.16.4. The DV90 for a distribution represents the drop diameter for which 90% of the total fuel volume of all drops

exists as smaller drops. The converse statement is conceptually even more informative; that is, 10% of the liquid fuel volume (and mass) of a spray exists as drops larger than DV90. It is becoming accepted that DV90, or some alternative metric for the large-droplet content of a spray, may be a parameter that is superior to SMD (D32) in correlating the HC emissions of different fuel sprays in G-DI combustion systems. When examining the results in Fig. 4.2-5, which shows a comparison of the droplet size distributions between swirl-type and hole-type injectors [446], it is evident that, although the

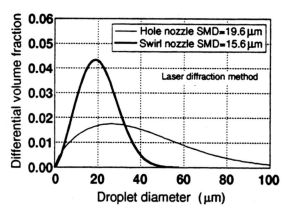

Figure 4.2-5 Comparison of drop size distributions for sprays from swirl-type and hole-type injectors [446].

difference in the mean droplet size (SMD) between the sprays from these two injectors is only 4 µm, the hole-type nozzle produces a wider droplet-size distribution having many larger droplets that are theorized to be a primary contributor to the observed increase in engine-out HC emissions.

Even though a design fuel rail pressure of 5 MPa is widely utilized, and seems to be adequate for producing an acceptable G-DI spray, a higher fuel rail pressure may be beneficial for some of the reasons listed below [487]. For a typical outwardly opening G-DI injector, it is found that the spray SMD is reduced from 15.4 to 13.6 µm as the fuel pressure is increased from 5.0 MPa to 10 MPa. Such a small incremental reduction in SMD would not seem to be significant given the magnitude of the increase in the fuel rail pressure; however, the total surface area for an injected quantity of 14 mg of fuel is increased by 13%, which should lead to a corresponding direct improvement in the fuel vaporization rate. More importantly, an elevated injection pressure may be required to reduce the key statistic for the maximum droplet size of the spray, namely the DV90 parameter. When the fuel rail pressure is increased from 5.0 MPa to 10 MPa, the DV90 of the spray at 30 mm downstream from the injector tip is reduced from 40 µm to 28 µm for a typical outwardly opening G-DI injector. In addition to the favorable effects on the spray characteristics, an elevated fuel rail pressure also increases the rate of injection and reduces the injector flow rate sensitivity to pulse-to-pulse variation in injector stroke.

4.3 Fuel Spray Classification

The three major classifications of fuel sprays from automotive fuel injectors are depicted schematically in Fig. 4.3-1. These are the automotive diesel spray, the PFI spray and the G-DI spray. For purposes of illustration a typical dual-stream PFI spray for a four-valve SI engine is depicted, although there are many spray configurations available. The diesel spray depicted in the schematic is used for automotive diesel engine applications, with contemporary designs utilizing a high pressure, common-rail system. Pressure levels of 50 to 80 MPa were common in the recent past, but diesel common-rail pressures of 120 to 160 MPa are now being utilized. These pressure levels yield a spray having an atomization level of 8 µm SMD or better. The diesel sprays are typically obtained from 5 to 8 individual holes that inject on a very wide cone (125° to 170° included angle). The PFI injector is by far the most common automotive fuel injector, with nearly a hundred million used annually in new production vehicles. This spray is also produced by a common-rail system, but uses a low rail pressure in the 0.27 to 0.45 MPa range, with typical atomization levels in the 85 to 200 µm SMD range. The cone angle of the individual PFI spray may range from 9° to 28°. The third major classification is for the spray obtained from a G-DI injector. As illustrated schematically, this spray is normally obtained from a common-rail system with a swirl-type fuel injector having a fuel pressure in the range of 5 to 13 MPa. This provides an atomization level of from 14 to 24 µm SMD.

The major classification of "G-DI fuel spray" may be further subdivided in many logical ways, but the categories selected for detailed discussion in this chapter have been designated on the basis of atomization mechanism and nozzle configuration. These six additional categories are depicted in Fig. 4.3-2, with the top row being designs that use a swirl plate, and the bottom row being non-swirl configurations. The swirl spray categories to be discussed are those of the inwardly opening pintle, the outwardly opening pintle and the shaped-spray (Casting-Net) swirl design. The non-swirl spray categories to be detailed are the slit type, the multihole type and the air-assisted type. The appearances of the fuel sprays from these six categories are illustrated schematically, with the wetted footprint that accompanies each spray type also shown. All of the spray types will be discussed in detail in subsequent sections.

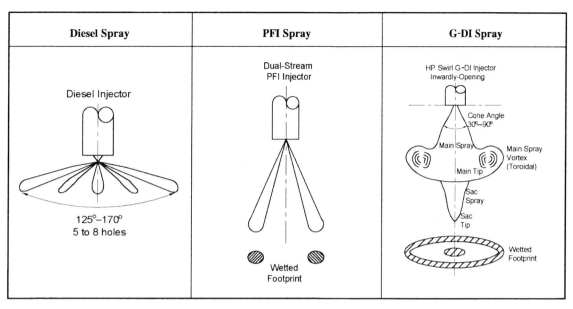

Figure 4.3-1 Broad classification of sprays from automotive fuel injectors.

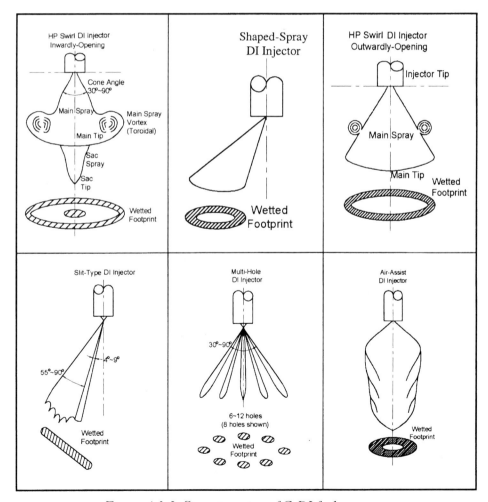

Figure 4.3-2 Six categories of G-DI fuel sprays.

The basic nomenclature that is commonly utilized to describe a transient G-DI fuel spray is illustrated in the schematic diagram in Fig. 4.3-3. In general the spray has a number of fairly identifiable features that may be made evident by imaging using a short-duration light flash. An important feature that is quite prominent at mid-injection is the main spray geometry, which for many nozzle designs has a form that approximates a cone. The angle subtended by the main spray is a descriptive parameter that is immediately obvious, but its use is associated with a number of caveats. First, as will be discussed in detail later in this chapter, this is not necessarily the cone angle that is used by injector manufacturers to specify an injector; nor can the cone angle be definitively represented in a schematic, as there is no single standard or universal definition for this most important parameter. Each corporation has an in-house definition of cone angle for purposes of classifying sprays into degrees of wide and narrow dispersion.

The G-DI spray has a leading edge, called the "main spray tip," that progresses, or penetrates, away from the injector nozzle tip as a function of time, normally penetrating on the order of 50 mm in less than 2 ms. A toroidal vortex that also moves away from the injector tip may also be attached to the periphery for some types of main sprays, particularly those from swirl-type nozzles. For many injector designs the leading edge of the spray may contain a separate sac spray that is comprised of residual fuel that is first to exit the nozzle orifice, and which generally continues as the most rapidly penetrating portion of the spray. If there is an interposed surface in the path of the spray within a penetration distance of up to 75 mm, a wetted footprint is normally formed during the injection process. If a sac spray is present, the wetted footprint may contain two distinct portions, as depicted in Fig. 4.3-3, one from the sac spray and one from the main spray. As will be discussed in Chapter 5, the entire spray cross-section does not necessarily wet the surface; thus the wetted-footprint most often exhibits a geometry that differs from the cross-section of the spray as illuminated by a pulsed laser sheet.

Two after-injections are depicted in Fig. 4.3-3 as small areas of fuel droplets that appear close to the nozzle tip in images of the spray development. If the spray is imaged at the correct time following the closing command, such droplet clouds will be evident if the pintle bounces when closing. Depending upon the design, after-injections may or may not be present, and may or may not be captured in the image unless the timing is carefully incremented. Such after-injections normally occur over a small time period of about 80 to 180 μs after the initial closure, thus they may be easily missed even if they are occurring. The atomization level is always degraded when after-injections are present, and may become so degraded that one or more ligaments are formed (usually two helical ligaments for a swirl-type nozzle). These ligaments are actually thin strings of fuel that are unatomized, and which are the last fuel to exit the nozzle. Thus they would appear as the final stage of the last after-injection, and are generally observed within 5 mm of the nozzle tip. The spray finger that is depicted as protruding from the main spray is shown schematically as similar to a ligament, but it is not. In contrast to a ligament, the spray finger is fairly well atomized (approximately 19–25 μm SMD), but is not as well atomized as the rest of the main spray. The presence of a spray finger (or multiple fingers) is generally an indicator of injector deposit formation, and is a region of the

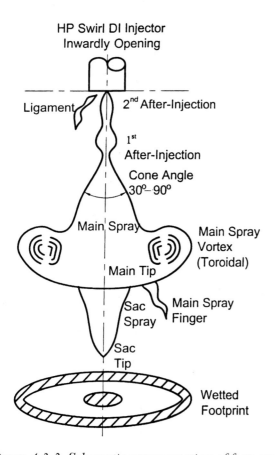

Figure 4.3-3 Schematic representation of features in a typical G-DI fuel spray.

main spray that has a moderately larger drop size and higher drop velocity, hence the greater penetration.

The laser-light-sheet photo in Fig 4.3-4 clearly illustrates the spray development for a typical swirl-type G-DI injector having a substantial sac volume. The larger drops associated with the sac spray are very evident, as is the sac penetration that occurs at a significantly greater rate and maximizing value than that of the main spray. Other classic features of the sprays from inwardly opening swirl-type injectors are the larger drops that are evident at the leading edge of the main spray, just ahead of the toroidal vortex, as well as the larger drops associated with the pintle closing ramp. The latter is, of course, not specific to swirl-type injectors, but is generally present for all G-DI injector designs except for two-solenoid, air-assisted types. The much narrower cone angle of the sac spray is obvious, with 12° to 20° being typical, even for an injector having a main spray cone angle of 70° to 90°. In fact the sac spray is often not a cone at all, but is perhaps best described as a column, which could be quantified by stating a column diameter rather than a sac cone angle. For the large sac spray in Fig. 4.3-4 the column diameter is about 16 mm. The leading boundary of the main spray, near the spray periphery, normally has larger drops present that result from the transient flow processes at injection initiation. High-speed images of the first 20 μs after the pintle lifts from its seat show that some poorly atomized fuel from each swirl channel exits at an increasing cone angle, rapidly attaining the swirl-channel exit area or pintle exit area, depending on the design. This small volume of sparse droplets, which is part of the total sac volume of fuel, continues at the maximum cone angle trajectory, and is little affected by the entrainment flow field that is being established. At slightly greater times in the pintle lift the angular momentum and mass flow rate of the exiting fuel is increased, and the fuel droplets begin to move back toward the injector axis, as is shown in the apparent curvature of the main spray boundary.

The finer level of atomization of the main spray during the time interval between the opening and closing transients is also shown in Fig. 4.3-4, and is evident even without phase-Doppler or laser-diffraction measurements. This mid-injection portion of the injection pulse generally produces the finest atomization, and it may be accompanied by a moving toroidal vortex containing a droplet size distribution that is devoid of drops over 20 μm in diameter, as they are centrifuged from the vortex. The spray cone angle in the vicinity of the

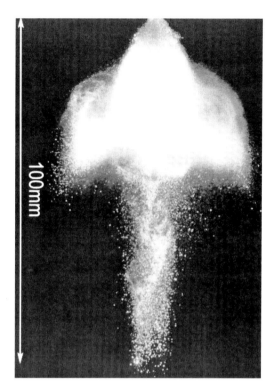

Figure 4.3-4 Features of a complete spray development event for a typical G-DI spray from a swirl-type injector.

injector tip is relatively constant over most of the injection event, normally oscillating only slightly over a range of a few degrees; however, as shown in the photo, the spray angle begins to increase and the level of atomization is degraded as the pintle begins to close. The fuel angular momentum and the rate of mass injection begin to decrease markedly, and the trajectory of the fuel droplets once more approaches the pintle exit angle.

As is shown in the detailed close-up view in Fig. 4.3-5a, the oscillations in exit spray angle are normally less than 5°, but in some cases can be as large as 20°, and can easily be detected by imaging. These oscillations can result from the interactions of internal fuel pressure waves, the spring/mass constants of the armature assembly and the decreasing flow rate of fuel through the exit orifice. Also obvious through imaging is the degradation in atomization as the fuel flow rate is reduced to zero. A very good closure will not have any after-injection, and will increase the SMD by less than 5 μm, whereas a very poor closure may yield many droplets of 45 to 65 μm, or even larger, and may have multiple after-injections. The details of the spray within 10 mm of the injector tip at, and just after, the time of

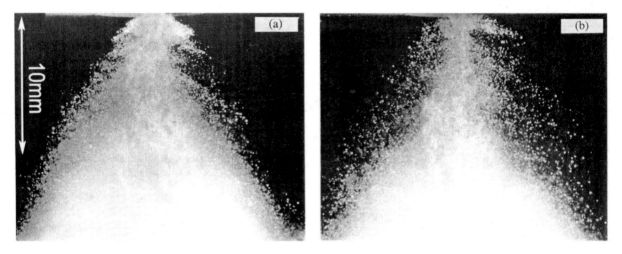

*Figure 4.3-5 Details of spray generation during the pintle closing event:
(a) exactly at the time of pintle closure; (b) pintle closure, plus 100 μs.*

pintle closure are clearly evident in the laser light sheet photos in Fig. 4.3-5. The spatial distribution of fuel drops at precisely the time of pintle closure is shown in Fig. 4.3-5a for a swirl-type injector that exhibits a very clean closure, with no associated ligaments and no pintle bounce. The modulation of the spray angle as the pintle moves toward the completely closed position over a time interval of 320 μs is fairly evident, as are the last, larger diameter drops that exit the tip at a greater included angle. The appearance of the spray 100 μs after closure is shown in Fig. 4.3-5b, and a comparison of the two panels illustrates the radial velocity of the drops that exited during the pintle closing ramp. The drop radial velocity can be as much as three times the axial velocity at the time of closure. All of the droplet velocities are very low at and after the time of closure, and the last drops remain in the vicinity of the tip for many milliseconds. By means of time-windowing the measurement of the spray in Fig. 4.3-5, the SMD of the drops that exited during the entire closing ramp was found to be 18 μm, whereas for the earlier main spray SMD was 15 μm. The closing that is depicted is to be considered a near-best-practice G-DI injector closure.

4.4 Sac Spray Considerations

The initial fuel spray from an inwardly opening G-DI injector is, in most cases, dominated by a small, identifiable quantity of fuel, and two distinct fuel sprays may be identified. These sprays are normally designated as the sac and main sprays, and have completely different spray characteristics, including cone angle, SMD, DV90 and rate of penetration. The initial, or sac, spray always

exhibits a higher penetration rate, poorer atomization, significantly more pulse-to-pulse variability, and generally has a much narrower cone angle. A photograph of a prominent, highly penetrating sac spray was shown in the previous section in Fig. 4.3-4. Many swirl injector designs yield a sac spray at less than 50% of design fuel pressure that is not axial, but is essentially a series of spray fingers at the exit angle. For a swirl injector having twelve swirl channels, this yields twelve individual initial sac sprays along a conical surface. The sac spray should ideally have a separate characterization from the main spray; however, this is seldom done in practice because of the measurement complexity and the requirement for precise time-windowing of the drop size and velocity data. This requires pre-testing with spray imaging to ascertain the proper timing window and spatial location for the size and velocity measurements. It is important to realize that if the sac spray is not characterized separately, which it generally is not, then the two sprays will be integrated into any measured characterization, thus yielding an average set of properties. The main spray almost always dominates this average set, as a sac spray volume (and mass) is normally only a few percent or less of the total fuel delivered. Hence, the fuel spray properties of the combined main and sac sprays normally differ only slightly from those of the individual main spray, and generally exhibit a few percent higher mean velocity and a few percent larger mean diameter. The sac spray, however, may disproportionately affect the DV90 value of the combined spray, as the DV90 value is controlled by the largest drops, which are almost always in the sac spray.

Whether or not the inclusion of sac spray drops in an integrated set of characteristics is valid or misleading depends upon the eventual use of the data. For engine combustion interpretation they should be included, but for spray simulation tuning the two sprays should ideally be separated. It should be noted that the volume of fuel constituting the sac spray is a constant geometric factor that is fixed by the particular injector tip, delivery channel and pintle design. It follows directly that the influence of the sac spray on the characterization values resulting from integrated main/sac measurements will increase as the amount of delivered fuel decreases. Thus, for engine operation near full load the sac spray may be only 1% of the fuel delivered, whereas at idle it may be 10%. For split-injection strategies, or for very brief fuel makeup pulses, the sac spray could constitute as much as one-third of the total. As the fuel pulse width is reduced, the main spray fraction decreases, finally reaching a pulse width where there is no main spray, just a sac. This low-pulse-width approach can be utilized to obtain the characteristics of the separate sac spray using either phase-Doppler or laser diffraction measurement techniques. It should be noted that this small pulse width is never used in practice, as injection-to-injection variability in the delivered fuel quantity and spray geometry generally becomes excessive for nearly all types of solenoid-activated G-DI injectors long before the main spray is suppressed. This is not necessarily true for piezoelectrically activated injectors, however, at least to the degree that it is observed for solenoid-actuated injectors. As an example, Fig. 4.4-1 shows the axial penetration for both the sac spray and the main spray as a function of fuel injection duration for a swirl-type DI gasoline injector. For an injection duration shorter than 0.5 ms, the sac spray is clearly dominating, and the main spray is not measurable.

Many terms for the sac spray may be encountered in the field of gasoline direct injection. Some are more correct than others, but none is totally accurate, including the term "sac." Commonly encountered descriptive terms for this initial spray from a G-DI injector are "slug spray," "pre-spray," "initial spray," "core spray," "leading mass," "center spike" and "sac spray." These are, for the most part, intended to be synonymous, and all refer to the separately identifiable spray that occurs at the initiation of the injection process in many G-DI injector designs. The terms "sac" and "core" have been adopted, and perhaps corrupted, from the terminology of diesel spray characterization. The term "core spray" is perhaps the least accurate, as the term "core" is generally considered to be descriptive of a spray breakup distance, and the term "sac spray," if used correctly, would only apply to a volume of fuel that was downstream of the pintle sealing surface and, hence, not at the fuel rail pressure when injection is initiated. For some types of G-DI injectors, including swirl designs, the lower-quality spray that is first observed upon pintle opening is due mainly to some of the fuel exiting the swirl channel without undergoing the full increase in angular momentum. Any true sac volume downstream

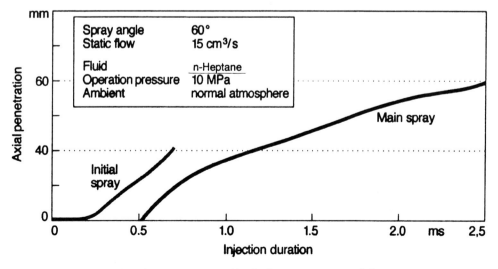

Figure 4.4-1 Axial penetration of both the sac spray and the main spray as a function of fuel injection duration for a swirl-type DI gasoline injector [369].

from the pintle in this type of injector merely adds to the degradation of the initial spray, but it is extremely difficult to separate the two effects. In this book the term "sac" has been selected to identify the separate, initial spray, regardless of its source, but perhaps the least misleading descriptors are pre-spray and slug spray, which assign a name to the observed spray without any associated implication as to why it occurs.

Figure. 4.4-2a shows the time history of the spray SMD from an on-axis swirl injector having a sac spray [357]. Clearly the SMD is initially large as the leading fuel droplets from the sac spray first enter the measurement location of 38.75 mm from the injector tip on the centerline of the spray. A peak SMD value is recorded for this initial sac volume before the instantaneous mean droplet size decreases rapidly for the main spray plume. As previously noted, this initial sac spray is composed of fuel that was in residence within the sac volume and within the lower portion of the swirl channels when the pintle moved off the seat. For the fuel pressure range of 3.45 to 6.21 MPa for this particular swirl injector, the initial axial velocity of the sac spray near the injector tip is found to be in the range of 68 to 86 m/s, which is higher than the velocity of 45 to 58 m/s for the main body of the spray. Time-resolved patternator measurements of the spray volume flux distribution for this spray are shown in Fig. 4.4-2b, and reveal that the sac portion of the spray mass is captured in the 1.5 ms cut-off time of the patternator. For a patternator cut-off time of 2.0 ms, the outer regions of the main spray begin to be collected, and for 4.0 ms the mass collected off-axis is larger than that collected on-axis. Visualization of the initial fuel slug, or sac spray, associated with the sac volume of a centrally mounted, high-pressure swirl injector inside a firing engine, reveals that this typical sac consists of relatively larger droplets of high velocity. These drops penetrate tens of millimeters prior to the later formation of the main spray cone, and can impact directly on the piston crown for early injection. For an injection timing of 80° after top dead center (ATDC) on the intake stroke, no major impingement of the fuel spray upon the piston crown is detected except for this initial slug of fuel from the sac volume [382].

The presence of a sac spray is generally detrimental to the mixture formation process, and contributes to an increase in smoke and HC emissions in many applications. However, there is a contrary view held by some that the sac spray can be beneficial, and that a relatively large sac volume can be used to stabilize the stratified-

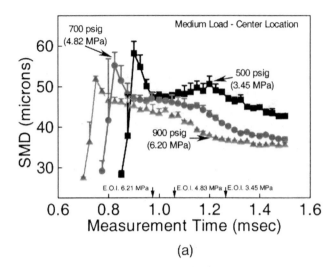

(a)

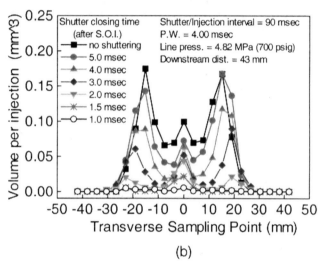

(b)

Figure 4.4-2 Time-resolved SMD variation for a G-DI swirl injector: (a) time histories of SMD for three fuel injection pressures; (b) fuel mass distribution inside the spray [357].

charge combustion, due to the associated high momentum [182, 183, 440]. In this regard it is well established that different G-DI operating modes require different spray characteristics for optimum performance. For homogeneous operation, a spray with a moderate to wide cone angle is generally optimal for air utilization, whereas a narrower-cone spray may be more effective in creating a highly stratified charge near the spark gap. The compromise between spray penetration and cone angle needs to be considered in optimizing air utilization. There is a viewpoint that a proper selection of the initial sac spray can be utilized to satisfy these two conflicting cone angle requirements without significantly degrading mixture preparation and the engine-out HC emissions [182, 183, 340, 440]. It is well known that the main spray from an inwardly opening swirl injector collapses to a narrower effective cone angle when injected into ambient air with elevated densities corresponding to late injection. This becomes more pronounced as the fuel quantity in the sac spray is increased, resulting in a further reduction in effective cone angle. Figure 4.4-3a shows the difference between the collected masses for two sprays having different sac volumes. The sac spray in this investigation was defined on the basis of the amount of

fuel collected within an angle of 20 degrees from the spray axis by a 37-ring patternator. The 5% COV limit map between ignition timing and end of injection (EOI) timing is illustrated in Fig. 4.4-3b. It is evident that, for some combustion system designs, increasing the quantity of fuel in the initial sac spray can enhance the engine combustion stability, and that the region of stable combustion can be broadened. An injector with a relatively large cone angle (70 degrees) and an appropriate quantity of initial center spray has been developed in an attempt to accommodate the conflicting requirements of homogeneous-charge and stratified-charge operation. However, the increase in overall mean droplet size and the associated increments in smoke and HC emissions with increasing sac volume must be carefully evaluated before this strategy is invoked. There is no clear consensus on this topic, but the considered view adopted in this book is that the disadvantages of a significant sac spray regarding wall wetting, HC and smoke emissions and injection-to-injection variability generally outweigh any advantages regarding main spray cone control or air utilization. Thus the recommendation is that sac spray be minimized or eliminated.

4.5 After-Injection Dynamics and Atomization

In many G-DI injector designs the closing is not nearly as good as in the example shown in Fig. 4.3-5, and ligaments of fuel may exit the injector during closure. In addition, the pintle may bounce one to three times, each time introducing a small amount (<500 µg) of quite poorly atomized or even non-atomized fuel. This small quantity of poorly atomized fuel has a minor negative effect on engine-out hydrocarbons, but has a more pronounced detrimental effect on smoke emissions. For this reason after-injections are to be avoided in any G-DI injector configuration. A triple after-injection that is associated with the closing event of a 65-degree G-DI swirl-type injector is shown graphically in Fig. 4.5-1. Figure 4.5-1a shows the initial after-injection that is associated with the maximum lift of the first bounce of the pintle, which normally occurs within about 100 µs of the initial closure. The fuel introductions from the second and third after-injections are shown in Figs. 4.5-1b and 4.5-1c, respectively. The bounce frequency is determined by the spring/mass constants of the system, and is modified slightly by the pintle elasticity and the hydraulic damping due to the extrusion of fuel from the seating area.

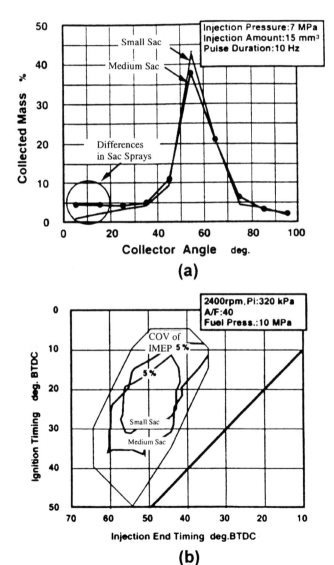

Figure 4.4-3 Effect of sac spray on combustion stability: (a) difference in collected fuel mass for two sac volumes; (b) 5% COV map of ignition timing and injection timing with different sac sprays [440].

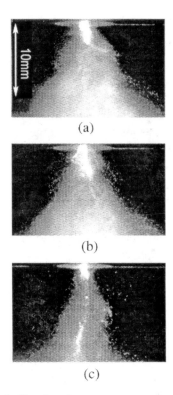

Figure 4.5-1 Triple after-injection associated with an injector closure: (a) early stage of first after-injection; (b) latter stage of second after-injection, with ligaments; (c) third after-injection and dissipation of the second.

A single small-to-moderate after-injection is very common in G-DI injectors; in fact, it occurs more often than not. Close attention is required in injector design to ensure that the pintle does not bounce on closure, as the normal response of a spring/mass system subjected to rapid impact closure is to rebound. Multiple after-injections can and do occur in some injector designs, with the observed limit being three. The height of each subsequent needle bounce is reduced, and the amount of fuel introduced is less, with the level of atomization being significantly degraded. The poor atomization associated with the second after-injection is quite obvious in Fig. 4.5-1b, with large droplets and some ligaments entering the combustion chamber. The third and final after-injection is always a very small quantity of fuel, thus it may not appear to be as poorly atomized as the second is, but it actually is. It may be noted that the remnants of the ligaments from the second after-injection are visible when the third occurs. Unlike the situation associated with a clean closure, in which the last droplets injected are at a very low velocity, the large drops and

ligaments associated with an after-injection can have fairly substantial velocities. There is also a correlation between injector noise and after-injection quantity and number, particularly for air-assisted injectors.

Detailed development of the sprays associated with multiple after-injections of a G-DI injector may be determined using a distance microscope coupled with high-speed imaging. Figure 4.5-2 shows a frame sequence of a very small region (3×6 mm) near the injector tip for two consecutive after-injections from a swirl-type injector. In this example of a very poor closing event the after-injections result from bounces of the pintle during the closing event. The first three frames in the top row clearly illustrate the degradation in atomization quality as the pintle initially seats. Frames four through seven show the initial formation and development of the first after-injection, which has moderate atomization without obvious ligaments. The second after-injection, however, is seen in frame eight to have very poor atomization, with a number of ligaments and very large fuel drops being evident. It is experimentally observed that the number of after-injections that occur for bench testing does not change as the injector operating temperature is increased to 90°C. The atomization level of each after-injection is somewhat improved, but the number of pintle bounces is unaffected.

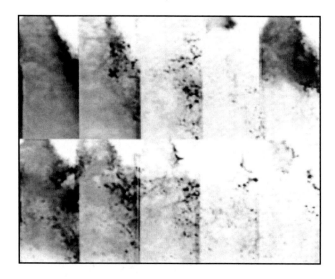

Figure 4.5-2 Time sequence of swirl injector closing dynamics showing details of two after-injections; distance microscope; frame rate: 25,000 frames/s; field of view: 3 mm × 6 mm.

4.6 Fuel Spray Penetration and Cone Angle Considerations

In all G-DI combustion systems the fuel spray-tip penetration characteristics are very important, and in some system designs are critical, in the matching of an injector to combustion chamber geometry. Unlike a PFI fuel spray in which the time of flight to the back face of the intake valve is not as important as the targeting of that face, the progress of the G-DI spray plume relative to the positions of the spark gap and moving piston crown or cavity is a primary concern. This is tantamount to stating that the spray-tip penetration characteristics of each injector must be well known in any G-DI combustion-system-development program. The influences of the spray penetration characteristics on the injection timing window, the smoke and the COV of IMEP, and also on the overall robustness of the combustion system are not to be underestimated. The fuel cloud must reach the vicinity of the spark gap and have sufficient time to form an ignitable mixture in that region regardless of whether the system is wall-guided, spray-guided or air-guided. For injector designs that have a sac, or pre-spray as it may be alternatively called, the penetration characteristics of both the sac spray tip and the main spray tip should be measured and correlated. Such correlations are usually obtained by imaging the spray development as a function of time as part of either a room bench test or an optical spray chamber test, but may also be obtained from imaging in an optical engine.

Although the concept of measuring spray-tip position as a function of time and optionally converting to engine crank angle seems relatively straightforward, in practice it is more involved. This is because the penetration characteristics of all G-DI sprays are influenced by many operating parameters, primary among which are the downstream ambient density/pressure, the fuel pressure, the injector tip temperature and, for hot operation, the volatility of the fuel being utilized. For G-DI fuel sprays from a particular injector there is, in general, a different penetration curve for each point on the engine operating map as the downstream ambient air density changes. For the common case of hot operation, with the injector and fuel cavity at a temperature in the range of 75 to 90°C, typical pump gasolines yield changes in spray cone angle and penetration rate as the ambient back pressure is changed. In addition to the well-documented, sudden collapse of spray cones from swirl-type injectors as the ambient density is increased, the sprays from many types of injector nozzles are also observed to collapse as

the back pressure is decreased for hot operation. The latter phenomenon, which has been called flash boiling in the literature, occurs even on injector types (nozzle designs) that do not exhibit spray collapse with increasing downstream density at room temperature. Thus, for fully warmed-up operation on real field gasolines, *both* the downstream ambient density and the downstream ambient pressure can significantly influence the effective spray cone angle, the rate of spray penetration and the spray wetted footprint.

Multihole injectors, air-assist injectors and swirl injectors all exhibit significant decreases in the effective spray cone angle and wetted footprint for hot operation, usually for ambient back pressures of less than 0.15 to 0.20 MPa. This decrease in cone angle is always accompanied by an increase in the penetration rate, with the spray behaving as a narrow, gaseous jet. Injectors having a slit-type nozzle exhibit changes that are much less pronounced than for other types, thus the task of correlating the spray penetration over the engine operating map is easiest for this design. For the other injector/nozzle types the task is more involved, as the loci of points where pressure collapse and temperature collapse occur are interrelated, are dependent on the volatility of the particular fuel being utilized, and may correspond to common points on an engine operating map. Thus, the use of a simple spray penetration curve that is determined for room conditions on the test bench *can be very misleading*. For the key uses of spray penetration data listed in Table 4.6-1, the extension of a simple penetration curve obtained at 20°C and one bar ambient could introduce errors in any of the four areas.

Penetration characteristics are comprised of a number of descriptive parameters that are used to quantify the progress of the fuel into the combustion chamber. The terminology and the corresponding physical parameters should be well understood, as a number of different parameters are sometimes described in the literature by the generic term "penetration," which can lead to errors

Table 4.6-1
Key applications of spray penetration information

- Injector selection
- Interpretation of combustion data
- Initial tuning of CFD models
- Establishment of final injector specifications

of interpretation. The terminology that is used to describe a transient G-DI fuel spray for any downstream ambient conditions is illustrated in Fig. 4.6-1. The generic case of both a main and a sac spray is shown. Spray penetration plots represent the loci of a defined portion of the spray as a function of time. The spray portion of interest is most commonly the main spray tip, but may additionally be the sac spray tip, the main spray tail, or the after-injection spray tip. The time may either be the time that has elapsed since the injector was first commanded to open, which is designated in the figure as SLP, or may be the actual time of flight of the spray portion being measured. If the main-tail or after-injection penetration were of interest, then the flight times from the first appearance of those particular spray portions would be utilized. The flight time differs from the time measured from the start of the injector pulse by the sum of the driver capacitive-charging delay and the mechanical opening time of the injector. Ordinarily this total dwell time would be subtracted to obtain the actual flight time, unless there is an interest in relating the spray position to particular engine crank angles. In this common scenario the start-of-injection timing of the engine corresponds to the start of the logic pulse and many crank angles are traversed before any spray appears at the nozzle tip. For this case the relationship between engine crank angle and spray-tip location would be incorrect if the opening dwell were subtracted. In general it is very easy to accurately determine the time of first appearance of the fuel spray at the nozzle tip, but it is usually quite difficult to establish the precise starting time for the main spray if a sac spray precedes the main spray. In this common case the main spray becomes very evident as a spray that is separate from that of the sac when the sac spray has penetrated some distance (6 to 12 mm). There may be a very slight difference in the starting times for the two sprays (50 μs or less), however, it is very difficult to quantify and is almost always ignored. The flight times of both the main and sac spray tips are usually measured from the first appearance of any spray, which will be the sac spray if a sac is present.

Although the tip positions of both the sac and main sprays are relatively easy to image and track, it should be noted that these regions are not comprised of a constant set of droplets, that is, there is not necessarily a drop that has the same penetration-time curve as that measured for the spray tip. The tip of the fuel spray is in fact comprised of a changing assemblage of drops that are passed by other drops and either evaporate or

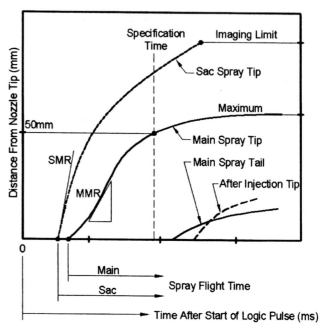

Figure 4.6-1 Terminology for G-DI spray penetration characteristics (MMR: main-spray maximum rate of penetration; SMR: sac-spray maximum rate of penetration).

move back from the tip. Other complicating factors include pulse-to-pulse variability in penetration and spray tip geometry, which are particularly prevalent for sac sprays. The only method of accounting for pulse-to-pulse variability in penetration is to acquire a set of positions for each operating condition that will constitute a statistically valid sample, and that will provide the mean and standard deviation of the penetration to the desired confidence level. A faster, but less accurate, approach that is commonly employed is to obtain five images at each operating point, then plot the mean penetration and the minimum and maximum limits. With regard to determination and interpretation of the spray front, this can be challenging for the main sprays from wide-cone injectors. Irregularities and spray fingers are often present, and can change significantly from injection to injection. For such conditions there is always a degree of subjective interpretation.

The penetration distance is most commonly the distance along the injector axis, which is the axial component for an offset spray. It is normally not measured and plotted as the actual flight path length of a drop, as most sprays have drops that penetrate over a wide range of angles and flight distances. For example, an oval pattern from an eight-hole multihole-type nozzle tip has eight individual sprays with four different offset angles relative to the injector axis. For an impact plane such as the floor of a piston cavity there will be four different impact distances or eight if the injector is inclined to the bowl floor. The most accurate, but again more time consuming, method of providing penetration data for multiple spray fronts is to measure each spray front along the individual flight path, indicating the angle from the injector axis. For the example spray of the eight-hole nozzle discussed previously, this leads to eight penetration curves at four different angles from the injector axis. It is evident that a saving in test time and plot complexity is achieved by assigning one maximum axial penetration curve to a complex spray pattern with multiple spray components or leading edges; however, care must be taken in using such curves to back-calculate times of impact to inclined surfaces.

If the time-imaged locations of spray tips are plotted, the generic spray characteristics that are represented in Fig. 4.6-1 are obtained. Each curve is monotonically increasing, and exhibits a maximum slope. This maximum slope, or rate of penetration, illustrated as MMR for the main spray and SMR for the sac spray, represents the maximum penetration rate that is obtained for a spray at any time in the penetration process. They correspond to the maximum tip velocities of the main and sac sprays. For the sac spray this almost always occurs very close to the nozzle tip, thus the slope of the curve plotted for the sac spray will continuously decrease. At any arbitrary point on the sac penetration curve, there will be a corresponding instantaneous rate of penetration, which is the tip velocity at that time and position. The same holds for the main spray, with the only difference being that the maximum rate of penetration usually occurs 5 to 12 mm downstream from the tip. Additional important points on the penetration curve correspond to the flight time at which the main spray tip penetrates to the distance at which drop size measurements are obtained, or to a defined point on the main curve that constitutes an injector procurement specification. The distance specified for both cases is usually 50 mm.

In many injector designs the absolute sac volume is very low, hence the sac spray becomes difficult to image when it has penetrated 60 to 90 mm and dispersed. The G-DI sac spray commonly does not reach a maximum penetration within practical distances, but behaves as depicted in Fig. 4.6-1: it continues to penetrate, but reaches a sac imaging limit. However, the main sprays from many G-DI injectors do attain a maximum

penetration limit, particularly for elevated levels of ambient density. There is a very important point of interpretation to be made here. The maximum penetration rate and the maximum penetration are two nearly independent parameters for the spray. It has been observed that two different sprays can have equal maximum penetration rates, but different maximum penetrations, and vice versa. If the use of the term "penetration" is not clearly explained in a technical paper, misinterpretations can occur. It is recommended that the modifiers of "maximum" and "maximum rate" be used whenever spray penetration characterizations are obtained and provided. This will distinguish between the maximum rate of penetration, which has the units of velocity, and the maximum penetration distance, which has the units of distance. For the characterization data set both the curve and the two numerical values should be provided for the main spray at each operating condition. Two curves and three numerical values should be provided for those cases where a sac spray is present, as there is normally no maximum penetration distance for the sac. For sprays having multiple spray fronts at various angles of offset from the injector axis, two options are available. The axial penetration of any spray front may be obtained and plotted for each time. This corresponds to the initial crossing of a virtual plane that is orthogonal to the injector axis and is placed at some axial distance from the nozzle tip. The second option entails much more work, and is described above. In either case it should be clearly indicated whether the plotted penetration distance is measured along the injector axis or an individual spray flight path.

A representative family of spray-tip axial penetration curves is presented in Fig. 4.6-2. These curves are for three injectors having swirl-type nozzles designed to deliver three different nominal spray cone angles. The designations of 70°, 80° and 90° refer to the values assigned by all injector manufacturers to the nominal angle subtended by the spray near the injector tip for a mid-injection time frame. For hot injector operation on volatile fuels this angle can change significantly, and even if it does not, the actual included angle of the downstream spray plume may be substantially narrower, which can result in increased penetration. The tip penetrations that are shown in Fig. 4.6-2 correspond to a very meaningful point in G-DI stratified-charge operation: that of hot, steady-state engine operation at light load using late-injection. The fuel volatility and quantity of fuel delivered for these penetration tests are

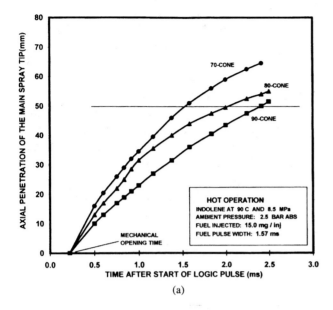

(a)

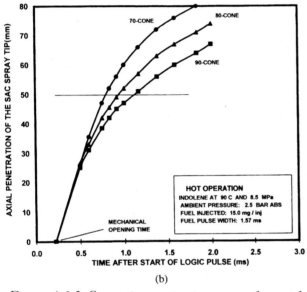

(b)

Figure 4.6-2 Spray tip penetration curves for swirl injectors of three nominal cone angles: (a) main spray tip penetration curves; (b) sac spray tip penetration curves.

also representative. It may be seen that the spray-tip penetrations of both the sac and main sprays do vary with the nominal cone angle of the spray for the swirl injector, with the maximum penetration rate generally increasing as the nominal cone angle is narrowed. The sac spray is seen to have a significantly higher rate of penetration and maximum axial penetration distance than the main spray. The maximum penetration rate of the sac spray for the test shown is 92 mm/ms, independent of cone angle, whereas it decreases from 63 mm/ms to 40 mm/ms for

the main spray as the nominal cone angle is increased from 70° to 90°. For the main spray the maximum axial penetration distance occurs later than the 2.5 ms time shown, but is less than 80 mm for all three sprays. Most sac sprays typically penetrate as shown in the example data, and do not exhibit a maximum penetration distance within any reasonable time frame. The sac spray instead disperses and become difficult to image at distances of 80 to 150 mm downstream. The trends that are shown and discussed are the same for injector operation at room temperature, although the values are somewhat altered.

The variations in the axial tip penetrations of sprays from a multihole injector, for both hot and cold injector operation, are shown in Fig. 4.6-3. The axial penetration along the direction of the injector axis is plotted. It should be noted that each inclined spray from a multihole nozzle would have an actual penetration distance and rate along the centerline of the spray that will be larger than the axial component. As a first approximation, this individual penetration curve may be obtained by applying a simple cosine correction using the angle of inclination of the particular spray relative to the injector axis. If all individual plumes in a spray do not break a virtual axial plane at the same time, then each plume must have a separate penetration curve. As is the case for all other types of G-DI injectors, the spray-tip penetration characteristics are altered by changes in the fuel rail pressure, the ambient conditions, the fuel volatility and the injector operating temperature. Both the ambient

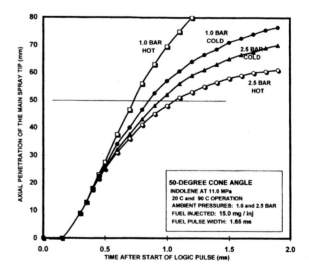

Figure 4.6-3 Variation in spray penetration for hot (90°C) and cold (20°C) operation of a multihole injector.

density and the ambient back pressure can *independently* influence the penetration characteristics, with the density influencing the drag, and the ambient back pressure influencing the degree of flash boiling. Combinations of test conditions can produce a variety of spray tip penetration curves that may be difficult to correlate without a comprehensive injector model. For a multihole injector, spray tip penetration using a typical gasoline is affected to varying degrees depending on the combination of operating temperature and ambient back pressure. For a 2.5 bar back pressure there is a small decrease in maximum penetration for hot operation versus room-temperature operation, whereas at 1.0 bar there is a significant change in the opposite direction. This illustrates the important fact that families of spray penetration curves may suggest simple trends, as in Fig. 4.6-2, or may convey complex behavior related to significant alterations in the spray geometry, such as is represented in Fig. 4.6-3.

As compared to a simple room-temperature bench test, it is certainly more difficult and time consuming to experimentally obtain a representative family of penetration curves that provides quantitative information on the effects of all possible changes in ambient density and back pressure, fuel pressure, operating temperature and fuel volatility. Thus the complete penetration operating map will not be known, but a knowledge of some areas, such as the spray penetration curve for 90°C injector operation on a fuel having a representative Reid vapor pressure (RVP) may be necessary in a critical development program. Imaging in an optical spray chamber or optical engine is required, and both should have the capability of using real gasolines and heating the injector tip, body and fuel inlet line to temperatures that correspond to those in the steady operation of the engine. It is also time consuming to account for cycle-to-cycle variations in spray-tip penetration, which can be quite pronounced, particularly for sac sprays. Such variations are best quantified by obtaining multiple images (5 to 12) at a fixed operating point and obtaining the sample mean and sample upper/lower limits of penetration. For most development programs the mean values of the penetration characteristics are usually sufficient, but the standard deviation of the pulse-to-pulse variability could be a factor in interpreting the causes of high values of the COV of IMEP. The ultimate predictive tool that could account for all of the environmental and operating effects on the spray plume geometry and penetration would be a computer model. The difficulty here is that a superior model of

both the injector and spray development is required, and even if such models existed they would have to be tuned based on experimental data. Such a model would also have to accurately account for the flash boiling of a multi-component fuel within the nozzle cavity.

Some operating parameters are not as influential as might be first surmised, and generally can be ignored in the spray penetration data matrix. The first parameter of little influence on the penetration is the repetition rate of the injector, which is the period of time in milliseconds between injections. The term "rep rate" should not, of course, have units of time, but that is now the commonly accepted terminology of the injector industry. It is found that whether the bench test repetition rate is 100 ms, 50 ms or 33 ms (10Hz, 20Hz or 30Hz), the spray penetration curve is only slightly altered. In contrast to a single injection event into a true quiescent ambient, there is a very small increase in spray penetration rate and maximum penetration as the injection frequency increases due to the steady, oscillatory air flow field that is set up by the multiple injections. This reduces the drag on the fuel droplets within the tip, thus slightly increasing the values of the two penetration metrics. The second parameter is the fuel pulse width, which may have an effect on the maximum penetration distance, but has only a small effect on the maximum penetration rate (maximum tip velocity). The maximum penetration distance increases with increasing fuel pulse width, particularly for elevated ambient densities/back pressures, as illustrated in Fig. 4.6-4; however, the maximum rate of penetration is altered only

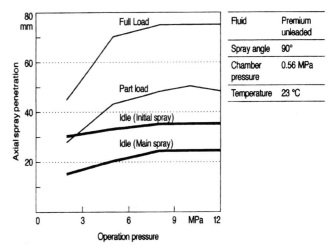

Figure 4.6-4 Spray axial penetration as a function of fuel injection pressure over different engine loads for a swirl-type injector [369].

for very short pulse durations (generally less than 1 ms). This is because the maximum penetration rate for the spray tip occurs at or near the nozzle tip for both sac and main sprays, and normally occurs very early in a fuel pulse. The application of fuel pulse widths of longer duration helps to sustain the early penetration rate, but does not increase it above the maximum value that was achieved close to the tip. If the major area of interest regarding spray penetration is the maximum penetration distance, then the fuel pulse width is a parameter of some influence, and the value must be specified. It should be noted that the fuel pulse width among various injectors and drivers is not universally correlated with fuel delivery, mainly because of design ranges in driver delay, static flow and mechanical opening time. Thus a better combination of parameters to use to unambiguously define the penetration test is the fuel mass per injection and the fuel pulse width. This information should be placed on every spray penetration plot.

Because spray cone angle changes with an increase in ambient pressure, there is an inherent difficulty in defining and correlating the spray cone angles that are obtained for a range of back pressures, particularly considering the definition(s) of spray cone angle that are employed by the injector manufacturers. Although each manufacturer has a unique in-house definition for G-DI cone angle, all are somewhat similar in that they are metrics for the angle subtended by the spray near the injector tip, and nearly all are obtained by imagery of the backlit spray taken at room bench conditions. The metric used by injector manufacturers for establishing the spray cone angle is illustrated in Fig. 4.6-5. A backlit image of the spray at some time, t, into the injection event

is used to define the angle subtended by the spray near the injector tip. Each manufacturer and end-user company has an in-house method of defining the image time, t, and the algorithm used to determine the subtended angle. The equation for image time normally is an expression for the time of mid-injection. A virtual line, which may be at 5, 10 or 15 mm from the tip, is used to define the intersection with the left and right spray boundaries. Some companies then use the tangent lines at these two points to define the subtended angle, with others projecting lines back to the point of the tip on the injector axis. Without knowledge of the in-house formulas for establishing the time, t, and the points in space for drawing the angle, it is virtually impossible to verify a stated spray cone angle using a bench test, as not enough detailed information is available for such a test to be conducted and the data reduced. Whether the angle is obtained from the images at a distance from the tip of 5, 10 or 15 mm, or some combination of two distances, the fact remains that this cone angle *is not the parameter that is indicative of a spray collapse* when dealing with the spray development process at high ambient densities. This will be discussed further in Section 4.12.

This is truly a situation where confusion reigns. The important realization is that the cone angle for which G-DI injectors are designed, classified and procured does *not* change significantly with ambient density, and does *not* collapse. Actually, the cone angle as currently defined by the manufacturers generally goes in the opposite direction to the narrowing of the spray, with the spray exit angle at the nozzle exit becoming slightly larger as the spray downstream is narrowed. A proposed shorter-term solution to this problem is to use the waist diameter of the spray plume, along with spray penetration, as an alternative to cone angle to characterize the spray under high levels of ambient pressure [195]. The waist diameter of the spray is what actually collapses in a collapsing spray; hence this is a reasonable suggestion. The schematic representation of such a cone angle is shown in Fig. 4.6-6. Many specific details of obtaining the overall and waist images must be defined, but the concept is straightforward. The boundaries of the spray waist may be used in combination with the distance from the tip for which the waist is imaged to compute a cone angle. This will yield an angle that is indeed significantly smaller when the spray is collapsed, as opposed to the cone angle as defined by the manufacturers, which may only change by 2° during a collapse event.

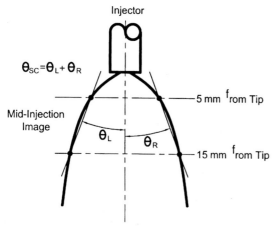

Figure 4.6-5 Metric for establishing the spray cone angle based on imagery of the backlit spray.

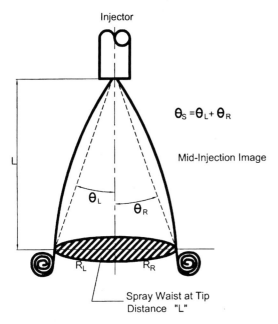

Figure 4.6-6 Schematic representation of spray cone angle determination using spray waist diameter along with spray penetration.

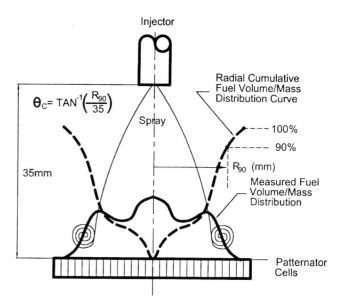

Figure 4.6-7 Schematic representation of spray cone angle determination based upon the measured cumulative radial distribution of fuel using the patternator.

Much of the problem is related to obfuscation associated with the use of the term "spray cone." When high-resolution patternation becomes routinely used to obtain the G-DI cone angle based upon the cumulative radial distribution of fuel (as has been routinely done for PFI sprays since 1985), the industry will then have yet another conflicting definition of "cone angle." As illustrated in Fig. 4.6-7, the ultimate definition of spray cone angle should be based upon the time-integrated mass distribution of the fuel, not upon imaging at an instant of time. In this mass-based technique, the liquid droplets from a G-DI injector are collected in a high-resolution patternator, which normally contains on the order of 250 to 260 individual collection cells. The liquid is normally collected for many consecutive injection events, until one cell has attained 90% of maximum capacity. To minimize losses due to vaporization, liquids such as solvents or n-decane have been used, although patternation tests using n-heptane are also quite common. In principle, the cumulative radial distribution curve of collected fuel mass/volume is generated, and a point on the curve is selected as representing the conical envelope for the fuel mass in the spray. Normally this is the 90% cumulative fuel point, although 80% has also been used. Even though a mass-based cone angle is considered to be preferable to an image-based cone angle, this technique is currently not being used by the majority of injector manufacturers for

measuring G-DI cone angles. It should be considered as a research technique that is under development, and will likely replace imaging for the determination of G-DI cone angles within the next six to eight years. Algorithms (cell assignment accounting) still must be defined for using the volumes of hundreds of cells in reducing the data, and each patternator can thus provide slightly different cone angles even for the same injector, particularly when patternators have different cell geometries and cell center spacings. Each algorithm can account for slight spray asymmetries in a unique way, thus yielding a slightly different cone angle on each patternator. Another restriction is that real pump gasolines cannot be used for such tests, as more than a third of the injected fuel is typically lost to evaporation, which obviously skews any measured mass distribution. It will obviously require a number of years before standards on patternator grid spacing, test fluids, injector test distance and reducing algorithms can be developed and agreed upon. This will happen, however, and mass-based cone angles will supplant those obtained from imaging as a more meaningful metric. The SAE Gasoline Fuel Injection Standards Committee has proposed a solution that restricts the use of the term "cone angle" to mass distribution determinations, with the current name for the imaged initial angle being changed to "spray angle." Until that is universally adopted, and G-DI injectors are referred to and ordered by spray angle, the confusion and obfuscation will continue.

4.7 Characteristics of Offset Sprays

Offset or angled sprays are quite common in many G-DI injector designs, with the basic reason being that they provide additional options in relation to packaging the injector and fuel rail within the total engine hardware configuration. The optimum positioning of the spray axis may have very limited flexibility within a combustion chamber design, but if the injector axis can be oriented up to 25° from the spray axis in any direction, then there are many more options for the injector mounting angle. Thus, the addition of a spray offset decouples the injector axis from a direct alignment with the spray axis. In this manner the injector and fuel rail can be more readily integrated with the other engine hardware such as the port runners, valves and the intake manifold. This helps to ensure a system configuration in which design compromises are minimized and injector serviceability is reasonable. Offset fuel sprays are typically available as an option in all types of G-DI injectors except single-fluid and air-assisted units incorporating an outwardly opening design. As an example, Fig. 4.7-1 shows the spray from the swirl-type injector of the Toyota first-generation D-4 engine. The spray from the slit-type injector that is used in the Toyota second-generation D-4 engine is illustrated in Fig. 4.7-2. Both injectors are designed to provide a fuel spray that is offset from the injector axis by 20°

The offset spray has a spray axis that is at an angle to the axis of the injector body. The offset is normally available for equal increments of offset angle such as 10, 15, 20 or 25 degrees. In swirl and shaped-spray (casting-net) injectors the turning is external, and the spray quality is slightly degraded as compared to what can be obtained with zero offset. For multihole and slit-type injectors there is little or no effect of offset angle, as turning is accomplished internally by modifying the orientation of the delivery holes or slit. For external-turning designs the mass distribution of fuel within the spray is not the same as for the zero-offset case. The spray symmetry that may be obtained with a zero offset is generally significantly compromised, with a greater concentration of fuel mass normally being present on the side toward the injector axis, although this is dependent on the particular method of external turning. Thus, for swirl-type injectors the offset spray plume cannot be assumed to be equal to that of a zero-offset spray plume that is transposed by the offset angle. The wetted footprint, SMD, DV90 and penetration characteristics will all be moderately altered in turning the spray, and the asymmetry should be accounted for by means of a separate spray characterization.

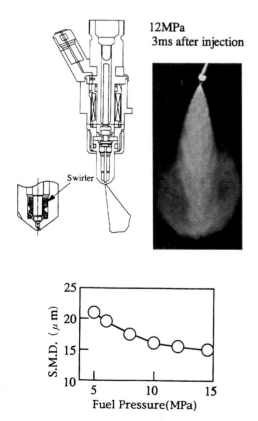

Figure 4.7-1 Offset spray from a swirl-type injector of the Toyota first-generation D-4 engine [156].

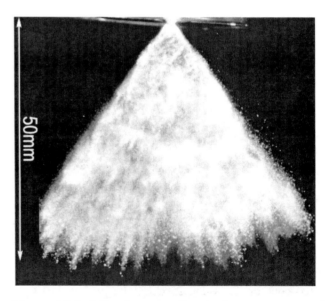

Figure 4.7-2 Offset fan spray from a slit-type injector of the Toyota second-generation D-4 engine (plan view).

The typical offset spray depicted in Fig. 4.7-3 shows turning of the spray axis through 15 degrees, and the asymmetry relative to that offset axis. As verified in the phase-Doppler data in Fig. 4.7-4, larger drops are obvious in the spray development photo at the spray periphery farthest away from the injector axis, as are the larger drops associated with the pintle closing. The latter has little to do with the offsetting of the spray, but the former is directly related to the details of the turning mechanism. Some designs have an SMD that is larger on the inside edge than the far edge, and others, like the example in Figs. 4.7-3 and 4.7-4, show the opposite trend.

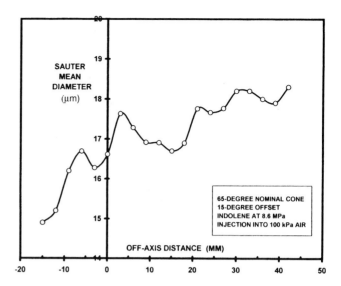

Figure 4.7-4 Profile of the mean drop diameter for an offset swirl spray (at 50 mm from the tip).

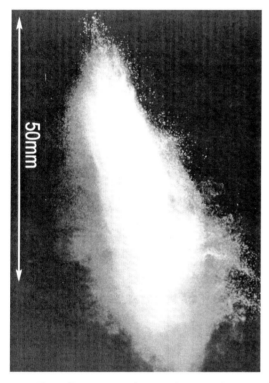

Figure 4.7-3 Geometry of a swirl spray that is offset by 15°.

4.8 Split Injection Considerations

There are two basic questions to be addressed in considering a split-injection strategy. As this almost always implies two injections per cycle, not three or four, the first question may be posed as follows: Is the second pulse degraded in terms of spray quality or fuel delivery because of the occurrence of the first pulse a number of milliseconds earlier? The second question relates to the pulse widths that must be used. If it is desired to inject 9.0 mg of fuel during an engine cycle, then the pulse width

for a single injection might be 1.30 ms; however, if that fuel is to be introduced by means of two equal injections, then the fuel delivered will be 4.5 mg per injection. This could require a pulse width of 0.82 ms, which may be below the lower linear range of the injector. In fact, it may require a fuel pulse width for which the delivered fuel quantity and spray geometry are much less stable than that obtained for a single-injection pulse width. The piezoelectric actuator stack is ideal for controlled multiple pulses, as the lift, opening time and fuel delivery are very accurate and well controlled compared to the typical G-DI unit. However, the piezoelectric actuator may be coupled with any number of tip designs, and some may have sac volumes. Thus some, but not all, of the problems associated with double pulsing of small fuel amounts could also be encountered with piezoelectrically actuated injectors. As the pulse width decreases for these small fuel deliveries, the contribution of the sac spray increases (assuming that the particular injector design has a sac volume), thus the spray quality could degrade markedly. Therefore, even if there were no pulse-to-pulse variability in fuel delivery or spray geometry, the SMD and DV90 will approach that for the sac spray as the fuel logic pulses become very brief. For double pulsing at light loads, it is indeed possible that two sprays of 27 μm SMD could be substituted for one of 17 μm.

The above phenomena are clearly illustrated in Fig. 4.8-1 for an actual G-DI injector and fuel spray. Figure 4.8-1a shows the spray for a fuel delivery of

9.0 mg, whereas Fig. 4.8-1b shows two consecutive sprays of 4.5 mg, each part of a double pulse of the same injector. It may be seen for this actual situation that the single-pulse spray is better atomized than the sprays of the double pulse. This is not at all related to the proximity of the two pulses of 4.5 mg, which were 15 ms apart, but is only related to the pulse width that must be used to deliver half of the total fuel mass. In fact, the two sprays look remarkably similar, even though they are degraded relative to the longer pulse. For heavier loads both the single-pulse and double-pulse sprays have fully developed main sprays and nearly equivalent levels of atomization. It is clear that the sprays generated from very short pulse widths appear to be more like a sac spray, and never develop the appearance of a full main spray. For example, the 9.0 mg pulse exhibits an obvious main spray with a toroidal vortex entraining air. The 4.5 mg pulse never achieves a developed main spray, but instead shows a transition spray appearance that is between a sac spray and the initial stages of a main spray. Also evident are larger drops and ligaments. If a double-pulsing strategy is to be invoked, the lower FPW threshold at which spray degradation becomes severe should be established, and

the use of a minimum pulse width that is less than this value should be avoided. For many injectors this will set a minimum engine load for which the double-pulsing strategy is an option. In the near future this minimum pulse metric could very well become a measure of goodness of the injector performance capabilities.

With regard to the first question on multiple-pulsing capabilities, most of the contemporary G-DI designs that are available for use in combustion system development can provide very closely spaced injection events that are essentially independent, even under elevated injector operating temperatures. Somewhat surprisingly, even 0.75 ms fuel pulses that are only 3.5 ms apart yield sprays that are basically independent, with the second event exhibiting little or no memory of the first injection. Of course, for such close spacing in time the residual droplet cloud from the first injection will be present, and the second injection will occur into that cloud, very much like a diesel pilot and main spray. As shown in Fig. 4.8-2, a second injection only 5.0 ms after the start of the first shows a very similar spray to the first, with the only difference being the presence of the dissipating cloud of droplets from the first injection.

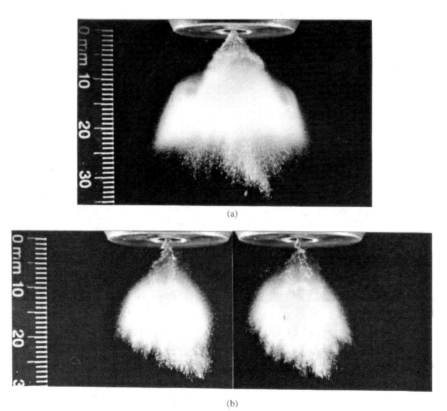

(a)

(b)

Figure 4.8-1 Comparison of single versus double-pulsed sprays for low fuel delivery: (a) single-pulsed spray for 9.0 mg fuel delivery; (b) double-pulsed sprays for 4.5 mg fuel delivery each.

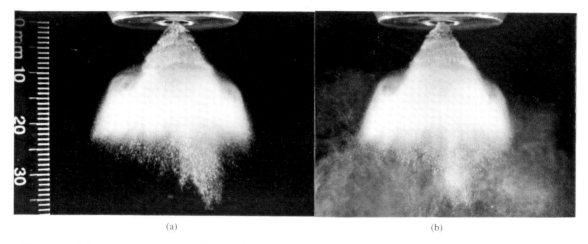

(a)   (b)

Figure 4.8-2 Spray formation for single and multiple pulsing: (a) single spray event; (b) second spray event at 5.0 ms after identical event.

4.9 Spray Characteristics of Single-Fluid Swirl Injectors

4.9.1 Swirl Nozzle Flow Dynamics and Effects of Design Parameters

The spray characteristics of swirl-type G-DI injectors are strongly influenced by nozzle design parameters such as swirl chamber geometry, the sharpness of the inlet to the discharge orifice and the exit radius of curvature of the discharge orifice. The flow processes within a pressure swirl atomizer involve the imparting of a swirling motion to the fuel within each swirl channel, with a liquid film being formed inside the discharge hole of the nozzle, leading to the formation of a thin, conical sheet at the exit plane [26, 27, 67, 72, 257, 278]. Experimental investigation of the internal dynamics of such a nozzle has been a challenge due to the complex transient nature of the flow, coupled with the very small dimensions.

CFD analyses have been conducted with a goal of understanding the fuel flow process inside the swirl-type nozzle [26, 27]. A CFD model of two-phase flow has been used to calculate both the location of the liquid-gas interface using the "volume of fluid" (VOF) method, and to compute the transient formation of the liquid film on the boundary of the discharge hole. The latter occurs due to centrifugal force acting on the fuel once it has been imparted with an increased angular momentum. A schematic representation of a swirl nozzle is presented in Fig. 4.9-1. This is for a design in which the maximum pintle lift is half the width of the conical slot. The radius of curvature at the inlet of the discharge hole influences the film development process, whereas the radius of curvature at the nozzle exit is considered to be one of the important parameters that affect the initial spray formation [26]. A one-dimensional model that calculates the pressure wave dynamics within a common-rail fuel injection system was used to generate the input data for the two-dimensional CFD simulation.

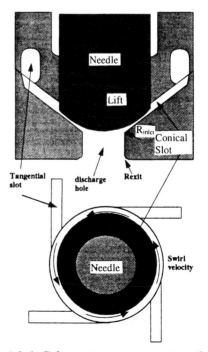

Figure 4.9-1 Schematic representation of a swirl nozzle [26].

The operation of the high-pressure pump was not modeled and, as a result, fuel pressure was assigned as an input to the model. Two different input signals for a fuel rail pressure of 5 MPa are presented in Fig. 4.9-2a, and Fig. 4.9-2b shows the calculated injection pressure at the gallery of the nozzle just upstream of the tangential swirl slots. As can be seen, following the pintle opening the pressure in the nozzle gallery decreases substantially. In fact this pressure loss can be as high as one-half of the nominal rail pressure. Although pressure-wave dampers are utilized within all G-DI fuel rails, the pressure fluctuations caused by the operation of all four injectors connected to a rail for a four-cylinder engine may still influence the formation of the liquid film inside the swirl nozzle, and could influence the subsequent spray characteristics. The predicted fuel injection rate is based on the assumption of a variable nozzle discharge coefficient as a function of pintle lift, with the result shown in Fig. 4.9-2c. It is evident that the instantaneous rate of fuel injection varies markedly during the injection process, which has a significant impact on transient spray development and the subsequent fuel mixing and vaporization processes.

The sequence of events that occurs in the development of the liquid-gas interface during the initial pintle opening is illustrated in Fig. 4.9-3. Some fuel initially enters the discharge hole with near-zero swirl velocity, thus yielding a poorly atomized pre-spray that is concentrated at the center of the injection hole. Following this initial stage, a gradual increase in swirl velocity occurs in the conical slot, forcing the bulk of the liquid to rotate and to move toward the wall of the discharge orifice. Concurrent to this swirl generation, a low-pressure recirculation zone develops at the center of the discharge orifice. Interestingly, two main liquid streams are normally observed to be present within the injection orifice. One stream is generated by the main flow entering through the pintle seat curtain area and moving toward the hole exit, while the second stream is formed from the liquid within the low-pressure recirculation region [27]. A more detailed examination of the predicted swirl distribution is presented in Fig. 4.9-4a, and reveals that the liquid trapped in the recirculation zone has almost zero swirl velocity, and has a much smaller axial velocity than the liquid that is forming the film on the walls of the discharge orifice. The liquid forms a lamella at its front as it moves parallel to the axis of symmetry of the nozzle and downstream of the hole exit. It is understandable that this portion of liquid generates a different spray pattern

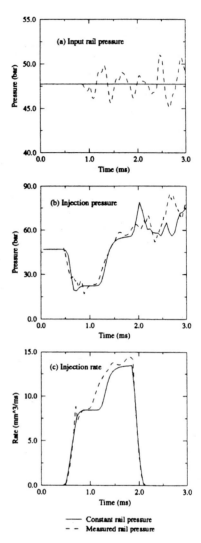

Figure 4.9-2 One-dimensional simulation of the flow in a common-rail G-DI injection system under full load conditions: (a) input rail pressure; (b) predicted injection pressure in the nozzle gallery; (c) predicted injection rate [26].

during the early stages of injection than that associated with the liquid film on the walls. An examination of the pressure distribution as depicted in Fig. 4.9-4b reveals that the pressure decreases along the conical slot and becomes equal to the back pressure in the area of the discharge orifice that contains the liquid film. However, depending on the magnitude of the pintle lift, the minimum values of the calculated pressure distribution may also be located at the seat curtain area where a pressure lower than that of the vapor pressure of the flowing liquid exists. This indicates that cavitation is possible in this location. The pressure decrease is larger at low pintle

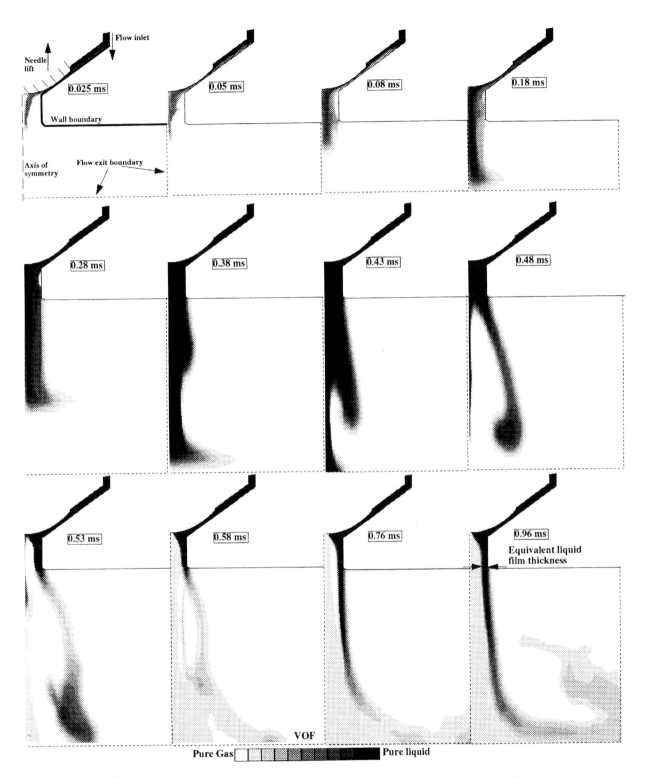

Figure 4.9-3 VOF distribution inside and in the vicinity of the injection nozzle hole at different time steps as predicted by the two-phase CFD model [27].

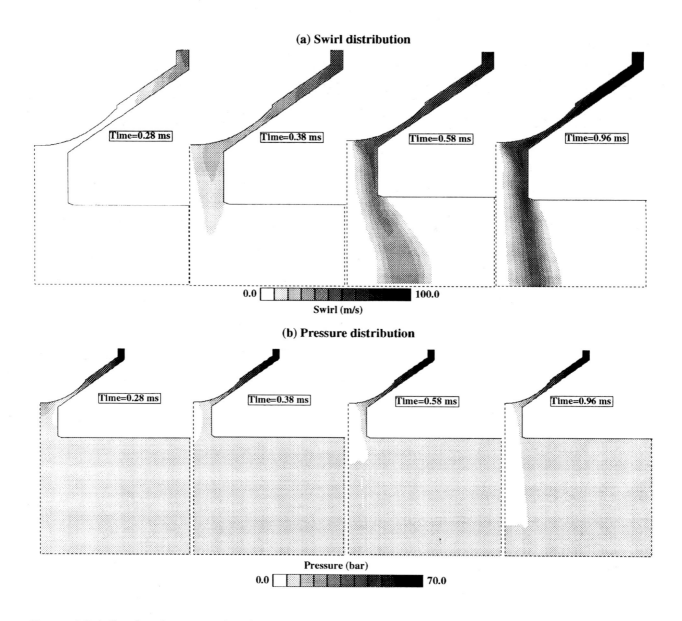

Figure 4.9-4 *Swirl and pressure distributions inside and in the vicinity of the injection nozzle hole at different time steps as predicted by the two-phase CFD model: (a) swirl distribution; (b) pressure distribution* [27].

openings, yielding an increased pressure differential across the seat curtain. The flow field prediction indicates that a vortex is created at the exit of the orifice by the exiting fuel displacing the air located in this region. The air vortices formed at the orifice exit interact with the injected liquid, and contribute to the breakup of the film.

Analyses of the effect of reduced initial swirl velocity on the swirl nozzle flow characteristics reveal that the film develops much more slowly, while the resultant spray attains a relatively smaller cone angle. It is also possible that, as the swirl level is reduced, either a liquid core or a cylindrical sheet will be formed instead of a liquid film. Because the level of swirl velocity is a function of the geometry of the tangential slots, it is predicted that nozzles having the same discharge hole but different swirl-channel geometries will yield considerably different sprays [26]. Increasing the flow rate through the nozzle by increasing the fuel rail pressure not only increases the injection velocity but also creates a thicker liquid film. In addition, the time required to establish a steady-state film is decreased, and the initial spray angle is narrowed. Therefore, it is expected that flow rate variations that occur during the injection process will also influence the spray pattern. Increasing the ambient density generally yields a reduction in the spray angle for the same fuel flow rate and swirl velocity, due to the increased shear forces acting on the moving liquid.

In a swirl atomizer the pintle lift is usually less than the width of the conical slot, resulting in a sudden contraction at the needle seat area that is slightly upstream of the discharge hole. Investigations have shown that for a reduced maximum pintle lift the liquid velocity at the needle seat curtain increases significantly, and that the injection pressure required to maintain the same fuel flow rate is increased correspondingly. A reduced pintle lift tends to promote the formation of a liquid core at the center of the nozzle instead of promoting the formation of a liquid film on the walls of the discharge hole. It should be noted, however, that each injection event is a full transient, thus the delivered spray from a swirl-type nozzle will be influenced not only by the instantaneous flow rate and swirl velocity, but also by the instantaneous pintle position.

Another important design parameter related to liquid film formation inside the nozzle is the radius of curvature at the inlet to the discharge hole. It has been determined that for a sharper inlet corner the

fluid-mechanical losses associated with redirecting the fluid inside the discharge hole are higher. This will result in the accumulation of a larger quantity of fuel in the transient recirculation zone that is formed at the center of the discharge hole. For a nozzle design having a sharper inlet corner, more time is required until pseudo-steady flow conditions are achieved. Interestingly, it has been found that for a sharper inlet corner a larger spray cone angle is initially formed, due to the smaller mass and lower inertia of the liquid contributing to the conical sheet at the nozzle exit. However, at a later time the spray cone angle becomes equal to that of the nozzle exit angle. Thus, the radius of curvature at the nozzle exit exerts a significant influence on the delivered spray cone angle. This design parameter, combined with the centrifugal forces acting on the liquid stream, determines the spray cone angle of the initially hollow-cone sheet spray that is formed at the nozzle exit [27]. Increases in the swirl inlet port area and/or decreases in the exit orifice diameter lead to an enhancement of the discharge coefficient of the swirl port, yielding a wider spray cone angle. Discharge coefficient and spray cone angle are found to be relatively insensitive to injection pressure differential over the range of 3.5 to 10 MPa, although this should be verified for each nozzle design. An air core is present at the needle tip, and may be considered as an obstruction to the central portion of the exit orifice. As the orifice diameter decreases, the size of this air core decreases significantly, which increases the discharge coefficient of the exit orifice [371–373]. It should again be noted that the understanding of these internal flow processes is almost exclusively derived from computational fluid dynamics, as experimental validation is extremely difficult.

4.9.2 Effect of Fuel Swirl Ratio on Spray Characteristics

One advantage that is claimed for the swirl-type nozzle is that the design can be customized to achieve a desired penetration curve by altering the fuel swirl ratio, thus providing one level of optimizing the spray configuration. The measured effect of the fuel swirl level on the droplet size and spray structure is summarized in Figs. 4.9-5 and 4.9-6 [259]. When the fuel swirl level is increased, a comparatively lower fuel pressure suffices for achieving a specified level of atomization. As illustrated in Fig. 4.9-5, increasing the fuel swirl level also promotes air entrainment, with the toroidal vortex ring that is initially generated near the injector tip growing to a larger scale vortex during the latter portions of the injection event. As shown in Fig. 4.9-7, droplet velocity measurements using phase-Doppler anemometry (PDA) verify that the axial velocity component of the drops in the spray decreases with distance from the injector tip, whereas the swirl-component remains fairly constant. The decrease in axial velocity is caused by drag on the fuel droplets that are moving relative to the ambient air. In contrast, the swirl component of drag is less than the axial component because the ambient air tends to rotate with the fuel droplets, as depicted in Fig. 4.9-5.

An increase in the level of fuel swirl leaving the nozzle orifice is found to yield a smaller mean drop size in both the initial spray and main spray regions, however, the mean drop size in the vortex region is not significantly influenced. It is theorized that the size of the droplets entrained into the vortex region is determined by the ambient air properties. As the ambient air density is increased, the droplet size in the vortex cloud region also increases due to changes in the entrainment characteristics of the ambient air. The fuel pressure has been found to have little or no effect on the mean size of drops in the vortex, as the ambient air properties are not altered [100].

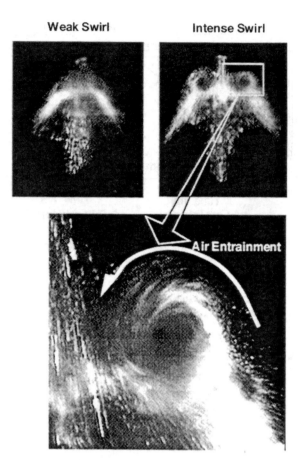

Figure 4.9-6 Effect of nozzle swirl intensity on spray structure [259].

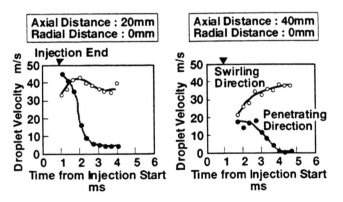

Figure 4.9-7 Time histories of droplet velocities in both axial (penetrating) and swirling directions at two measurement locations [196].

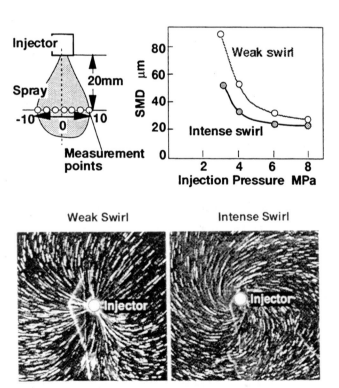

Figure 4.9-5 Effect of nozzle swirl intensity on droplet size and spray-induced air motion made visible using microballoon tracer particles [259].

Although a range of levels of angular momentum can be added to the injected fuel by changing the design of the swirl nozzle, it is found that the various designs tend to generate similar spray characteristics when the swirl Reynolds number is maintained the same [196, 285].

The swirl Reynolds number is defined as the product of the velocity in the swirl channel and the swirl radius, divided by the fuel viscosity [196]. The effect of swirl Reynolds number on the resulting spray cone angle and spray tip penetration of an injector with a tangential-slot swirl plate is illustrated in Fig. 4.9-8 [196]. The strong interrelationships between the spray characteristics such as mean droplet size, spray cone angle and penetration and the swirl intensity makes the custom design of a swirl nozzle for any single spray parameter fairly complex. It is worth noting that, even though swirl Reynolds number has been used to quantify and correlate the fuel swirl intensity of the swirl injector, it should not be used to correlate injection-to-injection spray variations or spray asymmetry. Such correlations are more correctly based on the measurement of the swirl torque from individual injection events [195].

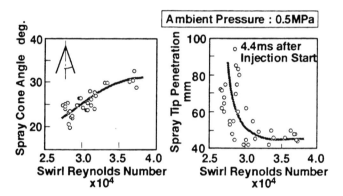

Figure 4.9-8 Effect of fuel swirl Reynolds number on the spray cone angle and penetration [196].

4.9.3 Spray Characteristics of Inwardly Opening, Single-Fluid Swirl Injector

For the fuel spray from a typical inwardly opening swirl injector, as depicted in Fig. 4.3-4, several distinct regimes of the transient spray process have been identified and discussed in the literature [100, 167, 174, 191, 206, 405]. These are designated as the following phases: delay, leading-edge, wide-spray-cone, fully developed and trailing-edge, and have some benefits in assisting in the understanding of the spray development process. The delay phase refers to the period of time between the sending of the fuel logic pulse to the injector driver and the appearance of the first fuel at the nozzle exit. This delay phase varies significantly with the injector design. The leading-edge phase results from the fuel in

the sac volume and the inital fuel in the swirl channel that has near-zero angular momentum, and refers to the time interval between the first appearance of fuel and the emergence of the main spray. After the pintle is opened fully the fuel attains a steady velocity, and a conical region of small droplets is formed [444, 472]. This generally yields the widest cone angle of the entire event, and is denoted as the wide-spray-cone phase. A vortex-cloud is subsequently formed by entrained air that carries small droplets from the spray periphery. As the injection pulse ends, the vortex-cloud continues to develop and move downstream. The continued entrainment of the ambient air causes a contraction of the angle subtended by the spray, which continues at the smaller angle for the fully developed spray. This period is denoted as the fully developed phase. The trailing-edge phase is the time period of pintle closure, and includes any time interval of pintle bounce [100, 405].

The initial atomization of the thin sheet of fuel exiting an inwardly opening swirl injector occurs at or very near the orifice exit. The degree of atomization achieved is a function of nozzle design factors such as the swirl-plate and orifice geometry, pintle opening characteristics, and fuel pressure. The second stage of spray atomization occurs during the spray penetration process, which is dominated by the interaction of the fuel droplets with the surrounding air flow field [293, 326, 332, 433]. The dynamics of an initially hollow-cone spray yields a transient toroidal vortex that forms, grows and moves downstream at an axial velocity that is generally slightly less than that of the spray tip. This vortex aerodynamically separates the fuel drops by centrifuging the larger droplets, and moves downstream with the smaller drops in containment. Phase-Doppler measurements show that the drop diameters within the toroidal vortex are typically all under 10 µm, even though the main spray may have drops of up to 40 µm in diameter.

The spray cone angle is an important parameter that is nominally determined by the injector design; however, in actual application the spray cone angle of a swirl injector also varies with in-cylinder air density and, to a lesser degree, with fuel injection pressure [510, 511, 523, 524]. With the pressure-swirl injector, the spray cone angle generally decreases with an increase in ambient gas density until a minimum angle is reached. Ambient gas density also has a strong influence on the minimum atomization level produced by pressure-swirl atomizers, with increased coalescence at elevated ambient densities yielding a lower atomization quality. In contrast, non-swirl atomizers generally yield a slightly wider spray cone angle as the ambient density is increased. This results from increased aerodynamic drag on the droplets, which produces a greater deceleration in the axial direction than in the radial direction [108]. Under conditions of elevated in-cylinder air density, corresponding to late injection at part load, a more compact droplet plume is required for a higher degree of stratification. Thus, for some combustion systems the phenomenon of spray cone collapse for the high-pressure swirl injector may be considered to be of some benefit for G-DI late-injection applications.

The predicted characteristics of hollow-cone sprays that are generated by a high-pressure swirl injector with a cone angle of 70° are shown in Fig. 4.9-9 [504]. These predictions were made for a swirl angle of 40° and an initial droplet velocity of 60 m/s, to simulate a spray injected into air at one atmosphere using a fuel pressure of 7.0 MPa. The interaction between the

droplets and the gas flow is found to be more pronounced for smaller droplets. It is interesting to note that for the case corresponding to a monodisperse spray of 40-μm droplets, analysis indicates that the droplets do not form a toroidal vortex. For sprays with different droplet size distributions but the same fuel flow rates, the spray structures are predicted to be considerably different. The fuel swirl component is found to influence spray development, and the spray shape at the transition between cone growth and toroidal vortex formation is quite different with and without fuel swirl. The spray cone angle for the case with swirl is significantly larger than that predicted for the non-swirl case. As is schematically illustrated in Fig. 4.9-9c, droplets in the mid-size range are found within the coarse droplet region, while droplets smaller than 10 μm do not form a hollow cone, but tend to concentrate near the injector axis. At 0.4 ms after the start of injection, all droplets near the injector tip are predicted to concentrate in a small region near the injector axis, verifying that the cone angle at the start of fuel injection is very small. At 0.6 ms after the initiation of injection, droplets near the injector tip are concentrated in an angular ring, indicating the development of a wider cone angle. The instantaneous spray cone angle increases from nearly zero to the steady value in proportion to the pintle opening. The measured droplet number density at the end of injection confirms that apparent vortex formation occurs only for droplets in the range of 10 to 25 μm, not for other droplet diameters. Moreover, the total number of droplets in the diameter range of 10 to 25 μm exceeds the numbers in the other ranges. The absolute maximum size of droplets forming the toroidal vortex is found to be about 20 μm, whereas the droplets that move directly down the hollow cone have diameters of up to 50 μm [504, 505].

Additional CFD studies [334] conducted on a 70° inwardly opening swirl injector show that the SMD of drops at a cross-section 50 mm from the injector tip is larger than the total spray-averaged SMD during the injection event, indicating that many of the smallest droplets near the injector tip do not penetrate 50 mm. The measurement of spatial mass distribution on a line 30 mm from the nozzle exit using spray patternation revealed that 99% of the liquid mass within that spray is located in an angular ring with an inner radius of 5 mm and an outer radius of about 20 mm [172]. The highest liquid flux is at a radius of about 15 mm. The thickness of the walls of the spray cone increases with time, while the outer boundary of the spray, which would

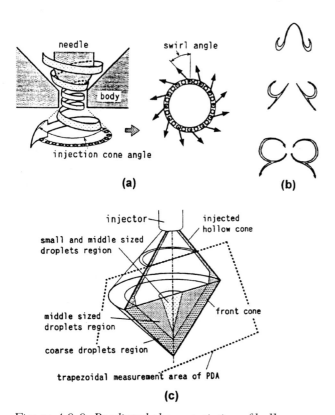

Figure 4.9-9 Predicted characteristics of hollow-cone sprays generated by a high-pressure swirl injector with a cone angle of 70°: (a) nozzle geometry; (b) schematic presentation of the three development stages for a hollow-cone spray; (c) schematic of the droplet size distribution inside a hollow-cone spray at 0.9 ms after the start of injection [504].

typically be used to identify the spray cone angle, does not change at all.

Measurements of the drop size distribution along the injector axis show a gradual increase in mean droplet diameter with increasing distance. The possible mechanisms for this are the coalescence of droplets and the complete vaporization of the smallest droplets. Drop coalescence is likely not a significant factor in this observation, as there is increased spray dispersion downstream of the injector; that is, the geometry of the cone structure causes the spray to be even less dense farther from the nozzle, making coalescence even less likely. The vaporization of smaller droplets definitely yields a shift toward a distribution having larger droplets. Even for non-evaporating sprays, selective aerodynamic drag is known to separate the drops in a distribution, yielding a different distribution of droplet sizes at different downstream locations. Many small drops attain terminal velocity and do not penetrate to a downstream measurement station. For an evaporating spray the measured mean drop size first increases with distance from the tip as the tiniest drops evaporate and are unavailable to be sampled. The mean drop size

eventually decreases as the spray is measured much farther from the nozzle tip. Drops that are significantly farther from the nozzle tip will have had a much longer time for vaporization to occur, which reduces the diameters of all penetrating drops [7]. Droplet measurement results using phase-Doppler anemometry for a range of ambient back pressure, fuel rail pressure and spray cone angles are shown in Figs. 4.9-10 and 4.9-11. The SMD is seen to decrease as the fuel pressure and spray cone angle increase, but increases as the ambient density increases. It is theorized that the use of a wider spray cone angle increases the spray dispersion, which decreases the droplet number density. This, in turn, reduces the probability of droplet coalescence [368, 369].

It is well known that the entraining air flow interacts with the spray droplets and directly influences spray cone development for all G-DI fuel sprays, not just for swirl sprays. The interaction between a transient fuel spray and the ambient air was investigated by injecting the fuel into a chamber filled with tracer particles of polymer micro-balloons [196]. The air flow structure was mapped by the trajectories of the tracer particles, which are shown in Fig. 4.9-12. Clearly an intense, transient,

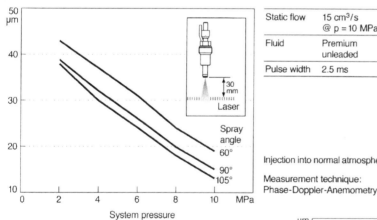

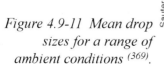

Figure 4.9-10 Mean drop sizes measured for a range of fuel pressures and spray cone angles [369].

Figure 4.9-11 Mean drop sizes for a range of ambient conditions [369].

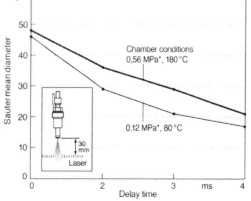

83

turbulent air flow field is generated due to the movement of the millions of drops in the fuel spray and the subsequent transfer of fuel droplet momentum to the ambient air.

4.9.4 Spray Characteristics of Outwardly Opening, Single-Fluid Swirl Injector

The swirl injector design that uses an outwardly opening pintle offers some unique advantages, but there are also some areas of design and operation that require careful monitoring, including the requirement for a very high quality microfinish on the pintle and a sensitivity to small accumulations of deposits on that surface. The immediate advantage of the outwardly opening design is the elimination of any classic sac volume downstream from the sealing line. This will eliminate the poorly atomized initial spray that invariably accompanies the sac spray, and will enhance the pulse-to-pulse variability of the spray, as sac sprays tend to be very unstable. The pseudo-sac volume that results from the initial fuel in the lower portion of the swirl channel not achieving the full angular momentum increase will still be present, and will yield some degradation of the initial spray, but even this is less than is encountered with inwardly opening swirl injector designs.

Spray development for an outwardly opening swirl injector for two extremes of ambient conditions is illustrated in Fig. 4.9-13. The hollow-cone spray structure that is produced at ambient pressure is clearly demonstrated in the left panel of the spray photo. In terms of measured performance, drop-sizing measurements show very good atomization characteristics for a new unit. An important contribution to this atomization is the elimination of the classic sac volume and the associated sac spray. In this manner the outwardly opening swirl injector is able to provide a spray having an SMD of less than 15 mm as measured by laser diffraction at 30 mm downstream from the injector tip, a DV90 of less than 40 mm, and a maximum limiting penetration of 70 mm when injected into air at one atmosphere [487].

4.9.5 Spray Characteristics of Shaped-Spray Injector

Figure 4.9-14 shows the spray characteristics of a typical shaped-spray injector for two operating temperatures. An asymmetric fuel distribution with a greater fraction of the total fuel being concentrated on the right side of the spray envelope is evident. With an increase in the fuel and injector operating temperature the maximum spray tip penetration increases. A comparison of the

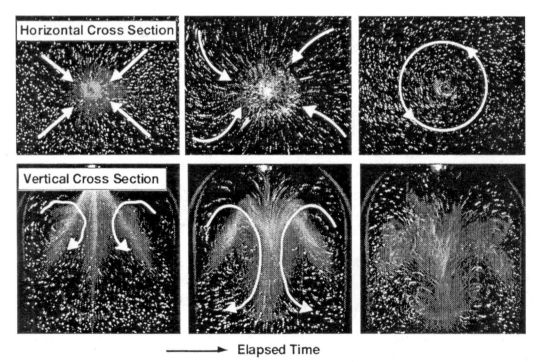

Figure 4.9-12 Mapping of the air-entrainment flow field of a transient spray using tracer particles [196].

84

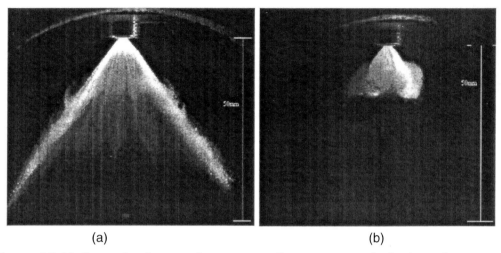

(a)   (b)

Figure 4.9-13 Spray development for an outwardly opening, single-fluid, swirl injector under different ambient conditions: (a) 1 ms after SOI; ambient backpressure: 0.1 MPa; (b) 1 ms after SOI; ambient back pressure: 1.5 MPa [487].

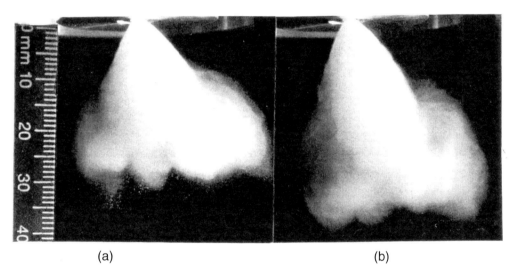

(a)   (b)

Figure 4.9-14 Spray characteristics of the shaped-spray injector: (a) spray derived at the fuel temperature of 20°C with injection into 0.25 MPa ambient; (b) spray derived at the fuel temperature of 90°C with injection into 0.25 MPa ambient.

conventional swirl nozzle and two shaped-spray nozzle designs (L type and taper type) is shown in Fig. 4.9-15 [312]. The spray pattern of the shaped-spray injector can be tailored to produce an inclined hollow-cone spray by varying the symmetric cavity shape at the tip of the orifice [238, 312]. An advantage of this class of spray nozzle is that the spray can be customized by modifying the nozzle tip, while other components are the same as those of a conventional swirl-type G-DI injector. As illustrated in the figure, the spray angles a1 and a2 are determined by the ratio of the velocity component in the orifice axial direction to that in the rotating direction. It should be noted that the air is entrained into the orifice cavity, forming an air core. The amount of air flowing into the orifice depends upon the orifice shape, and this flow strongly influences the spray angle. The L-type nozzle effectively produces a strongly inclined spray pattern due to a more asymmetric orifice shape.

Typical spray structures of the L-type and taper-type nozzles at the same nozzle L/D of 0.35 are illustrated in Fig. 4.9-16 [312]. Clearly the spatial distribution of the spray pattern for the L-type nozzle is more

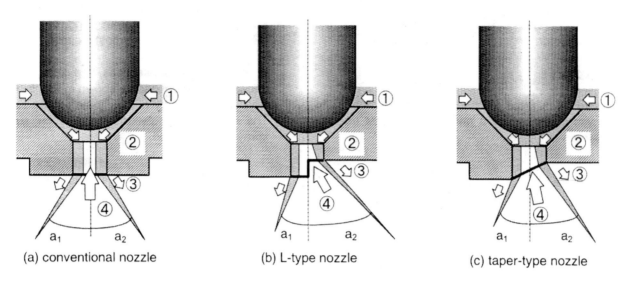

(a) conventional nozzle   (b) L-type nozzle   (c) taper-type nozzle

Figure 4.9-15 Spray formation mechanism of shaped-spray injectors: (a) conventional nozzle; (b) L-type nozzle; (c) taper-type nozzle [312].

inclined than that of the taper-type nozzle, as is illustrated in the vertical cross-section. The spray pattern in the horizontal cross-section at 40 mm below the nozzle tip indicates that the spray from the L-type nozzle is in a V-shape. In contrast, the spray cross-sectional pattern for the taper-type nozzle is nearly circular. The effect of ambient pressure on spray cone angle and penetration, as defined in Fig. 4.9-16, is illustrated in Figs. 4.9-17 [312]. The spray cone angle and penetration characteristics for the L-type nozzle are nearly constant

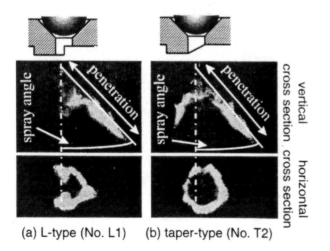

(a) L-type (No. L1)   (b) taper-type (No. T2)

Figure 4.9-16 Comparison of spray characteristics for L-type and taper-type shaped-spray injectors [312].

over a wide range of ambient pressure. On the other hand, both the spray cone angle and maximum penetration rate decrease with an elevated ambient pressure for the taper-type nozzle. For these nozzles the fuel pressure is found to have little effect on the spray pattern. The nozzle design parameter, L/D, as defined in Fig. 3.5-5, has a significant impact on the spray pattern, with this trend plotted in Fig. 4.9-18 [312]. The trend shown is similar to that for a conventional nozzle, in that both the spray cone angle and maximum penetration rate decrease as the L/D increases.

The results of a CFD analysis of the flow within different types of shaped-spray nozzles are shown in Fig. 4.9-19 [312]. The white region within the nozzle marks the air core where the velocity vectors are opposite to the fuel flow direction. It is clearly shown that the velocity vectors in the conventional nozzle are symmetrically distributed, and that the air flow, indicated by the solid arrow, is in the direction of the nozzle orifice axis. In contrast, both the fuel flow and the air flow directions inside the L-type and taper-type nozzles are significantly inclined, with the L-type having the greater inclination. Tests on a combustion system that was retrofitted with this type of injector nozzle in combination with an optimal piston cavity indicate that the amount of fuel wetting the piston bowl surface can be reduced by 30%, as compared to a conventional high-pressure swirl injector [238].

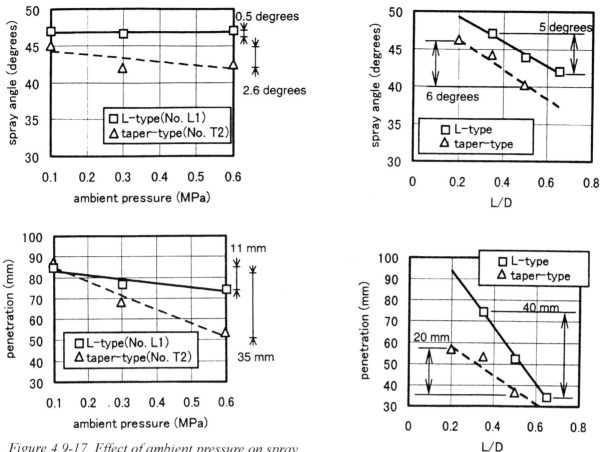

Figure 4.9-17 Effect of ambient pressure on spray cone angle and penetration (defined in Fig. 4.9-16) of shaped-spray injectors [312].

Figure 4.9-18 Effect of L/D on spray cone angle and penetration (defined in Fig. 4.9-16) of shaped-spray injectors [312].

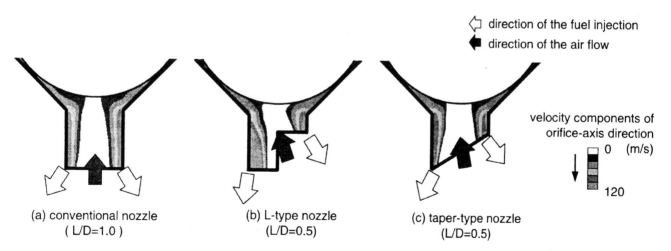

Figure 4.9-19 Predicted velocity contours inside different types of shaped-spray nozzles; ambient pressure of 0.6 MPa; fuel injection pressure of 7 MPa [312].

4.10 Spray Characteristics of Single-Fluid, Non-Swirl Injectors

4.10.1 Spray Characteristics of Slit-Type Nozzle

The slit-type injector nozzle generally has a single rectangular orifice, although in principle multiple or tapered slits could be incorporated. The spray from a slit-type injector has a cross-section that reflects the slit geometry, and, for a single rectangular orifice, expands downstream to form a spray in the shape of a fan. The level of atomization that is achieved is the coarsest among the nozzle types when measured at an equivalent fuel rail pressure, which must be at least 9 MPa to avoid spray-tip SMD values that exceed 22 μm. Of even more importance, elevated fuel rail pressures must be used to avoid DV90 values exceeding 45 μm. The penetration rate for a slit-type injector is also the highest among the nozzle types, even slightly surpassing that of a spray from a multihole nozzle at the same pressure. These penetration characteristics are neither inherently good nor bad, as requirements vary depending on the particular combustion system configuration. Obviously the characteristics of the fan spray from a slit-type injector have been used to significant advantage in the Toyota second-generation D-4 engine.

An example of the development of the fan spray is shown in Fig. 4.10-1 [223]. A circular segment, or fan-type geometry, is observed in the plan view, with a thin (5–6 mm) spray observed in the orthogonal end view. Little or no sac spray is observed from this nozzle, thus the fuel must be nearly completely purged from the slit cavity following an injection event. CFD analyses indicate that the fuel mixes with air relatively uniformly over the spray in the downstream portions of the fan, especially at elevated ambient pressures. As a result, improved homogeneous-charge operation can be achieved, and undermixing during stratified-charge operation can be minimized. The fan spray angle in the plan view is not significantly altered by changes in the ambient density, although the narrow angle in the side view does increase. Some recent studies also suggest that there may be two distinct sheets of fuel leaving the injector in the early spray development stage, as illustrated in Fig. 4.10-2, possibly suggesting that a fan spray forms along each internal surface of slit cavity [409]. Further studies are required to confirm this initial double-sheet observation. As will be discussed later in Section 4.13.2, some contraction of the fan-included angle does occur as the operating temperature is increased from 20°C to 90°C. Figure 4.10-3 shows a comparison of spray penetrations between the fan spray and selected conical sprays from swirl-type nozzles [443]. The fan spray from a slit-type nozzle is seen to have a larger penetration than a conical spray from a swirl nozzle, even for conditions of high ambient density.

As shown in Fig. 4.10-4, the important design parameters that influence the spray characteristics are the angle between the two sides of the nozzle cavity, θ_f, the distance between the center of the sac volume and the cross point of the extrapolated side, B, and the angle between the thin slit axis and injector axis, α. CFD analyses of the nozzle flow field reveal that a shorter B dimension results in a higher flow rate at the center of the spray, with no apparent flow along the sidewall, thus forming a center-projected spray, as shown in Fig. 4.10-4b. In contrast, an increased B dimension yields an increased rate of flow along the sidewall, which results in a center-recessed spray, as illustrated in Fig. 4.10-4d. For a constant B dimension, the flow rate distribution is nearly independent of the angle θ_f, thus θ_f can be varied to meet the requirements for various engine bore diameters without altering the fuel flow rate distribution. The CFD analyses also show that the fuel flow rate vectors at the nozzle exit are independent of the offset angle, α. This is advantageous, as it enables the injector mounting (packaging) constraints to be met by adjusting the spray offset angle, α, without altering the spray characteristics [443].

4.10.2 Spray Characteristics of Multihole Nozzle

As discussed in Chapter 3, the multihole G-DI injector certainly has significant advantages in that it offers flexible spray patterns, but it is not without disadvantages.

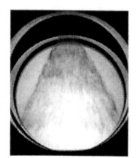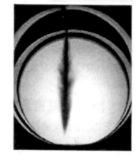

Figure 4.10-1 An example of the development of the fan spray from a slit-type nozzle [223].

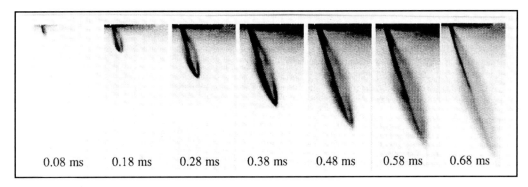

Figure 4.10-2 Side view of the fan spray derived from averaged Mie scattering images (laser light sheet is introduced from the right side) (409).

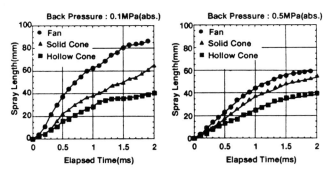

Figure 4.10-3 Comparison of penetrations for a fan spray from a slit-type nozzle and conical sprays from a swirl-type nozzle (443).

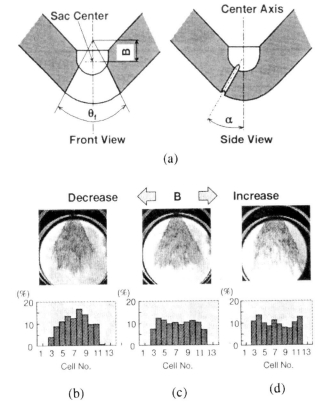

Figure 4.10-4 Effect of nozzle design parameters on the spray characteristics of the slit-type nozzle (443).

Basic, non-enhanced, pressure atomization is used to obtain the fuel droplets with this type of injector, hence relatively high fuel rail pressure (9.5 to 12.0 MPa) must be employed. As with the spray from a slit-type nozzle, the atomization degrades rather rapidly as the fuel pressure is reduced below 9 MPa. The mean droplet size increases as the fuel rail pressure is reduced, but the DV90 increases even more markedly. The maximum penetration rate, maximum penetration and wetting fraction, as with the spray from a slit-type nozzle, are slightly higher than those from swirl or air-assist nozzles are. This is because there are higher-momentum droplets in relatively narrow individual spray plumes that are both larger and of higher velocity. The relative drag on these larger droplets is less than that for a swirl-nozzle spray, hence the higher velocity at the target and the greater impact fraction.

As an example, Fig. 4.10-5 shows the structure of a typical spray from a multihole nozzle. The multiple spray plumes from a multihole nozzle are more complex than

may first be thought. A spray issuing from a single hole is indeed about as simple as can be conceptualized for G-DI injection conditions; in fact, it is a common example in textbooks dealing with atomization. However, the complicating factor is that the individual spray plumes interact in terms of air entrainment and spray plume blending, with the interaction being influenced by

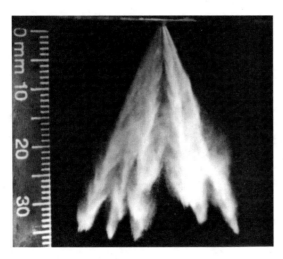

Figure 4.10-5 Typical spray structure of the multihole nozzle; 50° multihole; fuel temperature: 20°C; fuel injection into 0.25 MPa; fuel injection pressure: 11 MPa.

the spray-axis spacing, the downstream density and the temperature of the fuel in the nozzle. In fact, the spray plumes from a multihole nozzle can and do collapse under a number of conditions, although not at all the same conditions for which a swirl-nozzle spray collapses. The "collapse" or partial collapse for a collection of spray plumes from a multihole nozzle is actually a combining of the individual sprays into a lesser number through plume interaction. Depending on the initial spray pattern, this does not always have to result in just one plume. One critical parameter in the combining of individual plumes is the plume spacing, or from a geometric standpoint, the angular spacing of the axes of the individual holes. Four plumes on a 30° conical pattern may not interact and combine, whereas six plumes on a 40° conical pattern may. If the angular separation between plumes exceeds a threshold value, the plumes may not combine and collapse, although this is dependant upon the cone angle of the individual plumes for the operating condition that is being utilized. Thus, for a spray pattern that is a nominal oval in the normal, non-collapsed condition, the footprint of a collapsed spray may approximate a narrow rectangle (a line). The collapse is accompanied not only by a more compact wetted footprint, but also by a more rapidly penetrating spray tip. A faster, narrower, single plume is commonly observed when six or eight individual multihole sprays combine, although nominal crescent and oval wetting patterns may transmute into

two wetting spots, then into one. The causes of multihole spray collapse are twofold; the first is a decrease in the downstream ambient density, which is chiefly determined by the ambient pressure. If the absolute ambient pressure is decreased to less than about 200 kPa, multihole sprays of less than 65° will begin to exhibit collapse. At a level of 70 kPa to 90 kPa, the collapse is complete. This means that the spray geometry will normally be different for early injection than for late injection, as is the case for sprays from swirl injectors. It should be noted that the multihole spray behavior is completely opposite to that of the swirl-type nozzle, where the full cone is observed for early injection and collapse occurs for the high ambient pressures (greater than 200 kPa) that are associated with late injection. Secondly, the collapse of the multihole sprays into one plume also occurs for flash-boiling conditions, which can be at normal, steady operating temperatures of the engine and injector body using common pump gasoline. The individual spray plumes are much wider for hot (90°C) operation, even without collapse, and merge into one plume and one wetted footprint at low ambient pressures less than 200 kPa.

It is worth noting that for a multihole nozzle that is used in a G-DI application, the hole spacing and the spray collapse are critical in ensuring a good spray dispersion and reliable flame propagation between the spray plumes, particularly for air-guided and spray-guided systems. A poor spray distribution can result from a non-optimum spacing of holes and undesirable level of spray collapse, leading to an unstable flame kernel when ignited by a single fixed spark plug. With multiple sprays the rich mixture zones are inherently close to the lean mixture zones, thus the flame may not propagate uniformly unless the level of dispersion is quite good. The hole pattern is one important parameter in obtaining acceptable levels of homogeneity within the mixture cloud [118].

The results of a study of the combustion and emissions characteristics of swirl injectors (A) and multihole injectors (B), which were retrofitted in wall-guided, air-guided and spray-guided combustion systems, are shown in Fig. 4.10-6 [354]. When comparing the results in Fig. 4.10-6a obtained using a wall-guided combustion system, the soot reduction was more than 50% with the multihole injector due to the enhanced air entrainment characteristics. With the air-guided combustion system, however, both injectors yield very low levels of soot emissions as a result of reduced wall impingement

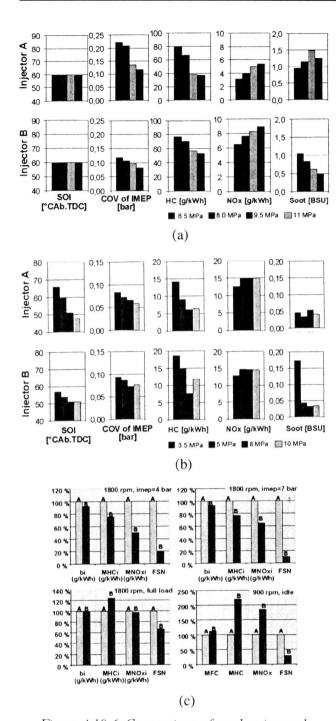

(a)

(b)

(c)

Figure 4.10-6 Comparison of combustion and emissions characteristics between swirl (A) and multihole (B) injectors: (a) wall-guided combustion system (the bars from left to right correspond to fuel pressures of 6.5, 8.0, 9.5, and 11.0 MPa, respectively); (b) air-guided combustion system (the bars from left to right correspond to fuel pressures of 3.5, 5.0, 8.0, and 10.0 MPa, respectively); (c) spray-guided combustion system (bi: ISFC; FSN: smoke emissons) [354].

with this type of combustion system. The pressure dependence of the atomization quality is significantly higher with the multihole injector, which is why soot emissions are quite high with the multihole injector at a fuel pressure of 3.5 MPa, as illustrated in Fig. 4.10-6b. It is well established that the swirl spray cone collapses when the ambient density exceeds a threshold value. This means that the swirl-injector-based, spray-guided combustion system is very sensitive to the spray characteristics of the swirl injector. Figure 4.10-6c shows a comparison of combustion and emissions characteristics between swirl and multihole injectors for a range of engine loads. The improvements in fuel consumption and emissions are significant for the multihole nozzle at partial loads of up to 7 bar IMEP. However, the improvement is quite small at engine idle with the spray-guided combustion system, due to the small amount of fuel injected at this load point. It is evident that the multihole injector has some advantages when utilized in a spray-guided combustion system. This is mainly due to the stable spray characteristics for elevated ambient density.

4.10.3 Spray Characteristics During Cold Crank and Start

The spray performance expected for a G-DI cold-crank-and-start is shown in Fig. 4.10-7 for three typical fuel rail pressures that might be encountered in the range of cold start strategies. The spray geometry and overall appearance is shown for a typical single-fluid, four-hole multihole injector, with Fig. 4.10-7a being for the lowest pressure that would likely be used (that obtained from a standard tank-mounted vane pump, or 0.4 MPa). Figure 4.10-7b is for an intermediate rail pressure of 1 MPa, which represents the type of spray that might be obtained with either a moderate-pressure boost pump or by partial priming of the main pump. The spray that would be obtained at the upper limit of cold-start fuel pressure is shown in Fig. 4.10-7c. The pressure level of 2 MPa represents the limit that might be obtained in an extended crank, coupled with a rapid-priming pump. It should be noted that these photos and comments apply to single-fluid injectors, not to PPAA injectors that require a fuel rail pressure level of only 0.6 to 0.8 MPa. The deterioration in the level of atomization is evident, with only 90 to 100 μm SMD being achieved at the lowest pressure of 0.4 MPa. This is not unexpected, as the design fuel pressure is in the range of 8.5 to 12.0 MPa,

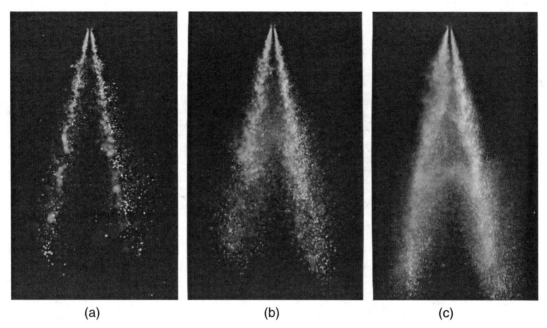

Figure 4.10-7 Spray photos of the four-hole, multihole injector at the ambient temperature of 20°C and atmospheric pressure: (a) fuel rail pressure of 0.4 MPa; SMD: 96 μm; (b) fuel rail pressure of 1.0 MPa; SMD: 62 μm; (c) fuel rail pressure of 2.0 MPa; SMD: 47 μm.

or a factor of 30 higher. Even at 1 MPa the SMD is only moderately improved to the 60 to 70 μm range. For the upper limit for cold starting of 2 MPa, the spray SMD is still in the range of 40 to 50 μm, which is well above the 18 to 22 μm that is required for good combustion. Additional factors to consider are the significantly extended fuel pulse width and reduced spray penetration rate that are associated with a rail pressure that is 15 to 30 times lower than the design level. The degradation in spray characteristics is very comparable for swirl-type and slit-type nozzles, thus graphically illustrating the problem of engine-out unburned HC emissions that accompany a G-DI cold-start at a reduced fuel rail pressure.

4.11 Characteristics of Pulse-Pressurized, Air-Assisted Sprays

Much work has been conducted over the past twelve years on the development and implementation of air-assisted, direct-injection fuel systems for automotive applications. Much of the earliest work was related to gasoline direct injection for two-stroke SI engines, and a significant number of references exist for this application [61, 74, 81, 88, 89, 90, 98, 129, 184–186, 205, 236, 241, 271, 280, 297, 317, 390, 465, 464]. Many methods of air-assist may be invoked in obtaining enhanced pneumatic atomization of the injected fuel, but what has evolved, as a production-intent injection system is a PPAA design. In pulse-pressurization two independent sole-noids are employed to first meter the fuel into a holding cavity, then to purge the cavity through the injector exit orifice into the engine cylinder using compressed air from an external source. The pressure level of the com-pressed air, typically less than 1 MPa, is sufficiently large to obtain a sonic blowdown into the cylinder, even for moderately late injection. This entrains the liquid and vapor fuel that resides in the holding cavity, and significantly enhances the atomization, vaporization and mixing of the injected fuel.

A very important point to emphasize regarding fuel spray formation with PPAA injection is that pneumatic atomization is utilized instead of pressure atomization or swirl-sheet breakup. In a PPAA system the fuel pressure level is *not* an important factor in the atomization that is achieved; in fact the fuel rail pressure must only be high enough to ensure that liquid fuel can be metered into the holding cavity in the time available. Normally a fuel pressure of less than 0.7 MPa is required. Conceptually the entire amount of metered fuel could be considered as a sac volume, as the fuel dwells at low pressure and is a blockage that must be purged by the air blast. The fuel is in fact blown from the cavity, and some

characteristics of a typical sac spray are indeed evident in time-resolved drop-size measurements. The mean drop size early in the blowdown is not as good as is achieved later in the purging process, which is similar to what is obtained in a swirl injector that has a moderate sac volume. Another characteristic of the purging of fuel from extended cavities is also present; that of an extended time of fuel exiting the cavity. It is difficult to achieve an efficient purging of the fuel holding cavity in a very brief time, hence short pintle opening times tend to yield fuel "hang-up," which is metered fuel that does not enter the cylinder on the intended cycle. Longer pintle opening times than are used in a single-fluid, high-pressure system must generally be employed to reduce the fuel-hang-up fraction, which results in an extended time period during which decreasing amounts of fuel are exiting the injector. Thus the injection of fuel is not as sharp, and the overall (integrated) mean rate of injection is not as high as can be obtained on single-fluid systems, which could require some compromises for light-load, stratified-charge operation. The level of pneumatic atomization is quite good for the air pressure levels that are utilized, with SMD values being comparable to those obtained for single-fluid swirl injectors using 10 MPa fuel. Another important point is that the peak velocity and associated momentum of the air-fuel charge being injected are quite substantial in such a system, and significant penetration rates and maximum penetration distances are normally achieved.

From a design standpoint, PPAA injectors could incorporate either outwardly opening or inwardly opening pintles, however, the units that are commercially available are outwardly opening. In the application of the PPAA injector to an engine system the introduction (metering) of fuel and the introduction of atomizing air may be sequential processes, with a dwell period for the fuel, or may alternatively be "nested," with a proscribed overlap of the fuel and air flow events. The particular strategy that is used depends upon whether compressed-air consumption is being minimized, or whether emissions are being optimized. In concept, many variations of fuel and air pressures and pulse timings may be utilized to optimize the spray performance for a specific application. It is unfortunate that the number of publications dealing with the application of PPAA injection systems to four-stroke DI gasoline engines is fairly limited [162, 176, 177, 308, 309], although it is should be noted that a substantial portion of the basic information in any of the PPAA publications may be applicable to

G-DI fuel system development. The spray from a PPAA injector for both the stratified-charge (late injection) and homogeneous-charge (early injection) modes are illustrated in Fig. 4.11-1 [60]. A comparison of the spray characteristics provided by typical PPAA and single-fluid, high-pressure swirl injectors is shown in Fig. 4.11-2 [177]. The spray from the PPAA injector is found to have a large number of small diameter drops. The maximum penetration of the PPAA fuel spray is lower than that of the swirl spray, with this difference becoming more pronounced as the ambient pressure is increased [177].

For the PPAA injector, the start of the air injection process and the pulse width for air injection are the parameters that determine the timing of fuel entry into the cylinder. These parameters control the mass of air being injected into the cylinder. The pressure level of the compressed air supply is generally not a critical parameter as long as it is above the threshold required for obtaining sonic flow at the most retarded injection timings, and is sufficient for effectively purging the fuel holding cavity. The pressure differential between the air supply and the engine cylinder does establish the air flow rate through the injector, and does determine the degree of fuel atomization that is achieved. As a result, the significant increase in ambient back pressure that accompanies late injection can degrade spray quality markedly if the air supply pressure is too low. It is this factor that constrains the latest injection timing that may be used. For example, if the spray from a PPAA unit is injected rather late into an ambient back pressure of 0.5 MPa using an air injection supply pressure of 0.65 MPa, the pressure differential across the nozzle

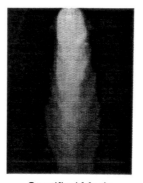 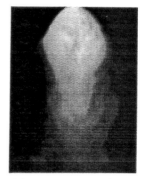

Stratified Mode   Homogeneous Mode

Figure 4.11-1 Spray characteristics of PPAA injector under stratified-charge and homogeneous-charge modes [60].

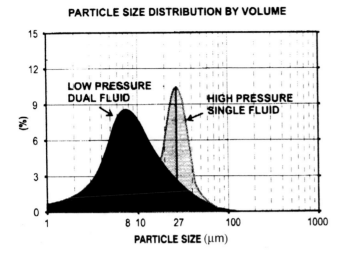

PARTICLE SIZE DISTRIBUTION BY VOLUME

LOW PRESSURE DUAL FLUID

HIGH PRESSURE SINGLE FLUID

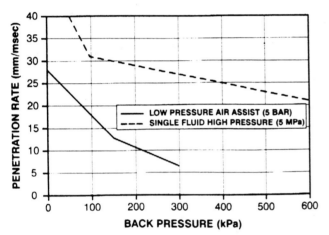

Figure 4.11-2 Comparison of spray characteristics for the PPAA injector and the single-fluid, swirl-type injector [177].

exit orifice is at most only 0.15 MPa. The size of the fuel spray plume for this condition is found to be reduced significantly due to the combination of a lower initial spray velocity and the decreased penetration that is associated with the elevated back pressure. In addition, phase-Doppler measurements of the drop-size distribution of the PPAA spray indicate that the mean drop diameter increases as the cylinder pressure is increased, thus reducing the number density of small drops in the spray [7]. In contrast to the level of air supply pressure, the timing of fuel injection into the holding cavity, the fuel rail pressure, the delay time dwell between the insertion of fuel into the holding cavity and the injection of air have been shown to be parameters that are not generally of critical importance to engine combustion [508].

The results of an analytical study of the spray structure derived from a prototype PPAA injector are presented in Fig. 4.11-3 [315]. For purposes of analysis and interpretation three distinct regimes of PPAA spray development are identified, which are denoted as the unsteady, steady and stagnant-flow regimes, respectively. In the unsteady regime, the flow is dominated by a starting vortex moving downstream. In the steady regime, a fixed vortex is formed just downstream of the pintle face, and air is entrained from outside the spray cone boundary. In the stagnant-flow regime, small droplets form a solid-cone structure, with the trajectories of small droplets being significantly influenced by the air flow field. The largest droplets in the spray tend to maintain their trajectories due to a decreased drag relative to their larger inertia, yielding a hollow-cone structure. Thus,

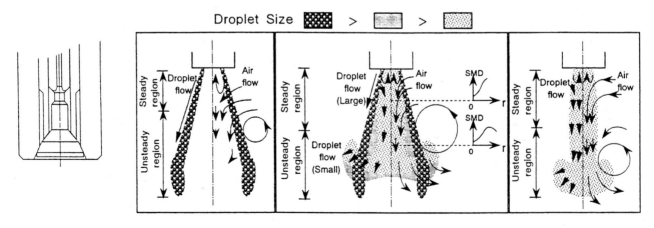

Figure 4.11-3 Spray structure of an air-assisted injector [315].

the mean droplet size is larger at both the spray tip and near the surface of the spray cone, and is smaller within the interior of the cone. A decrease in the nominal spray cone angle from a pulse-pressurized, air-assist injector generally yields a slight improvement in the level of atomization, which is counter to what is generally observed for single-fluid swirl sprays.

4.12 Ambient Density Effect on Spray Development

For conditions of low fuel pressure and low ambient density, such as occurs for injection into room air using half of the design fuel pressure, the spray from a swirl-type injector is indeed a true, hollow, conical spray. The individual flows from the swirl channels move out of the injector nozzle on a straight trajectory along a conical surface, with little or no flow in the cone center except for the initial sac spray. As either the fuel pressure or the ambient pressure/density is elevated, the spray boundary begins to develop a curvature, and the spray transitions from a true hollow cone to a partially filled bell shape. At an even higher fuel rail pressure and a moderate ambient density a toroidal vortex forms at the spray periphery, and becomes associated with the main spray. This vortex, coupled with the sac spray, gives an even more bell-shaped appearance. Many small to moderate-sized drops are entrained from the spray boundary to the interior of the developing spray, and very small drops are entrained into the moving toroidal vortex. For near-design fuel rail pressures and for elevated ambient densities exceeding about 2.5 times that of room conditions the entrainment mass flow attains a critical threshold and the spray cross-section and footprint become much smaller. As briefly mentioned earlier, this phenomenon is normally designated as the "collapse" of the swirl spray. As discussed in detail in the next section, this collapse also occurs if the injector operating temperature (body plus tip, including the fuel in the tip) is elevated to a threshold temperature that is dependent upon the volatility of the fuel being used. For a particular design of swirl injector there is a rather complex map of collapsed versus non-collapsed spray formation, with fuel pressure, ambient density/pressure, injector/fuel temperature and fuel volatility being the operating variables. Spray collapse to a narrow spray envelope is normally accompanied by an increased main-spray penetration rate, a more compact wetted footprint and a less prominent toroidal vortex. The overall SMD of the spray is little changed from

that of the non-collapsed spray for pressure collapse, but is a few microns smaller for temperature collapse. The latter occurs with hot injector operation (75°C to 110°C) and is basically the result of flash boiling within the injector tip cavity and discharge orifice. The spray that has undergone temperature collapse exhibits a very narrow and penetrating main spray, no separate identifiable sac spray and a significantly smaller SMD (10 to 12 μm).

An increase in the ambient density results in an increase in the droplet size within the spray plume. As illustrated in Fig 4.12-1 for the case of reduced ambient density during early injection, the fuel spray from a typical swirl injector exhibits a wide, hollow-cone structure. With an increase in the ambient density, such as by increasing the back pressure, the higher drag force and enhanced air entrainment can transform the spray geometry into a narrower, solid-cone shape [196]. The spray-induced air entrainment flow is much higher for the condition of elevated ambient density, and results in a significant transport of droplets to the interior of the spray, thus causing the spray cone from the swirl nozzle to collapse as depicted in Fig. 4.12-2 [369].

The structure of most G-DI fuel sprays changes substantially over the applicable operating ranges of in-cylinder density and fuel rail pressure. Fig. 4.12-3 contains a matrix of laser sheet photographs of the spray from a typical G-DI swirl injector. A hollow-cone structure of the swirl spray is present at low levels of fuel pressure and ambient pressure. At higher fuel rail pressures, the

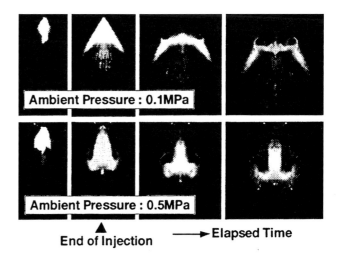

Figure 4.12-1 Effect of the ambient pressure on the spray characteristics of a swirl-type G-DI injector [196].

Mechanism **Numerical Simulation**

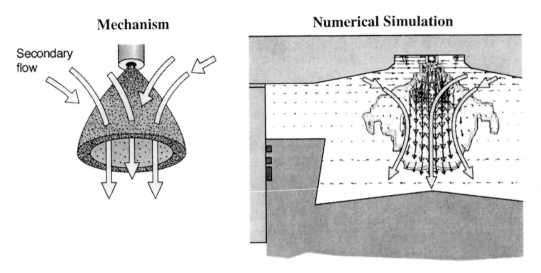

Figure 4.12-2 Mechanism of air entrainment and spray cone collapse at elevated ambient pressure [369].

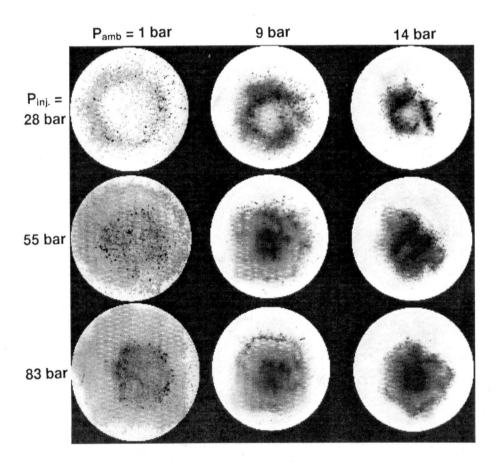

Figure 4.12-3 Effects of ambient back pressure (P_{amb}) and fuel rail pressure
(P_{inj}) on the cross-sections of G-DI sprays from a swirl injector
(images taken at 40 mm from tip and 4 ms after start of injection).

main effect is to fill the cone with more drops, decreasing the hollowness of the spray. Increasing the ambient back pressure yields a more compact cross-section, which is indicative of a narrower spray, although the spray cone angle near the tip, which is the definition utilized by injector manufacturers, may not be significantly altered. At elevated fuel pressures and back pressures the swirl spray is quite compact, with the highest fuel flux occurring on the injector axis. For an increasing ambient density the shape of the fully developed swirl spray transitions from a hollow cone to a bell, and finally to a bulb shape. Consequently, the spray cone angle obtained under atmospheric bench-test conditions may have very little similarity to the spray cone angle that may be formed at elevated ambient densities. This leads to difficulties in interpreting combustion results if the spray cone angle is obtained at atmospheric pressure and temperature, as previously discussed in Section 4.6.

As illustrated by the PDA data in Fig. 4.12-4 for swirl-type injectors having three different cone angles, the level of atomization is moderately degraded as the downstream ambient density increases. For the six cases represented in the plot the trends may be conveniently plotted versus ambient back pressure, as the ambient temperature is fixed and the downstream ambient density is increased by raising the back pressure. The SMD for the swirl-type injector is found to increase monotonically from about 16 mm to 22 mm as the ambient back pressure is increased from 0.1 MPa to 0.5 MPa.

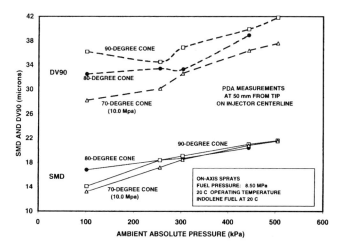

Figure 4.12-4 Measured effect of ambient pressure on SMD and DV90 for three swirl-type G-DI injectors (fuel pressure for 80- and 90-degree cone is 8.5 MPa; for 70-degree cone, it is 10.0 MPa).

The more universal conclusion for swirl-type injectors is that the SMD of room-temperature G-DI fuel sprays increase about 8% per bar from the baseline value at one bar, whereas it is about 6% per bar for hot fuel and injector conditions. For all G-DI injectors, not just the swirl-type, the SMD and DV90 are found to increase when injecting into a higher ambient density. The trends for other types will be discussed in subsequent sections. As is also shown in the plot, the DV90 statistic for the swirl spray also increases with ambient density, but it should be noted that the trend is not nearly so universal; in fact a different rate of increase is obtained for each injector type, operating temperature and fuel type. The increases in both the SMD and DV90 statistics are attributed to an increase in the rate of droplet coalescence for higher ambient densities, rather than to the diminished pressure differential across the tip orifice. For high-pressure, single-fluid injectors the change in pressure differential is negligible, although it can be a subtle effect for low-pressure, air-assisted G-DI injectors. The coalescence rate is normally an adjustable parameter in CFD spray modeling, thus experimental data on such SMD and DV90 trends can be matched.

4.13 Injector Operating Temperature and Fuel Volatility Effects on Spray Development

Not only does the injector operating temperature influence the spray development geometry, but it can also have a significant beneficial effect on the level of atomization and vaporization that is achieved. The injector operating temperature as used in this discussion, and in the associated tests, is actually a uniform temperature for the injector body, injector tip and the fuel within the injector, such as would be obtained with steady operation on a fully warmed engine. The geometric parameters of the spray that can be significantly altered by the operating temperature are the spray cone angle, the spray tip penetration rate, the maximum spray penetration distance, the sac spray, the toroidal vortex and the wetted footprint. The effect is very non-linear with operating temperature increase, and depends upon the volatility of the fuel, the ambient back pressure and density, and the value of the operating temperature. For pure hydrocarbons the important parameter is the boiling point at the downstream pressure, whereas for more realistic multicomponent fuels the standard metrics for volatility, such as RVP, driveability index or the T10 or T20 points of the distillation curve, may be used to correlate changes in the G-DI spray geometry with operating temperature. There is a valid debate in the current literature as to how much the changes in spray development are traceable to flash boiling. Regardless of the eventual outcome of this debate, there is no doubt that changes do occur in the sprays of all G-DI injectors at threshold levels of operating temperature, and that these thresholds are related to metrics for the volatility level of the fuel being utilized.

4.13.1 Injector Operating Temperature Effect on Spray Development of Swirl-Type Injector

Some injector designs, such as the swirl-type, exhibit a greater change than other designs, such as the slit-type, but all injector designs exhibit some changes in the spray as the temperature/volatility factor is elevated to critical thresholds. For fuels or fluids of very low volatility such as n-dodecane or test solvents, little or no effect on the fuel spray development and characterization parameters is observed in elevating the operating temperature from 10°C to 100°C. For these laboratory cases the threshold for change is above the temperature that is normally encountered during engine operation. If fuels of moderate volatility such as n-heptane or iso-octane are utilized, the threshold for spray geometry changes is in the vicinity of 70°C for a one-bar back pressure, and is slightly higher for elevated back pressures. For volatile hydrocarbons such as iso-pentane or n-hexane, or for multicomponent fuels having a low T10 temperature, the lower threshold of operating temperature can be as low as 45°C to 50°C.

The important point to be made is that spray characterizations that are obtained for bench room conditions *will be significantly in error* when the lower threshold of operating temperature is exceeded. The lower and upper threshold temperatures are a function of the fuel volatility and the cylinder pressure at the time of injection. If the injector body and the temperature of the fuel to be injected are both at 85°C to 100°C for market gasolines, the spray delivered to the combustion chamber will have quite different characteristics from those of the bench test at room conditions. As the lower threshold temperature is approached, the spray angle very near the tip (which is currently designated by the injector industry as the cone angle) *increases* slightly, but the spray waist downstream is narrowed. For a swirl injector the sac spray and toroidal vortex become much less prominent, finally disappearing altogether at the upper threshold temperature. The spray becomes more penetrating, and appears

to emulate a narrower, gaseous jet, with both the spray cross-section and the wetted footprint area being reduced. As the operating temperature nears the upper threshold the wetted footprint nearly disappears.

The changes in swirl spray development resulting from an increased injector operating temperature are very evident from the volume-illuminated (not light-sheet) spray images in Fig. 4.13-1. In this series for a 75°-cone-angle swirl injector operating on iso-octane fuel, the bench-test spray at 20°C is shown on the left. This is seen to exhibit the classic toroidal vortex and highly penetrating sac spray. The crossed lines in the images are PDA laser beams, which are utilized to make simultaneous drop-sizing measurements at 50 mm from the tip. Near the lower threshold temperature of about 70°C the cone begins to narrow and sac atomization is enhanced. The initial stages of spray transition with operating temperature are evident for the spray in the middle panel, which is for 75°C. Between 75°C and 100°C the transition to a very different spray progresses. The upper threshold temperature for the progression is about 120°C for this injector and fuel. At the upper threshold temperature the spray narrowing and penetration increase are complete.

It should be emphasized that the spray in the right panel exhibits a set of spray characteristics that is very different from those associated with the bench-test spray on the left. It should also be noted that the same test series performed for dodecane, a very low-volatility fuel, shows little or no change in the spray between 20°C and 100°C. If the fuel is changed to indolene, which is a blend of hydrocarbons that emulates gasolines, and which contains light ends that are not present in iso-octane, the spray transition is shifted to lower temperatures, and the performance highlighted in Fig. 4.13-2 is obtained. In this case the lower threshold temperature is about 60°C, corresponding to the more volatile fuel, and the spray is altered significantly for an operating temperature of 75°C. In fact the spray at 75°C using indolene is almost identical in characteristics to that at 100°C using iso-octane. At 100°C using indolene the spray characteristics have completely transitioned, meaning that the upper threshold temperature has been attained, and further changes with operating temperature increases are not prominent. Note that for injector operation at the upper threshold temperature there is no noticeable sac spray or toroidal vortex, and that the main spray plume is very narrow and penetrating. The spray cross-section and the associated cone angle are much

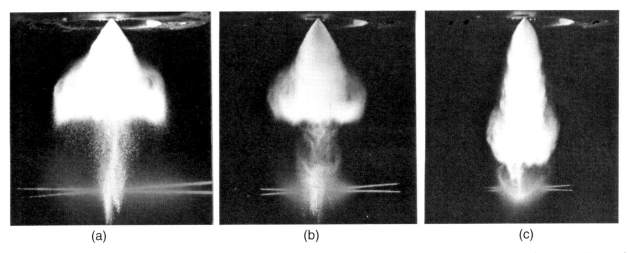

(a)   (b)   (c)

Figure 4.13-1 Effect of swirl-injector operating temperature on spray development (images taken at 1.5 ms after SOI; fuel: iso-octane; fuel pulse width: 1.5 ms; ambient pressure: 0.1 MPa): (a) 20°C; (b) 75°C; (c) 100°C.

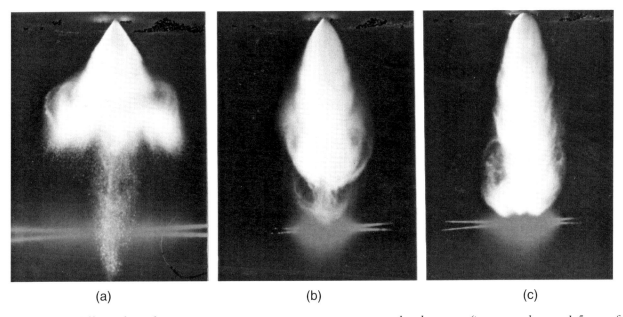

(a)   (b)   (c)

Figure 4.13-2 Effect of swirl-injector operating temperature on spray development (images taken at 1.5 ms after SOI; fuel: indolene; fuel pulse width: 1.5 ms; ambient pressure: 0.1 MPa): (a) 20°C; (b) 75°C; (c) 100°C.

reduced, and the spray cone angle of the fuel exiting the tip is very large, even approaching 130°.

The corresponding atomization performance for the three operating conditions in Fig. 4.13-2 are plotted in Fig. 4.13-3. These data were obtained at 50 mm from the tip using a 2-D, phase-Doppler, real-time system analyzer, with points measured every 2 mm from one edge of the spray to the opposite edge. Immediately evident from this typical data for indolene fuel are the enhancement in mean drop-size and the narrowing of the spray as the operating temperature is increased from

20°C to 100°C. On the injector axis, which also corresponds to the spray axis for this injector, the SMD is reduced from 16.6 μm to 13.9 μm. At the spray periphery the improvement is the most pronounced, changing in form from having the largest SMD (18 μm) in that position to having the smallest SMD (9 μm). This is tempered by the fact that there is not a very large percentage of the fuel mass flux near the spray periphery as compared to that near the injector axis.

Significant curvature of the spray boundary occurs near the injector tip for operation at temperatures above

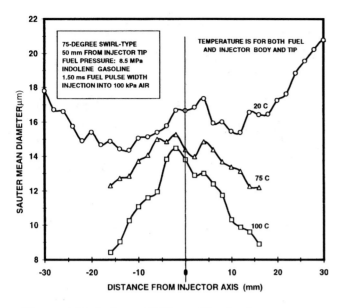

Figure 4.13-3 Spray SMD profiles for three injector operating temperatures.

the threshold, which will result in very large variations in the measured cone angle depending on the distance from the tip that is used to define the cone. This is an excellent illustration of the shortcomings of the current *ad hoc* method of defining G-DI cone angles by the use of the subtended angle of the spray near the injector tip. As is evident from the very narrow (20°, based on the waist diameter at 50 mm from the tip) spray shown in Fig. 4.13-2c, at 5 mm from the tip the "cone angle" would be measured and reported as 95°. At 10 mm from the tip, which is the defined distance used by some injector manufacturers to determine the cone angle, the value of the cone angle for this same spray would be measured as 55°, even though it may easily be recognized that the spray plume is effectively much narrower than this.

4.13.2 Combined Effects of Injector Operating Temperature and Ambient Back Pressure on Spray Development of Non-Swirl Injectors

The collapse characteristics of sprays from swirl-type injectors are fairly well known; however, those of other types of G-DI injector nozzles are much less investigated and documented. The fan sprays that are obtained from slit-type G-DI injectors exhibit the most robustness with regard to spray geometric alterations that occur due to changes in operating variables, such as fuel pressure, ambient density/pressure, injector body temperature and fuel temperature in the injector tip. The

sprays from all types of nozzles do change in appearance as the downstream ambient density is increased. The penetration rate and maximum penetration distance are generally reduced, thus at any fixed time after start of injection (SOI) the spray will appear less developed. This, however, is not what is designated as spray collapse. As is clearly illustrated in Fig. 4.13-4, even for the extreme case of conditions corresponding to injection on an engine that is operating at a steady, heavy load (Fig. 4.13-4b), the fan included angle is only reduced by about 15%, with the width dimension only increasing approximately 4–8 mm. For these conditions, which are for an injector/fuel operating temperature of 95°C and an ambient back pressure of 0.1 MPa, the resultant fan spray and the associated wetted footprint remain as a fan and a rectangle, respectively, with only slightly altered dimensions. This is compared to the fan spray geometry that is delivered at the more moderate operating conditions that correspond to light-load, late-injection operation (Fig. 4.13-4a). These conditions yield an injector/fuel operating temperature of 75°C and an operating back pressure of 0.3 MPa. For this change in operating condition the fan spray angle decreases from 79° to 67°, or 15%, with a corresponding increase in penetration at 1.40 ms from 45 mm to 58 mm.

As with the fan spray, the spray from a PPAA injector is only slightly influenced by changes in operating and ambient conditions. The sonic flow of a gas through the delivery orifice for critical and supercritical pressure ratios helps to isolate the process from the downstream conditions, and the presence of either liquid or vapor fuel in the delivered air stream creates only minor changes in the resultant spray geometry. The spray from a PPAA injector may be considered to be reasonably robust with regard to such changes. For the multihole-type nozzle this is definitely not the case. As with the swirl-type nozzle, the multihole spray geometry and wetted footprint change markedly with extremes of injector operation, although the boundary limits for geometric transition (collapse) are generally broader, making the multihole spray more robust than the swirl spray. As is clearly evident in Fig. 4.13-5, the spray from a multihole nozzle is influenced by the ambient operating conditions, with a transition from individual spray plumes to a single plume occurring in specific regimes of ambient density/pressure and injector body/fuel operating temperature. Whereas the spray from the swirl nozzle collapses as the ambient back pressure is increased, the sprays from most multihole nozzles tend

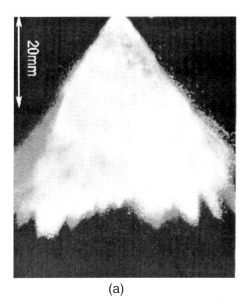 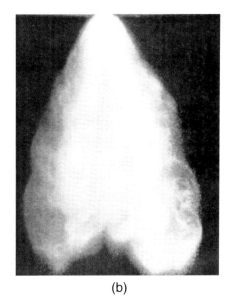

(a)   (b)

Figure 4.13-4 Variation in the fan spray from a slit-type injector for engine operating extremes: (a) light load operation; (b) heavy load operation.

to collapse as the ambient back pressure is decreased, particularly for operation with the injector body and fuel at 75°C or hotter. Some collapse of the multihole nozzle spray does indeed also occur at elevated back pressures, but this is only for levels exceeding 0.45 MPa. It is important to note that the transition region for the multihole spray provides more spray stability than is observed for a swirl spray. For example, it is not uncommon to observe many swirl spray images in an optical engine in which the spray geometry randomly oscillates between collapsed and non-collapsed. This is due to the narrow transition boundary for the swirl spray. Such random fluctuations are not generally observed for the sprays from multihole nozzles.

For bench-test room conditions the spray from a multihole nozzle is a collection of narrow, individual spray plumes, with one plume for each nozzle hole. There is also one individual wetted spot for each spray plume. As the injector body and gasoline within the injector are heated to a normal engine operating temperature (75°C to 90°C), the individual spray plumes become wider, with less well-defined spray boundaries. As the ambient back pressure is decreased from typical late injection values of 0.25 to 0.45 MPa, the plumes broaden and begin to combine, eventually becoming one plume at approximately 1.5 bar back pressure. At this point the spray appears to be very much like a swirl spray, complete with a small toroidal vortex. The total spray cross-section is decreased, as is the total area of

the wetted footprint, although these decreases are only about 15% as compared to the sum of the individual plumes. A further reduction in ambient back pressure to 0.1 MPa yields a fully-collapsed, highly penetrating spray such as is obtained for a swirl-type injector under the same operating conditions (compare to the photo in Fig. 4.13-2c). Some of the increase in penetration rate is due to decreased droplet drag at the lower back pressure, with the remainder being attributed to the decreased drag of one narrow plume versus multiple plumes, a spray cross-sectional area effect. With regard to spray geometry, the multihole injector is somewhat more robust than the swirl injector. This is also true regarding atomization changes with injector operating temperature and back pressure, but the multihole spray nonetheless seen to be affected.

To some degree, the characteristics of fuel sprays from all G-DI injector designs are affected by changes in injector operating temperature, fuel volatility, back pressure, ambient density or any combination of the four. In general the individual effects of changes in operating temperature and back pressure, such as are presented in Figs. 4.13-1 to 4.13-5, do not add linearly, but must be evaluated on the temperature and back-pressure operating map. This is complicated by the fact that there is an operating map for each fuel. It should be expected, however, that with additional research and modeling this would all be eventually correlated using a dimensionless operating map. The parameters could certainly

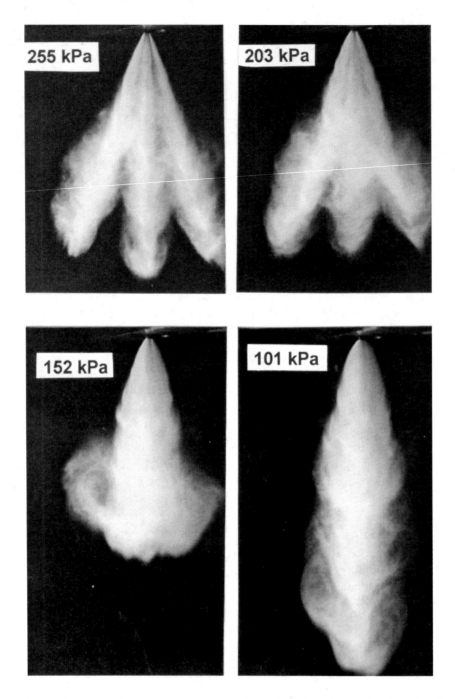

Figure 4.13-5 Effect of hot operation of a multihole injector for four levels of ambient back pressure; six-hole nozzle; 50° spray; fuel: indolene; fuel temperature: 90°C; fuel injection pressure: 11.0 MPa; amount of fuel injected: 10 mg per injection.

include the ratio of the ambient back pressure to the vapor pressure of the fuel, as well as the ambient density. A summary of the sensitivity of spray geometry to changes in the ambient and operating conditions for the spectrum of nozzle types is provided in Table 4.13-1.

Figure 4.13-6 illustrates the effect on spray atomization characteristics of a change from room-temperature operation to engine-temperature operation for seven levels of ambient back pressure. The phase-Doppler data are for a 50°, six-hole injector, and show that there is a moderate increase in SMD as the back pressure is increased, although not as significant as is observed for swirl units. This trend for the sprays from hole-type atomizers is also attributed to an increased rate of drop coalescence at higher ambient densities. Also evident is an improvement in the level of atomization with operating temperature, which is quite significant except for very high back pressures. The six individual spray plumes began to blend at the operating condition indicated by the dotted line, and form a single, central footprint at the leftmost point. For injection into 1.0 bar ambient the SMD is reduced by half in heating the injector and fuel from 20°C to 90°C, whereas there is only a 1.9-μm improvement at 0.6 MPa. Figure 4.13-6b shows the associated variation in the DV90 statistic for the same data runs in Fig. 4.13-6a. Within the repeatability

constraints of measuring DV90, which are discussed in Section 4.16.4, the enhancement in the spray DV90 is found to be very significant with heating, with many fewer large drops present, and an 18–28 μm improvement exhibited over the entire range of back pressures. The DV90 value is found to be fairly large for the multihole class of injector when spraying into low back pressures, but this performance is improved substantially as the back pressure is increased just one or two bar. For multihole and slit injectors the SMD and DV90 values are slightly larger than those of swirl and air-assist units, although the sac spray of a swirl-type injector, if measured separately, has values that are larger yet.

For multihole sprays the individual spray plumes become wider at the lower threshold temperature, and begin to blend with adjacent plumes, with a key parameter being the included angle between adjacent spray axes. A wider angular spacing requires a more elevated threshold temperature to merge the individual plumes. A circular multihole injector with multiple sprays at 50° will therefore blend into one plume at a lower operating temperature than a 70° design, or they may not blend at all if sufficiently separated. In terms of the wetted footprint, an oval pattern may become a line pattern as the upper threshold temperature is approached. For an air-assist or slit-type injector only moderate changes

Table 4.13-1

Degree of sensitivity of spray geometry and wetted footprint to changes in ambient and operating conditions

Injector Type	Degree of Sensitivity	Comments
Slit	Very robust	• Very little effect • Slight narrowing of fan angle at high temperature
Air-Assisted	Robust	• Small effect only • Slight narrowing of cone at high temperature
Multihole	Moderately sensitive	• Collapses at reduced (quite low) back pressures when hot • Exhibits effect of flash boiling
Swirl	Very sensitive	• Collapses at increased back pressures • Exhibits effect of flash boiling

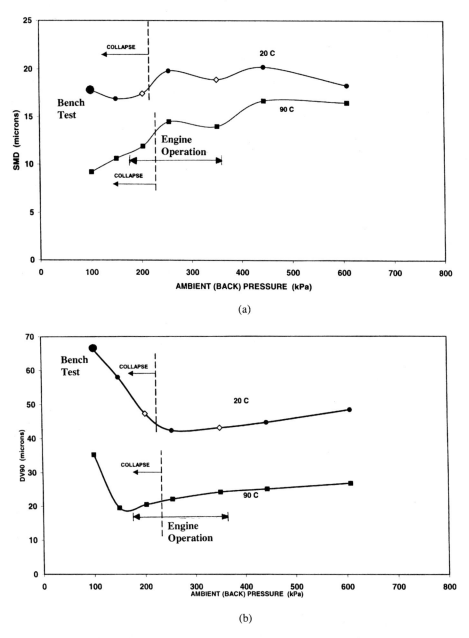

Figure 4.13-6 Effect of operating temperature and back pressure on the atomization of a 50°, six-hole, multihole nozzle spray: (a) trends for SMD; (b) trends for DV90.

occur with typical operating temperature increases, thus these designs are the most robust regarding alterations in spray geometry. Spray narrowing occurs for the air-assist design and for the major spray axis of the slit design, with a moderate thickening of the spray minor axis occurring simultaneously.

A number of investigations have confirmed that the injector operating temperature has a significant effect on spray characteristics because the fuel present

inside the injector nozzle may be nearly at the same temperature as the nozzle cavity wall, particularly for a low fuel flow rate or a low frequency of injection. This applies particularly to stratified-charge operation [454–458]. An experimental investigation on the effect of fuel volatility revealed that the spray changes from a hollow to a solid cone when the cylinder head bulk temperature is increased from 30°C to 90°C [171]. Such an increase in nozzle tip temperature also leads to an increase in spray-tip penetration in the later stages of spray formation. This is associated with a spray cone angle contraction that occurs at the higher operating temperature, which results in a delayed spray deceleration [24]. At an elevated injector operating temperature and a low back pressure, acetone-doped iso-octane experienced flash boiling, which caused the spray to change from the hollow-cone structure observed under cold conditions to a solid-cone distribution. The transition in apparent structure was observed to occur at around 70°C. As the ambient pressure increases, the transition into a solid cone is less pronounced. For example, at 0.06 MPa back pressure an intermediate change is observed, but this change is not observed at higher back pressures for which flash boiling is considered not likely to occur, or when the low-boiling-point dopant is absent. Experiments with indolene showed results similar to that of acetone-doped iso-octane. This indicates that the light components in the indolene experience boiling under the high-temperature, low-pressure condition. The observed solid-cone structure is explained by rapid evaporation and flash boiling of the volatile species, followed by transport of the smaller droplets to the center of the jet. It is interesting to note the pronounced influence of the cylinder head and injector operating temperature on the spray characteristics.

Because normal injector operating temperatures and tip-cavity fuel temperatures would be expected to be at least equal to the coolant temperature, and possibly 20°C hotter, fuel species may thus be expected to have a direct impact on the spray development process. This will influence the subsequent mixture concentration distribution, particularly for stratified-charge operation. For a common field gasoline with a multi-component blend, low-boiling-point components will be present that will alter the spray characteristics. This implies that the local equivalence ratio may not only be a function of the fuel vapor concentration, but also of the local composition [1, 295]. An analysis of the effects of fuel components on the structure of a spray from a high-pressure swirl injector has been conducted for the case of late injection using room temperature fuel. This KIVA-II study [295] found that the initial vaporization rate of the multicomponent spray is slightly higher than that for the single-component case, and that the vaporization rate decreases quite rapidly toward the end of the vaporization process. This is attributed to the more volatile components vaporizing slightly more rapidly at first, while at later stages the heavier components take longer to vaporize. The predictions for a single component also show a

predominantly hollow-cone liquid spray, with very few liquid droplets present near the spray centerline. This does not hold true for the multicomponent case, for which many smaller droplets are predicted near the spray center. For a multicomponent fuel, more rich-mixture areas are observed during the initial stages of spray development. A decrease in the area of maximum equivalence ratio is predicted for the multicomponent case as compared with the single-component case.

In addition to the effect of injector operating temperature on spray development, there are two noteworthy non-spray effects that should also be considered. It is well known that an increased cylinder-head operating temperature tends to promote increased formation rates for injector deposits. In addition, the amount of fuel delivered by the injector will, in general, also vary slightly with increased injector body temperature, which can lead to an air-fuel metering error if the bench-test flow curve is used without correction. Thus, the standard test specification for the thermal flow shift of the G-DI injector must also be taken into account.

4.14 Spray Atomization Ranges for Design and Operating Variables

The level of atomization of the fuel delivered by a G-DI injector is influenced by the injector type through the particular design of the fuel passages within the nozzle tip. The design may incorporate fuel swirl channels, a single delivery orifice, multiple holes, a slit, an offset guide to produce an angled spray, or may provide for air assist. The general atomization performance levels of the various classifications of G-DI injectors are listed in Table 4.14-1. This table summarizes the average performance of each class of injectors at the design fuel rail pressure. In general, the atomization level of sprays from swirl, air assist and piezoelectric-actuated injectors is somewhat finer than that obtained from multihole and slit injectors, in spite of the fact that slit and multihole injectors normally employ a design fuel rail pressure that is 1 to 2 MPa higher than that of swirl injectors. Technically, piezoelectric should not be in Table 4.14-1, as the term refers to the method of actuation rather than to an injector tip design. In fact, this type of actuation could be used with many of the other nozzle types that are listed. What piezoelectric actuation does offer to all nozzle types is a significantly more rapid and uniform opening ramp, as well as enhanced injector dynamic and working ranges. Another key advantage is the ability to open against much higher fuel

Table 4.14-1
General atomization performance levels of
injector types at the design fuel rail pressure

Injector Type	SMD*
Swirl (on-axis)	14–17 μm
Swirl (off-axis)	16–19 μm
Air assist	15–16 μm
Piezoelectric	13–18 μm (dependent on tip design)
Multihole (on-axis)	16–19 μm
Multihole (off-axis)	16–19 μm
Slit (on-axis)	17–20 μm
Slit (off-axis)	18–21 μm

* Measured for indolene clear under room bench
conditions at the maximum fuel flux location in the
spray at 50 mm from the tip.

pressures (15–20 MPa) than would be possible with a
solenoid. The more rapid opening translates into less
time spent in the period of low pintle lift and large pressure losses across the pintle seat curtain, which in turn
provides improved atomization levels during injector
opening. An atomization penalty is normally observed
for offsetting the spray from the injector axis, with the
SMD penalty being larger for greater offset angles. This
is true for swirl and slit injectors, but there does not
appear to be an atomization penalty associated with offset sprays from multihole injectors.

4.15 Current Best-Practice Performance of G-DI Injectors

With the rapid development that is occurring in G-DI
injector design and performance, it is obvious that a compilation of the best values for each of the key performance
parameters, which constitutes a table of best-practice
performance, is a living document. The injector performance envelope is continually being pushed as new
and improved injectors are designed, built and tested.
Table 4.15-1 contains a tabulation of contemporary
best practice for a number of key injector performance
parameters. Because injectors have a range of design
fuel rail pressures, four categories are listed, with
three pressures for a single-fluid injector and one narrow pressure range for an air-assisted injector. Each

category has current best-practice values for the performance metrics. All of the droplet size statistics presented
in the table were obtained by a two-component, phase-
Doppler, real-time-system analyzer for a very wide range
of injector hardware. All injectors were measured for
operation on indolene fuel at 20°C using a standard test
protocol. It is very important to realize that for a particular injector type and design the fuel atomization level that
is actually achieved is significantly influenced by many
operating variables, as summarized in Table 4.15-2. The
effects listed are averages based upon tests of many swirl-
type and multihole injectors, including both on-axis and
off-axis designs, for the range of operating parameters
listed in the table.

4.16 Issues with G-DI Fuel Spray Characterization

4.16.1 G-DI Fuel Spray Measurement Considerations

A number of similar characteristics among G-DI fuel
sprays in free space are revealed by precise phase-
Doppler measurements of time-resolved droplet arrival.
Whether it is a spray from a swirl, slit, air-assist or
multihole injector, certain generic features are noted.
This includes the initial dead zone due to the flight time
from the nozzle tip to the measurement location. As shown
by the actual G-DI spray data in Fig. 4.16-1 for a swirl-
type injector at a distance of 35 mm from the tip, the first
drops to arrive at this measurement location have, in general, the highest velocities and the largest diameters. The
7200 drops plotted are coincident drops, which are the
highest quality phase-Doppler measurements, meaning
that the Doppler bursts from each drop that passes through
the measurement volume simultaneously satisfy both
diameter and velocity validity criteria.

In this illustrative but real example the first drops,
which are from the sac spray, arrive at about 1.05 ms
after the initiation of the square-wave logic pulse. The
1.05 ms delay period without any drops being recorded
results from a 0.25 ms built-in driver capacitive delay
(OEM optional), a 0.22 ms injector mechanical opening time and a 0.58 ms minimum flight time from the
injector tip to the measurement location for even the
fastest drop. It may be seen that the fastest drops have a
velocity of about 50 m/s at 35 mm from the tip, although
they exited the injector with a velocity of nearly 80 m/s.
This information is not obvious from measurements at
35 mm from the injector tip, but can be ascertained from

Table 4.15-1
Current (2002) best practice bench-test performance for key injectors and spray parameters[a] (spray cone angle for single-fluid injectors: 55–80; amount of fuel per injection: 15 mg; indolene fuel at 20°C; injection into 1 atm)

Injector and Spray Parameters	Single Fluid			Air-Assisted
Fuel Rail Pressure (MPa)	5.0	8.5	11.0	0.5–1.0
SMD Main Spray (μm) [b]	16.7	15.0	13.9	16.4
DV90 Main Spray (μm) [b]	32.5	28.7	25.0	29.6
DV80 Main Spray (μm) [b]	31.6	27.7	24.1	28.5
SMD Sac Spray (μm) [c]	21.9	19.4	18.3	18.8
DV90 Sac Spray (μm) [c]	35.1	30.3	29.0	32.5
DV80 Sac Spray (μm) [c]	33.6	28.8	27.6	30.8
SMD Closing Spray (μm) [d]	19.0	17.3	16.4	17.2
DV90 Closing Spray (μm) [d]	33.7	29.1	27.6	31.7
DV80 Closing Spray (μm) [d]	32.2	27.5	26.1	30.0
After Injections	None	None	None	One small
Main Spray Skew (°)	0.0	<1.0	<1.2	<1.0
Sac Spray Skew (°)	<1.0	<2.0	<2.3	<1.0
Opening Time (ms), 12V Volt Design	0.26	0.27	0.29	0.26
Opening Time (ms), 24 to 70 Volt Design	0.16	0.17	0.18	–
Closing Time (ms)	0.36	0.34	0.32	0.41

[a] Including swirl, multihole, slit and pulse-pressurized air-assist
[b] Fuel-volume-flux weighted PDA across entire spray at 50 mm from injector tip
[c] PDA at 50 mm from injector tip on sac spray centerline over sac arrival time
[d] PDA at 20 mm from injector tip on injector axis for 3 ms following actual closure

Table 4.15-2
Measured variation in spray atomization level for G-DI fuel system operating variables

Operating Variable Change	SMD Increment or Decrement
Normal field gasoline variation	±0.9 μm
Range of normal hydrocarbons/alcohols: n-decane to n-pentane	−1.7 μm
E85 versus iso-octane	+2.2 μm
Decreasing fuel rail pressure (11 to 6 MPa)	+3.8 μm
Increasing fuel and injector temperature (15 to 90°C at 1 bar ambient)	−3.2 μm
Increasing ambient back pressure (1 to 5 bar at 20°C)	+5.1 μm
Decreasing ambient pressure (2.0 to 0.8 bar) at 90°C	−6.0 μm
Double pulsing within six milliseconds (second pulse relative to first)	<0.5 μm

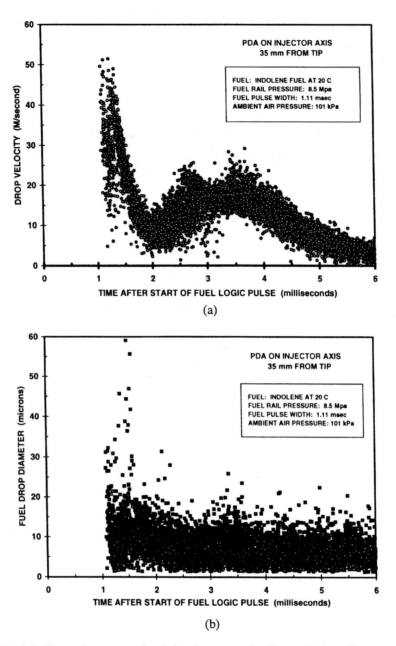

Figure 4.16-1 Typical time-resolved droplet arrival pulse at 35 mm from injector tip:
(a) drop velocity arrival pulse; (b) drop diameter arrival pulse.

spray penetration photographs showing the spray-tip position at 50 μs intervals near the start of injection. The sac spray drops are seen to be the largest of the entire pulse, with diameters of 15 to 58 μm being measured at the earliest times. The sac spray only passes through the measurement location for about half a millisecond, as is evident from the drop diameter arrival plot. The main spray then transits the phase-Doppler measurement volume, exhibiting a mean velocity maximum and minimum as the main spray pulse builds up

and decays. For this typical 1.1 ms fuel logic pulse (idle) the peak velocity of the main spray pulse occurs at this downstream location at approximately 3.3 ms. By 5.5 ms the pulse has decayed significantly, with all drops at 35 mm from the tip having velocities under 6 m/s, which may be considered as the start of the spray tail. The spray tail contains nearly all small, low-velocity, entrained drops that will continue to pass through the measurement volume for an additional 30 to 40 ms in a free-space test.

The trend in droplet size that is documented in Fig. 4.16-1 is typical of the sprays from many G-DI injectors. The largest drops are almost always the initial drops, with a rapid improvement in mean drop size occurring as the sac spray exits the measurement location. Following this is a continuing trend toward a smaller mean drop size as the main spray and spray tail pass through, although the change is relatively slow compared to that associated with the sac spray. The initial drops are generally the largest and fastest because they are either associated with the sac spray or, for injectors with little or no sac volume, they are still associated with an initial opening process in which the atomization is not as good as is achieved at full pintle lift. These largest drops experience the least relative drag, and pass the smaller drops that are decelerating more rapidly. In the 0.58 ms minimum flight interval to the measurement location, the largest and fastest drops move to the front of the spray (the spray tip) and maintain their velocity to a greater degree. This aerodynamic separation is very evident from the drop arrival pulse data. The difficulties inherent in defining and determining particular values for the mean drop size and mean drop velocity for such a transient spray are evident from the plots. The pulse is highly transient, and every spatial position will record a different droplet arrival history. If the measurement by phase-Doppler or laser diffraction is not time-windowed, which is a restriction of the acceptable measurements to a proscribed time window relative to the start of the fuel logic pulse, then a large percentage of the droplets acquired will likely be from the spray tail. This will bias the reported SMD, DV90 and mean velocity to smaller values, although it may be argued philosophically that these drops are also from the associated injection event and should perhaps be included. The contrasting viewpoint expressed here is that these diameters and velocities are not representative, particularly at many tens of milliseconds after the injection event. Some drops may even be measured more than once at later times if recirculating vortices are present. If this recognition of a possible bias is accepted, then time-windowing of the measurements should obviously be employed with either phase-Doppler or laser diffraction sampling. A complicating factor, however, is that there is no single, fixed time-window value that is correct for all sprays and fuel pulse durations. For example, a 5 ms time-window is not appropriate if a 6 ms fuel pulse is being utilized. It is obvious that as the time window for accepting drop data is decreased

from 10 ms down to 6 ms, and then further reduced to 2 ms, the measured drops will become more representative of the sac spray, and less representative of the main spray. In the limit of a data time-window cut-off of 1.6 ms in the example plots, only the sac droplets would be measured, as the leading edge of the main spray has not yet arrived at 1.6 ms.

Normal drop-sizing measurements typically do not account for atomization that is associated with injector closure, mainly because the measurements are almost always made well downstream from the injector tip (35 to 50 mm). This is done for many practical reasons, as measurements within 5 to 10 mm of the tip with either laser diffraction or phase-Doppler almost always result in significant data dropout during the injection pulse. A special test using time-windowing near the tip for only the 100 µs time interval of the last stages of pintle closure can be conducted, which reveals that the SMD and DV90 can exceed 28 µm and 50 µm, respectively, if the closure is not well done. Standard downstream measurements of drop sizes never include these drops because they are introduced with such low velocities (0.5 to 2.0 m/s) and such wide trajectories that they do not cross the laser measurement volume in a reasonable time frame, if ever. This is evident in the laser sheet photographs of the spray associated with injector closure that were discussed earlier in Section 4.3. The fuel droplets photographed near the injector tip in Fig. 4.3-5 will *never* penetrate 35 mm to a drop-sizing measurement location, and will *never* be sampled. A 60-µm drop introduced at 1 m/s requires 35 ms to travel the 35-mm distance down to a phase-Doppler or laser-diffraction measurement location, even if it has no radial velocity component and does not decelerate. In reality these drops are moving mostly radially, and do decelerate. Thus, it is important to be aware that the atomization characteristics associated with injector closure, whether good or bad, almost always escape being included in spray quality measurements, and that the spray quality at closure should be evaluated separately either by reviewing laser-sheet spray images taken at that time or by a special time-windowed test that monitors the late drops near the injector tip.

In the absence of a standard or recommended practice document on G-DI spray measurement and reporting, the spray measurement protocol that is utilized within each OEM or end-user group will necessarily be *ad hoc*. The results reported for a given injector will necessarily be influenced by the *ad hoc* choice of tip distance and time-window cut-off, in addition to the choice of the physical operating variables that are tabulated in this chapter. In establishing meaningful in-house test protocols for G-DI sprays it should be noted that the measurement distance of 50 mm from the injector tip is the most common, but distances of 35 mm and 40 mm have been employed. A time-window cut-off for data acquisition that yields an 85% reduction in the maximum drop velocity is recommended as a good rule of thumb. In the example in Fig. 4.16-1 this represents a reduction from 52m/s at 1.25 ms to 7.8 m/s, which translates into a window cut-off time near 6.0 ms. With regard to time-windowing it should be noted that the very latest phase-Doppler and laser-diffraction instruments allow for virtual time-windowing after the test, which has some significant advantages in test setup time. This would permit a broad, generic time window of 10 to 12 ms to be used for each test regardless of the injector and spray, with the specific, shorter time window that provides an 85% reduction in drop velocity invoked during post-processing. The only penalty here is that with the broad time window, many drops will be measured that will not be used. This will require additional testing time.

4.16.2 Spray Characterization Issues

The necessity of characterizing the key areas of G-DI spray development and performance were discussed in detail earlier in this chapter. The focus areas of a combustion-system development project that require such information were also addressed. However, the applicability and relevance of specific spray tests and characterizations must also be addressed, as the limits of a particular characterization may not be obvious. With regard to the understanding and meaningful characterization of fuel sprays from G-DI injectors, there are two very important considerations to be discussed in detail:

- G-DI fuel sprays generally have completely different characteristics for engine operating conditions than are exhibited in bench tests at room conditions.
- There are currently no industry-wide spray measurement and reporting standards for G-DI fuel sprays, therefore reported values for any injector and spray parameters are very difficult to verify or reproduce.

The injector manufacturers do supply some injector and spray data, but much of the data is related to the flow characteristics. This is certainly important information, however, it is normally not sufficient for a full combustion system development program. The detailed testing outlined below normally must be performed in any G-DI development program. Injector manufacturers may supply some specialized spray or injector data, but they generally are not able to provide data for real fuels or for hot injectors, fuels and ambient conditions. With very few exceptions the data supplied with G-DI injectors or injector prototypes are for room bench conditions; that is, for 20°C ambient fuel and injector, and for injection into one bar ambient using a liquid other than pump gasoline, indolene or iso-octane. The common test fluids are normally solvents, but may be n-heptane. These standard data are certainly not to be denigrated, as they have a number of valid uses, the first of which is to place the injectors into critical initial classifications by spray cone angle, spray offset angle and flow capacity. The data should be considered as very necessary, but sometimes not sufficient.

It is when interpreting combustion data, optimizing a chamber design or setting up a CFD spray model that much of the bench spray data is found to not apply directly. As was demonstrated graphically in Sections 4.12 and 4.13, the spray characteristics of many types of G-DI injectors are significantly altered if the ambient density or operating temperature is changed from that of the standard bench test at room conditions. If the ambient density is either lowered or elevated, with elevated being more meaningful for late injection, most of the spray parameters are changed, and may be changed significantly. For conditions corresponding to those that exist in an engine for late injection, the spray cone angle, mean drop size, DV90 and penetration rate are all found to be substantially different. If the injector body, tip cavity and fuel that is resident within the injector are then heated to a typical steady-state engine operating temperature of 75°C to 90°C, even more striking changes in the spray development may be observed. If a real field gasoline is then substituted for the usual test solvent, the spray transformation is complete.

At bench test conditions the spray is relatively insensitive to the test fluid, and solvents yield results that do not differ greatly from those obtained with pump gasolines or pure hydrocarbons. However, for conditions that correspond to real G-DI engine operation, the fuel spray that is delivered by many, even most, G-DI injectors is very different from the spray that was characterized on the test bench at room conditions using a low-volatility test fluid. If such standard bench data are used to interpret combustion measurements or to initialize spray models and tune penetration predictions, then erroneous conclusions may be drawn. Moreover very significant changes in the wetted spray footprint, the collapsed spray cone angle, the SMD and DV90, and the spray-tip penetration rates *do* occur, with significant alterations in drop size distributions and transient spray development being exhibited. The bottom line is that the two extreme test conditions almost always yield two completely different spray appearances and characterization data sets. There is no easy accommodation to this observed fact, other than to be aware of possible errors in interpretation if standard bench data are used where not applicable. Bench testing at 0.1 MPa ambient for a 20°C injector and test fluid is to be contrasted with hot testing using 90°C gasoline and a 90°C injector that injects into a 0.3 MPa ambient. The largest changes in spray development between these two tests are exhibited by swirl-type injectors, with multihole injectors showing moderate changes, and slit-type and air-assisted injectors showing minor changes. If laboratory test facilities are available for providing improved spray characterizations under more realistic conditions, such data should be obtained and utilized. The standard bench data would seem to be directly applicable to cold starting conditions, and may apply to early injection for either stoichiometric or lean operation. However, even for test evaluations of cold cranking and starting, additional bench spray data at reduced fuel rail pressures may be required, as the high-pressure fuel pump may not be fully primed.

Although there are some worldwide standards on the testing and reporting of the characteristics of PFI injectors and sprays, there are currently none for G-DI injectors and sprays. SAE International has a number of recommended practice documents for PFI injection hardware, including J1832, J1862, J1537 and J1541, all fully updated since 2000, but the only spray parameter for which a standard testing and reporting procedure exists is the PFI spray cone angle. This test using a low-resolution patternator is documented in SAE J1832, and has been the accepted standard since 1985. The cone angle is defined as the angle that includes 90% of the total mass of the PFI fuel spray at the specified collection radial distance. All manufacturers of PFI injec-

tors have gone well beyond the low-resolution patternator test for spray cone angle that is described in SAE J1832. Manufacturers currently utilize internal *ad hoc* test protocols that are based upon high-resolution patternation. As was illustrated in Fig. 4.6-7, high-resolution patternators employ 250 to 260 cells to capture the liquid fuel and define the mass distribution. Neither the low-resolution nor the high-resolution patternator tests have been applied directly to determining the mass-distribution cone angle of a production G-DI injector. The standard low-resolution cone angle test protocol and specified test hardware in SAE J1832 does not, and cannot, apply to G-DI fuel sprays, as the specified collection distance range is far too great (90 to 143 mm), and too much of the fuel spray is lost to evaporation and droplet terminal velocity attainment prior to collection. Some research experiments have been conducted to measure the G-DI fuel mass distribution using high-resolution patternators, but as yet no manufacturers are reporting G-DI spray cone angles that are based upon such measurements. This is to be considered as a prime area for research and development, as real mass distribution profiles and mass-based values of cone angle have many advantages over image-based cone angles, particularly for asymmetric sprays.

With regard to measurements of spray atomization, it may come as some surprise to note that there is *no* accepted national or international standard on how spray drop sizes are to be measured and reported, even for the well-established PFI fuel spray. ASTM E 799-92 and ISO/WD 9276-2 (Document number ISO/TC 24/SC4 N115E) address the standard practice for the reduction of data from particle size measurements, but do not deal with the specific test protocols of gasoline fuel system hardware. Part of the reason for this is the manner in which PFI engines operate, that is, from a fuel puddle on and around the intake valve, which results in a moderate insensitivity to the atomization level of the spray. The spray cone angle and spray targeting, which are interrelated, are known to be the most influential parameters in a PFI system. The PFI spray mean drop size, however, does have some effect on cold start hydrocarbon emissions and on the level of cold-start enrichment that is required, as it is a metric that is an indicator of vaporization rate.

With the move toward more stringent emissions standards and with the increasing emphasis on G-DI engines, there is an increasing awareness that test protocols for spray measurement and reporting are required for both PFI and G-DI fuel sprays. Until the standards are fully promulgated and approved, data that are obtained and reported for fuel sprays must necessarily be based on *ad hoc* standards, with each injector supplier and end-user company having their own in-house measurement protocol. This means that two different injector manufacturers may report a 16 μm SMD spray for their injectors, but one may be measured by means of line-of-sight laser diffraction at 40 mm from the tip using Stoddard solvent, whereas the other may be measured at a single location of 50 mm from the tip using phase-Doppler anemometry with n-heptane as the fuel. The measurements may or may not be time-windowed. This example illustrates the difficulty of comparing the spray performances of injectors from various manufacturers or of reproducing data that are reported in the technical literature. In the absence of a global standard the comparison of spray performance can be fully resolved only by additional testing of the two injectors using the same test protocol for both. This will likely be a third test procedure that differs from that of either of the two injector manufacturers in terms of drop sizing equipment, test procedures and test fuel.

4.16.3 G-DI Spray Measurement Techniques and Hardware

Characterizations of G-DI fuel sprays are normally obtained using a standard test protocol that may be specific to a company, or may be partially based upon the proven techniques discussed in this section. The test injector is mounted in either a holding fixture of large heat capacity on an optical spray chamber, or on an optical engine. The holding fixture should have the capability of heating both the injector and the inlet fuel, and should provide for automatically controlling the set temperature. This permits the stabilization of the injector operating temperature, which includes the temperature of the injector body, tip, fuel and fuel inlet tube. The fuel pressure may be obtained either by a motor driven pump and rail, or by using a high-pressure bladder accumulator system that is pressurized with nitrogen gas to the desired fuel inlet pressure, which is normally in the range of 5.0 to 13.0 MPa for current G-DI hardware. The fluid specified for the characterization tests may be any fluid, including solvent substitutes for fuel; however, standard comparative tests should ideally be performed using a fluid having physical properties such as specific gravity, viscosity, surface tension and volatility that do

not differ greatly from those of pump gasoline. In this regard, fluids such as n-heptane, indolene, indolene clear, iso-octane, Howell EEE and California Phase II certification fuel have been utilized and documented in the literature. For spray tests under hot operating conditions a narrow-boiling-range test fluid such as n-heptane or iso-octane will likely not yield the same spray changes that are observed for pump gasoline, and it may be advantageous to use one of the other multi-component fluids listed above.

An optical spray chamber should have at least three quartz windows, ideally UV grade, all of a minimum 85 mm clear viewing diameter. For G-DI drop sizing using a PDA, one of the quartz windows is normally oriented for the 30° forward scattering angle, whereas laser diffraction measurements will also require in-line windows for line-of-sight access. This general setup will facilitate both drop sizing and the introduction of a pulsed laser light sheet for either Mie scattering or planar laser-induced fluorescence (PLIF) imaging. The chamber should be designed to purge the residual fuel droplets from previous injections by using a very low dilution flow of nitrogen gas (or air for solvent sprays) from the top to a bottom exit. A low injector firing frequency such as 1 to 4 Hz will minimize the required dilution purge flow, as 250 ms to 1000 ms will be available for the purging of residual fuel droplets before a subsequent injection event occurs. It is important to be aware that, for G-DI driver boxes that do not have a built-in capacitive-charging delay interval, operating the injector at 1, 2 or 3 Hz may yield sub-standard sprays due to capacitor voltage decay. If the first fuel appears at the nozzle exit within 0.16 to 0.35 ms after the start of the fuel logic pulse, then there is very likely no built-in capacitive-charging delay. For these cases very low injector operating frequencies should not be used unless it is first verified that the spray quality is not degraded. An injector driver control system is normally used to trigger the injector, laser pulse, strobe and imaging system. The driver should be the specific driver box for the injection unit being characterized, and the correct pulse mode (high-true, low-true) *must* be used. For the facilitation of comparisons of injectors, it may be advantageous to fix the mass of fuel delivered per injection and adjust the fuel pulse width to provide it. This is because the fuel pulse width varies so widely among injectors and drivers, particularly if capacitive-charging delays are incorporated. Thus, standard test fuel deliveries can be selected for each class of test, such as 5 to 6 mg for idle, 10 to 15 mg for light load and 30 to 35 mg for heavy load.

The transient fuel spray may be characterized by determining a number of key spray parameters, including the SMD, DV90 (or DV80), the spray angle, the droplet arrival history, the wetted footprint, the sac and main spray penetration curves, and the Mie or PLIF spray-development photographs. The characterization of a fuel spray is actually a matrix of individual characterizations, as the spray test ideally would be performed for both 20°C and for hot operation at 0.1 MPa ambient, and at least one relevant, elevated, ambient pressure. For this elevated pressure, values between 0.25 and 0.45 MPa have been used. If an unheated ambient gas such as nitrogen is used in the optical spray chamber, then a pressure should be computed and utilized that provides the proper ambient density for the point in the compression stroke that is being emulated. It should be noted that for flash boiling effects it is the ambient pressure that is the key parameter, not necessarily the ambient density. If additional fuel pressures or fuel pulse widths are to be investigated, the test matrix may be expanded accordingly.

The development of the transient fuel spray may be documented in the standard test by Mie-scattering images of a pulsed laser light sheet, or, alternatively, by a strobe or flash lamp of no more than two microseconds duration. A typical thickness range for the laser light sheet is from 0.25 mm to 0.75 mm, with the sheet normally positioned in precise alignment with the injector axis. The electrical connector of the injector may be oriented at a right angle to the lens of the imaging system, pointing toward the right side of the resultant image. This is done for purposes of repeatability and standardization. For offset or asymmetric sprays the electrical connector is almost always aligned with the offset direction by the injector manufacturer, thus this orientation of the connector will provide a meaningful image. For asymmetric or special offset-spray injectors such as the offset slit-type injector that is used in the Toyota second-generation D-4 engine, either a special injector-tilt fixture or a special alignment of the light sheet are required to ensure that the sheet passes through the main portion of the developing spray. For example, a unit of an injector model that is designed to provide a spray being nominally 20° off the injector axis may, in fact, generate a spray that is 18.6° off-axis. For this particular unit the light sheet must be inclined at 18.6° and is normally positioned to pass

through the center of the exit cavity of the injector tip. For non-standard PLIF imaging of a spray, for example when both fuel liquid and vapor are to be imaged, the UV laser frequency may be optionally used to produce a sheet. UV grade windows must be utilized in the optical spray chamber.

The wetted footprint for the spray may be obtained by using a sheet of special fuel-sensitive paper (such as Japanese Origami paper) that is positioned on a plane at the location of the drop size measurements described above. This sensitive paper turns bright red whenever even micro amounts of fuel impinge, and this footprint may be imaged one second after a single injection event. In approximately 10 seconds the fuel in the footprint vaporizes, leaving no trace of the original outline. This is ideal for optical spray chamber work, as it avoids removing sets of windows to replace the paper for the next test. The footprint should ideally be obtained prior to any drop size measurements to ensure that the measurement point or line location is optimum. Corrective image-processing software is required to correct for photographic distortions due to perspective and an oblique viewing angle.

Spray-tip penetration data may be obtained from Mie scattering images, with a series of images at intervals of 100 μs being typical. For each ambient pressure and temperature condition a reference scale should be included in the images. Because injection-to-injection variability in penetration is fairly common and noticeable in most G-DI fuel sprays, the tip-penetration data from multiple repeat images (5 to 12) should be used to generate tip velocity data points, which may then be mathematically smoothed and integrated to obtain smoothed and monotonically increasing penetration curves. The spray cone angle of the transient G-DI spray is normally obtained by means of spray imaging using a high-intensity flash lamp and backlighting. For the time corresponding to mid-injection the cone widths at both 5 mm and 15 mm from the tip may be measured, and the included angle between two lines passing through these four points may be computed as the spray angle. It is important to note that this angle is here denoted as the spray angle, not as the cone angle, even though the latter is the *ad hoc* nomenclature used in the literature. It is recommended for future consideration that the term "cone angle" be reserved for the result of patternation measurement techniques in which the fuel mass distribution is measured, and in which the cone angle that includes 90% of the total delivered fuel mass can be computed. This is the standard SAE recommended practice for determining the cone angle of a PFI fuel spray, but it has not yet been standardized for G-DI sprays due to the relatively recent emergence of G-DI spray patternation measurement.

An external-input, digital clock system may be optionally utilized to obtain a complete fuel droplet arrival history for each test point and condition. In this particular test the driver pulse generator also provides a reset pulse to reset and start a digital clock that is integrated into the drop measurement system. In this manner the droplet arrival time relative to the start of each injection event is recorded along with the diameter and velocity for each fuel drop in the database. As the number of drops acquired during a single injection event is only in the range of 50 to 180, this technique permits overlaying the data from many consecutive injection events in order to obtain a detailed time history of the transient. A data time-window of 0–6 ms, 0–7 ms or 0–8 ms is recommended for all measurements, with the value that most closely corresponds to an 85% decay in the peak droplet velocity being selected. This means that no drop will be accepted into the sampling database if it arrives more than 6, 7 or 8 ms after the initiation of the fuel logic pulse. The basic data in each droplet-arrival-history data set so obtained may be analyzed in a number of useful ways. For example, the data can be easily subdivided into bins of arrival time such as 1–2 ms, 2–3 ms, etc. A set of drop size distribution statistics may then be generated for each bin, thus providing the time variation of any particular distribution statistic. The basic data can also be selectively screened for drops above or below any specified threshold of size, velocity, or combinations of velocity and size, such as momentum, Weber number or Reynolds number. This is particularly useful in the development and improvement of spray models and sub-models.

The measurement location for G-DI phase-Doppler or laser diffraction measurements is most often within a plane that is orthogonal to the injector axis and 35, 40 or 50 mm from the tip at its closest point. This distance range has been determined to be a reasonable compromise for G-DI fuel sprays, as measurements closer to the tip normally result in increased data dropout for phase-Doppler and laser-diffraction systems during critical periods of the pulse, although laser-diffraction systems are more robust in this regard. The data dropout is the result of high drop number densities and laser beam obscurations that are associated with the early stages of spray development. Attempts to measure in regions of very high number density or very high fuel flux (which normally correspond to tip distances of 5 to 25 mm) generally result in very low data acceptance rates during those brief time intervals when such conditions exist at the measurement location. Unfortunately, this usually corresponds to a time of significant interest, such as the first 500 ms after main spray arrival.

4.16.4 DV80 versus DV90 Measurement Accuracy

The relative position of any statistical parameter such as DV90 or DV80 on the cumulative volume distribution curve will be determined by the specific shape of the curve. This will in turn depend upon the droplet number distribution for the entire range of droplet diameters that is measured. Mathematically, the cumulative volume curve is generated by simply generating the total volume contained in the number of drops in each size bin, from the smallest to the largest, and summing the accumulated sub-total of liquid volume as the bin number is incremented. Optical equipment for droplet sizing normally divides the measured range into 35 to 60 separate measurement bins, depending upon the device, and the number of validated droplets in each size bin constitutes the raw distribution data from which all spray statistics are derived.

Figure 4.16-2 illustrates a wide range of size distributions that could be encountered, although more extreme cases, such as bi-modal or tri-modal, could be encountered, particularly if such distributions are obtained for special cases such as after-injection. Figure 4.16-2a illustrates a typical number distribution curve for a total sample size of 7200 accepted (coincident) droplets. A G-DI spray normally contains droplets in the 1 to 50 μm interval, with only relatively few droplets in the 30 to 50 μm range. The most probable drop size is almost always in the range of 5 to 10 μm, with more than half of the total drops in the spray having a diameter in that window. In the example in Fig. 4.16-2a the most probable drop size is 6 μm. The sampling of the spray using only 7000 to 20000 fuel droplets is assumed to be representative of the near-infinite population of drops (normally exceeding 8×10^6), although there are known biases that are associated with the various laser-diffraction and phase-Doppler configurations that can be corrected. If the measured proportion of drop sizes is indeed representative of the entire spray, which it may

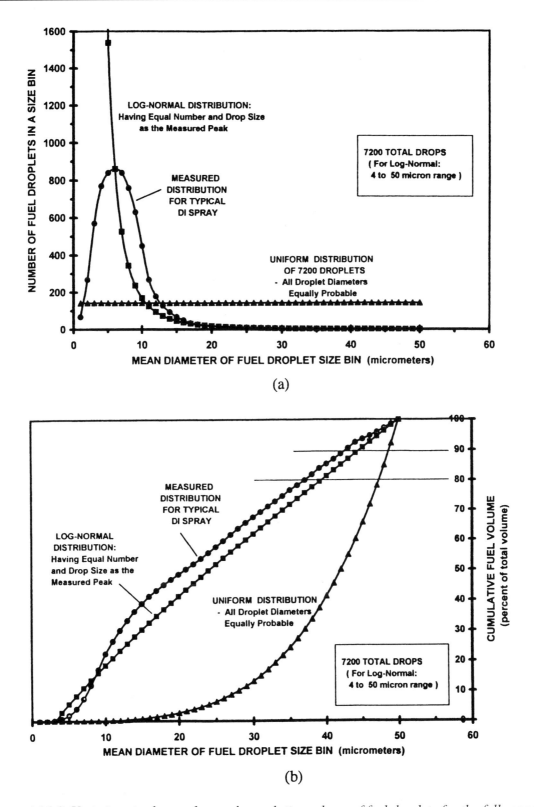

Figure 4.16-2 Variations in the number and cumulative volume of fuel droplets for the full range of droplet size distributions: (a) number of fuel droplets; (b) cumulative fuel volume.

not be for a phase-Doppler measurement at a single location, then a volume distribution for the spray may be generated. Figure 4.16-2b illustrates the cumulative volume curves that are obtained for a measured number distribution and for two mathematical number distributions: log-normal and uniform. The log-normal is very commonly invoked in mathematical representations of spray distributions. Even though only a few large droplets are measured, the cubic relationship with droplet diameter holds very disproportionate fuel volumes that are contained in drops over 30 μm in diameter. The thousands of droplets that are measured in the size range less than 10 μm are seen to contain only about 20% of the injected fuel. Thus, the largest 5% of the measured droplets, or 360, contain more than 60% of the injected fuel. The log-normal representation is seen to have an even greater proportion of small drops, and some minimum-size cut-off is generally required if a log-normal distribution is used to represent a real G-DI fuel spray; otherwise a tremendous number of very small drops will be predicted. The lines representing DV90 and DV80 are also drawn on Fig. 4.16-2b, and represent 90% and 80% of the total fuel volume in the spray, respectively. For a spray having a maximum drop diameter of 50 μm, a DV90 value in the vicinity of 43 μm is to be expected, with a DV80 value about 5 μm smaller. If the maximum drop size in the distribution is reduced to 40 μm, the expected DV90 and DV80 values would be approximately 34 and 29 μm, respectively.

Although it is a parameter that is often referenced in the field of spray measurement, DV90 may not be the ideal variable to use as an indicator of the large-drop content in a G-DI spray. This is not because the drop diameter corresponding to 90% cumulative volume (and also mass) is not a valid parameter, but because it is difficult to experimentally obtain an accurate and repeatable value for the 90% cumulative volume diameter for a highly atomized spray. As illustrated in Fig. 4.16-3 for sprays in which the most frequent drop diameter is typically near 8 μm, a few drops of 50 to 60 μm in diameter contain a very disproportionate fraction of the total fuel volume in a measured distribution. Because a factor of seven on the diameter is equivalent to a factor of 343 on mass or volume, only nine of these larger drops contain as much fuel as about 3000 of the smaller drops.

What happens in actual measurements of the G-DI drop-size distribution using either laser diffraction or PDA is that a total sample of 12000 to 20000 drops will be obtained, with typically many thousands

of drops in the 8 to 10 μm diameter range, and just a few drops measured in the 50 to 70 μm range. The statistical algorithms are then applied to this measured distribution, and key parameters such as D32 (SMD) and DV90 are computed. This is all mathematically correct; however, the problem that arises is that a repeat run will yield an SMD value very close to the initial result, but the repeat DV90 value can differ from the first by more than 30%. A careful investigation of why this occurs shows that one test of 15000 drops may record 8 drops of 50 to 60 μm, whereas the repeat test may contain up to 7 more or 7 less drops. This typical statistical sampling result for a real spray yields greatly different DV90 values. More than 120000 drops would have to be sampled in every measurement in order to reduce the statistical uncertainty of the DV90 values to a reasonable level. This is impractical, and there would appear to be only two alternatives: either continue to report DV90, but clearly state the error band that results from multiple runs, or, as an improvement, phase out the use of DV90 as an evaluation metric for G-DI sprays and substitute a parameter that can be measured much more repeatably with reasonable sample sizes. DV80 has proven to be a good compromise candidate for typical G-DI sprays, yielding greatly improved repeatability for repeat runs using sample sizes of 15000 drops. The DV80 diameter represents the drop diameter in a distribution for which 80% of the total volume of the spray is in smaller drops, with 20% present in larger drops. Assuming that the liquid density is constant, which is a very good assumption, these percentages apply to fuel mass as well. The DV80 parameter, which is on the order of 28 μm for a typical G-DI fuel spray, is much less sensitive than DV90 to small run-to-run differences in the number of very large drops, whereas DV90 responds directly. Therefore DV80 could be a meaningful spray metric.

4.17 Summary

The characteristics of the mean and sac sprays that are produced from the range of DI gasoline nozzles are very important in interpreting engine combustion and emission data and in establishing CFD models and sub-models. The characteristics of cone angle, mean droplet size, DV90 (or DV80), penetration rate, maximum penetration, and wetted footprint all vary with the design of the injector nozzle, the operating temperature, the ambient conditions, and the volatility of the fuel being used. The nozzle may incorporate a fuel swirl

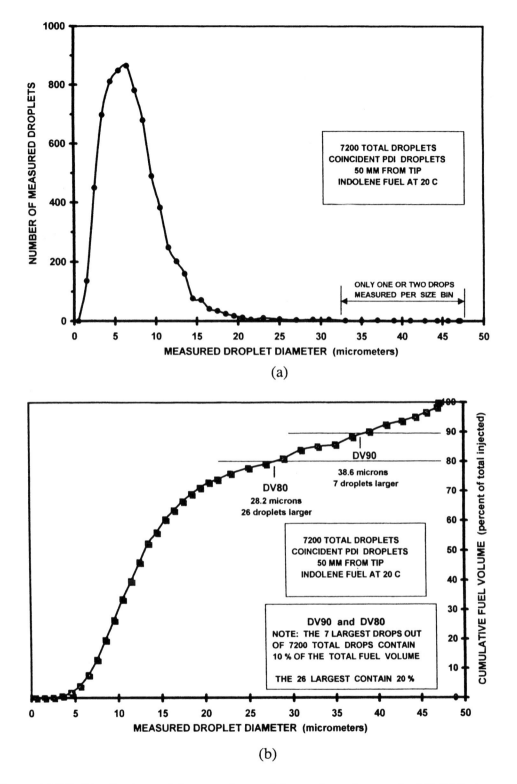

Figure 4.16-3 Distributions of drop number and drop volume for a typical G-DI fuel spray: (a) relative occurrence of droplet diameters; (b) cumulative fuel volume.

plate, a slit, multiple holes, a shaped single cavity or air-assist, and may be actuated by either an electronic solenoid or a piezo stack. Fuel pressures are typically in the range of 5 to 10 MPa; however, there is interest in injection hardware that can operate in the 10 to 15 MPa range. The spray orientation may be designed to align with the axis of the injector body (on-axis), or may be offset by a specified angle (offset spray). The development of a G-DI fuel spray is a highly transient event, normally lasting less than 6 ms for light-load operation, with the spray generally being slightly degraded at both the beginning and the end of injection. This is particularly true if either a substantial sac spray or an after-injection is present. The spray may result from either an inwardly opening or an outwardly opening pintle, with the latter being less common, but generally having the advantage of little or no sac volume. Specialized optical equipment and procedures are required to characterize the transient fuel sprays from a G-DI injector, and worldwide standardization of terminology, techniques and recommended practice has not yet been achieved. Hence caution must be employed in comparing the spray characteristics of injectors from various sources, as it is most likely that the spray parameters were measured using different fluids, ambient conditions, positions in the spray and optical techniques.

As a class of G-DI injector the swirl-atomized unit performs quite well, achieving SMD values on the order of 15 to 16 µm for fuel rail pressures in the range of 8.5 to 10.0 MPa. The DV90 values, which represent the drop size for which 90% of the fuel volume is in smaller drops, range from 28 to 35 µm for the same conditions. The spray performance of both slit-type and multihole injectors is nearly comparable to that of swirl-type injectors, being only 1 to 3 µm larger in SMD and 4 to 12 µm larger in DV90 at the same level of fuel pressure. This is normally compensated for by specifying a slightly higher level of operating fuel pressure, typically 10 to 12 MPa. The droplet velocities and penetration rates tend to be lower for the swirl-type spray than for the multihole, slit and air-assist types of injectors. The slit injector has quite uniform spray properties along the spray in the direction of the slit, with an SMD of about 18 to 19 µm, and only degrades slightly at each end. The sprays from the holes of a multihole injector are fairly independent under room-bench-test conditions, much like those of the VCO diesel nozzle, but can interact at elevated operating temperatures and low ambient pressures. A spray SMD of 16 µm may be

obtained from a pulse-pressurized, air-assisted injector for fuel rail pressures under 1.0 MPa. Droplet velocities and penetration rates for the spray from a pulse-pressurized, air-assisted injector tend to be slightly higher than even those of sprays from either the slit or multihole injectors, because the drops in a pulse-pressurized, air-assisted spray are basically entrained in a sonic, gaseous jet.

The near-term research and developmental topics for the areas of G-DI injectors and fuel sprays will include the expansion of fuel rail pressure limits, spray shaping and optimization, multiple pulsing and piezo-electric actuation. The current practical operational boundaries of fuel rail pressure, which are approximately 5.0 to 12.0 MPa, will be investigated with the goal of expanding both limits. There is certainly an interest in operating a single-fluid G-DI fuel system at 2.5 MPa, as it reduces the pump load and priming time and can provide reduced pump wear. New atomization techniques will have to be invoked to obtain single-fluid mean drop sizes of less than 20 µm with this level of fuel rail pressure. In contrast, there is also an interest in developing fuel injectors, fuel rails, fuel pumps and pressure regulators that operate in the 15 to 20 MPa range. As with the diesel fuel system, there is a continuing trend over many years toward higher operating fuel pressure levels, mainly because advantages regarding combustion and emissions benefits are realized. Improvements in smoke, HC and rate of injection capability have obviously outweighed the disadvantages of a higher pump load and an extended priming time. Even at 10 to 12 MPa the limit of this trade-off has obviously not been reached.

The area of spray shaping, which mainly applies to multihole and slit-type injectors, will experience expanded interest, as this provides a combustion-system design option for controlling the spatial distribution of the fuel. Options for a range of fuel distributions in space will be developed by programming hole and slit configurations. Hole patterns such as circular, oval, line, half-circle, L-shape, crescent and rectangular will be added to the portfolio that is available to combustion-system developers, which will permit enhanced optimization.

The development and refinement of multiple-pulse injection is yet another area that will be the subject of heightened interests in the near term. Both the capabilities of injection systems, as well as the knowledge base of the effects of multiple pulsing, will continue to be expanded with increased utilization in G-DI control systems to be expected. This development area is weakly coupled to that of piezoelectric actuation, in that the use of such actuation is an enabling technology for an entire range of multiple-injection options. The extension of injector dynamic range and the associated capability of invoking very brief fuel pulses combine to make the piezoelectric actuation of injection a near-ideal tool for incorporating multiple-pulse strategies into an engine control system. This is tempered by the observed degradation of spray quality for very brief pulses, which may limit the use of more than two injections per cycle regardless of the actuation source. If it is demonstrated by research investigations that complex strategies of multiple injections can indeed enhance either combustion or emission parameters for some operating condition or environment, then piezoelectric actuation will greatly facilitate the system implementation. Two, three or even four distinct injection events per cycle could be conducted, with precision make-up pulses, catalyst preheat injections, NOx storage catalyst regeneration events and particulate control pulses also included in the repertoire of strategies that can be invoked. These near-term R&D activities will lead to new advances, improved hardware and enhanced strategies, and will facilitate the continuous improvement of G-DI engines over the next decade.

Chapter 5

Mixture Formation Process and Approaches

5.1 Introduction

The transient in-cylinder flow field that is present during the intake and compression strokes of a G-DI engine is one of the key factors in determining the operational feasibility of the system. The magnitude of the mean components of motion, as well as their resultant variations throughout the cycle, are of an importance that is nearly comparable to that of the fuel injection system. On a microscopic scale, a high level of turbulence is essential for enhancing the fuel-air mixing process; but additionally, a controlled mean or bulk flow is generally required for the stabilization of a stratified-charge mixture plume. The optimum flow field for G-DI engines depends on the type of combustion system and the injection strategy that are being used, which is a difficult compromise for full-feature G-DI engines that operate with multiple injection strategies. The turbulence velocity fluctuations near TDC on compression for SI engines can attain the same order of magnitude as the mean velocity, and the turbulent diffusive transport and convective transport can be of equal influence in determining the initial development of the combustion process [103, 479]. Also, the integral scale of the mixture concentration fluctuation inside the combustion chamber can be as large as that of the velocity fluctuation [514]. This can result in strong concentration fluctuations at a fixed position, such as the spark gap location, which can lead to difficulties in obtaining a stable flame kernel.

The preparation of the fuel-air mixture is, by definition, one of the most important processes in ensuring a successful G-DI combustion system. The spray-air-wall interactions and the spray-induced air motion all must be considered in order to optimize the mixture formation process and exploit the full potential of gasoline direct injection. The conditions inside the engine cylinder, as defined by the gas composition, temperature, pressure, and air flow field exert a very substantial effect on the spray atomization and dispersion, the air entrainment in the spray plume, and on the subsequent fuel-air mixing process. The complex and time-dependent spray/air-flow-interaction process determines the rate of fuel-air mixing and the resultant degrees of charge cooling and mixture stratification. In addition, unintended fuel impingement on the combustion chamber surfaces, impacting either the cylinder wall, head or the piston crown or both, can be a source of HC emissions, combustion chamber deposit (CCD) and engine oil dilution. The mixture formation process and the associated control strategies will be described in detail in this chapter.

5.2 Relation of In-Cylinder Flow Characteristics to G-DI Combustion

5.2.1 Typical In-Cylinder Flow Characteristics in SI Engines

In general, a rotating, off-axis, flow structure exists within the cylinder and the combustion chamber, and this coherent structure has an instantaneous angle of inclination between the cylinder axis and the principal axis of rotation. The rotational component having an axis that is parallel to the axis of the cylinder is denoted as swirl, and the component having an axis that is perpendicular to the axis of the cylinder is denoted as tumble. The magnitudes of both the swirl and tumble components are very dependent on the particulars of the intake port design, intake valve geometry, bore/stroke ratio, and the shape of the combustion chamber. It should be noted that it is almost impossible to generate a pure, on-axis, swirling flow inside the modern internal combustion engines during the intake stroke. Any swirling flow that has been observed during the intake stroke from modern engines contains a certain level of tumble. In fact, when the tumble component of a swirling flow is high, this type of charge motion is sometimes called inclined swirl, namely combined swirl and tumble, in order to differentiate the flow field from that of a swirl-dominated charge motion. Another air motion that persists and must be taken into account in combustion system development is squish. This flow is generated in the radial direction in the piston-to-head clearance when the piston approaches the compression TDC.

When comparing the effects of tumble and swirl on the combustion process, the turbulence intensity at TDC on compression correlates better with tumble than with swirl. The key point regarding turbulence generation is when and to what extent the bulk motion decays into turbulence. Many flow studies have revealed that tumble is more effective in turbulence generation than swirl for a given angular momentum. In fact, it is generally agreed that it is easier to develop a conventional tumble-based combustion system for a two-intake-valve engine than to develop a swirl-based system. In the case of swirl, the rotating motion is more effectively conserved during the compression stroke, which tends to delay the turbulence generation that results from the breakdown of swirl. This is because of the circular boundary formed by the cylinder walls, which is more conducive to maintaining a swirling flow. The influence of tumble on combustion may be limited to the early stages, as swirl more effectively stores kinetic energy past TDC on compression. As the tumble ratio increases, the breakdown of the mean tumble motion into turbulence occurs earlier in the compression stroke. In comparison, the location of the center of swirl rotation has a significant impact on the characteristics of turbulence at the spark location. The transition from inclined swirl to horizontal swirl during the compression stroke is progressively delayed as the tumble ratio is increased. Research has shown that tumble enhances the burn rate to a greater degree than could be expected from the turbulence intensity that it generates. This is considered to be the result of the shear flow increment that expands the flame area.

In addition to turbulence, it has long been recognized that the mean velocity above a certain level has a direct impact on flame kernel growth and convection. When there is a strong mean velocity, the flame kernel will be constantly convected in that direction, and will thus be less sensitive to the random convective effects of the largest scales of the turbulence. As a result, cycle-by-cycle variations in combustion can be reduced, and more stability in early flame growth, and eventually the overall combustion process, can be achieved. However, cycle-by-cycle variation of the mean velocity is considered to degrade combustion stability, even though it is a random fluctuation of air motion in the cylinder just as is turbulence. The swirl motion generally has a lesser degree of cycle-by-cycle variation of mean velocity than is associated with tumble motion.

5.2.2 G-DI Flow Field Characteristics and Considerations

There are four key features of the in-cylinder flow field: (1) the mean flow components, (2) the stability of the mean flow, (3) the temporal turbulence evolution during the compression stroke, and (4) the mean velocity near the spark gap at the time of ignition. For homogeneous combustion in an SI engine, the combination of high turbulence intensity and low mean velocity at the spark gap is desirable. This is generally achieved for PFI engines, and also for G-DI engines that operate exclusively in the early-injection mode. Therefore, a flow structure that can transform the mean-flow kinetic energy into turbulence kinetic energy late in the compression stroke is considered desirable for the homogeneous combustion case. The G-DI engine using late injection, however, operates best with a flow field having an elevated mean velocity and a reduced turbulence level, which aids in obtaining a more stable stratification of the mixture. This indicates that the optimum flow field depends upon the injection strategy that is being used, which is an optimization problem for G-DI engines operating with both strategies. For G-DI combustion systems, control of the mixing rate by means of the bulk flow is generally more important than the scheduling of turbulence generation. This is not to imply that turbulence is not important to the combustion process; in fact, turbulence is known to be an important factor in entraining EGR into the local combustion area [262].

Both swirl-dominated [302] and tumble-dominated [259] flow structures are used to achieve stratified-charge combustion in G-DI engines. For the tumble case, the fuel plume is deflected from a shaped cavity target in the piston, and the vapor and liquid fuel are then transported to the spark plug. For the swirl-dominated flow field, the mixture cloud is generally concentrated at the periphery of the piston cavity [73]. In combustion system development, the effects of squish must be carefully evaluated. Some of the features of swirl, tumble and squish and the associated influences on the G-DI combustion system are summarized in Table 5.2-1.

Because of the more favorable geometry, the swirl component of in-cylinder motion generally experiences a lesser rate of viscous dissipation than the tumble component. Therefore it is preserved longer into the compression stroke, and is of greater utility for maintaining mixture stratification. The swirl flow is usually combined with a squish flow that imparts a radial

Table 5.2-1
Principal features of key in-cylinder charge motion patterns

Swirl	• Yields less viscous dissipation and is preserved longer into the compression stroke • Good for maintaining stratification • Intensified when combined with squish • Engine speed dependent, yielding a limited operation zone for adequate fuel-air mixing • Lower cycle-by-cycle variation
Tumble	• Can be transformed into turbulence near TDC by tumble deformation and the associated velocity gradients • Only totally transformed into turbulence with a flat pancake chamber • Incomplete transformation into turbulence may lead to an elevated mean flow • Effective in creating high levels of near-wall flow velocities for promoting wall film evaporation • Effective in enhancing mixing by turbulence generation • Yields larger cycle-by-cycle variations in combustion than does swirl • Tend to decay into a large-scale secondary flow structure, making stratification more difficult
Inclined Swirl	• Tumble combined with swirl • Combined flow characteristics of tumble and swirl
Squish	• Not pronounced until piston is near TDC • Only changes the bulk flow, intensifying the swirl or tumble • Effect of reverse squish must be evaluated

component to the air motion as the piston approaches TDC on compression. A shaped cavity in the piston may also be utilized to assist in achieving the required turbulence production late in the compression stroke. The combined effects of squish and swirl normally yield enhanced swirl and an augmented turbulence intensity during the early portion of the combustion period. It should be noted that the employment of swirl to promote fuel-air mixing is considered to have operational limits. This is because the momentum of the swirling air increases in proportion to the engine speed, whereas the additional momentum imparted by the fuel spray is independent of the engine speed. As a result, the engine speed range in which operational levels of fuel-air mixing can be realized may be limited [3, 196].

The tumble component of the in-cylinder flow field is transformed into turbulence near TDC by large velocity gradients that are associated with tumble deformation, and can be totally transformed only if the combustion chamber geometry is sufficiently flat. Otherwise, an incomplete transformation of tumble kinetic energy will occur, which generally results in an elevated mean flow velocity at the spark gap. Further, tumble-dominated flow fields in G-DI engines generally yield larger cycle-by-cycle variations in the mean flow than those obtained for swirl-dominated flows [38]. These variations influence both the centroid and the shape of the initial flame kernel following ignition, but normally do not produce significant changes in the combustion period or flame speed [438]. Furthermore, the tumble component of the motion tends to decay into large-scale secondary flow structures due to the effect of the curved cylinder wall, which makes maintaining a stable mixture stratification more difficult. With regard to turbulence generation, the presence of a significant tumble component is effective in enhancing the turbulence intensity at the end of the compression stroke. This enhancement is essential to compensate for the reduced flame speed of a lean, stratified-charge mixture. The tumble motion that is present early in the compression stroke rapidly decays into multiple vortices that have a size on the order of the turbulence length scale. This rapid transformation of kinetic energy into turbulence is not generally observed for swirl-dominated flow fields. The swirl flow continues to rotate relative to a center point that generally precesses in a complex path around the vertical cylinder axis for the entire time period from the beginning to the end of the compression stroke. It should be noted, however, that high-swirl-ratio flows can centrifuge the largest droplets from the fuel spray onto the cylinder wall, causing an increase in fuel wall wetting.

Another feature of tumble motion results from its inherent rotational acceleration during compression. Elevated levels of near-wall flow velocities are created, and continue to be created, even relatively late in the compression stroke. This can promote the evaporation of a wall film that is formed by an impinging fuel spray. Thus the transport of fuel vapor to the point of ignition may be enhanced by this flow structure. The reverse-tumble-dominated flow field, in conjunction with a specially designed piston cavity, is utilized in the Mitsubishi GDI combustion system to create a stratified charge near the spark gap. The piston cavity is designed to control the spray impingement and flame propagation by enhancing the reverse tumble flow throughout the compression stroke, with the assistance of squish flow from the exhaust to the intake side of the chamber. Reverse tumble as the dominant in-cylinder air motion may indeed be effective for designs in which the spark plug is centrally located and the injector is positioned below the intake valve. In such designs reverse tumble can be effective in moving the vapor and liquid fuel toward the spark gap subsequent to spray impingement on the piston cavity surface.

It is worth noting that almost all G-DI combustion systems have more stringent requirements regarding charge motion than are needed for PFI engines in order to prepare the stratified-charge mixture [105, 106].

Because the cone spray created by the swirl-type G-DI injector has a relatively low penetration rate, the role of charge motion to transport the fuel plume to the spark gap becomes very important. However, the specially designed, fixed port geometry that has been optimized for stratified-charge operation may lead to a significant flow loss during WOT operation. As a result, the theoretical benefit of volumetric efficiency improvement that can be offered by G-DI engines for improving full-load performance may not be fully realized, and a compromise has to be made between part and full load requirements. With a variable charge-motion-control (CMC) device, intake port flow characteristics that are comparable to those of stoichiometric PFI engines can be obtained even for a system tuned for best stratification and homogeneous-charge dilution capability. Variable charge motion can be applied to both tumble and swirl systems, with deactivation of one intake port being the most simple and direct implementation method. Therefore, excellent port flow characteristics can be obtained even considering an optimum injector location [482]. It should be noted that one of the major drawbacks with increased charge motion is the increased heat loss to the cylinder wall, which is generally associated with a fuel economy penalty. All of these disadvantages become more pronounced at increased engine speeds [250]. Fortunately, the heat loss associated with a high charge motion is less with stratified-charge operation, as the flame may have less contact with the wall surface.

For a combustion system incorporating a swirl control valve (SCV), a high-swirl-ratio flow can be generated during the latter portion of the intake stroke. As a result, the swirl ratio in the top portion of the cylinder is generally of a greater magnitude than that in the vicinity of the piston crown. Such a swirl ratio differential will result in a pressure difference along the cylinder axis, which will yield a net flow from the piston crown to the spark gap, countering the mixture dispersion in the cylinder axial direction. This pressure difference could help to maintain a highly stratified mixture around the spark gap [318–320, 323, 325]. Similar to the vertical flow generated by the aforementioned phenomenon, vertical air motion along the cylinder axis in the Nissan NEODi combustion system has been observed [333, 334]. This particular air motion is called liftup swirl, and has been confirmed by both CFD analysis and laser-induced fluorescence (LIF) experimental measurements. The mechanism of liftup swirl is explained by the following reasoning: two swirl ratios are identifiable in a bowl-in-piston chamber, one in the main chamber and one in the bowl. The two swirl ratios can differ significantly when one of the two intake valves is closed to induce swirl. The pressure at the swirl center is lower than that at the periphery, thus a pressure gradient exists from the bowl to the spark gap, with an upward flow generated from the edge of the bowl to the center of the chamber. CFD analysis shows that the presence of horizontal swirl during the compression stroke generates a strong vertical flow inside the piston bowl, thus enhancing the transport of the mixture plume from the bowl to the spark gap.

A large number of G-DI investigations have been dedicated to assessments of the effects of charge motion on mixture formation, combustion and emissions. A comparison of the engine performance of a G-DI engine that operates using both the swirl and reverse-tumble concepts shows that these two types of flow fields provide similar light-load engine performance for the overall air/fuel ratio in the range of 35:1 to 40:1. However, for the high-load region in an air/fuel ratio range of 20:1 to 30:1, problems of combustion stability and smoke emissions may be encountered with swirl-dominated engines. Also, the required control system for this type of engine is generally more complex in order to accommodate engine load transients [181]. A G-DI concept using an "inclined swirl" that contains equal swirl and tumble flow components has been proposed; this ostensibly combines the best features of these two flow structures [489]. An inclined swirl at an angle of 45° significantly enhances the turbulence intensity and provides an associated reduction in the COV of IMEP [123], although the extent of the reduction would certainly depend upon the specific combustion chamber geometry being utilized.

A study of the influence of the in-cylinder air flow pattern (tumble vs. swirl) using a Mazda prototype G-DI engine under stratified-charge operation revealed that a swirl-flow predominance exhibits an improved light-load ISFC, whereas a tumble-flow predominance results in a marked combustion instability that requires additional intake throttling [503]. In-cylinder HC measurements obtained near the spark gap by fast FID indicate that for the reference swirl case a stable combustible mixture is formed near the spark plug gap at the time of ignition, yielding low HC fluctuations. In contrast the air/fuel ratio was found to be about 24:1 for tumble flow, with the associated in-cylinder HC emissions fluctuation being much larger. It was also found that an increase in the swirl ratio can further reduce the ISFC, which is attributed to improved fuel transport, more rapid evaporation of fuel in the film on the piston-bowl surface and reduced fuel dispersion. It should be noted that the combustion system used in this study was originally developed and partially optimized for swirl flow.

The stratified-charge operating requirements of G-DI combustion systems can conceptually be achieved with either swirl or tumble air motion concepts, as a number of studies have indicated [109, 110, 481]. Usually the performance differences between swirl and tumble systems are smaller than is observed among different methods of establishing the stratified charge, such as spray-guided versus wall-guided approaches. With a flat combustion chamber, intake-generated charge motion can be achieved quite effectively by a swirl port that introduces a swirling intake flow. With a pentroof-shaped combustion chamber, however, tumble flow is generally found to be more effective. Variable charge motion is recommended to achieve good performance at both part and full load, which favors swirl flow, as variable swirl by means of port throttling or full deactivation can be generally achieved with less complexity than variable tumble. As a result, variable swirl is used in a number of current production G-DI engines [259, 440, 492]. However, interest in the variable tumble concept is increasing, and several variable-tumble-based G-DI combustion systems have been proposed and developed [49, 127]. Figure 5.2-1 provides a comparison of the flow characteristics of different charge-motion-based G-DI concepts [42]. It is evident that the variable charge motion system provides the flexibility of generating a strong air motion when required, without compromising charging efficiency when strong charge motion is not necessary.

For a four-valve G-DI system the valve cross sections necessary for achieving competitive full load characteristics leave little design flexibility in the pentroof geometry. In general, tumble-based combustion systems require relatively deep piston bowls with medium or smaller valve angles. However, the use of deep bowls is known to degrade homogeneous combustion significantly. In comparison, swirl flow can be preserved more efficiently due to reduced momentum dissipation through transformation into small-scale flow. Therefore, the use of swirl flow can provide good results even with relatively shallow piston bowls. In such a case, combustion chamber height can be reduced to accommodate designs having a small valve angle.

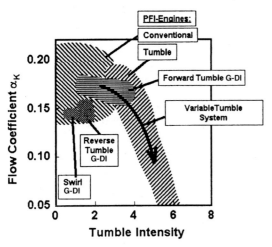

Figure 5.2-1 Comparison of flow characteristics of different charge-motion-based G-DI concepts [42].

Some systems that employ squish as the dominant motion for charge stratification are illustrated in Fig. 2.1-2d [108]. The principle is to use the squish to generate turbulence and improve mixture preparation. The squish area must be carefully determined for these systems in order to minimize fuel encroachment into the squish area. Figure 5.2-2 shows the time histories of calculated non-dimensional squish velocities for combustion chambers with two different clearance heights [201]. The calculation is based solely on volume changes during compression. A maximum squish velocity is

observed around 10° before or after compression TDC. It is worth noting that the effects of both inward and outward squishes on G-DI combustion processes must be carefully evaluated. The combustion characteristics and engine performance with these two different clearance heights were also compared in an engine test by simply raising the cylinder head and liner relative to the crankcase. This had the effect of reducing the engine compression ratio from 12.7:1 to 10.4:1. It was concluded that the effect of squish flow on mixture preparation and combustion is quite limited. It was also found that NOx levels are similar, which suggests that there is little or no change in the air/fuel ratio in the vicinity of the spark gap. The similarity in the response of HC emissions to combustion phasing further suggests little change in mixture transport time. The observed change in fuel consumption is less than would be expected from the decrease in the compression ratio, which is due to a reduction in engine friction resulting from the lower cylinder pressures.

5.2.3 Effect of Fuel Injection Event on In-Cylinder Flow Field

In addition to the intake-induced bulk air motion, another important issue that should be addressed in G-DI combustion system design is that the spray-induced flow field can exert a significant influence on the in-cylinder air flow field [69]. The secondary flows

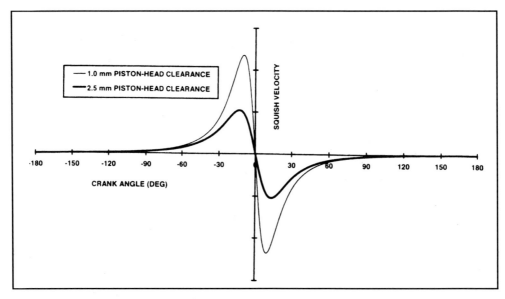

Figure 5.2-2 Time histories of calculated non-dimensional squish velocities for combustion chambers with two different clearance heights [201].

induced by the transient spray itself may promote spray contraction, or "collapse," with the thresholds being dependent on the droplet size and the in-cylinder ambient conditions [432]. Analytical studies can predict the effect of a spray-induced flow field on the flow structure inside the combustion chamber of a G-DI engine. As shown in Fig. 5.2-3, a KIVA code analysis for the early injection case of a centrally mounted injector predicts that the momentum generated by the injected stream of liquid droplets is partially transferred to the surrounding gases, which increases the kinetic energy of the charge soon after the fuel is injected. This spray-induced flow enhances in-cylinder air-fuel mixing; however, the increased kinetic energy rapidly decays as the piston moves up during the compression stroke. Hence, the increase in TDC kinetic energy over the non-injection case is relatively insignificant. For injection later than 150°ATDC on intake, the turbulence intensity as enhanced by fuel injection is substantially higher than that of the non-injection case. About 10% extra turbulence intensity is generated by the typical G-DI spray when the initiation of fuel injection occurs later than 150°ATDC on intake. For operation in the early-injection, homogeneous-charge mode, the later the start of the injection, the higher the turbulence intensity at TDC on compression [150], although mixture homogeneity may be degraded.

Spray-induced motion does indeed affect large-scale, in-cylinder flow structures. In particular, it increases the mean velocities of the gases in the spray region, and significantly suppresses the intake-generated bulk flow for a wide range of fuel injection timings [150]. This indicates that the local mean velocity in the case of fuel injection may be increased or decreased, depending on the measurement location inside the combustion chamber. That is why some experimental studies on spray/air and spray/piston interaction processes during the early injection mode [4, 148, 422] reveal that there is a slight decrease in the tumble intensity caused by the injection event. Figure 5.2-4 shows a comparison of the resultant tumble between the injection and no-injection cases for two locations at the radial distance of 15 mm off the cylinder axis and 38 mm from the cylinder head [422]. During the intake stroke following the fuel injection event, velocities between injection and no-injection cases are very similar. After bottom dead center, a tumble motion develops and the velocity traces for the injection and no-injection cases begin to diverge. The tumble velocity level is lower for the injection case between 180° and

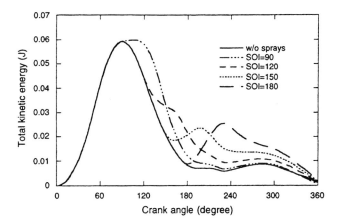

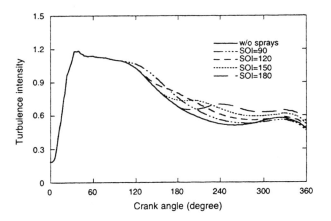

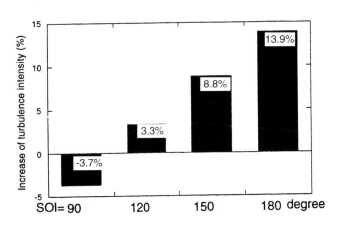

Figure 5.2-3 Effect of fuel injection on the total gas mean kinetic energy and gas turbulence intensity— normalized by the mean piston speed [150].

250° ATDC, as indicated by the difference in vertical velocity between the two locations; however, the tumble component is mostly dissipated by 270° on intake for both cases. This could be because the tumble center is moved upward. The velocity for the no-injection case becomes positive at an earlier point in the cycle. This indicates that the downward momentum that is added by the injection event influences the in-cylinder velocities for a significant portion of the cycle.

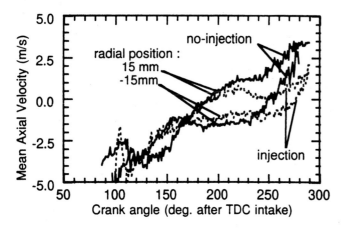

Figure 5.2-4 Measured axial velocities as a function of crank angle at two in-cylinder locations: 15 mm from the cylinder axis and 38 mm from the cylinder head; centrally mounted injector; 750 rpm; fuel injection timing: 90–100°ATDC; ensemble-averaged data [422].

Tests of a second-generation Toyota D-4 engine reveal that in-cylinder fuel injection during the compression stroke significantly increases the mean velocity and turbulence intensity of the in-cylinder flow inside the piston cavity [223]. Figure 5.2-5 shows the measured time histories of flow velocity at the spark gap with and without fuel injection. In this combustion system a stratified charge is created using a fan-shaped fuel spray from a slit-type nozzle, with a shell-shaped piston cavity [223] (illustrated earlier in Fig. 2.3-10). A strong increment of bulk motion is generated by the process of fuel injection into the piston cavity. As compared in Fig. 5.2-6, the turbulence intensity generated by the fuel injection and averaged over the time period from 30°BTDC to TDC is even greater than that generated by the SCV of a lean-burn engine having a helical port. This indicates that aligning the fuel injection direction with the in-cylinder bulk air motion could increase the in-cylinder turbulence intensity significantly.

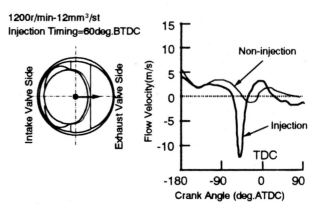

Figure 5.2-5 Time histories of measured flow velocities with and without fuel injection [223].

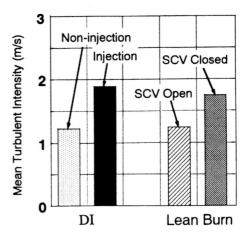

Figure 5.2-6 A comparison of turbulence intensities between the Toyota second-generation D-4 engine with and without fuel injection, and the lean-burn PFI engine with and without SCV [223].

5.3 Fuel-Air Mixing Process

5.3.1 In-Cylinder Charge Cooling

In an ideal G-DI engine, gasoline is injected directly into the cylinder as a well-atomized spray that is vaporized completely by absorbing heat only from the in-cylinder air. This is a very efficient mechanism for cooling the air charge that resides within the cylinder. As shown in Fig. 5.3-1, the charge temperature is decreased as the latent heat of vaporization is transferred from the air to the liquid fuel. This can, in fact, yield an increased mass of air in the cylinder if injection occurs during the induction process, thus increasing the volumetric

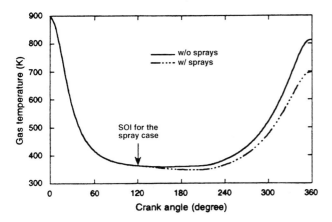

Figure 5.3-1 A comparison of time histories of gas temperature with and without in-cylinder fuel injection during induction [150].

efficiency of the engine. A reduced charge temperature at the start of compression also translates directly into a lower compression temperature, thus enhancing the mechanical octane number of the combustion chamber. In contrast, fuel droplets of 100 µm to 200 µm SMD are injected into the intake port in a typical PFI engine, with a liquid film being formed on the back of the intake valve and on the wall of the lower port. This liquid fuel vaporizes under the influence of concentration gradients, port vacuum, and by absorbing heat from the valve and wall surfaces. The latter makes it more difficult to achieve an efficient charge-cooling effect in a PFI engine. As a result of the thermodynamic effects of in-cylinder charge cooling, the ideal G-DI engine exhibits the advantages of both greater maximum torque and a higher knock-limited compression ratio as compared to the PFI engine. The cooling of the in-cylinder induction air due to direct injection of a vaporizing fuel spray during the induction event has been found to provide four major advantages, listed in Table 5.3-1.

Table 5.3-1
Four key advantages of in-cylinder charge cooling

- Increased volumetric efficiency
- Reduced compression temperature and heat losses
- Reduced autoignition tendency at the same compression ratio
- Higher knock-limited compression ratio

The decreased charge temperature that is associated with direct fuel injection during the induction process provides an increase in the trapped cylinder mass over the non-injection case; however, as would be expected, the gain in volumetric efficiency disappears when injection is retarded to the end of the intake stroke [150]. It should be noted that some fraction of the beneficial charge cooling effect and knock suppression due to fuel vaporization still persists even though fuel is injected into the cylinder after the intake valve is closed. At the other extreme, if the injection timing is advanced to the early stages of the intake stroke, fuel impingement on the piston crown can occur. The increased wall wetting is effective in cooling the piston, but will reduce the effective charge-cooling rate, resulting in a smaller temperature decrease for the charge and a smaller gain in trapped mass. Therefore, for injection circumstances that result in significant wall wetting, the benefits of charge cooling will be diminished. The level of atomization does indeed influence the volumetric efficiency gain, as rapid vaporization of fuel droplets in the air stream during induction is required. Droplet sizes must be sufficiently small that the bulk of the injected fuel mass vaporizes during the time available for the induction event. The cooling effect continues as fuel continues to vaporize; however, the volumetric efficiency gain is directly related to the fraction of fuel that is vaporized by the time that induction is complete. The cooling of the intake charge also modifies and improves the engine heat transfer process, particularly for the early injection associated with high-load operation. On the compression stroke the charge density is monotonically increased, and the charge temperature is elevated to values higher than the wall temperature, with heat being transferred to the walls. This heat transfer advantage diminishes rapidly as the injection timing is retarded toward the end of the intake stroke [150].

It is instructive to compare the charge cooling effect on volumetric efficiency for two extreme situations. One case is similar to that of the PFI engine where the fuel is vaporized only by heat transfer from the intake port and valve surfaces. The other case corresponds to the ideal G-DI engine where the fuel is vaporized only by absorbing thermal energy from the air. For an assumed initial intake air temperature of 100°C and a fuel temperature of 50°C, it is found that for direct injection the volume of the mixture after vaporization at standard conditions is about 5% smaller than the volume of intake air [9]. Under the same inlet conditions, however, if fuel is vaporized and heated to the intake air temperature by heat transfer from the wall only, the mixture volume will increase by 2% due to the volume of the fuel vapor. Thus, the total difference in the mixture volume of the two extreme cases can be as large as 7%. The standard volumes may also be converted directly into differences in trapped mass. However, it should be noted that these extreme cases are not totally descriptive of real G-DI and PFI processes, because some vaporization of fuel in air occurs in PFI engines, and some wall film vaporization of fuel occurs in actual G-DI engines. Moreover, for the cold-start case, the fuel, air and engine cylinder wall have similar temperatures. As a result, the actual difference in the engine volumetric efficiency between PFI and G-DI engines can be significantly less than that computed from the ideal limiting cases, and is quite dependent on the specific engine design, fuel characteristics and operating conditions. Dynamometer testing for a particular combustion system has indicated that volumetric efficiency improvement is approximately 1/3 of the theoretical maximum difference, or about 2.5%, and exhibits a strong dependence on injection timing [9]. At a constant pressure the difference in the calculated charge temperature for the two extreme cases can be as large as 30°C, depending on the assumed intake air and fuel temperatures. The charge temperature for the case of injection during intake was found to decrease by 15°C at the end of the induction process due to fuel evaporative cooling [9]. This translates to a significant decrease in gas temperature at the end of the compression stroke. Therefore, the G-DI engine utilizing mid-induction injection of fuel will exhibit an enhanced resistance to autoignition. As shown in Fig. 5.3-2, the knock-limited spark timing can be substantially advanced for this operating mode. Alternatively, the knock-limited compression ratio can be increased by as much as 1.5 ratios for 91 research octane number (RON) fuels, thus achieving a significant gain in engine thermal efficiency. The benefit of charge cooling resulting from early fuel injection is shown in Fig. 5.3-3 as an octane number improvement of 4 to 6, which allows an increase of compression ratio of up to 1.5 [201, 274]. Invoking the maximum permitted increase in compression ratio will directly enhance the power and torque characteristics of the G-DI engine, and will result in a direct, substantial improvement in engine BSFC.

The study using the Nissan prototype DI engine reveals that for a stoichiometric air/fuel ratio, charge

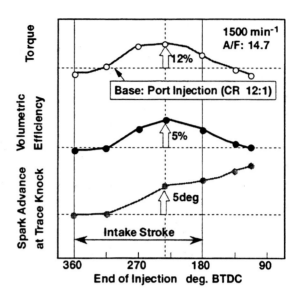

Figure 5.3-2 Improvements in G-DI engine torque, volumetric efficiency and spark advance with mid-induction injection; 1500 rpm and WOT [196].

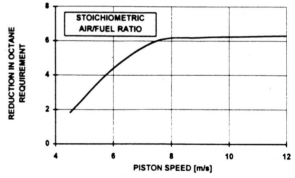

Figure 5.3-3 Octane requirement improvement of G-DI engines at stoichiometric air/fuel ratio [274].

cooling has the effect of lowering the charge air temperature by approximately 20°C. As a result, even under WOT operation at low engine speed, a G-DI test engine exhibits a 6% higher power output than a comparable engine using PFI. This improvement is attributed to the ignition timing advance made possible by the reduced knock tendency [438]. The elimination of fuel vapor in the intake port, which displaces air in the PFI case, is another factor in the improvement of volumetric efficiency [203]. It is also worth noting that the optimal SOI timing for maximum volumetric efficiency may not be the same as that for maximum torque due to the limited time for achieving charge homogenization [19]. For the Fiat DI combustion system, maximum torque is obtained at an injection timing that is about 20 to 30 crank angle degrees earlier than that for achieving maximum volumetric efficiency [20].

5.3.2 In-Cylinder Fuel-Air Mixing Characteristics

The fluid-mechanical state within an engine cylinder, as represented by the temperature, pressure and air flow field, exerts a very substantial effect on spray atomization and dispersion, air entrainment in the spray plume and the subsequent fuel-air mixing process. The complex and time-dependent spray/air-flow and spray-wall interaction processes determine the rate of fuel-air mixing and the degree of mixture stratification [39, 68, 77, 80, 96, 97, 218, 270, 349, 397, 404, 463, 507]. It is difficult to generalize the characteristics of the in-cylinder fuel-air mixing process, as mixture preparation is highly dependent on the details of the combustion system configuration, spray geometry, in-cylinder flow structure, and fuel injection strategy; however, the process can be described in three basic stages [301, 505]. For a wall-guided combustion system with an axisymmetric spray impinging upon a piston cavity, the first is the free-spray stage in which fuel droplets initially form a hollow cone and generate a toroidal vortex. The in-cylinder flow field has a direct effect on both the trajectory of the spray plume and the fuel vapor distribution [358]. The second stage begins with the interaction of the spray with the piston surface. During this stage the spray angle and piston cavity shape control the boundary of the spatial distribution of air-fuel mixture. The spray-wall interaction process will be discussed in detail in the following sections of this chapter. After the completion of spray impingement on the piston cavity, the mixture formation process enters the final stage, namely the dispersion and convection of the air-fuel mixture. In this stage, the number of fuel droplets decreases significantly due to evaporation that is enhanced by a steep rise in the ambient gas temperature near the end of compression. The rich-mixture region rapidly diminishes in extent, and the air-fuel distribution within the mixture cloud becomes more uniform. Throughout the entire mixture formation process, the interaction between the spray and the in-cylinder flow field is primarily determined by the relative momentums of the in-cylinder flow and the free and impinged sprays [367, 375]. Further, the impinged spray trajectory is greatly

influenced by the angle of impingement relative to the piston surface. The final distribution of fuel vapor concentration at the time of ignition is significantly influenced by this interaction with the cavity shape.

Experimental investigations [148] indicate that for a centrally mounted swirl injector, the greatest fuel droplet density is observed along the axis of the spray even though the spray is ostensibly a hollow cone. The bulk entrainment flow field and the recirculation of small droplets induced by the toroidal vortex are largely responsible for this. For a tumble system the tumble momentum is reduced by the fuel drops that are entrained in the flow, while a downward air flow component is induced by the spray event. This momentum deficit is detectable well into the compression stroke.

It is worth noting that the entrainment of air into the fuel spray plume is a very important element of mixture preparation. Even for minimal in-cylinder air flow fields, a substantial flow field is created by the typical G-DI injection event. This transient pulse results from the exchange of momentum between the millions of fuel droplets and the in-cylinder gases, and can significantly alter the mean air flow field and the turbulence characteristics. During this 4 to 6 millisecond exchange, air is entrained into the developing and dissipating spray plume, which promotes mixing of the fuel and air, and can alter the spray geometry. The air entrainment characteristics are mainly determined by the spray geometry and injection event rate history, such as would be observed in a bench test with an infinite environment. However, the transient entrainment flow field is greatly constrained and influenced by combustion chamber geometry, particularly for wall-guided systems. A substantial portion of the known dependence of combustion characteristics on fuel spray geometry is traceable to the air entrainment history that results from the interaction of a plume geometry with the air flow field. This entrainment occurs until all of the spray momentum is dissipated, including the time periods both prior to and after wall interaction. For some sprays such as those from wide-cone swirl injectors, much of the entrainment occurs prior to any wall interaction, whereas for a fan spray from a slit-type nozzle the entrainment after wall interaction may predominate.

KIVA-3 CFD studies of the effect of the mean flow components of swirl and tumble on the in-cylinder fuel-air mixing process of a G-DI engine indicate that injected hollow-cone sprays can be significantly deflected by the intake flow, with the smaller droplets

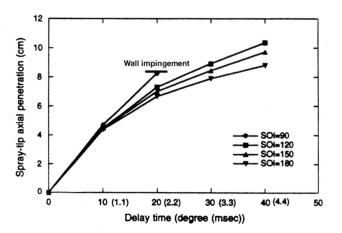

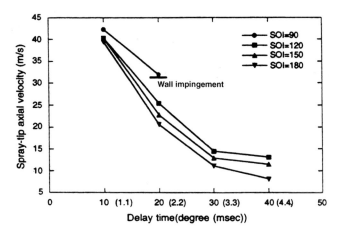

Figure 5.3-4 Computed spray-tip axial penetrations and velocities (1500 rpm; 0.1 MPa MAP; iso-octane)[150].

tending to follow the air flow stream lines. For a simulated combustion system having a centrally mounted injector that injects fuel axially into the cylinder during the intake stroke, the spray-tip penetration for the tumble-dominated flow field is found to be greater than that for either the swirl-dominated or the quiescent flow fields. Spray-wall impingement is found to occur by bottom dead center (BDC) on intake, after which some of the liquid fuel is predicted to remain as a wall film while other droplets are entrained by the air flow field. A rich-vapor region remains near the piston surface during the compression stroke in the relatively quiescent, non-tumble flow field. For both the tumble-flow and swirl-flow cases, the rich-vapor region is found to move with the main in-cylinder flow field and disperse. However, some pockets of rich mixture are predicted to remain near the piston surface [149].

For the centrally mounted injector position, confirmation that a tumble flow field is likely to deflect the spray toward the cylinder wall is provided by an experimental study of the effect of various charge motions on G-DI mixture preparation [347]. A swirl flow tends to concentrate the spray plume at the center of the cylinder, and is effective in reducing cylinder wall wetting. The lean limit can be extended markedly when the spray tip velocity is reduced and the spray cone angle becomes narrower. For a solid-cone spray with a spray cone angle of 45°, a G-DI test engine could be operated at an overall air/fuel ratio of 40:1.

The hollow-cone spray structure that occurs for injector bench tests in a low-pressure, quiescent environment is also obtained in engines with a centrally mounted fuel injector when injection occurs during the intake stroke [150, 151]. However, the intake-generated flow field does influence the trajectory of the injected spray significantly, with the spray being deflected and the spray-tip axial penetration being slightly increased. Due to the combined effect of deflection and increased penetration, spray impingement on the cylinder liner occurs when fuel is injected between 90° and 120° ATDC on the intake stroke, even though the spray is injected axially. The computed spray-tip axial penetrations and velocities of the main spray versus the delay time after the start of injection are shown in Fig. 5.3-4. The velocities were computed from the spray-tip penetration data. The intake flow is found to have the largest influence on the spray for injection timings earlier than 90° ATDC on the intake stroke. The details of the spray-wall impingement depend not only on the injection

timing, which changes the phasing between the spray and piston velocities, but also on the instantaneous flow field. Figure 5.3-5 shows the computed time histories of the liquid fuel fraction on various combustion chamber surfaces. In the worst case, for this configuration, the amount of liquid fuel that impacts all wall surfaces is as high as 18% of the total fuel injected, leading to the formation of relatively rich vapor regions near the piston surface late in the compression stroke. The vaporized fraction of injected fuel is shown in Fig. 5.3-6 for an injection timing of 120° ATDC. It is predicted that almost 90% of the fuel is vaporized at a crank angle of 330° ATDC on the intake stroke. The distributions of the mixture in the three different equivalence ratio ranges ($\phi > 1.5$, $1.5 > \phi > 0.5$, $\phi < 0.5$) are not significantly different for an injection timing later than 90° ATDC on the intake. This is considered to be because the remaining liquid in these cases, which is about 3% of the total

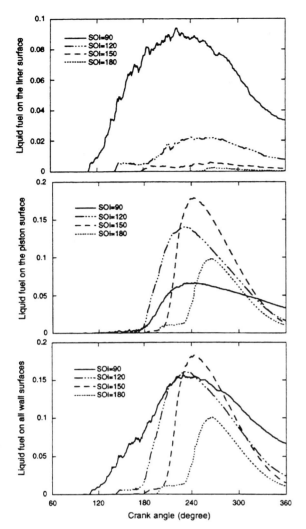

Figure 5.3-5 Computed time histories of the liquid fuel fraction of the totally injected fuel on the cylinder liner, the piston crown and the sum of the two cases (1500 rpm; 0.1 MPa MAP; iso-octane) [150].

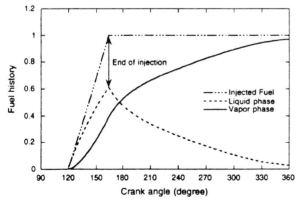

Figure 5.3-6 Computed vaporization history of fuel injected during the induction stroke; injection timing: 120°ATDC; normalized by total amount of fuel injected (1500 rpm; 0.1 MPa MAP; iso-octane) [151].

fuel injected, is located near the piston surface, and the rich mixture in this region is less affected by the in-cylinder flow field. Although variations in injection timing result in differing levels of charge stratification, the general trend with regard to the locations of rich and lean regions is not significantly modified. The gross features of the charge distribution are primarily determined by the injection and flow-field orientations. For the early-injection cases considered, the mixture ratio is generally leaner in the main-chamber region and richer in the squish region, with the air/fuel ratio ranging from 8:1 to 24:1.

Numerical predictions of the fuel-air mixing process with a side-mounted injector indicate that for early injection the initial development of the spray is largely unaffected by the air motion due to the high spray momentum [274]. For the early-injection operation, with an injection timing of 170°ATDC on intake, spray impingement on the cylinder wall occurs at 185°ATDC, causing a rich mixture to remain near the piston crown throughout most of the compression stroke. At a crank angle of 25°BTDC on the compression stroke, the bulk tumble motion begins to decay rapidly and a fairly homogeneous stoichiometric mixture is produced. Over 90% of the injected fuel vaporizes by 20°BTDC on the compression stroke.

CFD analyses of the fuel-air mixing process of the reverse-tumble-based G-DI combustion system with a side-mounted fuel injector reveal that, for early-injection operation, the homogeneity of the mixture is degraded by increasing the average equivalence ratio [86]. For overall lean operation the mixture is nearly homogeneous, while a rich region exists above the center of the piston for overall stoichiometric operation. A similar analysis for stratified-charge operation shows that the fuel initially ignited by the spark comes directly from the injector and is not guided by the piston bowl. The main function of the bowl geometry would appear to confine the fuel cloud during the flame propagation process. Some improvements in the description of the interaction between hot surfaces and the spray (the wall-film sub-model) are required even for early injection timing, as the fuel distribution in the cylinder directly affects flame propagation. Interactions between potentially impinging liquid droplets and the piston cavity are also important for stratified-charge operation. An experimental investigation of mixture preparation for the same reverse-tumble-based G-DI combustion system indicates that a large cycle-by-cycle variation of air/fuel ratio is present near the spark gap when operating in the stratified-charge mode. These variations correlate with fluctuations in the start of combustion, and with IMEP. For injections that occur during stratified-charge operation, particularly for injectors having spray cone angles wider than 60°, part of the fuel reaches the spark gap on a direct path from the injector without any redirection or impingement on the piston wall. This is normally referred to as the short-circuiting of the wall-guided spray. The other portion of the spray plume that does interact with the piston surface or cavity may or may not form a wetted area. If a film is indeed formed, then liquid fuel deposited on the piston surface may contribute directly to smoke emissions [62].

An investigation of the interaction of the fuel spray with the in-cylinder air flow field reveals that significant bowl wall wetting occurs when the fuel spray direction aligns with the air flow velocity vector [252]. The fuel cloud is dispersed by the air flow, and a lean mixture zone is formed downstream of the injector. For the case of injecting fuel against the air flow direction, the tip of the spray is rapidly decelerated. When injecting fuel in an orientation that is orthogonal to the air flow direction, the spray development and penetration characteristics are not significantly influenced by the air flow. This demonstrates the important relationship between the in-cylinder air flow field and the orientation of the fuel spray axis.

The effects of the in-cylinder swirl ratio and nozzle type on the resulting penetrations of G-DI fuel sprays have been investigated extensively [157]. Six nozzle types of one, two and three-hole configurations were used to obtain spray penetration data for three in-cylinder swirl ratios and four in-cylinder pressure levels. The spray-tip penetration and trajectory are found to be strongly dependent on both swirl ratio, nozzle geometry and orientation of the spray axis relative to the air velocity vector. Measurements of variations in the spray-tip velocity with distance and time for in-cylinder injection of gasoline show that time histories of spray penetration predicted from individual droplet drag correlations are not accurate, as the spray tip is not composed of a single collection of droplets during the injection event, and does not experience the same drag force history as any one individual droplet [158]. A three-regime method of spray penetration and velocity correlation is necessary for accurate predictions: an early-time regime that is determined by the injector opening characteristics; a middle-time region that is controlled by spray-tip drag; and

a late-time region that is dominated by the in-cylinder flow field.

In achieving a stable stratified-charge mixture, the important considerations are the combustion chamber geometry, the air flow field, and the shortening of the interval between injection and ignition. The feasibility of creating a stable stratified-charge mixture using two high-flow-rate fuel injectors have been investigated for a Toyota prototype combustion chamber [446]. In this study the injectors were a hole-type nozzle and a swirl nozzle, each with a piezoelectric actuator with the capability of rapidly opening and closing the needle valve at fuel pressure levels exceeding 15 MPa. The concept of this limiting case is to control the degree of fuel vaporization from the liquid film on the piston by means of the hole nozzle, and to control the fraction of fuel vaporization in the air flow field by means of the swirl nozzle. The injectors selected had the capability of injecting fuel at very high pressures (>20 MPa) in a short duration to permit the flow rates to be set independently. The hole nozzle provides a spray with a narrow cone and a high penetration, thus it is suitable for achieving a stratified-charge mixture around the spark gap at light load. Because the fuel vaporization rate achieved using the hole nozzle depends mainly upon heat transfer from the piston, the delivered droplet size distribution is found to be of less importance than that of the swirl nozzle. Visualization of spray development indicates that vaporization of the fuel spray from the swirl nozzle is initiated well before any wall impingement occurs. For this nozzle the vaporization rate does not substantially depend on heat transfer from the heated wall. For the hole-type nozzle the vaporization rate is suddenly enhanced as the fuel spray interacts with the wall; therefore, effective utilization of the thermal energy of the piston is very important if a hole nozzle is to be used to achieve a stratified-charge mixture. For the low-load, low-speed condition (approaching idle), the two nozzle types exhibit opposite trends of BSFC variation with the fuel injection pressure. For the hole nozzle the BSFC increases with a lower fuel pressure up to a limiting value of 8 MPa, whereas the swirl nozzle yields an improvement in BSFC. For the medium-load, high speed condition, both nozzles require an elevated fuel rail pressure to maintain a low BSFC. It may be assumed that if the injection-to-ignition time interval is reduced, the fuel pressure would have to be elevated in order to shorten the injection duration. In general, because the fuel spray of the hole nozzle has less entrainment and dispersion than that of the swirl nozzle, a suitable stratified-charge mixture can be more readily achieved at light load. At higher loads, an optimal strati-fied-charge mixture can be achieved most readily with a swirl nozzle, due to its higher rate of spray dispersion.

Spray characteristics are also of particular impor-tance for emissions associated with homogeneous-charge operation [473]. The effects of spray characteristics on engine HC emissions during early injection have been determined for both a solid-cone spray (A) and a hollow-cone spray (B). The measured spray characteristics and the resultant HC emissions are tabulated in Figs. 5.3-7 and 5.3-8. The use of a hollow-cone spray yields a decrease in HC emissions as the injection timing is var-ied from 20°ATDC to 150°ATDC on the intake stroke. At an injection timing of 20°ATDC the measured HC difference between the solid-cone and hollow-cone injectors is quite small, but becomes significant at later injection timings. This is because earlier injection tim-ings generally lead to severe wall wetting, which gener-ally has a greater effect on the HC emissions than the spray quality. For the solid-cone spray the HC emissions increase as the injection timing occurs later during the intake stroke. For the hollow-cone injector, however, the HC emissions decrease as the injection timing is retarded within the experimental range.

Fuel composition, through the associated effect on fuel volatility, can exert a significant influence on spray characteristics, especially under conditions of low ambient pressure and high injector operating tempera-ture. This was discussed in detail in Chapter 4. Except for these conditions, the effect of fuel composition on parameters such as viscosity, specific gravity and sur-face tension produces only a small effect on the result-ant spray. The effects of fuel composition on the distribution of fuel within the combustion chamber were investigated at different injection timings for both cold and warm conditions [78]. This study included three dif-ferent fuels of commercial gasoline, iso-octane, and a three-component research fuel. For combinations of injector, fuel temperature and fuel volatility that result in flash boiling, the spray dispersion and droplet size are significantly influenced. The effect of flash boiling on mixture homogeneity at the time of spark is found to be sensitive to injection timing. Mixing between the air-borne droplet/vapor cloud and the surrounding charge is observed to be incomplete for some of the fuels and injection timings studied, which suggests that adequate droplet dispersion in the initial phases of fuel injection is

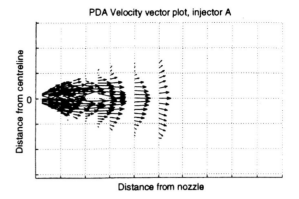

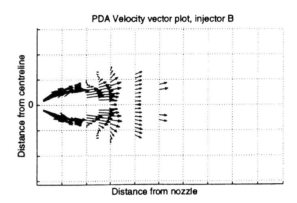

Figure 5.3-7 Spray velocity fields for a solid-cone spray (A) and a hollow-cone spray (B); phase-Doppler spray measurement [275].

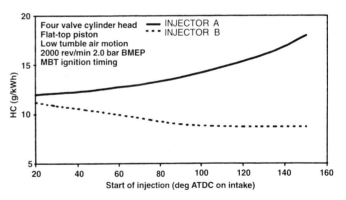

Figure 5.3-8 Effect of the spray characteristics on HC emissions for early injection [275].

a critical requirement for the formation of a good homo-geneous charge in a G-DI engine. The spray characteris-tics typically produced by a hollow-cone swirl injector may not be ideal for generating a fully homogeneous charge under medium to full-load operating conditions, even though it may be excellent for stratified-charge operation [(247, 248)]. As a summary, Table 5.3-2 provides the guidelines for an optimal mixture preparation strategy.

5.4 Spray-Wall Interactions

Within the rather small confines of a combustion chamber, millions of fuel droplets having more than a 50:1 range in both velocity and diameter are introduced with each G-DI injection event. A substantial fraction will achieve terminal velocity and vaporize well before any chamber boundary surface is reached. These are primarily the smaller and slower drops. Another fraction that is substantial for wall-guided systems, and hopefully very small for well-designed spray-guided and air-guided systems, will eventually come into proximity to a chamber

boundary surface. These are generally the largest and fastest drops. It is the latter class of drops that will be addressed in this section. These drops experience some effect of the chamber surfaces within a boundary zone, with some fraction being deflected and the remainder actually impacting.

The impingement of a fuel spray on the surface of a piston cavity is intentionally utilized to establish a stratified-charge mixture in wall-guided G-DI combustion systems. For such systems, and for spray-guided and air-guided systems, some degree of unintended fuel wetting on the chamber and cylinder surfaces is inevitable. With such wall wetting expected, an improved understanding of the spray-wall interaction process is

Table 5.3-2
Guidelines for an optimal mixture preparation strategy

Injector and Spray Characteristics Requirements	• Appropriate spray cone angle to insure good air utilization for early injection as well as to avoid fuel wetting of the cylinder wall and the piston crown outside of the bowl • Appropriate spray penetration characteristics to insure good air utilization for early injection while avoiding wall impingement • Minimum sac volume and no after-injections • Clean closure without 60+ μm drops or ligaments • Sufficient injection rate and low-pulse-width stability to reliably inject small fuel quantities
Spray Targeting	• Optimized spray-axis angle to achieve a combustible mixture at the spark plug gap while avoiding unintended wall impingement
Injection Timing	• Injection timing for the early injection, homogeneous operating mode should be advanced to take full advantage of in-cylinder charge cooling • Spray tip should "chase" the piston to minimize spray/piston-crown impingement during early injection • Injection timing for the stratified-charge operating mode should be as retarded as possible to minimize excessive fuel diffusion • Injection timing should be advanced enough to enable reliable ignition and avoid smoke emissions
Chamber Design	• Combustion chamber geometry and piston crown/cavity shape should be optimized to match the spark plug location, injector location, spray geometry, and spray orientation • Swirl and/or tumble should be optimized for the entire range of operation

extremely important in the G-DI development process to either better utilize the intentional spray-wall impingement in the wall-guided concept or to adjust parameters to avoid the unintended fuel-wall wetting.

5.4.1 Interactions of G-DI Sprays with Interposed Surfaces

The generic interaction of a transient G-DI fuel spray with an interposed surface is depicted schematically in Fig. 5.4-1. The interaction dynamics are, of course, very time dependent, and are illustrated for a time that is well into the injection event. In such interactions the upper plume of the spray toward the injector tip is basically unaffected in terms of its geometry and spray characteristics; that is, an image of the spray taken without the impact surface present can be overlaid on that region, as can drop-size and velocity profile data. Although the schematic diagrams represent 90° and 45°, for the generic interaction the impact angle of inclination and impact distance along the injector axis (or spray axis for an offset spray) may be any values. The surface may be initially wet or dry, hot or cold, rough or smooth, moving or stationary, and may or may not have deposits that adsorb and desorb liquid and vapor fuel. Each of these variables may have any value within a wide range for a particular interaction.

The impact zone, or more correctly, the interaction zone, is first manifested as a small toroidal vortex that expands radially outward from the initial impact point or points and transfers a portion of the momentum of the original spray. This vortex maintains a near circular form relative to the surface even if the impact is on an angle. For an inclined spray, however, the center moves down the plate from the initial impact point. If the injector produces any form of sac or pre-spray or multiple sprays, then the leading edge of the sac spray, main spray finger or individual multihole spray each produces a toroidal vortex that expands radially along the surface. This vortex is not necessarily the result of an impact, but only momentum exchange, as the toroidal vortex occurs even for a transient, gaseous jet. The overall appearance of the interaction event, either the side view or the top view, is little influenced by the details of the surface condition. The bulk motion of the deflected spray plume and the position in space at a given time is only very slightly affected by surface wetness, temperature or roughness. For example, very precise measurements indicate that the spray plume velocity along the surface is about two to four percent higher if the surface temperature is 130°C instead of 20°C. Such slight effects are difficult to measure, as they may be on the order of the typical injection-to-injection variability of the spray itself.

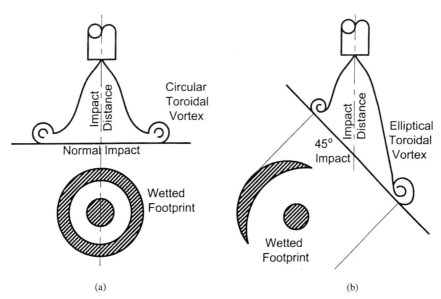

Figure 5.4-1 Schematic of the generic interaction of a transient G-DI fuel spray with an interposed surface (room temperature; 0.1 MPa ambient back pressure): (a) normal impact; (b) 45°-inclined impact.

A schematic representation of the range of phenomena that can occur for a G-DI spray interacting with an interposed boundary is provided in Fig. 5.4-2. For this representation an infinite flat plane is placed in the path of a transient G-DI spray that would otherwise be developing in a free, unconstrained environment. The side view of the interaction in Fig. 5.4-2a illustrates the zone discussed above in which there is little or no effect of interposition of the blocking plane. This is found experimentally to be in the range of 5 to 8 mm above the surface. The spray geometry, droplet sizes and droplet velocities measured at points above this second vertical plane are unchanged when the blocking plane is alternatively inserted and removed. A number of possible phenomena that occur within the zone of influence of the blocking plane are illustrated. The side view shows both an upstream and a downstream vortex, but these are merely a two-dimensional, cutting-plane representation of the toroidal vortex that is moving radially outward with time. There may be one or more wetted areas in which a liquid film is present. The pattern shown is representative of that obtained for a swirl-type injector under bench-test conditions; however, the shape of the wetted footprint will differ significantly among injector types. A multihole injector having eight holes can yield eight individual toroidal vortices and wetted areas for bench-test conditions. Each wetted area has the potential to provide splashed drops, generated when drops of sufficient momentum and Weber number impact a liquid film. A portion of the film may be reatomized, producing drops

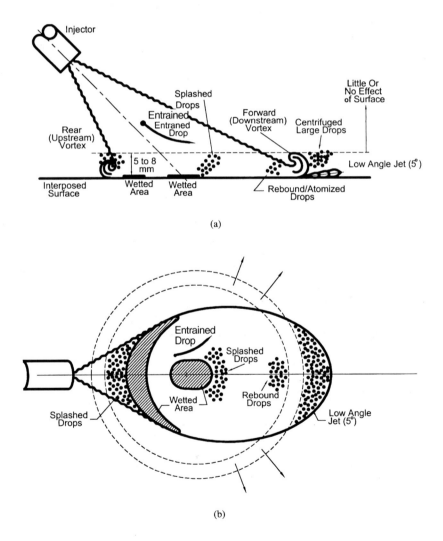

(a)

(b)

Figure 5.4-2 Schematic representation of the range of phenomena that can occur for a G-DI spray interacting with an interposed boundary: (a) side view of the interaction; (b) top view of the interaction.

having size and velocity distributions that are distinctly different from those of the original spray. Other drops in the spray may be entrained in the air flow field that is established by the injection event. All momentum normal to the plane is dissipated, and the drop moves parallel to the plane in the entrainment flow field. Those drops that retain some momentum normal to the plane will impact it, and if the impact is in a location having no liquid film, three additional outcomes are possible, depending upon the droplet Weber number at that time. First, the drop could adhere to the surface, thus initiating a new site for a film. Second, the drop could rebound without breaking up, maintaining some portion of the original momentum. Finally, the drop could break up into a set of smaller rebound drops, which is a reatomization that results from the impact event.

If the interaction produces any actual impact of liquid contacting the surface—and it is important to note that it may or may not—then two additional generic features will be present. The first is a low-angle (approximately 5°) secondary stream of generally larger droplets that is experimentally observed to occur at the leading edge of the deflected spray plume. These drops are distributed in the form of a fan that is centered in the direction of the impact. Most of the droplets are located at or near the center of the fan, with the droplet number density decreasing at increasing off-center distances. The second generic feature is a collection of larger droplets at an intermediate angle (30° to 40°) to the impact surface. These features are generally present for all impacting G-DI sprays, regardless of the impact angle or whether the surface is wet or dry, hot or cold, or rough or smooth. However, the droplet number density and size distribution for each of the two sets of droplets are significantly influenced by the above parameters, with larger and more numerous droplets generally resulting from wetter, colder and rougher surfaces. For G-DI sprays that impact at angles of 50° or less, the low-angle jet is usually detected at the leading downstream edge of the toroidal vortex. This thin, fan-shaped jet and the cloud of larger leading-edge drops is observed for both initially-dry surfaces and for surfaces with pre-existing liquid films. The jet and the large drop cloud are more pronounced for the case of an initially wet surface, but are readily detectable even if a dry surface is impacted. The presence of reatomized droplets having sizes that exceed those in the original spray is *ad hoc* evidence of splashing, and, when present for an initially dry surface, is evidence that later-arriving droplets impact the liquid film created by earlier-impacting droplets. More research needs to be conducted to identify both the physics and the thresholds that will fully explain the observed phenomena. This will permit accurate correlations to be made, and will provide valuable information that is needed to enhance CFD submodels.

The interaction of the spray from a 60° G-DI injector with a piston bowl of a wall-guided combustion system is illustrated in the laser sheet photos in Fig. 5.4-3. The injector operating temperature and the indolene fuel are at 90°C, and the ambient back pressure is 0.25 MPa. The injector axis is targeted at the cutaway bowl floor at 45°. The impact distance along the spray axis is 35 mm and the bowl floor is initially dry. At 0.7 ms after the start of the fuel logic pulse the image in Fig. 5.4-3a shows that the spray front is just

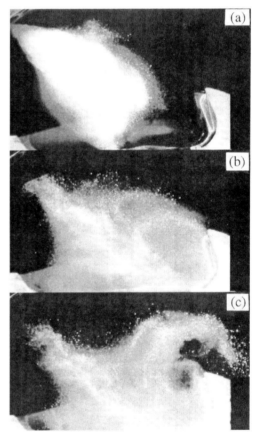

Figure 5.4-3 Interaction of the spray from a 60° G-DI injector with a piston bowl of a wall-guided combustion system (fuel temperature: 90°C; ambient back pressure; 0.25 MPa; spray inclination angle to the bowl floor: 45°): (a) 0.7 ms after the start of injection; (b) 2.2 ms after the start of injection; (c) 4.0 ms after the start of injection.

contacting the bowl floor. The initial contact is not on the spray axis, but is displaced toward the injector, with the maximum wall wetting occurring there. In the image in Fig. 5.4-3b, the forward portion of the bowl is filled at 2.2 ms, and injection has been over for 0.4 ms. As shown in Fig. 5.4-3c, at 4.0 ms the fuel droplets are arcing out of the bowl under the influence of a complex flow pattern that has two counter-rotating vortices: one traceable to the toroidal vortex of the spray and the other due to motion up the bowl lip. In this example the interaction is not optimum, as the spray moves over and down the outside of the bowl lip. Also evident are the larger droplets at the leading edge of the spray exiting the bowl, which is the first fuel to arrive at the spark gap.

There are a number of important considerations regarding the interactions of G-DI fuel sprays with wall boundaries and piston bowl cavities, with the first

being that G-DI spray wall interactions are, for the most part, spray deflections rather than true impacts [272]. This is true even for wall-guided systems in which spray-wall interactions are designed to occur. When a boundary surface is interposed in the path of a G-DI fuel spray, the droplets very near to the surface, of course, do not have the same velocity and diameter history that are exhibited at that same point in a free spray. Figure 5.4-4 shows the velocity history for the arriving drops at a point in space as measured both in free space (Fig. 5.4-4a) and with an interposed surface plate (Fig. 5.4-4b). The drop velocity measured and plotted is the velocity component normal to the plate, with a positive velocity being downward, which is into the plate. For the data in Fig. 5.4-4b the plate is placed 3 mm below the phase-Doppler measurement point. It may be seen that the characteristics of the early portion of the spray are affected very little whether

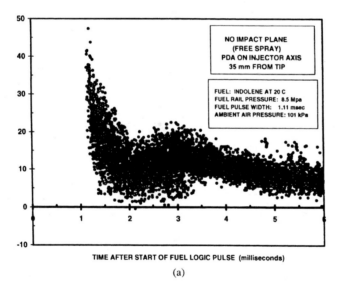

TIME AFTER START OF FUEL LOGIC PULSE (milliseconds)

(a)

Figure 5.4-4 Velocity histories for the arriving drops: (a) free spray; no inserted plate; (b) same spray and measurement location; spray interacting with 45°-interposed plate.

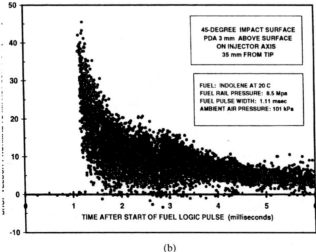

(b)

the plate is interposed or not. The latter half of the inter-action pulse does exhibit an effect of the presence of the plate, with larger drop velocities being measured. This is the result of a different entrainment flow field that develops when a plate is present. None of the measured drop histories, nor any photographs of sprays interacting with plates, verify either impact or non-impact on the interposed surface. Of the thousands of drops represented in Fig. 5.4-4b, it is not known which ones impacted the plate and which ones did not. If there is no wetted foot-print extending to this particular measurement location then it is indeed known that none of the drops wetted; however, if some wetting occurs then further analysis and correlation are required to estimate the thresholds that apply. Correlations resulting from research on the phys-ics of the interaction process are used to establish the momentum and Weber number criteria that must be sat-isfied for a drop to impact and either wet the surface, rebound or reatomize.

One simple misinterpretation regarding G-DI spray-wall interactions is that the bulk of the spray impacts the piston cavity surface, forming a substantial liquid film, and that this liquid film moves along the surface under the influence of the in-cylinder air flow field. In reality, sprays from nearly all types of G-DI injectors at a typical mean impact distance of about 35 mm wet the surface very little. Even for bench tests at room temperature and ambient pressure, with injection normal to the surface, only about 5% to 14% of the injected fuel mass actually forms a liquid film on the interposed surface. For injec-tion at an inclined angle relative to the interposed sur-face, even less wetting may occur. Less wetting also occurs if the injector and fuel temperature are elevated, or if the ambient density is increased. Even at room ambient conditions the vast majority of the millions of fuel droplets, most under 9 µm in diameter, do not actu-ally contact the interposed surface, but are instead entrained in the injection-generated flow field, and move as particles suspended in the transient gaseous flow along the surface. Very small droplets have high relative drag, and rapidly transfer their momentum to the surrounding air. This means that a very substantial transient air flow field is initiated within about a millisecond, and that the droplets rather rapidly obtain terminal velocity relative to that flow field. In fact, for the typical distributions of droplet velocity and droplet diameter only the very larg-est, fastest and earliest arriving fuel drops impact the sur-face. These large droplets contain a disproportionate amount of fuel for such low numbers of impacting drops.

The vast majority of the droplets (99.9%) do not have sufficient momentum normal to the interposed surface, and do not wet the surface, but are instead deflected at some time during a 5 ms time period by the entrainment air flow field, all the while losing mass and diameter due to vaporization.

Analytical studies of fuel-wall impingement asso-ciated with droplet over-penetration in G-DI engines have been conducted by computing the drag coefficients of the droplets [82, 83]. The penetration distance for impact was assumed to be 20 mm for late injection and 80 mm for early injection. For the case of early injection of a spray having an SMD value of 15µm, most of the droplets in the spray were predicted to decelerate to a very low velocity prior to reaching the piston crown. Simi-lar results were also found for late injection, in spite of the significantly reduced penetration distance that is avail-able before the spray impacts the piston. The rapid drop-let deceleration is due mainly to the higher air densities that are encountered for the late injection condition, which results in increased droplet drag and enhanced vaporiza-tion rates. It was predicted that an increase in droplet size and a decrease in spray cone angle increases the amount of fuel impinging on the piston crown.

For a typical G-DI fuel spray in a wall-guided sys-tem the approximately 8 million droplets in a 1.5 ms injection pulse interact with the interposed boundary sur-face, and the majority are influenced by it over a period of about four to six milliseconds. The overall features of such an interaction at 45° to the impact boundary are shown in the images in Fig. 5.4-5 for a 60°, swirl-type injector and a 35 mm impact distance. At 0.70 ms the main spray is just approaching the plate, which is dry and unheated. At 1.30 ms the spray plume is interacting with the boundary surface, and wetted areas are formed. Drop-let size and velocity data in sprays just a few millimeters above the impact surface verify that the effects of the boundary are not propagated very far upstream into a tran-sient G-DI fuel spray. In fact the detectable zone of change from free space conditions expands only 5 to 8 mm above the surface. Measurements at the same point in the spray only 3 mm above the plate location made with and with-out the plate present show little or no difference for approximately the first 1.5 ms. At later times, of course, the droplets move in the diverted flow field and increas-ingly show the effect of the plate. The image at 2.20 ms shows the interaction continuing at the time of the end of injection, with an arc of the normal toroidal vortex being clearly evident. On the back edge the toroidal vortex is

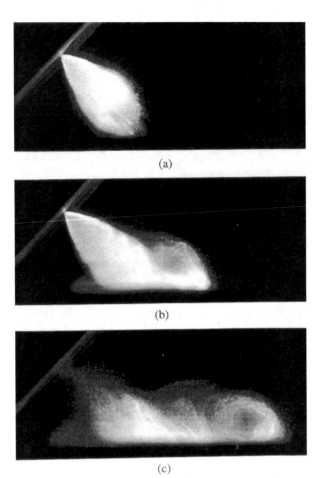

Figure 5.4-5 *Overall features of the spray-wall interaction: 45° to the impact boundary; 60°, swirl-type injector; 35 mm impact distance: (a) 0.7 ms; (b) 1.3 ms; (c) 2.2 ms.*

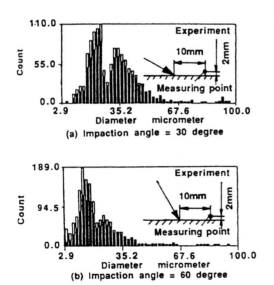

Figure 5.4-6 *Effect of spray inclination angle on droplet size distribution after impingement* [332].

not as prominent because the plate is encountered prior to full entrainment being established. It is worth noting that the orientation of injector axis relative to the plate surface has a significant effect on the droplet size distribution downstream from the impingement point, as illustrated in Fig. 5.4-6 [332]. These phase-Doppler measurements show a bi-modal drop-size distribution with larger drops measured for a shallower impact angle. All drops including original, splashed (reatomized) and rebound are included, and much additional rescarch is required to separate the droplet origins.

The magnified images of the tip region of a 60°-cone spray impinging on a cold flat surface are shown in Fig. 5.4-7. The images are taken using a distance microscope in combination with shadowgraph imaging. The liquid fuel shows up as fine dark dots

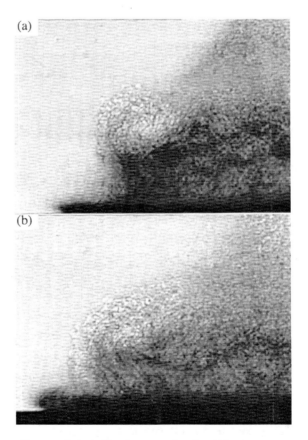

Figure 5.4-7 *Shadowgraph of a 60°-cone spray impinging normally on a cold, flat surface at a distance of 38 mm downstream; injection duration: 1 ms; ambient pressure: 0.1 MPa; image dimension: 1.2 mm × 1.5 mm; (a) image taken at 2.2 ms after start of injection; (b) image taken at 2.5 ms after start of injection.*

within the silhouettes, whereas the vapor phase is manifested as a layered eddy structure. It is interesting that closer to the wall, a layer of wall-jet-like spray with a thickness on the order of 0.1 mm is formed, which propagates along the wall at a speed of greater than 50 m/s. The circulation of the expanding toroidal vortex is evident in the images. Phase-Doppler measurements that are taken at a position such that this vortex moves through the test point show that the droplets are smaller near the center of rotation. This indicates that larger droplets are centrifuged within this propagating vortex, either moving to the periphery or being ejected completely.

A comparison of the spray-wall interactions for cold and hot impact surfaces is presented in Fig. 5.4-8. The interaction sequences are also obtained using a shadowgraph configuration, with the left sequence being for a 23°C surface temperature and the right sequence for 160°C. The imaging times for corresponding panels are identical. The sequences verify a point of discussion that was noted in prior sections: that the overall bulk interaction of a spray with an interposed surface is only slightly influenced by the surface temperature. The position of the deflected spray tip is seen to be somewhat further advanced for the heated plate, indicating a slightly higher propagation velocity along the heated surface. Otherwise, the time sequences of the interaction are quite similar [355].

The earliest droplets to arrive are the least influenced by the entrainment air flow field, as this field is just being established. Either with or without a sac spray present, they are generally the largest drops in the spray. The sac spray normally has a DV90 value that is significantly larger than that of the main spray (typically 45 μm versus 30 μm), and even without a sac volume the drops associated with the pintle opening process are larger than those that are generated later. In addition to being the largest, these early drops also have the highest velocities, as is clearly demonstrated in plots of time-resolved droplet arrival measurements. The sac spray thus has the largest and fastest drops that arrive at the potential impact surface before the entrainment air flow field transverse to the surface is fully established. The main criterion for impact versus no-impact is the droplet momentum normal to the surface in the developing transient air flow field. This momentum, if sufficient to overcome the developing air flow field along the plate, carries the droplets to the surface where, depending on the droplet Weber number and the angle of approach relative to the

surface, they may either form a liquid film, rebound or interact with the film from prior droplets to create splashed droplets. This was illustrated schematically in Fig. 5.4-2. Any splashed droplets will have a size distribution that differs from that of the original impacting droplets, and may contain drops that are larger than any of those in the original spray. Thus the main parameters for the minority of droplets that actually impact are the droplet Weber number, the angle of approach and the properties and thickness of any liquid film at the time of contact.

Another misconception is that there is very significant splashing and droplet rebounding from surfaces that are interposed in the path of the G-DI fuel spray. This viewpoint may at first seem logical based on the very elevated fuel rail pressures that are used; however, these fuel pressures produce very small drops having high drag, rapid vaporization rates and enhanced momentum transfer to the surrounding air. Rather than increasing droplet impact rates and splashing, an increase in fuel pressure generally diminishes these effects. An additional factor contributing to the misconception is the observation that single-droplet impact tests and PFI fuel spray tests *do* generally indicate splashing and rebounding, particularly for PFI fuel sprays that impact a pre-existing liquid puddle. For G-DI fuel sprays, however, there is no pre-existing liquid film prior to injection, and the vast majority of the droplets attain terminal velocity in the flow field established by the injection event. The only liquid film that could be impacted is that formed less than 2 ms earlier by prior impacting droplets in the same injection event. Figures 5.4-9 and 5.4-10 show the momentums and Weber numbers, *We*, for 7200 measured droplets from a swirl-type injector arriving at an on-axis point that is 35 mm from the tip. These are the same droplets for which the velocity normal to the plate was shown in Fig. 5.4-4b.

Although the Weber numbers of the individual spray droplets are plotted in Fig. 5.4-10, and thresholds of Weber number may be experimentally determined for G-DI spray rebound, splash, film entry and reatomization, correlations that include the Ohnesorge number, *Oh*, may also be developed. The dimensionless Ohnesorge number represents the ratio of the liquid viscous shear force to the droplet surface tension force, with a large *Oh* number corresponding to a very viscous droplet. The Weber number represents the dimensionless ratio of the distorting aerodynamic drag force on a droplet to the surface tension force that acts to maintain the droplet as a sphere. By combining the dimensionless

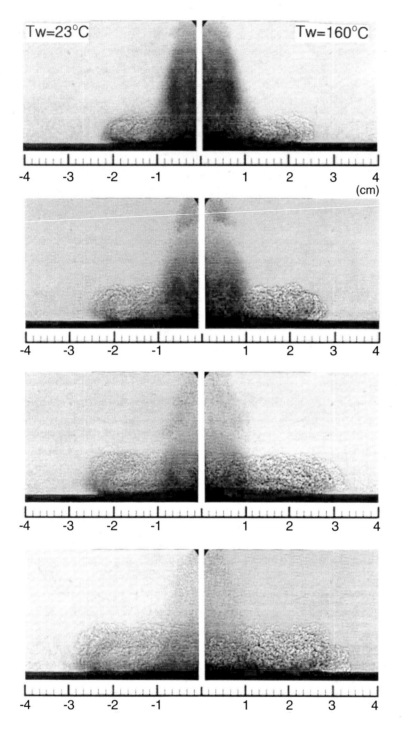

Figure 5.4-8 Shadowgraph of a 20°-cone spray impinging normally on a flat surface at a distance of 3.8 cm downstream; injection duration: 1 ms; ambient pressure : 0.1 MPa; left panel: $T_w = 23°C$; right panel: $T_w = 160°C$; (a) image taken at 1.9 ms after start of injection; (b) image taken at 2.1 ms after start of injection; (c) image taken at 2.3 ms after start of injection; (d) image taken at 2.5 ms after start of injection.

numbers, the ratio of the aerodynamic drag force (which is distorting the droplet) to the viscous shear force (which is resisting droplet deformation) may be expressed as $We^{1/2}/Oh$. Experimentally, the effects of droplet liquid viscosity on breakup have been correlated using a number of formulations, but $We/Oh^{1/4}$ has been successfully employed over a wide range of viscosities. For the typical range of fuel compositions and droplet temperatures at impact that are encountered in G-DI sprays, the Ohnesorge correction is small, thus Weber thresholds will suffice. In order to summarize the very important area of G-DI spray-wall wetting, it is important to be aware of the considerations in Table 5.4-1. These observations should be of significant benefit in understanding spray

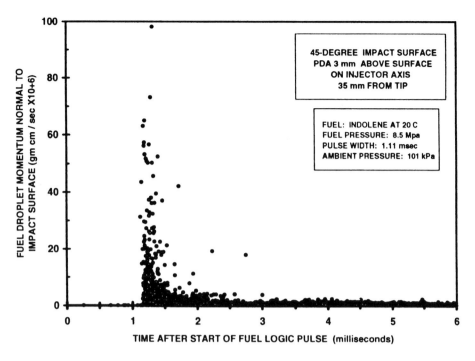

Figure 5.4-9 Time-resolved drop momentum for drops arriving from a swirl nozzle.

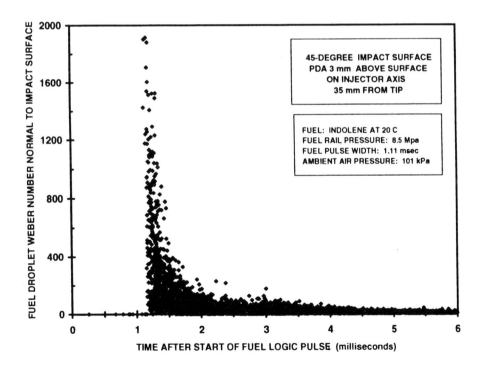

Figure 5.4-10 Weber numbers for 7200 measured droplets from a swirl-type injector arriving at an on-axis point 35 mm from the tip.

Table 5.4-1
Important considerations regarding the interpretation and modeling of G-DI spray-wall interactions and wall wetting

- In general, it is the largest, fastest and earliest droplets that wet an interposed surface.
- The vast majority of droplets never impact the surface.
- Most droplets are entrained into the spray-generated air flow field that exists for a total duration of 4 to 7 ms.
- The wetted footprint does not correspond to the spray cross-section at the impact location; it is a smaller sub-set of the cross-section.
- Typical engine air flow fields may move the apparent visual spray boundary, but have little or no effect on the location or geometry of the wetted footprint.
- Splashing is very much reduced as compared to PFI fuel sprays, as there is no pre-existing liquid film when injection is initiated.
- The droplet momentum relative to the surface and the droplet velocity relative to the transient entrainment flow field may be used to determine impact probability.
- For droplets that impact, the droplet Weber number may be employed to correlate experimental thresholds for rebound, splash, reatomization and film entry.
- A viscosity (Ohnesorge) correction is normally not required for G-DI applications.
- Weber number thresholds that are obtained from single-droplet impact tests cannot be used directly; further research is required to refine the correlations for G-DI sprays.

targeting in wall-guided systems, and unintended wall impingement in air-guided and spray-guided systems, and in obtaining detailed experimental data for use in Weber number correlations and spray-wall sub-model tuning.

In concept the general determinant of impact will be the momentum normal to an interposed boundary surface of an individual drop in the immediate proximity of the boundary. If this individual momentum is above a certain threshold value, then the drop will not be slowed to such a degree that it becomes entrained in the air that is moving along the boundary. Further, for only those selected drops that exceed the normal momentum threshold and contact the boundary, the determinants of whether the drop forms a liquid film, rebounds without volume change, rebounds with breakup, or enters an existing liquid film and splashes (reatomizes) is the droplet Weber number normal to the boundary surface and the local angle of incidence. Also required is a knowledge of the boundary properties, which includes the state of any liquid film that exists when the drop arrives. In concept, prediction of the fate of an impacting drop invokes a series of thresholds,

with each window between thresholds being indicative of a possible impact outcome. Such a threshold for momentum, coupled with a series of thresholds for Weber number and the surface conditions, could readily be applied to time-resolved PDA data, which provide the required starting information. However, a number of practical difficulties are encountered in applying such a sequential screening procedure to real G-DI fuel sprays. First, it is very difficult to establish the correct thresholds for transient interactions that involve millions of droplets and lasts only milliseconds. This type of droplet pulse and transient air entrainment flow field can significantly alter any thresholds that might be measured for a single drop approaching and impinging on a surface. The rebound of a drop upstream is much more difficult with a counterflowing entrainment flow and tens of thousands of close drops moving opposite to the rebound. For these and other reasons, the thresholds will change with time during the pulse, thus there is no fixed set of thresholds that can be applied for the entire injection event. The very first drop to arrive at a boundary experiences little or no entrainment flow along the surface due to the injection event, as the transient entrainment flow field is just being established. Two milliseconds later, however, the air flow field along the surface is much more established, and a higher drop momentum normal to the surface is required for impact. A drop having the same diameter and velocity as one that impacted two milliseconds earlier may now be entrained and made to move along the plate without impacting. These simple examples illustrate some of the difficulties that must be addressed in the near term if wall wetting is to be accurately predicted by spray-wall sub-models. Ideally, data from a spray in free space would be obtained for a location such as is depicted for one location in Fig. 5.4-9. Data obtained with a real impact surface in position could be used to check model predictions, but it is impractical to set up such a real impact test each time an impact is to be predicted. The free-space spray data could be used in combination with modeling to predict the transient entrainment flow field for a combustion chamber geometry, as well as the time-varying thresholds for momentum and Weber number. The early droplets, particularly the initial sac spray droplets, certainly have significant momentum, and *can and do impact* any surface that might be interposed at that location. It is evident that many of the early drops will *rebound* and *splash* in addition to forming a liquid film. However, the fraction of the 8 million droplets in the spray that actually contact the surface is minuscule, although the mass fraction can be significant due to the cubic relationship between droplet diameter and the fuel mass of a droplet.

5.4.2 Spray-Wall-Interaction Phasing Issues

The interpretation of G-DI combustion data will be facilitated by an appreciation and understanding of the phasing of spray and wall-interaction events that occur in the engine. This is important in applying the data for a specific injector to the optimization of system performance, and in interpreting the results of various test hooks that are run on the engine. Whether the system is wall-guided, air-guided or spray-guided, the combustion chamber boundaries represent a confining boundary that is changing slightly with time relative to the development of the spray. Characterizations that are based upon free-space tests, even those that are obtained with real fuels and actual operating temperatures and pressures, must subsequently be applied to an injection event that is very much more constrained. The free-spray penetration curve and spray cone angle are not perfectly applicable to in-chamber injection, since the spray decelerates as it approaches a boundary surface. The symmetric toroidal vortex that is observed for a swirl spray in free space may not develop at all on the side nearest a piston crown or bowl cavity, thus altering the entrainment field and spray development. If the CFD spray model is very good, the model may be used to match the transient spray characteristics in free space, then extended to predict the spray performance that is modified by the highly constraining boundaries. If such a model is not available, then the information that follows should be considered in applying spray characterization to interpreting combustion data.

The fuel injection event and the interaction of the spray with combustion chamber surfaces are multi-step events. Each step requires a certain amount of time, which translates to a crank-angle interval in the engine. As has already been described in Section 3.6, an understanding of these phase shifts begins with the command from the EMS for the injector to open. The start of this command is denoted as the start of the logic pulse (SLP) point. This square wave is sent to the injector driver, where an optional driver delay time may be built in to ensure full voltage on the capacitor. In practice this precise fixed delay may be any of the following: zero, 0.25, 0.50, 0.75 or 1.00 ms, and is determined by the driver manufacturer. Following the fixed delay a modified pulse (no longer a square-wave pulse) that is specific to each manufacturer is sent to the armature-actuation unit of the injector. Normally this is a solenoid, but may alternatively be a piezoelectric stack. The time required for the pintle to first lift from the seat after the driver pulse is initially sent to the armature actuator is denoted as the mechanical opening delay. This point in time is the true start of injection, or SOI, with the first fuel appearing at that time at the exit orifice of the injector tip. It may range from 0.15 to 0.48 ms for solenoid injectors, but is an order of magnitude shorter for piezoelectric actuators. Although the range of mechanical opening time is large among injector designs, for a given injector it generally varies only a few microseconds from injection to injection or with operating conditions. For inwardly opening injectors it generally increases up to 5% as the fuel rail pressure is increased by 15%, as the opening is opposed by the fuel pressure. It should be noted that from the standpoint of acquiring combustion data from an engine on a dynamometer, the SLP and/or ELP is almost always recorded, as it is an easily acquired electrical signal, whereas the true SOI is not known and is not recorded. The SLP is often assumed to be synonymous with SOI, which is definitely not valid. Simply, the ELP is non-synonymous with EOI.

As illustrated in Fig. 5.4-11 for typical light-load and idle conditions, the fuel spray exiting the injector is displaced significantly in time and crank angle from the

EMS command pulse, with 3 to 10 crank angle degrees being very typical. At high engine speeds, or with a substantial driver capacitive charge time, the phase difference between SLP and SOI can exceed 20 crank angle degrees. For closing, the process is reversed, except that there is no capacitive driver delay. The end of the logic pulse (ELP) is transmitted with negligible delay to the

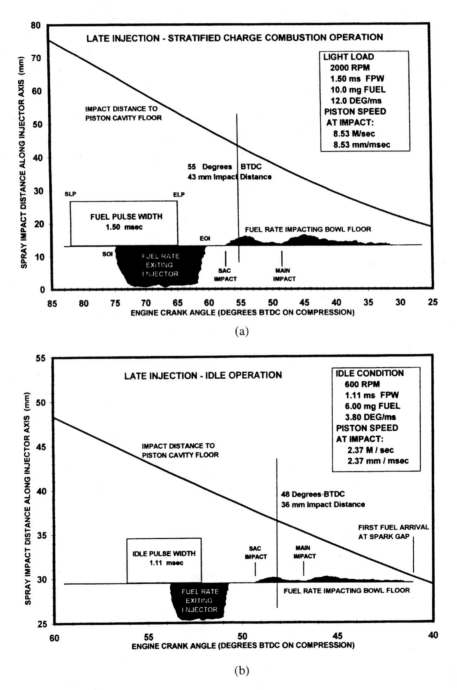

Figure 5.4-11 Light-load fuel injection timeline and impact distance variation: (a) light load at 2000 rpm; (b) idle at 600 rpm.

armature actuator, thus cutting off the current flow. For solenoid injectors the magnetic field then begins to collapse and, at a threshold level of magnetic force, the return spring force begins to move the pintle toward the seat. The point at which the pintle is initially seated is known as the end of injection or EOI point. The time from the ELP to the EOI point is known as the mechanical closing delay, and can range from 0.30 ms to as much as 0.80 ms for a solenoid-type G-DI injector. Most contemporary injector designs operate in the 0.30 to 0.40 ms range.

Two very interesting observations may be made regarding the typical data represented in Fig. 5.4-11a. The first is that the time interval during which injection actually occurs is *not at all* equal to the duration of the fuel logic pulse. In fact it is the FPW, minus the driver delay, plus the closing time, minus the opening time. The second observation is that at light loads the fuel is *still* in flight toward the piston cavity floor or spark gap, even though the fuel logic pulse and entire injection event are over. This is true in spite of the fact that the distance between the injector tip and the impact surface is decreasing rapidly due to the piston velocity of about 8.5 mm/ms. Injection is over at EOI, but the first droplets from the sac volume do not reach the bowl floor for another 3 crank angle degrees. It is very typical for light-load and idle operation that the injection event is completed before any fuel reaches the combustion chamber surface, even for a wall-guided system. In panel A, the sac spray is seen to interact with the piston bowl for about 8 crank angle degrees, with the largest and fastest droplets creating a wetted area. More than 30 crank degrees after the injector was first commanded to inject, the first droplets from the main spray arrive at the piston bowl floor. Droplets from the main spray continue to arrive at the bowl for the next 16 crank angle degrees, with the piston moving ever closer to the injector tip. The schematic representation shows the main spray continuing to impact and wet the bowl floor; however, nearly all of the smaller, lower-velocity droplets in the main spray are deflected, and wet very little, if at all. For a wall-guided system the spray plume traverses the piston bowl floor to the spark gap for another 1.0 to 1.5 ms, which corresponds to an additional 12 to 18 crank degrees.

Figure 5.4-11b illustrates the injection event phasing for idle operation, with the main differences from light-load operation being the lower quantity of fuel injected and a reduced engine speed. Even though injection is started later on compression, and the impact distance is less, the observations made for light-load operation are also found to be valid for idle. The first fuel does not arrive at the spark gap for 4 ms after the injector is first commanded to inject, which corresponds to 15 crank angle degrees. Thus, there are important phasing relationships to be considered in evaluating combustion data from hooks of injection timing or spark timing, and a basic understanding of the spray-event timings can be very beneficial. The use of ELP timing for combustion tests on engine dynamometers circumvents the accounting for some phase delays, such as the driver and mechanical opening delays, but there are still closing delays and flight times to be considered in interpreting combustion events.

5.4.3 Spray Wetted Footprints

The wetted footprint of a G-DI fuel spray is an important consideration for a combustion system development program, as it is a metric for spray targeting. As discussed in the previous section, spray droplets that have sufficient momentum when nearing the interposed surface will impact that surface. Here, "sufficient momentum" means that the droplet has sufficient initial velocity and is not slowed by viscous dissipation to a velocity level where it can be entrained in either the swirl or tumble air flow field or the transient entrainment air flow field that is initiated by the injection event. Two important considerations as to whether this is achieved for any droplet of any initial velocity are the flight distance and the time within the injection event (early or late). For all but the narrowest sprays there are significant differences in flight distance between the spray inside edge and outside edge, particularly when the spray axis is inclined to the interposed surface. As the change in penetration rate is normally very non-linear with the distance traveled, flight distance is a key parameter influencing the level of arriving droplet momentum.

With regard to the time within the injection event, droplets are exiting the nozzle at all times during the injection event. It is very important to note that early droplets do not experience the same air flow field as later ones, as the transient entrainment flow field undergoes rapid development when injection is initiated. This side-drag force, which tends to turn a droplet from a straight trajectory and move it transverse to the plate, is much different at 3 ms than at 1 ms. Late droplets that arrive at the potential impact surface are almost always of lower mean velocity than earlier droplets, partially due to injection dynamics and partially due to aerodynamic filtering of the velocities; that is, even if droplets having the same size and velocity distributions are introduced at the injector tip at two different moments, measurements at a downstream location will still show the larger and higher velocity droplets preferentially arriving first. The slower droplets will arrive downstream later, and even the smaller, faster droplets will be slowed to a greater degree than the large droplets, and will arrive later. Thus the droplets that arrive first at a downstream location near an interposed surface will almost always have a larger SMD, a higher mean velocity and a straighter trajectory. Even if the injector design does not produce a sac spray, the droplet arrival history at a downstream location can exhibit the appearance of a sac spray because of these effects. The transient entrainment air flow field will have minimum influence on these droplets, whereas the last arriving droplets are greatly influenced and have a lower chance of impacting. Thus, wall wetting preferentially occurs early in the injection event as opposed to later.

Perhaps contrary to perception, the existing in-cylinder air flow field due to inlet-generated swirl or tumble is found to have little or no influence on the wetted footprint. This perception perhaps has its source in in-cylinder spray videos that show the G-DI spray plume being substantially deflected by an air crossflow.

The fallacy is that the smaller, lower velocity drops are deflected, but these would not wet anyway. For typical G-DI targeting distances of 28 to 44 mm, neither the location nor the shape of the wetted footprint is altered to any significant degree by an existing crossflow or transverse flow. Even the extreme case of G-DI injection into the crossflow from a shop-air-nozzle blast only moves the wetted footprint about 2 mm to the side for typical G-DI sprays. Of course, the highly visible smaller droplets of the spray are deflected by such a crossflow, which gives the appearance of moving the entire spray to one side. The very fact that the typical G-DI spray-wetted footprint is little affected is verification that there is a selected subset of fuel droplets that have the required momentum and straight trajectory to impact the surface regardless of the crossflow field or the entrainment flow field. That this is a rather sharp cutoff is evident from the fact that the introduction of such a crossflow does not substantially alter the fringes of the wetted footprint; otherwise continuous increases in the crossflow velocity would be manifested in movement of the footprint fringes on the upstream side.

The spray wetted footprints presented in Fig. 5.4-12 illustrate the variations that can result from changes in injector type, angle of inclination and operating temperature and back pressure. The most significant

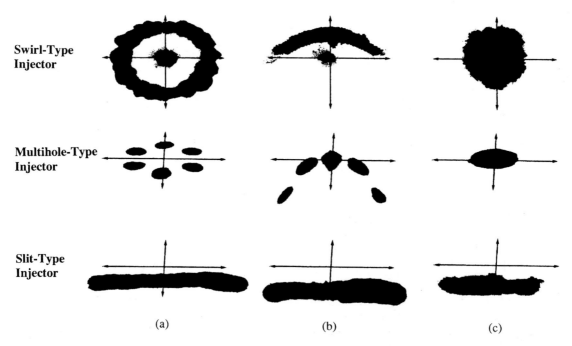

Swirl-Type Injector

Multihole-Type Injector

Slit-Type Injector

(a)   (b)   (c)

Figure 5.4-12 Examples of wetted footprints for sprays from swirl, multihole and slit-type injectors: (a) orthogonal injection; (b) 45°-inclined injection; (c) orthogonal injection from hot injector filled with hot fuel.

parameter is the flight distance to impact, with a secondary factor being the approaching angle. This is made evident by simply rotating the impact plane around the point of intersection of the spray axis, thus increasing the impact distance on one side and decreasing the distance on the other. For all G-DI fuel sprays the increase in wetted footprint area on the closer side is evident. Flight distances may be, for example, 35 mm on the spray axis, with 28 mm for droplets on the near edge of the spray and 50 mm for drops on the far edge. As may be seen in Fig. 5.4-12, this change is more than sufficient to drastically alter the wetting pattern, even though a higher velocity crossflow air does not. On the far side of the swirl spray there is no longer any wetting, whereas the near side experiences a much wider footprint. It should be noted that such footprints contain important information as to *where* wall wetting occurs or does *not* occur, and as to the exact pattern, but do not provide quantitative information as to the total fuel mass or its distribution within the footprint. If 2% to 12% of the total injected fuel mass for a light-load operating point wets the impact surface and forms a film, then the mass of fuel in the film will range from about 0.25 to 1.3 mg. The total weight can be measured by utilizing a very low volatility fuel and weighing the detection paper using an analytical balance, but this can be done practically only for a bench room test, not for tests in an optical spray chamber at elevated temperature and pressure. Laser interferometric measurements of the instantaneous film thickness would most likely have to be utilized in conjunction with the total footprint area to obtain an indication of the fuel mass that wets the impact surface for a given combustion chamber configuration.

Substantial changes in the wetting pattern are seen for all but slit-type injectors when elevated operating temperatures and back pressures are used. The effect of the operating environment on spray development was discussed in detail in Chapter 4, but with regard to wall wetting it may be stated that the wetted footprints are observed to change markedly with transitions of the overall characteristics of the spray. Thus the observed collapse of the sprays for a multihole injector from six individual plumes to one is tracked by a change in the wetted footprint from six individual spots to one central larger spot. This may seem obvious with hindsight, but it certainly could have been theorized that the small subset of drops that wet the surface might have been independent of the collapse process. When the spray is observed to collapse to a new, narrower geometry, the wetted footprint is also found to coalesce into a more central position. There is an effect of the angle of approaching that should also be noted. If the impact distance is maintained constant at any value such as 35 mm, and the angle of approach relative to the impact surface is varied, then some variation in wall wetting does occur for G-DI sprays. However, it does not appear to be correlated by a simple trigonometric conversion that adjusts the droplet velocity to its value normal to the impact plane. A very small local target placed at an angle within the G-DI spray will be wetted regardless of the angle of inclination, whereas if it is part of a larger impact plane at the same location and angle it may not be wetted.

The intersection of a cone with an interposed plane is, of course, a classical conic section. Because the majority of G-DI sprays are relatively axisymmetric and relatively conical, and because the majority of intended impact surfaces such as piston bowl floors are relatively flat, the interpretation of spray footprints is often facilitated by evaluating them as conic sections. This simple consideration can help to explain and correlate a number of ostensibly complex patterns, even without requiring flow modeling. For most low to moderate engine loads and speeds the FPW is brief (under 1.8 ms) and the piston motion during the impact event is slight (under 3 mm), thus the cone/plane geometry is almost static, even in a running engine. Hence there is good agreement between footprint data from a pressure chamber test in the laboratory and spray impact outlines on pistons taken from engines that have run on the dynamometer for many hours. If the ambient density, injector operating temperature, injector mounting angle and impact distance are emulated in a laboratory pressure chamber, the footprints so obtained will correspond to the outlines in the carbon on a removed piston. This gives confidence that some portion of G-DI spray targeting studies related to bowl design, spray cone angle selection, spray-bowl overshoot and smoke reduction can be conducted in a spray chamber, which is generally quicker than dynamometer testing. The value of the wetted footprint is that it shows the precise location of the fuel that must undergo a transition from a liquid wall film to a vapor while confined to that spot, whereas the remainder of the fuel in the spray is not in a film and is not stationary. Final tests would still have to be performed on a running engine to verify the system, but time may be saved by screening out combinations that have obvious problems, such as excessive, broad wetting or the overspraying of a target. Figure 5.4-13 shows a comparison of LIF-measured fuel film and the carbon deposit pattern on the piston crown with a centrally mounted 30°-cone-angle fuel injector [424]. Carbon deposits are observed on the piston top following late injection experiments. The deposit pattern shown in Fig. 5.4-13b bears a close relationship to the shape and location of fuel films as illustrated in Fig. 5.4-13a. The deposits accumulate rapidly and start to be noticeable only after a few hundred fired cycles, and are more pronounced for later injection timings. During initial stages the deposit appears to form an outline of the fuel film, as shown in Fig. 5.4-13b. With more fired cycles, the deposits fill in gradually and form a continuous layer that is similar to the pressure-chamber fuel film in shape and location. An image of fully developed carbon deposits on a piston crown is shown in Fig. 5.4-13c. Clearly these carbonaceous deposits are accumulated around the regions where substantial fuel films are formed due to spray impingement.

The caveat regarding the matching of ambient density and injector operating temperature is an important one. Bench testing at room conditions does indeed provide footprints, but, strictly speaking, these footprints only occur in the engine for cold crank and start, and possibly not even then because of the significant change in injection timing that may be invoked. Injection timing changes can significantly alter the impact distance, which is a key parameter. Spray geometry and wetted footprint change significantly with operating conditions for all but slit-type injectors, thus these conditions must be emulated if the footprint is to be representative.

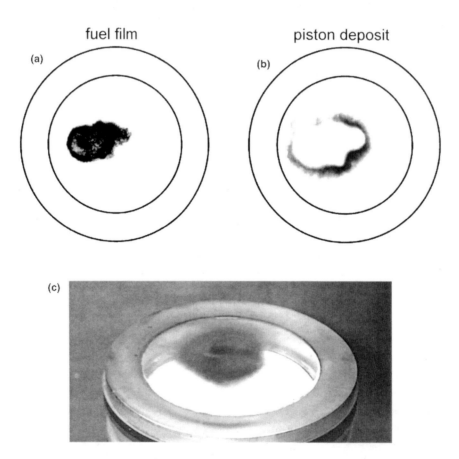

Figure 5.4-13 Comparison of an LIF-measured fuel film and the carbon deposit pattern on the piston crown with late fuel injection timing; centrally mounted, swirl-type fuel injector: (a) fuel film recorded at 65 crank angle degrees after the SOI of –90 crank angle degrees on compression stroke; (b) typical piston deposit pattern for late injection operation after approximately 100 fired cycles; (c) piston-top carbon deposits after several thousand fired cycles with late injection (SOI = –90 crank angle degrees) [424].

At room conditions a swirl-type injector of spray cone angle greater than about 45° will typically be a hollow-cone spray, and for orthogonal injection will provide a bull's-eye pattern for the wetted footprint. This will consist of an outer annular ring that corresponds to the swirl-channel or pintle-exit angle, and a central spot that is generated by the sac spray. In fact, the fuel from every individual swirl-channel is generally evident. This is clearly shown in Fig. 5.4-12a for a swirl injector. If the injector axis is inclined from being orthogonal to some other angle, such as 45° as shown in a perspective view in Fig. 5.4-12b, the wetted footprint will be totally changed. The pattern generally changes from a bull's-eye to a crescent, with the central circular spot changing to an ellipse. The ellipse is not centered, but will have the injector axis at one of the foci. The spray-cone boundaries do translate very well into elliptical boundaries on an inclined impact plane, as shown in the perspective view in Fig. 5.4-14. For a hollow-cone spray the wetted footprint will be bounded by ellipses corresponding to the inner and outer cone angles of the spray. Even without measuring the actual spray it may be surmised that the spray outer cone is about 81 degrees, the inner cone is about 60 degrees and the sac spray cone is about 12 degrees. It is helpful to consider that the wetted area is a subset of the spray intersection area of the impact plane. The spray must be present at a point in the plane to wet; however, the spray may be present but not wet. The minor exception to this is the wetting that may occur due to splashing from a liquid film that originates in the spray intersection area. This occurs in the footprint in Fig. 5.4-14.

The very essence of G-DI spray-wall interaction complexity is illustrated by the footprints in Figs. 5.4-12

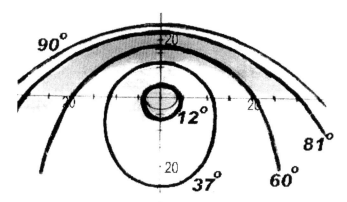

Figure 5.4-14 Spray inner, outer and sac cones derived from the wetted footprint for an inclined injection angle of 45° for a 90° swirl-type G-DI injector.

and 5.4-14. Even for the very simple case of a hollow-cone G-DI spray at room conditions, tilted from normal impact (90°) to 45° impact without changing the spray axis impact distance, the wetted footprint does not extend continuously along the space between the ellipses. Even though the spray is everywhere within the annulus, the spray wets in some regions but reaches a point at which no wetting occurs. The presence of the spray may be verified by passing a pulsed laser light sheet through the spray at the exact position where an impact plate would be. Thus, for significant areas of the spray there is no interaction with the wall, and no wall-wetting. The droplets in the spray are obviously present at those locations (at least very close to the wall) but do not wet the surface. It might be expected that the wetted footprint resulting from the spray from a hot injector with hot fuel would have a greatly reduced area, but this is generally not the case. For swirl, multihole, air-assist and slit-type G-DI injectors the total wetted area for hot operation is not much less than is observed for the same impact geometry at room conditions. The biggest difference due to hot operation is illustrated in Fig. 5.4-12c, which shows that the spray collapses for some injector types, and the spray footprint becomes a single centralized area. If the spray axis is inclined to the impact plane, the footprint will generally be displaced toward the injector, as that side of the intersection ellipse will wet, while the other side may not wet at all.

One of many necessities of modeling the spray-wall interactions is to be able to predict the wetting patterns shown in Fig. 5.4-12, which is much more involved than it may appear. An accurate prediction of both the pattern and the distribution of fuel within the pattern must be made for any spray targeting geometry, various fuel properties, and ambient conditions. For 90°C operation the spray spatial distribution and characterization will be completely different from that for 20°C operation, as will the rate of change of droplet diameter due to vaporization during the flight to the impact surface. This will alter the drag characteristics of the spray. From Fig. 5.4-14 it may also be seen that the footprint is broadened from the actual spray boundaries at the shortest impact distances, which is likely due to splashing from a thicker liquid film. For arbitrary points on an inclined impact plane, both the local flight distance and the compound angle of approach to the wall are changed, as are the local spray characteristics and the local time of arrival of the first droplets.

The properties of the impact surface, such as the roughness (microfinish), temperature, presence of deposits or the presence of a pre-existing liquid film can have a small effect on the amount of wetting or the wetted footprint, however, the main influences of these variables are on what occurs just after impact or after a film is formed. Mie-scattered or volume-illuminated imaging observations show that the bulk spray motion for the transient interaction with an interposed surface is little affected by the above parameters. Except for a few percent higher spray-front velocity along the impact plate for a hot surface (T > 120°C), it is very difficult to detect differences in the wall-interaction images between a rough and smooth surface, or between a hot and cold surface. The overall bulk motions of the spray plume and its spatial position at a given time are basically unaffected. However, the microscopic details of the outcome of the droplet impact events *can* be significantly influenced. Whether a droplet forms a film, enters a film, rebounds intact, shatters and rebounds or splashes within a film is not only determined by a series of Weber number thresholds for the drop and the angle of approach, but also will be influenced by the above surface properties. Whether a droplet reaches an impact surface will not be strongly determined by the surface deposits or surface temperature, but the details of the interaction and the subsequent inclusion of the wall-interaction fuel mass in the combustion event will be significantly influenced. The three general regimes that control the overall effect of spray-wall interactions on G-DI combustion are summarized in Table 5.4-2.

Table 5.4-2
Three general regimes controlling the overall effect of spray-wall interactions on G-DI combustion

Aerodynamic Regime	• Fuel mass reaching potential impact surface is determined. • Near-tip distributions of droplet size and velocity are important. • Flight distances to impact and ambient conditions are key factors. • Angle of inclination of target surface and the in-cylinder air flow field are of less importance.
Impact Regime	• Very small fraction of the total droplets in the spray reach the interposed surface and interact with it. • These droplets can wet, rebound or splash, with the actual fraction being dependent not only on the droplet Weber number and angle of approach at the time of impact, but upon the specific wall surface properties and state. • Whether the wall is wet or dry, or the degree to which it is smooth or rough, can significantly alter the outcome of the droplet interaction.
Post-Injection Regime	• Between the end of injection and the combustion event. • The small percentage of fuel mass that participates in the wall interaction is affected by the physical conditions within the combustion chamber. • This mainly involves the vaporization of that small mass of fuel, consisting of both the vaporization of droplets rebounded or splashed, as well as the evaporation of the liquid film remaining on the wall. • This may be a total or partial vaporization prior to the spark, and will depend upon many factors such as the in-cylinder flow field near the film, the wall temperature, the fuel distillation curve and the presence and structure of surface deposits at the film location.

The regimes discussed here are sequential; that is, the post-injection regime is very dependent on the outcome of the impact regime. If no droplets survive the aerodynamic regime, with all droplets being entrained in the injection-generated flow field and none reaching the wall, then the impact regime and post-injection regimes become moot. Tens of parameters and thresholds are influential in the overall process that incorporates all three regimes, which is the reason why this is a very difficult area to correlate and model. These thresholds are critical to spray-wall sub-models, and much research is being conducted to ascertain the limits of applying single-droplet data to transient G-DI sprays. Even within a single regime such as impact there is an ongoing debate regarding the proper Weber number thresholds for splashing, rebounding and wetting, and as to whether existing data for PFI sprays or single droplets can be appropriately applied. It is obvious that the mass, thickness, shape and time-history of any fuel wetted area in the engine, whether the wetting is intentional or unintentional, are parameters that can influence combustion and emissions. It is also evident that additional critical research needs to be conducted on G-DI spray-wall interactions to more fully understand these complex processes and enhance the predictive capabilities of CFD spray-wall interaction sub-models.

5.5 Unintended Spray-Wall Impingement

5.5.1 Effect of Spray-Wall Impingement on Combustion and Emissions

For a wall-guided G-DI combustion system fuel impingement on a specially designed piston cavity is utilized to create a stable, stratified-charge mixture at light load. Other than the intended spray impingement that is associated with guiding the spray, any unintended fuel wall wetting should be either eliminated completely or minimized to the maximum extent possible [294]. It has been well established that matching and optimizing the spray cone angle and spray-tip penetration rate to the piston cavity geometry is one of the most important steps in minimizing unintended fuel impingement for wall-guided G-DI systems. With substantial unintended fuel impingement on the combustion chamber surfaces, improved fuel atomization can only partially enhance mixture preparation. Pool-burning of the resulting wall film will occur, along with the associated negative effects of increased heat loss and elevated HC and soot emissions. Other negative impacts include the formation of excessive chamber deposits and engine oil dilution. The equivalence ratio at compression TDC becomes leaner when wetting of the piston crown outside of the bowl occurs [4].

Figure 5.5-1 shows fuel film formation as a function of fuel injection timing for a combustion system

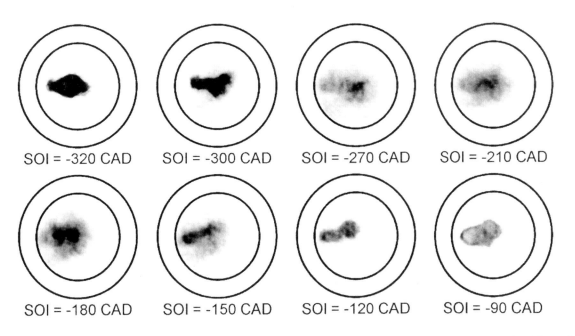

Figure 5.5-1   Fuel film formation as a function of fuel injection timing; centrally mounted, 30° swirl-type injector; images recorded at −30 crank angle degrees relative to the compression TDC [424].

having a centrally mounted, 30° swirl-type injector [424]. The images were recorded at the time of the spark (−30 crank angle degrees). The mass of fuel impacting the piston is significantly greater for both early and late injection when the piston is closer to the injector tip. Histories of visible pool fire luminosity for late injection (SOI = −90 crank angle degrees) are shown in Fig. 5.5-2 [424]. The pool burning of the fuel film is first visible at 20 crank angle degrees after the compression TDC as the piston descends below the fire deck. The pool fires are attached to the fuel film on the piston top and follow the piston as it descends. Toward the end of the cycle, pool fire luminosity weakens and the flames lift off the piston surface. At 300 crank angle degrees after the compression TDC the flames have extinguished, although some liquid film may still remain unburned due to the low volatility of the remaining liquid film and the low oxygen concentration.

The amount of fuel-wall impingement is known to vary significantly with injection timing and engine speed, thus the injection timing must be optimized in order to avoid spray over-penetration and the associated unintended wetting of chamber surfaces [9, 235, 259, 422, 423]. It is generally agreed that the timing for early injection should be adjusted so that the spray tip "chases" the descending piston, but does not significantly impact it. For this the injection timing is critical, as the penetration characteristics of the sac and main sprays must be matched to piston recession velocities for a range of engine speeds and crank angles. Figure 5.5-3 shows an example comparison of the phasing of spray-tip penetration history and piston crown position at 1000 rpm for various injection timings [259]. It is evident that the injection timing is critical to avoiding unintended spray impingement for early injection, and to achieving the desired spray-piston-cavity targeting for creating a stable

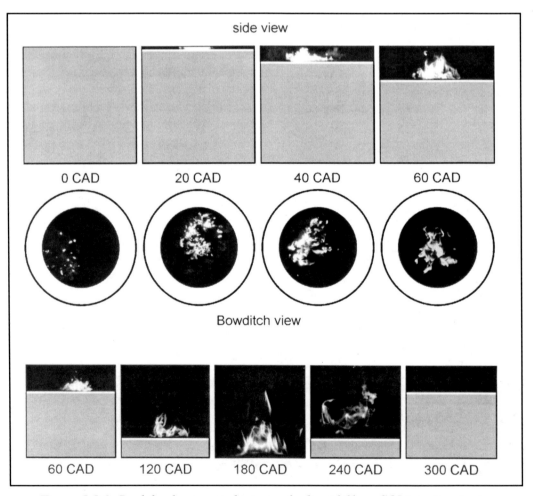

Figure 5.5-2 Pool fire luminosity histories for liquid films; SOI injection timing of −90 crank angle degrees [424].

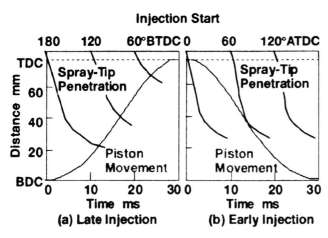

Figure 5.5-3 *Spray-tip penetrations relative to piston position at an engine speed of 1000 rpm (late injection timing is based on compression TDC; early injection timing is based on intake TDC)* [259].

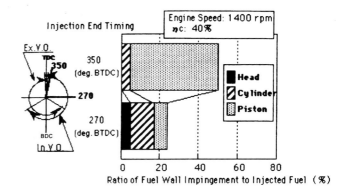

Figure 5.5-4 *Effect of injection timing on fuel wall wetting for a combustion system with an intake-side-mounted injector* [183].

stratified-charge mixture for late injection. Due to the time required for the physical processes of fuel evaporation and fuel-air mixing, each combustion system will exhibit a minimum time interval between the EOI and the occurrence of the spark that is required to avoid an over-rich mixture in the vicinity of the spark gap. By definition, the engine rotation rate in crank degrees per millisecond increases linearly with engine speed; as a result, the injection timing in terms of crank angle degrees must be advanced as the engine speed increases [302]. For purposes of monitoring the avoidance of spray impingement, SOI timing is the most meaningful, whereas for monitoring the mixture preparation interval, EOI timing is the most applicable injection timing parameter. Both should be recorded during G-DI engine development programs, and both have been utilized in engine control algorithms. The warnings given in Section 5.4.2 on the phase differences between indicated and actual SOI and EOI should be kept in mind.

A detailed investigation of the effect of injection timing on fuel wall wetting for a prototype DI combustion system shows a number of key points [182, 183]. Figure 5.5-4 illustrates the results for an engine speed of 1400 rpm. When injection occurs early in the intake stroke, namely 350°BTDC on the compression stroke, significant fuel wetting of the piston crown occurs, with little or no fuel wetting of the cylinder wall and head. For later injection timings the total extent of fuel wall wetting decreases markedly, although there is a trade-off between cylinder wall wetting and piston crown wetting. For an injection timing of 270°BTDC, the wetting

of the piston crown is diminished significantly, whereas the wetting of the cylinder wall is nearly doubled as compared with that obtained for an injection timing of 350°BTDC. It is worth noting that HC emissions are very sensitive to the specific fuel wetting location inside the combustion chamber. Figure 5.5-5 shows a comparison of transient HC concentrations that are obtained for the same amount of liquid fuel wetting different locations inside the chamber. The lowest HC levels are obtained using LPG (liquefied petroleum gas), which yields no wall wetting. Figure 5.5-6 shows the effect of the engine operating (coolant) temperature on HC emissions for different in-cylinder fuel wetting

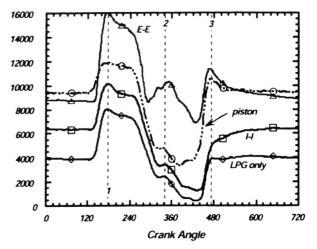

Figure 5.5-5 *Variations in transient HC concentration with the location of the wetted fuel film; constant film mass; part load operation (1500 rpm, 0.262 MPa BMEP); engine coolant temperature of 90°C (LPG; liquefied petroleum gas)* [291].

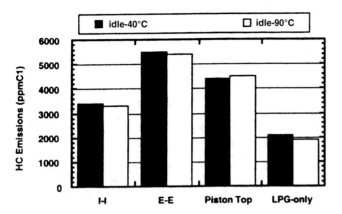

Figure 5.5-6 Effect of coolant temperature on HC emissions with different in-cylinder fuel wetting locations for idle operation [291].

locations. Wetting of the exhaust valves (E-E) by fuel produces the highest HC emissions, as compared to wetting the piston top and intake valves (I-I). This trend is also true for a range of coolant temperatures [291].

The presence of a film of liquid fuel on a chamber surface during the combustion event, whether intended or unintended, can have a significant effect on the measured smoke emissions. Figure 5.5-7 shows the effect of fuel impingement and fuel film thickness on smoke emissions. The smoke emissions data are obtained in engine tests conducted for a range of injection timings and injected fuel quantities, with the amount of fuel wall wetting and wall film thickness calculated by CFD for each test. It is evident that the level of smoke emissions is more dependent on the wall film thickness and less related to the total amount of fuel on the piston surface.

A thicker wall film exhibits a lesser degree of vaporization before ignition occurs, and the remaining fuel film yields a diffusion flame that produces smoke. This indicates that even a small amount of fuel impinging on the piston crown can produce a considerable amount of smoke if the resultant wall film is relatively thick. It should be noted that the temperature of the surface having the film also has a significant effect on fuel vaporization. A cold wall significantly diminishes the fuel film vaporization rate, and generally degrades the spatial uniformity of the fuel-air mixture, whereas a wall surface temperature in the range of 90°C to 130°C generally promotes fuel film vaporization [512]. In fact, the injection timing is purposely advanced in some production G-DI engines in order to obtain spray impingement on a hot piston to improve fuel vaporization under certain operating conditions [385]. However, the charge-cooling benefit will be significantly reduced by this operating strategy.

The general relationship between the initial spray-tip velocity and the spray SMD is shown in Fig. 5.5-8 for a wide range of G-DI injector types and fuel pressure levels [108]. It is evident from the data that diminishing returns occur for the spray-atomization benefit that can be obtained by increasing the fuel rail pressure. Any further increase in the initial spray-tip velocity may result in excessive spray penetration, but may not significantly enhance atomization. Currently most fuel systems for production G-DI engines utilize fuel pressures in the range of 5.0 to 10.0 MPa, although there are some recent systems utilizing variable fuel pressures that have the design authority to operate in the range of 10 to 13 MPa. The curve shows that selected levels of

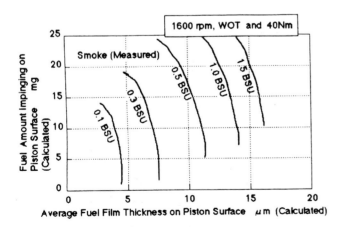

Figure 5.5-7 Effect of fuel impingement on the piston crown and fuel film thickness on smoke emissions [183].

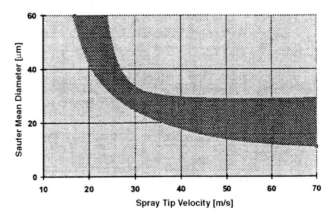

Figure 5.5-8 Measured relationship between the initial spray-tip penetration rate and mean droplet size for a wide range of injectors and fuel rail pressures [108].

atomization for production combustion systems are associated with initial spray-tip velocities in the range of 40 to 60 m/s, which is approximately twice the typical peak flow velocity of the in-cylinder flow field. It should be noted that some newer types of G-DI injectors such as offset multihole and offset slit-type injectors have initial spray-tip velocities in the range of 90 to 100 m/s. Depending upon the orientation of the spray axis relative to the in-cylinder flow field, the momentum of the fuel spray can be a substantial addition to the momentum of the flow field. Because the clearance height is generally quite small in G-DI engines, the fuel spray from G-DI injectors having high penetration rates could result in some degree of unintended wall impingement if targeting of the spray is not optimized. Such unintended impingement must be monitored carefully.

5.5.2 Effect of Cylinder-Bore Fuel Wetting on Oil Dilution

As discussed in the previous section, unintended fuel impingement on combustion chamber surfaces; either on the cylinder wall or the piston crown, or both, often occurs in today's G-DI engines. Non-optimal fuel injection strategies may also lead to fuel wetting of the cylinder wall, causing gasoline to be displaced by the moving piston ring and transported into the crankcase [120, 380, 436]. This can severely degrade the oil quality, and can adversely impact the cylinder bore wear. Oil dilution resulting from gasoline impingement on the cylinder wall is one of the factors to be investigated during the development of a production G-DI engine.

Experimental results from the study of oil dilution for a wide range of engine operating conditions are summarized in Fig. 5.5-9 [436]. As shown in Fig. 5.5-9a, moderate oil dilution is measured at high load with the

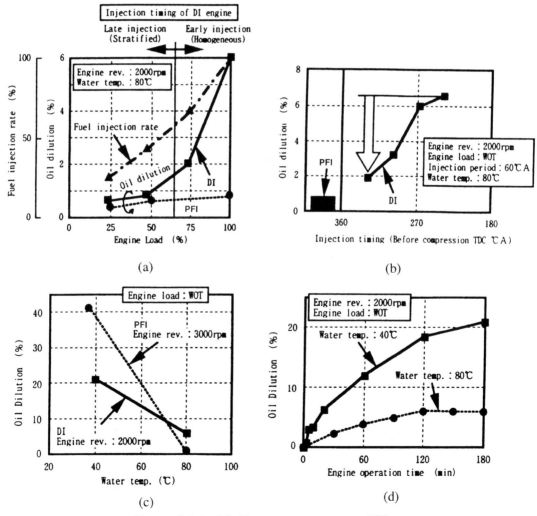

Figure 5.5-9 Oil dilution measurements [436].

G-DI engine, but is not observed in a baseline PFI engine. For late injection corresponding to stratified-charge operation, a similar degree of oil dilution is observed in both the PFI and G-DI engines. For this operating point, the fuel is well confined within the piston bowl and fuel wetting on the cylinder wall is effectively avoided. For a similar reason, earlier injection timing is effective in limiting the oil dilution, as shown in Fig. 5.5-9b. As illustrated in Figs. 5.5-9c and 5.5-9d, the coolant temperature and engine operating time are important parameters in minimizing oil dilution. Closing the flow control valve on this engine is found to reduce oil dilution, due to the fact that the particular air flow pattern provided by the flow control system affects the amount of liquid fuel that impinges on the cylinder wall. Significant oil dilution has been observed for high-load operation in a number of G-DI engines. In general, combustion systems incorporating a side-mounted fuel injector and a spray penetration associated with a high injection pressure (10 to 13 MPa) are expected to exhibit more cylinder-bore fuel wetting, particularly for early injection. However, vehicle test results indicate that the overall level of oil dilution as measured in the field only incrementally exceeds that observed for PFI vehicles. This is quite different from what has been observed in dynamometer tests, and further comparisons of engine oil and coolant temperature histories in the vehicle must be conducted to ascertain the reason.

Investigations of oil dilution in G-DI and PFI engines [120] via oil sampling from the cylinder bore periphery reveal that oil dilution for the PFI engine is more pronounced on the exhaust side of the engine, whereas oil dilution occurs across the cylinder bore for the G-DI engine. Both PFI and G-DI engines exhibit a linear correlation between oil dilution and increasing engine specific fuel consumption. There is evidence to indicate that oil dilution on the intake side results from fuel droplet entrainment through turbulent mixing, whereas oil dilution on the exhaust side results from direct impingement of the fuel spray on the cylinder bore with an intake-side-mounted injector. It is generally agreed that any operating parameters such as retarded injection timing and increased spray penetration that may increase fuel impingement on the cylinder bore are very likely to increase oil dilution. It has been found that the maximum extent of oil dilution is likely to occur between the two operating points of low-speed WOT and maximum torque. Studies that utilized crankcase oil sampling have indicated that oil dilution has only a minor dependence on the hot soak time, the engine air/fuel ratio or the amount of engine oil in the crankcase [380]. Instead, oil dilution for a fixed operating condition is found to increase monotonically with the T50, T70 and T90 points of the fuel. The key factors that affect the amount of fuel impacting the cylinder wall are summarized in Table 5.5-1.

Table 5.5-1
Key factors affecting the amount of fuel impacting the cylinder bore

Spray Characteristics	• Spray-tip penetration curves for the sac and main sprays; maximum penetration • Cone angle, mean drop size, maximum drop velocity • Sac volume, DV90
Combustion System Layout	• Injector mounting location and orientation of spray axis • Combustion chamber geometry (piston bowl, dam or baffle) • In-cylinder flow structure, especially during injection
Fuel Injection Strategies	• Injection timing and injection pressure
Engine Operating Conditions	• In-cylinder air density and temperature; engine load and speed

Investigations of the effects of the spray patterns from swirl-type and slit-type nozzles on oil dilution reveal that the fuel spray from the slit-type nozzle produces less oil dilution [189]. A study of the effect of oil dilution on the amount of top ring wear using the 3.0L DI engine with a slit-type nozzle shows that the position of the piston-ring end gap does not make a significant difference in the level of oil dilution. Fuel enters the second land through not only the end gap, but also the sliding face of the ring. Oil dilution not only decreases the oil viscosity, causing the oil film thickness to decrease, but also increases the amount of oil that adheres to the cylinder, causing an increase in the oil supply to the top ring. When these two opposing effects are not balanced the oil film becomes thinner, leading to an increase in piston ring wear. Therefore, it is important that oil dilution and ring wear characteristics be investigated for a wide range of engine operating conditions, not just for a high-power condition [189].

5.6 Wall Wetting During Cold Start

Achieving an air-fuel mixture of adequate homogeneity with low levels of engine-out HC emissions for the operating modes of cold crank, start and run is one of the noteworthy challenges in G-DI engine development. As discussed previously in Chapter 3, all current G-DI production engines are started at a fuel rail pressure that is substantially less than the design pressure. This generally yields a relatively poorly atomized spray having a degraded dispersion of fuel. Even with improvements in the priming and pressure-rise times of engine-mounted, high-pressure fuel pumps, it is unlikely that even 75% of the design pressure can be achieved during a crank and start time of acceptable duration. Other conditions that exist during cold crank and start that are not conducive to good combustion and low HC emissions are the relatively low charge motion and the low temperature of the fuel and combustion chamber surfaces. These conditions, coupled with the degraded atomization, significantly reduce the overall vaporization rate of the fuel. The larger, slower vaporizing drops tend to reach the chamber surfaces more readily, thus yielding a combination of an increased fraction of injected fuel wetting the walls and a slower rate of evaporation for the cold fuel on the cold surface. An additional factor working against the optimization of cold-start combustion is that the fuel must be injected over a very wide range of crank angle. The combination of low fuel rail pressure and very low

engine speeds during crank necessitates an injection event that can extend in some cases to a third of an engine revolution. The ideal G-DI engine would operate without cold enrichment however, the real engine is subject to the conditions just described, and does require some enrichment, although generally not as much as for a PFI engine.

A comparison of the measured fuel inventory between G-DI and PFI engines in the first cycle of an engine start with fuels having two different T50 points: 102°C (fuel A) and 82°C (fuel B), is illustrated in Fig. 5.6-1. With the PFI engine, almost half of the injected fuel remains as liquid in the intake port, and is carried over to the subsequent cycle. Even though port fuel wetting does not occur for the G-DI engine, the in-cylinder wall wetting is quite significant, causing an increase in engine-out HC emissions. It is interesting to note that the level of cylinder wall wetting with the PFI engine is not very sensitive to fuel volatility, whereas port fuel wetting exhibits a moderate effect. In contrast, in-cylinder wall wetting with the G-DI engine decreases as fuel volatility increases.

Figure 5.6-2 shows a comparison of the required amount of fuel for PFI and G-DI engines from the first to the ninth cycle during cold start, which represents the ignition limit of each cycle. For the PFI engine the required amount of fuel does not exhibit a continuous decrease. Instead, the required amount of fuel undergoes a step decrease for the second cycle. For the G-DI engine at a low initial fuel pressure of 0.4 MPa, the

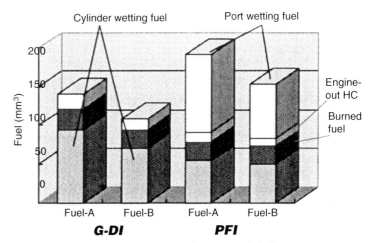

Figure 5.6-1 Comparison of measured fuel inventory between G-DI and PFI engines in the first cycle during engine start with fuels of two different T50: T50 = 102°C for fuel A; T50 = 82°C for fuel B [246].

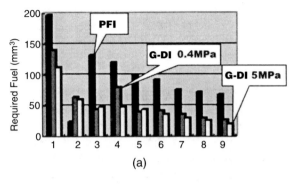

(a)

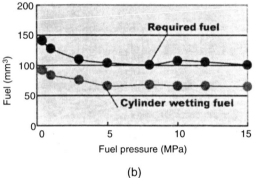

(b)

Figure 5.6-2 Comparison of the amount of fuel required for PFI and G-DI engine; first to ninth cycle: (a) required fuel for PFI and G-DI engines; (b) required fuel and cylinder wall wetting fuel as a function of fuel rail pressure [246].

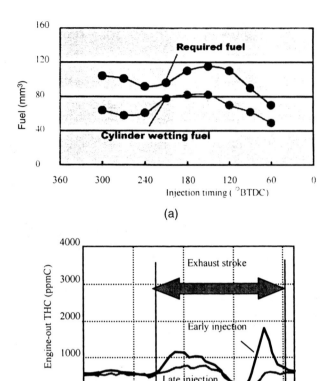

(a)

(b)

Figure 5.6-3 Effect of injection timing on in-cylinder fuel wetting and engine-out HC emissions: (a) required fuel and cylinder wall wetting fuel as a function of injection timing; (b) engine-out HC emissions as a function of injection timing [246].

required amount of fuel at the first cycle is about 25% less than that for the PFI engine. In contrast, the amount of fuel required for the second cycle is twice that for the PFI engine. On subsequent cycles the G-DI engine requires significantly less fuel than the PFI engine, even for a low fuel pressure of 0.4 MPa. If a fuel pressure of 5 MPa could be maintained from the first cycle during cold start, the required amount of fuel would exhibit a continuous decrease. An increase in fuel pressure improves fuel spray atomization, thus reducing the amount of fuel wetting the cylinder.

The effect of injection timing on cylinder wall wetting and engine-out HC emissions is illustrated in Fig. 5.6-3. Fuel wetting the cylinder wall and the amount of fuel required for the cycle of a cold start are higher when injection occurs near the intake TDC or intake BDC, and are minimized for injection at 270°BTDC (early injection) and 60°BTDC (late injection). This is consistent with the measured HC emissions. Late injection timing requires a higher injection pressure due to the limited time available to supply the required amount of fuel.

Injector heating has been utilized as a means to improve cold start performance, and the measured effect of injector heating on fuel evaporation rate and the associated in-cylinder fuel wetting are shown in Fig. 5.6-4. Clearly, heating the fuel injector significantly increases the fuel vaporization rate, and is effective in reducing the amount of fuel required for cold start.

For typical pump gasolines, it has been observed that some degree of flash-boiling will occur within many G-DI injectors when fuel is injected into a vacuum. Such flash boiling will improve the level of fuel atomization. The absolute cylinder pressure levels required (less than 75 kPa) can be generated by delaying the opening time of the intake valve. Figure 5.6-5a shows the time histories of in-cylinder pressure during the intake stroke with conventional and retarded valve timings. A much stronger air flow can be generated during the early stage

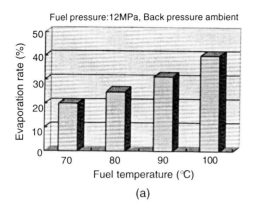

Fuel pressure:12MPa, Back pressure ambient

(a)

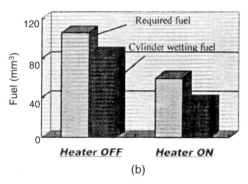

(b)

Figure 5.6-4 Effect of injector heating on cylinder wall wetting and engine-out HC emissions: (a) fuel evaporation rate as a function of fuel temperature; (b) effect of fuel heating on required fuel and cylinder wall wetting fuel [246].

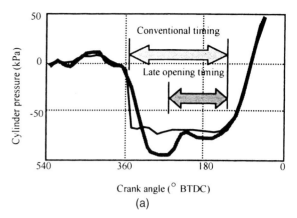

(a)

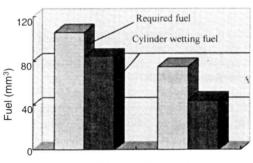

(b)

Figure 5.6-5 Effect of retarded intake valve opening on cylinder wall wetting during cold start: (a) cylinder pressure with conventional and late intake valve opening; (b) effect of intake opening on required fuel and cylinder wall wetting fuel [246].

of the intake stroke due to the low pressure inside the cylinder with the retarded intake valve opening. As illustrated in Fig. 5.6-5b, cylinder wall wetting may be reduced approximately 40%, compared to the standard valve timing, leading to a reduction in the required cold enrichment.

The strategies that may be invoked for reducing the engine-out HC emissions associated with a G-DI cold start are summarized in Fig. 5.6-6. The baseline HC emissions are derived from a G-DI engine operated in an early injection mode at a low fuel injection pressure for 10 seconds. This corresponds to the HC emissions level of current production G-DI engines. The engine-out HC emissions of such engines are still substantial as compared with the current PFI engine. It may be seen that each individual strategy has the potential to reduce engine-out HCs by 20 to 35%. A combination of strategies is required to reduce the G-DI HC emissions below that of the contemporary PFI engine, which is also a moving target.

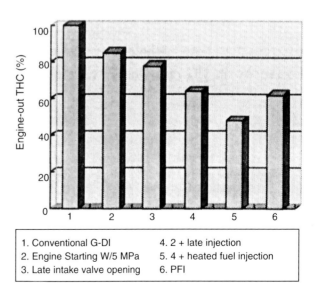

1. Conventional G-DI   4. 2 + late injection
2. Engine Starting W/5 MPa 5. 4 + heated fuel injection
3. Late intake valve opening 6. PFI

Figure 5.6-6 Comparison of strategies for reducing cold-start, engine-out HC emissions for G-DI engines [246].

5.7 Summary

The preparation of the required fuel-air mixture is one of the most important overall processes in ensuring a successful G-DI combustion system. The transient in-cylinder flow field that is developed during the intake and compression strokes is one of the key design factors in determining the operational feasibility of the system. The magnitude of the mean components of motion, as well as their resultant variations throughout the cycle, are as important as the characteristics of the fuel injection system in the success of a G-DI system. The optimum flow field depends upon the injection strategy that is being used, and is a compromise for full-feature G-DI engines that operate with both homogeneous-charge and stratified-charge modes. For G-DI combustion systems, the control of the mixing rate by means of the mean flow history is normally used in preference to the scheduling of turbulence generation, with the latter being the case for a conventional PFI engine. The spray-induced flow field is known to exert a significant influence on the overall in-cylinder air flow field, and should be evaluated.

For wall-guided combustion systems, spray-wall impingement is purposely utilized to achieve a stable stratified-charge combustion at light load. For other than this intentional bowl-cavity targeting, unintended spray-wall impingement generally results in less charge cooling, and in an increase in HC emissions, CCD and oil dilution, and should be avoided to the greatest extent possible. Therefore an improved understanding of the spray-wall interaction process is crucial to minimizing the effect on emissions. With regard to the interactions of G-DI fuel sprays with piston crowns or piston bowl cavities, detailed research has shown that, contrary to the usual schematic representation of such processes in spray-wall control systems, the G-DI spray is predominately deflected, not impacted. For most G-DI injectors at typical bowl cavity impact distances of 28 to 45 mm, generally less than 10% of the total injected fuel mass actually impacts the bowl and forms a liquid film on the impact surface, even for bench tests at room conditions. In fact, it is generally only the largest and earliest droplets that impact, with many of those being sac spray droplets for injectors that have a sac volume. Because most of the fuel mass is present in the largest droplets in a distribution, only a very small percentage of the total number of droplets in a spray actually impact. The vast majority (>99%) of the droplets in a G-DI spray are instead entrained in the transient air flow field that is established by the injection event. These droplets rapidly decelerate through drag and achieve a terminal velocity that corresponds to the entrainment flow field velocity at that location. These drops move as particles suspended in the flow field, which, of course, follows the impact surface contour. The time histories of the mass, shape and thickness of any spray-wetted area in the engine, whether created intentionally or unintentionally, are critical in G-DI engine combustion and emissions. Further research efforts are required to more fully understand these complex processes and enhance the predictive capabilities of CFD spray-wall interaction sub-models.

Chapter 6

Combustion Process and Control Strategies

6.1 Introduction

Over the past three decades a number of concepts have been proposed to exploit the potential benefits of direct gasoline injection for passenger car applications, but it is only in the last decade that any have been available in production vehicles. One important reason for this twenty-year delay was the lack of precise controllability of the fuel injection system. Early systems that were based upon diesel fuel pumps and pressure-activated poppet nozzles experienced significant limitations in performance and control. Although the engines developed in the initial attempts were relatively successful in achieving improvements in BSFC, the mechanical pumps that were utilized had inherent speed and timing limitations, and the specific power densities obtained with these engines were generally inferior to those of the diesel engine.

The concentrated development of the G-DI engine over the last decade has benefited significantly from the application of electronic, common-rail, fuel injection systems. It should be noted that this type of injection system, which has long been utilized on production PFI engines, is also currently being applied to diesel engines, and has seen some application on two-stroke DI gasoline engines. Such injection systems provide fully flexible timing over the entire speed and load range of current gasoline engines by enabling a strategy of optimum injection timing and metering at all engine operating points, which can provide improved catalyst light-off and regeneration characteristics, increased engine performance and reduced engine-out emissions.

A G-DI engine equipped with an electronic, common-rail system has the ultimate potential of achieving a significantly improved BSFC, while simultaneously achieving a specific power output that is equivalent or superior to that of a PFI engine [274]. In fact, a number of the potential advantages of the G-DI combustion system are directly attributable to the capabilities of the electronic fuel injection system. It is the flexibility of the contemporary G-DI fuel injection system that provides the freedom of fuel-air mixing and combustion process control, as illustrated in Fig. 6.1-1. Figure 6.1-2 shows a

typical operating map for a generic G-DI engine. A stratified-charge G-DI engine must also be capable of operating reliably for early-injection, homogeneous-charge conditions in order to provide the designed full-load performance. In addition, a successful G-DI combustion system must be capable of a smooth transition between operating modes. Therefore, all G-DI combustion systems, whether stratified charge or fully homogeneous, must provide good combustion characteristics over the full range of engine operating conditions. The combustion characteristics that are obtained for key control strategies will be discussed in detail in this chapter.

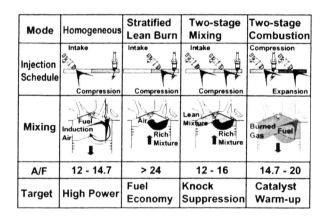

Mode	Homogeneous	Stratified Lean Burn	Two-stage Mixing	Two-stage Combustion
Injection Schedule	Intake / Compression	Intake / Compression	Intake / Compression	Compression / Expansion
Mixing	Fuel Induction Air	Air / Rich Mixture	Lean Mixture / Rich Mixture	Burned Gas / Fuel
A/F	12 - 14.7	> 24	12 - 16	14.7 - 20
Target	High Power	Fuel Economy	Knock Suppression	Catalyst Warm-up

Figure 6.1-1 Flexibility of fuel-air mixing and combustion process control of G-DI engines [501].

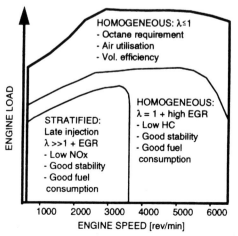

Figure 6.1-2 Typical G-DI engine operating map [275].

6.2 Engine Operating Modes and Fuel Injection Strategies

The operation of G-DI engines designed to include stratified-charge combustion can be classified into the four basic operating modes, listed in Table 6.2-1. The least complex, full engine load is obtained for stoichiometric, homogeneous-charge operation, with maximum air utilization required for a high specific power output. For medium load, lean, homogeneous-charge or stoichiometric (with EGR) operation is used, with control parameters optimized for a combination of fuel economy and NOx emissions. For NOx control, this regime will most likely be stoichiometric operation with EGR. Lean, stratified-charge operation is used for idle to medium-load operation in order to achieve maximum fuel economy. From a control standpoint the most difficult mode is generally that of load transition, in which the operating mode is changing from one of the above three modes to another. The details of these modes will be discussed in this section.

6.2.1 Early-Injection, Homogeneous-Charge Operation

The homogeneous-charge operating mode in the G-DI engine is designed to achieve a homogeneous mixture to meet the requirement of medium-to-high engine loads. Depending on the overall air/fuel ratio, the mixture can be homogeneous-stoichiometric or homogeneous-lean. When used in combination with complementary in-cylinder charge motion, this strategy has the potential to provide part-load performance that is comparable to that of the conventional PFI engine and full-load performance superior to that of the PFI engine [10]. Early injection and the associated charge cooling increases volumetric efficiency and also makes it possible for the engine to operate at a slightly higher compression ratio, which provides an additional incremental improvement in fuel economy. An early injection strategy with an essentially homogeneous charge can also achieve a number of benefits in the areas of cold start and transient emissions, as will be discussed in detail in Section 6.2.3.

6.2.2 Late-Injection, Stratified-Charge Operation

The injection of fuel relatively late in the cycle is normally invoked to obtain a stratified-charge mixture in the cylinder. Depending on the overall air/fuel ratio, the mixture cloud can be either stoichiometric-stratified, or lean-stratified. Generally speaking, the term "stratified charge" refers to lean-stratified operation, which is widely utilized for this G-DI operating mode.

Table 6.2-1
Operating mode classifications for full-feature G-DI engines

G-DI Operating Class	Engine Load	Injection Timing	Description
1	Very light load and idle	Late injection	Stratified, overall lean, some throttling
2	Light to medium load	Late injection	Stratified, overall lean, no throttling
3	Medium load	Early injection	Homogeneous, stoichiometric with EGR or lean
4	Full load	Early injection	Homogeneous, stoichiometric or richer-than-stoichiometric
	Load transition		Realize a smooth transition from one mode to the other

In contrast, the term "stoichiometric-stratified charge" normally refers to a special strategy that is used to reduce catalyst light-off time (to be discussed in Section 6.2.3). The objective of charge stratification in the G-DI engine with an overall-lean air/fuel mixture is to operate the engine unthrottled at part load at an air/fuel ratio that is leaner than is combustible with the conventional lean, homogeneous mixture. This is achieved by creating and maintaining charge stratification in the cylinder such that the air/fuel ratio at the spark gap yields stable ignition and flame propagation, whereas areas farther from the point of ignition are either very lean or devoid of fuel. In general, air/fuel mixture stratification is realized by injecting the fuel into the cylinder late in the compression stroke, however, it may also be possible to achieve stratification with early injection, and some success toward this goal has been obtained with an air-assisted fuel system [130, 131]. The use of charge stratification with an overall lean mixture can indeed provide a significant improvement in engine BSFC [50, 259]. This is due primarily to a significant reduction in the pumping work associated with throttling, but there are also additional benefits such as reduced heat loss, reduced chemical dissociation from lower cycle temperatures, and an increased specific heat ratio for the cycle, which provide incremental gains in thermal efficiency.

There is a consensus in the literature that achieving stable, stratified-charge combustion in a G-DI engine while controlling the engine-out HC emissions to a very low level is a difficult task. The interrelationships between injector location, spray characteristics, combustion chamber geometry, EGR rate, injection timing, and spark timing are quite complex, and must be optimized for each specific system. The engine operating range in which charge stratification can be effectively utilized to obtain the available thermodynamic benefits should be designed to be wide enough to encompass the most frequent engine operating points, otherwise mode transitions will occur too frequently [227]. In general, it is exceptionally difficult to achieve full air utilization and the associated high specific power using a late injection strategy. This is due to incomplete mixing, which not only degrades air utilization, but also may yield excessive smoke emissions. For most prototype G-DI combustion systems, obtaining efficient, part-load, stratified operation with acceptable values of COV of IMEP has proved challenging.

6.2.3 Stoichiometric-Charge Operation

Direct-injection engine operation using stoichiometric combustion has the distinct advantage of being able to utilize three-way catalysis, thus avoiding the dependence on lean-NOx catalysis [47, 138, 379, 399]. The attainable advantages of stoichiometric G-DI engine operation are summarized in Table 6.2-2. Homogeneous-charge, stoichiometric G-DI engines using fuel injection during induction provide increased volumetric efficiency and low tailpipe emissions. This is achievable due to the combination of a three-way catalyst, quicker engine starting and less cold enrichment, resulting in an elevated exhaust temperature and a more rapid catalyst light-off. A multiple injection strategy that incorporates late injection and combustion during the expansion stroke is also an option, and may further reduce the catalyst light-off time. Considering all of these factors, homogeneous, stoichiometric G-DI engines do exhibit the potential for meeting future emissions targets [279].

When fuel injection timing is significantly retarded, a stoichiometric-stratified mixture can be generated, which can be used to advantage in reducing engine-out emissions. The use of a stoichiometric, or near-stoichiometric, stratified charge is found to retard initial combustion phasing marginally, whereas main combustion is significantly retarded. Depending upon the degree of charge stratification, and upon combustion chamber geometry, EGR rate, charge motion, and fuel vaporization rate, the burn rate decreases more rapidly than is observed for homogeneous-charge operation. Thus, the use of relatively late combustion is possible even with fixed spark timing. As a result, HC and NOx emissions can be lowered and the exhaust gas temperature can be elevated [110]. Such a stratification measure constitutes an alternative to conventional catalyst light-off strategies. Figure 6.2-1 illustrates the advantages exhibited by the moderate stratification of a stoichiometric charge, first among which is the potential for ultra-low tailpipe emissions. It is certainly true that the additional energy supplied to the exhaust is associated with an increase in fuel consumption; however, this is in effect only until the catalyst threshold temperature is achieved; therefore, it only marginally influences overall fuel consumption. A significant concern with this strategy is the likelihood of increased soot emissions due to the non-homogeneity associated with retarded injection timing.

Table 6.2-2
Attainable advantages of stoichiometric, homogeneous-charge G-DI operation (relative to PFI best practice)

Engine Start	• More rapid cold start • Less cold-start enrichment required • Strategy for more rapid catalyst light-off applicable • Reduced cold-start HC emissions
Transient	• Improved transient response • Less acceleration enrichment required • More precise air/fuel ratio control • Ability to aggressively invoke fuel cut-off on deceleration
Combustion	• No combustion system compromise for different mode operations • In-cylinder charge cooling benefit • Reduced heat loss during compression stroke • Improved combustion stability • Enhanced EGR tolerance • Up to 2 ratios increase in knock-limited compression ratio • Reduced sensitivity to fuel volatility for driveability and cold start • Applicable to lean, homogeneous-charge operation without system modification
Fuel Economy	• Up to 5% gain in integrated fuel economy • Up to 5% improvement in volumetric efficiency • Ability to aggressively invoke fuel cut-off on deceleration • Ability to invoke engine shut-off instead of idle • Up to 2 ratios increase in knock-limited compression ratio
Performance	• Up to 5% improvement in volumetric efficiency • Up to 7% increase in peak torque and power • Engine downsizing possible while maintaining torque/power
System Flexibility and Complexity	• Less control complexity as compared to stratified-charge G-DI engine • Control flexibility for system optimization
Combined with Other Technologies	• Possible enabler of some other technologies such as boosting, ability to employ engine shut-off during idle, continuously variable transmission (CVT), and hybrid electric vehicle (HEV)
Emissions	• No lean-NOx after treatment system required • Three-way catalysis can be fully utilized • Reduced emissions as compared to lean, stratified-charge operation • Reduced emissions during transient operation

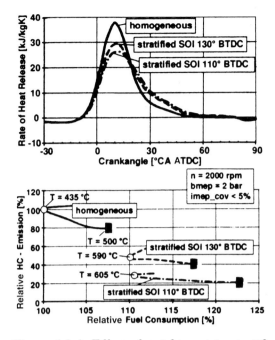

Figure 6.2-1 Effect of stoichiometric-stratified operation as a catalyst light-off strategy; close-coupled three-way catalyst [481].

6.2.4 Slightly Lean, Stratified-Charge Operation for Improving Catalyst Light-Off Characteristics

It is generally acknowledged that the catalyst light-off temperature for CO reduction is lower than that for HC reduction. CO oxidation can assist in heating the catalyst, leading to a reduction in the catalyst light-off time. Based upon this consideration a concept with a slightly lean, stratified-charge mixture has been proposed to form CO locally in the combustion chamber through fuel-rich combustion, as illustrated in Fig. 6.2-2 [501]. This is realized by retarding the injection timing for an overall slightly lean mixture. The CO and surplus O_2 will be dispersed within the overall charge during the expansion and exhaust strokes, and the CO-O_2 mixture will react on the catalyst surface to warm the catalyst. It has been found that in combination with post injection (also designated as two-stage combustion), which is another catalyst-light-off-enhancing technique and will be discussed in Section 6.3.3, the slightly lean, stratified-charge strategy can significantly reduce the catalyst light-off time. Figure 6.2.3 shows the effect of this operating strategy (in combination with extremely retarded spark timing) on the light-off characteristics of a closed-coupled catalyst with an

NOx trap catalyst mounted downstream. When the slightly lean, stratified-charge strategy is applied immediately after a cold start in combination with a spark timing of 15°ATDC, the exhaust gas temperature can be elevated rapidly to 650°C, and the close-coupled catalyst is warmed up quite quickly. As a result, catalytic oxidation of both CO and HC is initiated much more rapidly than would otherwise occur.

As discussed in Section 6.2.3, smoke emissions are generally elevated for operation with a retarded injection timing and a relatively rich mixture. Figure 6.2-4 shows the effect of fuel injection timing on burning angle, fuel consumption, smoke, HC, CO emissions, and O_2 concentration [501]. When the EOI is set within the range of 40–70°BTDC on the compression stroke, CO and O_2 concentrations increase significantly due to the strong stratification, whereas smoke emissions are suppressed to a very low level. Further retardation of the injection

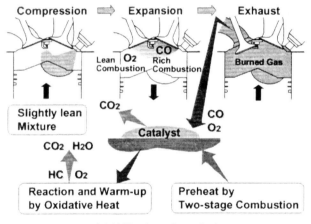

(a) Utilization of catalysis

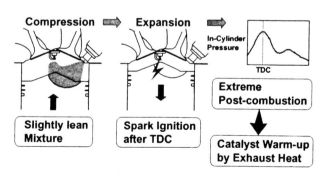

(b) Utilization of exhaust heat

Figure 6.2-2 Schematic of the slightly lean, stratified-charge operation strategy for reducing catalyst light-off time [501].

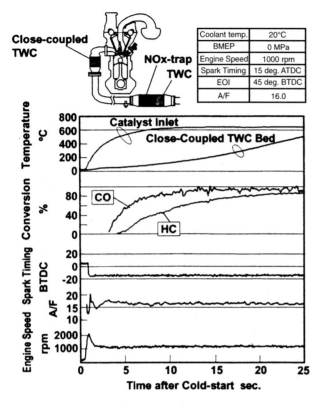

Figure 6.2-3 Effect of the slightly lean, stratified-charge operation strategy in combination with retarded spark timing on the light-off characteristics of a closed-coupled catalyst with an NOx trap catalyst mounted downstream. (501).

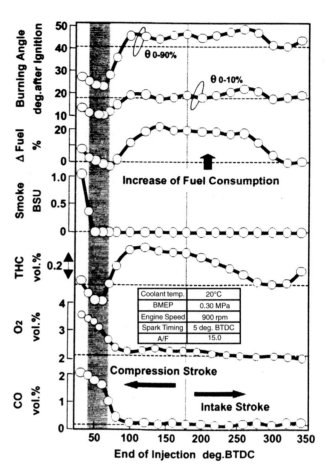

Figure 6.2-4 Effect of fuel injection timing on burning angle, fuel consumption, smoke, HC and CO emissions, and O_2 concentration (501).

timing, however, increases smoke emissions sharply, due to the more limited time available for fuel to mix with the air before ignition occurs. This problem can be overcome to some degree by extending the interval between the end of fuel injection and the spark timing, which will enhance the degree of fuel vaporization, dispersion and mixing. When this interval is too brief, the mixture is locally too rich due to insufficient fuel vaporization and mixing. In contrast, fuel dispersion will eventually degrade the stratification when the interval is too long. The slightly lean, stratified-charge strategy does permit a fairly wide interval for the optimization of the stratified mixture. Figure 6.2-5 shows the effect of the interval between injection and spark timing on CO and smoke formation with this approach. When the interval is short, smoke emissions increase with CO concentration. By optimizing the interval, however, the smoke level can be suppressed below an acceptable threshold level while a sufficient amount of CO can be produced to meet the requirement for reaction on the catalyst surface.

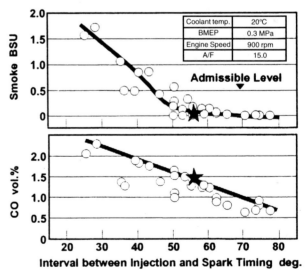

Figure 6.2-5 Effect of the interval between injection and spark timing on CO and smoke formation with the slightly lean, stratified-charge approach (501).

A possible disadvantage of the slightly lean, stratified-charge strategy is the fuel economy penalty associated with retarded spark timing. Figure 6.2-6 shows a comparison of the effects of spark retardation on the measured fuel consumption of both the slightly lean, stratified-charge strategy and the post-injection (two-stage combustion) strategy. It is evident that fuel consumption increases when the spark is retarded beyond the compression TDC; however, the fuel economy penalty associated with spark retardation is significantly lower than that of post injection. Therefore, if the slightly lean, stratified-charge approach is invoked, the post-injection strategy can be either avoided or reduced in duration, which can improve fuel economy significantly [501].

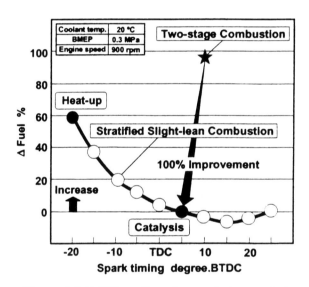

Figure 6.2-6 Effect of spark retardation on fuel consumption for the slightly lean, stratified-charge and post injection (two-stage combustion) strategies [501].

6.2.5 Operating Mode Transition

Both the early-injection mode for homogeneous-charge operation and the late-injection mode for stratified-charge operation are widely utilized in current G-DI combustion systems. This, however, requires a transition from one mode to the other without misfires, partial burns or noticeable torque fluctuations that can impact the driveability rating [286, 509]. Achieving this with an overall lean mixture can be a formidable task. It should be emphasized that the key element in achieving a smooth mode transition is to minimize torque fluctuation.

Electronically controlled throttling is effective in maintaining the necessary torque demanded by the driver or electronic controller, thus permitting the control system to maintain a smooth mode transition. This control-system technique is being increasingly utilized in G-DI engine applications. In addition, the use of a torque-based engine control system can effectively enhance the overall G-DI vehicle fuel economy. By avoiding close coupling of engine operation to driver input, the engine can be operated in the stratified-charge mode for an extended period of time during transients while maintaining acceptable vehicle driveability.

In addition to electronic throttle control, a number of other approaches derived from the flexibility of G-DI fueling and combustion control are also available for obtaining a smooth mode transition. The Toyota first-generation D-4 production engine utilizes a split-injection strategy, also called two-stage injection, to achieve mode switching. This two-stage injection strategy injects portions of the total fuel into the cylinder during both the intake and the compression strokes [156, 302]. The details of this technique will be discussed in Section 6.3.1.

With the Mitsubishi GDI engine produced in 1996, mode transition control is achieved by the accurate control of air flow rate using a conventional throttle valve that is controlled by the accelerator pedal, and by a bypass solenoid valve that is electronically actuated using pulse width modulation [196]. Figure 6.2-7 schematically illustrates both the air and air/fuel ratio control during the transition from the stoichiometric operation of early injection to the overall lean operation of late injection. In order to minimize torque fluctuation during the transition, switching should ideally be conducted only when the generated torques of both modes at the same engine air flow rate are identical. As is shown, early injection at an air/fuel ratio of 18:1, and late injection at an overall air/fuel ratio of 25:1, generate the same torque at the same air flow. The air/fuel ratio is gradually increased from 14:1 to 18:1 by increasing the amount of engine air flow while maintaining the injected fuel quantity. The combustion mode is then switched to late injection within one cycle. In the first cycle of the late-injection mode, the injected fuel quantity is adjusted to achieve an air/fuel ratio of 25:1, which is enleaned further to an air/fuel ratio of 35:1 on subsequent cycles by reducing the injected fuel quantity and increasing the air flow simultaneously. As a consequence, engine operation is managed to minimize the torque differential at any transient conditions.

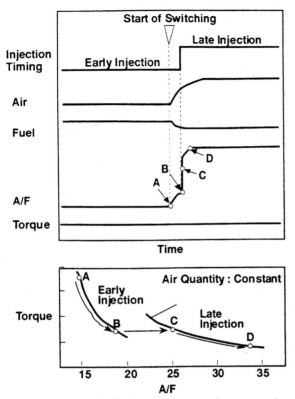

Figure 6.2-7 Air/fuel management during mode-transition operation of the Mitsubishi GDI engine [196].

During the transition between modes, two air/fuel ratio limits have been identified in some proposed G-DI control strategies [328]. One applies to stratified-charge operation with a low air/fuel ratio limit of approximately 24:1 to avoid soot formation. Another limit applies to combustion instabilities that are associated with homogeneous, lean operation and is encountered for air/fuel mixtures that are leaner than 19:1. This range of air/fuel ratios (19:1 to 24:1) is regarded as a forbidden window and it is recommended that all air/fuel ratios in this range be avoided during transitions.

6.2.6 Comparison of Operating Mode Complexity

Engine fuel consumption and NOx emissions for the three G-DI operating modes of homogeneous- stoichiometric, homogeneous-lean and stratified-charge are compared in Fig. 6.2-8 [109, 481]. It is evident that stratified-charge operation is only advantageous for low engine loads, even for the optimized charge-stratification concepts. If the engine BMEP exceeds 5 bar, a more efficient combustion process can be achieved alternatively using homogeneous-lean operation. This alternative has the benefit of comparatively low NOx emissions. Moreover, if the

stratified-charge mode is extended to high-load operation, smoke is likely to be produced in the over-rich regions near the spark gap. As the load increases, the injection duration increases nearly linearly with load, resulting in a fuel-air mixture cloud that increases in size and fuel concentraion, but is constrained by the piston and chamber geometry. The mixture cloud can exhibit a locally rich stratification as a consequence of being geometrically constrained, and there being only a short mixing time available for fuel introduced near the end of injection. When the fuel cloud associated with the first fuel injected is ignited, combustion occurs rapidly because of the locally rich mixture. The flame propagates rapidly through the late-injected portion of the mixture, possibly yielding soot and HC emissions. Soot may also be generated by the increased impingement of the fuel spray on the piston crown as the injection duration is increased. The soot formation threshold is considered to be a primary limitation that is eventually encountered as the engine load is increased and EOI approaches TDC [11]. In this case the amount of smoke can be reduced or alleviated by operating the engine in the homogeneous-charge mode using early injection. A carefully designed transition between low-load, stratified-charge operation and high-load, homogeneous-charge operation is important for achieving very low particulate emissions [361].

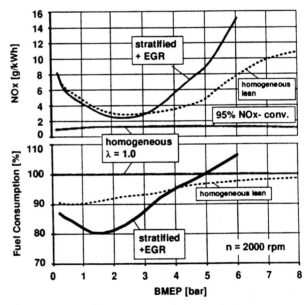

Figure 6.2-8 Comparison of engine fuel consumption and NOx emissions as a function of engine load for three G-DI operating modes: homogeneous-stoichiometric, homogeneous-lean and stratified-charge [481].

For the Mitsubishi GDI engine, the stratified-charge mode using late injection has been extended to 50% of full load without an increase in soot [259]. It is theorized that sufficient excess air exists in the cylinder for the soot to oxidize rapidly. When the load exceeds 50%, however, soot particles are not fully oxidized and begin to be emitted from the engine. Injection timing has been found to be an influential parameter in suppressing the formation of soot. For the DI diesel engine the amount of fuel that can be injected must be restricted so as to limit the soot emissions to permissible levels, resulting in lower performance. In the G-DI engine, however, this problem can be circumvented over the entire engine operating range, without degrading engine performance, by adjusting the injection timing [302].

It is worth noting that the degree to which the benefits of G-DI can be realized is directly associated with the extent to which the engine and control system are designed and configured to take maximum advantage of the best features of each operating mode. Table 6.2-3 shows a comparison of the degree to which the advantages of G-DI over PFI can be achieved for three G-DI operating modes. An analytical comparison between stratified-charge operation combined with homogeneous-lean operation, and stratified-charge operation combined with stoichiometric, high-EGR operation indicates that

Table 6.2-3
Extent to which the advantages of G-DI over PFI may be achieved for three operating modes

G-DI Advantages over PFI	Early-Injection Stoichiometric, Homogeneous-Charge Operation	Early-Injection, Lean, Homogeneous-Charge Operation	Late-Injection, Lean, Stratified-Charge Operation
Knock-limited compression ratio increase due to charge cooling	Full	Partial	None
Reduced octane number requirements due to charge cooling	Full	Partial	None
Increased volumetric efficiency due to charge cooling	Full	Partial	None
Fuel cut-off on deceleration	Full	Full	Full
Reduced cold-start HC emissions	Full	Full	Full
Rapid cold start	Full	–	Full
Reduced cold enrichment requirement	Full	–	Partial
Less acceleration enrichment required	Full	Full	Partial/Full
Reduced CO_2 emissions	Full	Full	Full
Improved transient response	Full	Full	Full
Reduced cylinder-to-cylinder air/fuel ratio variation	Full	Full	Full
Lower idle speed and reduced idle fuel consumption	–	Partial	Full
Fuel economy improvement due to reduction in throttling loss	None	Partial	Full

the latter has a slight advantage with respect to homogeneous-lean operation. The NOx conversion requirement shows a benefit for stoichiometric, high-EGR operation in the light-load range, with only a 60% conversion efficiency required, as compared to 85% for the homogeneous-lean strategy. The G-DI operating strategy that combines stratified-lean, light-load operation with stoichiometric, high-EGR, mid-load operation has substantial benefits in emissions, with only 1% of the fuel economy potential compromised [480].

6.2.7 G-DI Engine Operating Classes

As outlined in Table 6.2-1, current G-DI systems may be categorized using four operating classes. In this table the degree of complexity of the G-DI engine operating system is shown to progressively decrease, with the least complex system being the early-injection, homogeneous-charge, stoichiometric, G-DI engine. The incremental step from a homogeneous-charge G-DI system to a stratified-charge system does indeed offer the potential for significant fuel economy improvement, but there is a consensus that stratified-charge operation is not obtainable without a significant increase in system control complexity. The advantages of early injection, homogeneous G-DI (Class 3 or 4) could, in some applications, be more extensive than for full-option G-DI (Class 1 or 2), due to the more complex hardware and the associated development time required for the latter.

The development of advanced G-DI engines that meet future stringent emissions requirements could occur in two stages. First, prior to the emergence of a proven lean-NOx aftertreatment system, homogeneous-charge stoichiometric G-DI (Classes 3 and 4) can be developed with increased engine specific power at high load as a key design goal. These engines can be designed to achieve BSFC values that are slightly superior to current PFI engines. Exhaust emissions in this G-DI class can be controlled through the combination of a three-way catalyst and EGR. A reduction in HC emissions during cold start, coupled with improved transient response, will be the major advantages over the conventional PFI engine. The availability of an efficient and durable lean-NOx catalyst will enable the second stage of G-DI development, and engines incorporating ultra-lean stratified-charge operation for load control may then be developed and optimized. These engines will most likely be designed to provide both the specific power of the homogeneous-charge mode and a 10% to 15% increase in overall vehicle fuel economy [521].

6.3 Split Injection Strategy

Experiments have shown that certain performance and emissions advantages may be realized by invoking the strategy of split fuel injection. The flexibility afforded by electronic, common-rail systems permits split injection, which is the scheduling of more than one fuel injection event during a single engine cycle. Introducing portions of the total fuel mass at predetermined times in the cycle can indeed influence combustion and emissions, and provide another set of parameters with which to optimize. As discussed in detail in Chapter 4, the capability of an injection system to reliably deliver two, or even three, distinct spray events of acceptable atomization quality in the time period occupied by the intake and compression strokes is very much dependent on the specific characteristics of the fuel injector. The pulse widths for multiple injections are significantly shorter than for the equivalent single injection, thus a key parameter is the specific minimum pulse width at the low-flow end of the linear flow curve for a given injector. If a pulse width less than this threshold is utilized, a large portion of the pulse is devoted to the injector opening and closing ramps, and it is commonly observed that spray atomization quality is degraded. In addition, large injection-to-injection variations in fuel quantity may occur with some injector designs. This is the main reason why the most common approach to split injection is to inject fuel twice during one engine cycle, as opposed to three or four times. This strategy may also be denoted as two-stage or double injection in the context of some publications. The rapid actuation of a piezoelectric injector could certainly expand the viable option beyond double injection if combustion or emission benefits can be documented.

6.3.1 Two-Stage Injection Strategy for Mode Transition

For enhancing the smoothness of the mode transition, a two-stage injection strategy that injects portions of the fuel into the cylinder during both the intake and the compression strokes has been adopted, with the operating region being illustrated in Fig. 6.3-1 [156, 302]. With this injection strategy, a weak stratification is formed, with excessively rich or excessively lean mixtures avoided, and stable combustion being obtained from the overall-lean mixture. This two-stage injection does indeed facilitate a smooth transition from lean-stratified combustion to full-load, homogeneous-charge combustion. The potential of this two-stage injection for enhancing G-DI engine performance and exhaust emissions may be noted from the data in Fig. 6.3-2 [156].

Two studies have been conducted to explore the potential of a two-stage injection strategy [163, 314] in

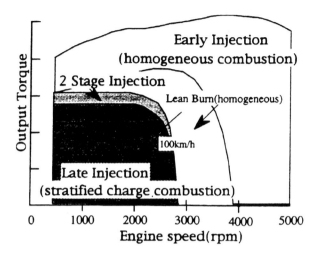

Figure 6.3-1 Operating map of the Toyota first-generation D-4 engine [156].

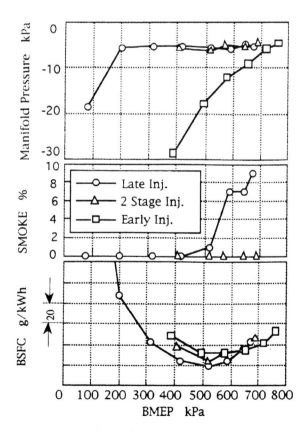

Figure 6.3-2 Effect of two-stage injection on engine performance and emissions during load transition [156].

which the main fraction of the fuel is injected during the intake stroke, ensuring the formation of a homogeneous, premixed, lean mixture, and a rich mixture is created in the vicinity of the spark gap. This is done by subsequently injecting the remaining fraction of the fuel just prior to the spark. This was found to provide more stable ignition and faster combustion. The lean combustion limit is definitively shown to be extended with this technique, while fuel consumption and HC emissions attain the same level as in the case of homogeneous combustion. This injection strategy has also been evaluated using a single-cylinder engine with a bore of 135 mm and a compression ratio of 9.6:1, and equipped with a multihole nozzle [314]. Stable combustion is realized in this two-stage-injection engine over a wide range of operation without detonation.

Although the injection hardware utilized in this study is not representative of contemporary production G-DI systems, there are some interesting observations that should be considered. As compared to operation with stoichiometric, homogeneous combustion, significant enhancements in fuel consumption and NOx emissions are achieved. Optimized combinations of spark timing, secondary fuel injection timing, the fuel split between the primary and secondary injections, and the number of nozzle holes are essential for improved ignition and combustion with this strategy. Both NOx and HC emissions are reduced by retarding the ignition timing; however, the BSFC increases due to the reduced degree of constant volume combustion. Retarding the timing of the secondary fuel injection yields similar trends for NOx and BSFC; however, HC emissions increase monotonically due to the decrease in available mixing time. Because the split fraction is altered by increasing the fuel quantity in the secondary injection, both HC emissions and BSFC are improved, but NOx increases slightly. As the load is reduced, the mixture formed by the first stage injection becomes extremely lean, and HC emissions exhibit an increase. The stratified mixture created by the two-stage injection exhibits a high flame speed and a short combustion duration,especially at higher excess air ratios. The use of two-stage injection enables the use of less volatile fuels and lower octane numbers.

The second study on two-stage injection [163] used a water-cooled, 4-cylinder engine with a square-shaped piston cavity, a bore of 102 mm, and a compression ratio of 12:1. As a consequence of the early primary injection a lean, homogeneous mixture is formed at the end of the compression stroke. This avoids the formation of an ultra-lean mixture, which could result in flame extinction, particularly at low engine speeds and light loads. The fuel from the later secondary injection creates a near-stoichiometric mixture near the spark gap, which improves the early development of the flame over a wide range of engine speeds and loads. The optimum fraction of fuel in the secondary injection has to be controlled carefully in order to improve both the lean limit and the BSFC. As noted earlier, a fuel injector that can accurately and reliably deliver small quantities of fuel is required if this strategy is to be successfully implemented. If the fraction of fuel in the primary injection is too low, the main pre-mixed homogeneous mixture becomes too lean and the normal flame of the lean homogeneous mixture is extinguished. A spark plug with an extended electrode is a beneficial addition in the implementation of this strategy, as it yields a substantially extended lean limit for a wide range of secondary fuel injection timings.

6.3.2 Split Injection for Improving Full-Load Performance

Another split-injection strategy, called two-stage mixing, was invoked to suppress autoignition and improve the low-speed torque on a G-DI engine [15]. As was the case in the studies discussed above, fuel is injected twice during the entire mixture preparation process, with the first injection performed during the early stage of the intake stroke, and the second injection occurring late in the compression stroke. Figure 6.3-3 schematically illustrates this strategy [264]. The optimal injection strategy and the resulting benefits are shown in Figs. 6.3-4 and 6.3-5. This strategy is effective when the majority of the fuel is supplied by the second injection, with the first injection being utilized only to create a mixture with an overall air/fuel ratio from 30:1 to 80:1. Significant knock suppression is achieved at low engine speeds, and it is theorized that the premixed mixture created by the first injection is too lean to achieve autoignition of the end gas, and that the stratified mixture created by the second injection does not have sufficient reaction time to autoignite. It is also theorized that the second injection may contribute to soot formation when the mixture is rich. However, soot emission is not observed when this strategy is utilized, except at very low engine speeds, even when the mean air/fuel ratio is 12:1 [15]. The soot generated during the early stage of the combustion process is later oxidized completely, as depicted

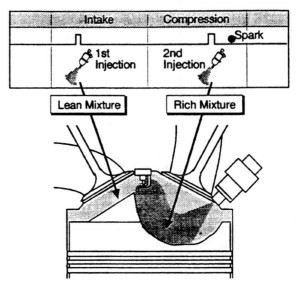

Figure 6.3-3 Schematic of the two-stage mixing strategy for enhancing low-speed torque [264].

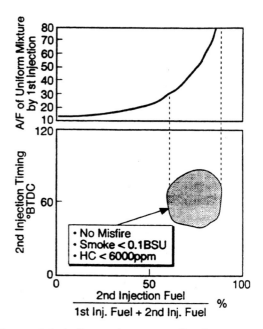

Figure 6.3-4 Optimal injection for the two-stage mixing strategy (600 rpm; WOT; air/fuel ratio: 12:1; first injection timing: 280°BTDC; ignition timing: 20°BTDC) [264].

in Fig. 6.3-6 [264]. Furthermore, the presence of soot is effective in extending the flame propagation limits to leaner mixtures. As a result, the low-end torque characteristics of the G-DI engine are improved significantly, even when using a compression ratio of 12.5:1.

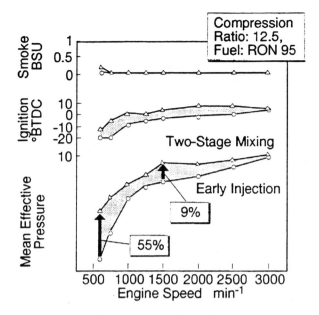

Figure 6.3-5 Combustion and performance improvements resulting from the two-stage mixing strategy [342].

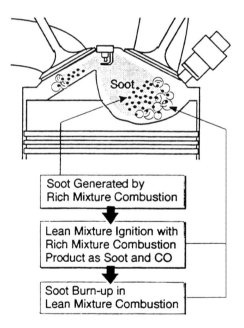

Figure 6.3-6 Schematic of soot formation and oxidation mechanisms for the combustion of a stratified-charge mixture [264].

A similar strategy was also proposed to increase full-load torque [506]. By splitting the total amount of delivered fuel into two separate pulses, both charge cooling and knock suppression can be enhanced. The timing

of the second injection is not necessarily restricted to the latter stages of the compression stroke, but may occur over a fairly wide timing window during the compression stroke. It is important to note that combustion stability is relatively insensitive to the timing of the secondary injection pulse when two-thirds of the fuel is injected in the first fuel pulse. For this strategy the highest IMEP is achieved when the secondary injection occurs near 60°BTDC. By comparison, if half or more of the fuel is delivered by the secondary injection, the start of the secondary injection should be no later than 150°BTDC if combustion instabilities are to be avoided. In order to improve the robustness of combustion, a ratio of 2:1 between the primary and secondary injected fuel quantities is an excellent starting point for optimization. This two-stage mixing strategy may also be utilized in combination with the slightly lean, stratified approach (Section 6.2.4) to minimize the amount of soot generated from the late injection of a substantial amount of fuel. Figure 6.3-7 shows the effect of the fraction of fuel injected during the second injection on the combustion duration, fuel consumption, smoke and HC emissions, and CO concentration for three engine loads. When the fraction of fuel in the second injection increases, the stratification level of fuel distribution is increased, leading to an increase in CO concentration and a rapid heat release. Somewhat surprisingly, HC emissions are the lowest when all of the fuel is injected during the compression stroke, possibly because fuel injected during the intake stroke may be emitted without complete combustion, because the mixture near the chamber wall is too lean. When the fraction of fuel in the second injection is increased, fuel consumption is improved at light load but deteriorates at heavy load, indicating that the advantages of stratification are lost with an increase in load. Through optimizing the fuel injection timings of both the first and second injections, smoke emissions may be kept from increasing.

6.3.3 Post Injection for Improving Catalyst Light-Off Characteristics

The process of split injection, when implemented as a two-stage combustion strategy, may also be utilized to rapidly heat an underfloor, three-way catalyst that is located downstream of the lean NOx catalyst [15, 266]. With this strategy the engine is operated lean in the late injection mode under cold-start conditions, as schematically illustrated in Fig. 6.3-8. A supplementary amount of fuel is injected during the latter stages of the expansion stroke,

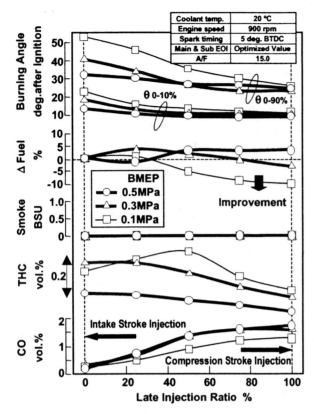

Figure 6.3-7 Effect of two-stage mixing on combustion duration, fuel consumption, smoke and HC emissions, and CO concentration over three different engine loads; slightly lean, stratified-charge strategy to reduce catalyst light-off time [501].

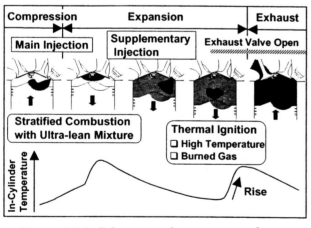

Figure 6.3-8 Schematic of two-stage combustion strategy for reducing catalyst light-off time [501].

resulting in an increased exhaust gas temperature. When the supplementary fuel is injected early in the expansion stroke, a luminous diffusion flame from liquid fuel

droplets results, whereas when fuel is injected during the mid-portion of the expansion stroke a slow pre-combustion reaction occurs and ignition is delayed. For the later injection timing the effect of this strategy on exhaust gas temperature and associated HC emissions is shown in Fig. 6.3-9. As illustrated in Fig. 6.3-10, this two-stage combustion strategy is only applied for the first 20 seconds after the initiation of cranking, which reduces the catalyst light-off time to 100 ms from the original 300 ms. This expansion-stroke-injection strategy does indeed yield a significant reduction in cold-start HC emissions. As has already been shown in Fig. 6.2-6, a significant fuel economy penalty is associated with this two-stage combustion strategy. Therefore, it should be implemented only after careful evaluation of the

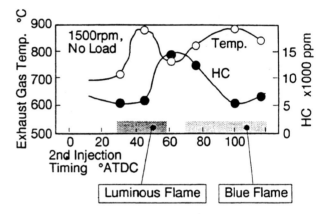

Figure 6.3-9 Effect of two-stage combustion on exhaust gas temperature and associated HC emissions [342].

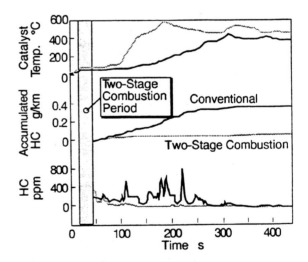

Figure 6.3-10 Effect of two-stage combustion on catalyst light-off characteristics [342].

advantages and disadvantages. Combining this strategy with other approaches such as the slightly lean, stratified-charge technique has been found to be effective in reducing fuel consumption while still meeting the requirement for reducing the catalyst light-off time.

This two-stage combustion strategy using post injection is a viable alternative to secondary air injection for obtaining a more rapid catalyst light-off during engine cold start. As shown in Fig. 6.3-11, the reduction in light-off time achieved with post injection (double injection in the figure) is comparable to that with secondary air, but with less hardware complexity. It should be emphasized that a non-trivial system optimization effort is required if the post-injection strategy is to be utilized, and optimization must include avoidance of the particulate emissions associated with sub-standard mixing and incomplete combustion [93].

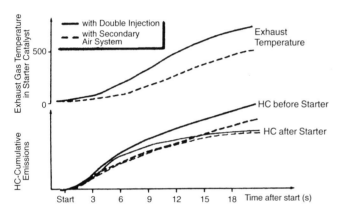

Figure 6.3-11 Comparison of exhaust gas temperature and HC emissions [93] for double-injection and secondary-air strategies.

6.3.4 Post Injection for NOx Storage Catalyst Regeneration

As will be discussed in Section 8.3.7, oxygen-deficient exhaust is required occasionally to regenerate the NOx storage catalyst. This is very difficult to realize in the exhaust stream during extended periods of time for part-load, lean operation since any fuel injected, if not precisely optimized in time and quantity, will lead to torque fluctuation. With a direct injection engine, this precisely controlled rich spike can be created in the exhaust stream anytime through demand by the engine management system for a post injection late in the expansion stroke. Through optimizing the air flow rate, fueling rate, fuel injection timing, EGR rate, and spark

timing, the NOx storage catalyst can be regenerated without any noticeable torque fluctuation.

6.3.5 Split Injection for Control of Homogeneous-Charge, Compression-Ignition Engine

Direct injection of fuel into the cylinder with flexible fuel injection timings also provides the opportunity to control the combustion process of a homogeneous-charge, compression-ignition (HCCI) engine. HCCI combustion is the process whereby a homogeneous mixture is autoignited through compression. This combustion process has the potential to significantly reduce NOx and particulate emissions, while achieving higher thermal efficiency and having the capability of operating with a variety of fuels. To a degree, the HCCI combustion process is able to combine the best features of an SI engine using gasoline fuel and a diesel engine using diesel fuel. In a manner similar to that in an SI engine, fuel and air are mixed to obtain a homogeneous mixture, which can eliminate fuel-rich diffusion combustion and can thus dramatically reduce particulate emissions. With an ignition process similar to that of a diesel engine and an extremly lean mixture to support a propagating flame, the HCCI engine undergoes an autoignition process that can eliminate the bulk flame propagation of the conventional SI engine. Therefore, NOx emissions from a gasoline HCCI engine can be very low when compared to those from a conventional SI engine using gasoline fuel. Furthermore the unthrottled operation of the gasoline HCCI engine is possible at an equivalence ratio of 0.3 and an EGR rate of 45% without misfire, thus yielding a high thermal efficiency and extremely low NOx. The HCCI engine is thus an attractive technology that can ostensibly provide diesel-like fuel efficiency and very low engine-out emissions without the requirement of lean NOx aftertreatment systems.

HCCI combustion is achieved by controlling the in-cylinder temperature, pressure, and composition of fuel-air mixture at the time of ignition so that the charge ignites spontaneously throughout the entire combustion chamber. Compared to the positive ignition mechanism of conventional internal combustion engines, such as a spark for SI engines and fuel injection for diesel engines, the lack of a distinct triggering event makes the control of this combustion process quite challenging. The local charge temperature and the air/fuel ratio are the key variables controlling the initiation of the HCCI combustion process. The injection of fuel directly into the cylinder provides the potential to control this combustion process through the altering of the local fuel concentration by varying the injection timing. The gas temperature may also be altered through charge cooling from fuel evaporation. Early fuel injection provides adequate time for fuel to vaporize and mix with the air to achieve a homogeneous charge. Late pilot fuel injection into the combustion chamber during the compression stroke can control HCCI combustion by increasing the local fuel concentration in some regions of the combustion chamber. The capability of split injection with G-DI engines can combine these two functions. Moreover, control of HCCI operation during transient requires an accurate fuel metering in order to avoid engine knock or misfire. Clearly direct injection of fuel into the cylinder makes this type of fuel metering requirement possible during engine transient operation.

Figure 6.3-12 shows a proposed concept to control the HCCI combustion process through a combination of early exhaust valve closing and in-cylinder fuel injection [476]. Early exhaust valve closing traps a significant amount of residual gas in the cylinder to provide a portion of the temperature required to initiate the HCCI combustion process. Fuel can be injected into the cylinder directly through a G-DI injector well before the opening of the intake valve to utilize the residual gas heat to promote fuel vaporization. There will also be sufficient mixing time available to obtain a homogeneous charge. It is apparent that the capability of the G-DI engine for multiple direct fuel injection provides the opportunity to

control the HCCI combustion process. With further development of the HCCI combustion strategy, the G-DI engine can be calibrated to incorporate an HCCI operating mode as one of the control options in order to take full advantage of this unique combustion process.

6.3.6 Implementation of Split Injection in Engine Operating Map

The operating map of an engine that effectively uses split injection is shown in Fig. 6.3-13. Further comparisons of split injection (double injection in the figure) strategy for a range of fuel-mass split ratios under various operating loads are summarized in Fig. 6.3-14 [261]. For the transition between stratified-charge and homogeneous-charge operation, double injection permits torque matching with minimized NOx emissions. Experimental investigations have indicated a split ratio of approximately 3:1 (primary:secondary) yields favorable results. The use of the double injection strategy during part-load, steady-state operation is beneficial for soot emissions reduction, and could provide an incremental fuel economy improvement at low engine speed in the transition area between stratified-charge and homogeneous-charge operation [260, 261]. At full load a torque increase of up to 5% can be achieved by the extension of the knock limit. The optimum split ratio for this operating mode has been shown to be approximately 3:7, but is dependent on the basic engine design. For the aforementioned strategies the typical SOI is 330°BTDC for the first injection and 80°BTDC

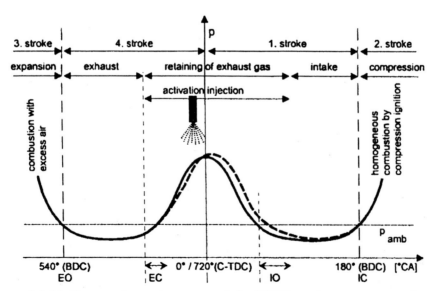

Figure 6.3-12 Proposed concept to control the HCCI combustion process through a combination of early exhaust valve closing and in-cylinder fuel injection [476].

for the second injection. When implementing double injection for fast catalyst light-off, the recommended SOI is 80°BTDC for the first injection and 50°ATDC for the second injection. It should be noted that the optimum split ratio is strongly dependent on the actual catalyst heating demand, and will vary with the volume and heat losses of the system [261].

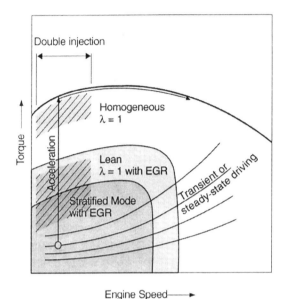

Figure 6.3-13 Engine operating modes that may effectively implement the split injection (double injection) strategy [261].

6.4 Combustion Characteristics

6.4.1 Homogeneous-Charge Combustion

The characteristics of G-DI combustion change significantly with the control strategy that is used [234, 321, 322, 324, 331, 447]. Engine combustion characteristics for very early injection, which is the precursor for forming a homogeneous mixture in the cylinder, are basically the same as those of standard, premixed-charge, PFI engine. Investigations of the combustion characteristics of a G-DI engine using early injection combined with a range of injection timings during the intake event reveal that, with a centrally mounted injector, the emissions and engine performance are indeed quite similar to that of a baseline PFI engine [407]. This indicates that the use of early injection timing achieves a homogeneous mixture because of the significantly longer time available for mixture preparation. Retarding the injection timing yields higher CO emissions because the mixture at the time of ignition is less homogeneous, hence the combustion is less stable. HC emissions are minimized with injection at BDC on induction due to a decrease in fuel impingement on the piston crown and other quenching areas. High-speed-video imaging of combustion reveals that early fuel injection results in a blue flame during combustion that is similar to that observed in the PFI

Double Injection Mode	Switch-Over Double Injection	Part-Load Double Injection	Full-Load Double Injection	Catalyst-Heating Double Injection
Operating Area	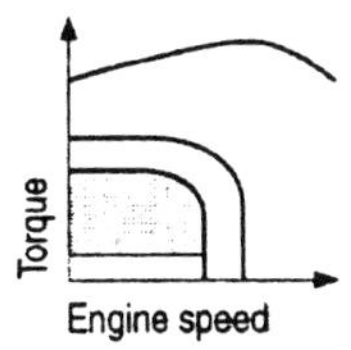	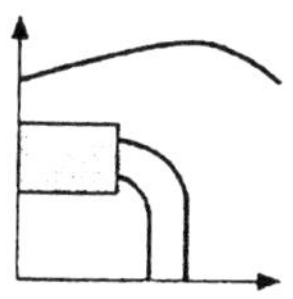	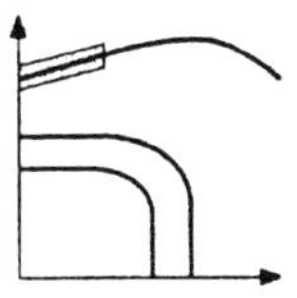	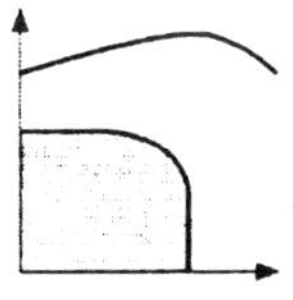
Reasons for Use	Optimized torque constant during switch-over with minimized NOx	Decrease in soot emissions and fuel consumption	Increase in torque output due to optimized knock limit	Active catalyst temperature control for elimination of sulfates, fast warm-up
Typical Timing of the 1st and 2nd Injection	300° BTDC 80° BTDC	330° BTDC 80° BTDC	330° BTDC 80° BTDC	80° BTDC 50° ATDC
Split Factor (%) (Note: 100 means 1st injection only; 0 means 2nd injection only)	75	75	Variable between 30 and 70	Variable due to heating demand

Figure 6.3-14 Principal applications of the split injection (double injection) strategy for optimizing combustion, emissions, performance, and driveability of G-DI engines [261].

engine; however, a yellow flame indicative of stratified-charge combustion is observed when the injection timing is retarded to the end of the intake stroke. For a G-DI engine operating in the homogeneous-charge mode at full load, the heat release curve obtained is nearly identical to that from a PFI engine. A slightly reduced heat release rate is obtained for some G-DI engines at full load, which is indicative of a detectable degree of charge inhomogeneity [108].

For injection during induction, retarded injection timing improves the start of combustion and advances the 50% heat release point due to the presence of a rich mixture zone around the spark gap. However, changes in injection timing do not significantly alter the combustion duration. For the late-injection, stratified-charge mode, an excessively rich mixture near the spark gap must be avoided in order to maintain stable ignition and minimize smoke. Therefore, it is imperative that a mixture preparation strategy that will accommodate a wide range of engine operating conditions be developed for each specific G-DI design [13]. For homogeneous charge combustion, a wide crank-angle window exists for the timing of early injection, but for stratified-charge combustion only a narrow operating window is available for the timing of late injection. At higher loads, a longer injection duration with the accompanying richer mixture results in a decreased sensitivity to injection timing [201–203].

The combustion characteristics associated with early fuel injection under cold start conditions have been investigated using a homogeneous, stoichiometric mixture [408]. For an early injection timing of 30°ATDC on the intake stroke, the heat release rate during the later stage of combustion is observed to be quite low due to the slow evaporation of liquid fuel from the piston crown. As a result, HC emissions and engine BSFC are degraded. Very early injection timing during induction delays initial combustion and extends the main combustion period due to the presence of a lean mixture resulting from the slow vaporization of the liquid film on the piston crown. Visualization of spray impingement on the piston shows that, for an injection timing of 30°ATDC on intake stroke, the relative spray impingement velocity is higher, and the impingement footprint is smaller, than for later timings. Due to the brief time available before spray impingement, the contribution of the in-cylinder air flow field to spray penetration and

evaporation is limited, resulting in a substantial liquid film being formed on the piston crown. For later injection timing, with the piston moving away from the injector more rapidly, the relative impingement velocity of the spray is reduced and the impingement footprint is generally larger. As a consequence, the fuel film is generally thinner and the evaporation of fuel droplets in the ambient air flow field is enhanced due to the increased time available prior to the initial impingement. All of these factors contribute directly to a reduction in the amount of fuel on the piston crown. Visualization of the associated combustion indicates that, for the case of injection at 30°ATDC on intake stroke, the entire chamber is initially filled with a blue flame, after which a yellow flame is observed at the center of the chamber. This yellow flame persists until the beginning of the exhaust stroke, and is attributed to pool-burning of the film of liquid fuel on the piston surface, as discussed in Section 5.5.1. For an injection timing of 110°ATDC on the intake stroke, the presence of a yellow flame is quite limited, and it occurs at the periphery of the chamber rather than at the center. The yellow flame observed at the periphery of the chamber results from a lower film evaporation rate due to reduced air velocities in this region. If high-velocity air can be directed to the area of spray impingement, the wall film evaporation and fuel transport can be enhanced.

The throttling losses of the gasoline spark-ignition engine are relatively small for high-load operation. For this operating mode, engine efficiency is determined primarily by the compression ratio and the specific combustion characteristics; however, increases in the compression ratio and advances in the ignition timing for best efficiency are limited by mixture autoignition in the end-gas region. Improvements in combustion chamber geometry, piston, charge cooling, and residual gas control to modify flame propagation at high load have proven to be effective means for knock reduction in both PFI and G-DI engines. Modification of charge motion in order to obtain symmetric flame propagation is an effective way to improve the inherent knock resistance of the chamber [484]. In general, the best compromise among HC, NOx, BSFC, and COV of IMEP can be obtained by a combustion process that offers a fast and stable initial phase, a moderate main combustion rate and a locally uniform and symmetric end of combustion that avoids flame quenching.

6.4.2 Comparison of Stratified-Charge and Homogeneous-Charge Combustion

When the air/fuel mixture is successfully stratified for an idle or low-load condition, the mixture surrounding the spark gap is slightly rich at the time of the spark. If this is achieved, the reaction rate will be high enough to sustain efficient and stable combustion [344–346, 376, 377]. For a PFI engine operating at idle condition the combustion rate is low, and combustion stability is generally marginal, primarily because of the large amount of residual gas. For the case of a G-DI engine at idle, the initial combustion rate is approximately the same as that for the full-load condition [448]. The G-DI engine, however, exhibits a significant advantage in both ignition delay and burn duration when compared to a PFI engine of equivalent geometry [199, 207, 459]. As illustrated in Fig. 6.4-1, the initial flame kernel develops rapidly in the rich mixture region near the spark gap; however, the rate of flame propagation is reduced in the lean outer region of the stratified charge. The significantly reduced combustion rate near the end of the combustion process is one of the causes of the observed increase in HC emissions [227]. The overall high flame speed does

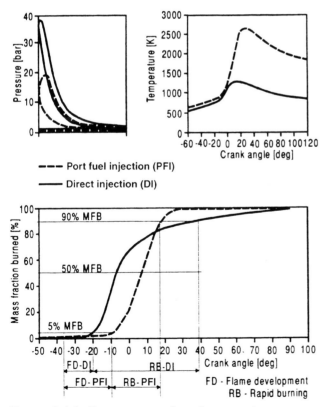

--- Port fuel injection (PFI)

—— Direct injection (DI)

Figure 6.4-1 Comparison of combustion characteristics between PFI and G-DI engines [413].

allow ignition timing to be retarded more for the G-DI engine than for the conventional PFI engine, and combustion rate and stability are enhanced rather than degraded with a reduction in idle speed [448].

A detailed comparison of pressure-volume diagrams for homogeneous-stoichiometric and stratified-charge operation of a production G-DI engine at 1500 rpm and 2.62 bar BMEP is shown in Fig. 6.4-2 [51]. Both are optimized at minimum spark advance for best torque (MBT) timing, and the EOI is optimized for best BSFC. Typical features of these two combustion processes are illustrated in Table 6.4-1 [51]. It may be observed that the peak motoring pressure at a manifold absolute pressure (MAP) level of 72 kPa for stratified-charge operation is nearly the same as the peak firing pressure for homogeneous, stoichiometric operation. From the log-log, pressure-volume diagram shown in Fig. 6.4-2 it is evident that stratified-charge combustion releases heat more rapidly at the beginning of the event. This is demonstrated by the rapid rise in pressure immediately following the spark, and by the slope of the curve in the vicinity of peak pressure. In contrast, homogeneous-charge combustion exhibits a smooth curve throughout the combustion event, thus indicating a slower initial burn rate and more uniform heat release.

The combustion characteristics of fast, early burn and slow, late burn with a stratified charge are illustrated in Fig. 6.4-3. If stratified-charge combustion were uniformly fast, further reductions in fuel consumption would be possible [51]. Combustion phasing with stratified-charge operation is seen to be quite different from that of homogeneous-charge combustion, with an overall air/fuel ratio of 30:1 burning much faster during the early and middle combustion periods than does a homogeneous stoichiometric mixture. At an air/fuel ratio of 30:1, the ambient density in the combustion chamber is approximately double that for stoichiometric operation at the time of spark, due to the induction of about twice as much air. If the mixture near the spark gap were stoichiometric for both cases, the flame speed would be nearly the same for both. However, with nearly twice the density for the 30:1 charge, about twice as much mass must undergo heat release per unit distance of flame travel. That is why the 0–2% mass burn time is 7.5° for stratified-charge combustion as compared to 13.3° for stoichiometric operation. This is a typical feature of the lean-stratified-charge combustion process, and is independent of the injection system (air-assisted or single fluid) and operating conditions (with or

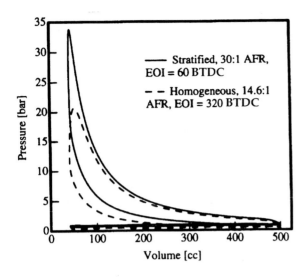

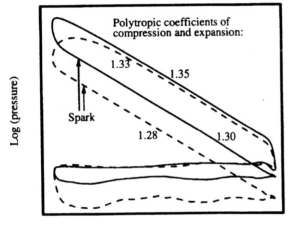

Figure 6.4-2 Pressure-volume diagrams for homogeneous and stratified operation; 1500 rpm, 2.62 bar BMEP and MBT spark advance [51].

Table 6.4-1

Comparison of combustion characteristics between homogeneous-charge and stratified-charge combustion at 1500 rpm, 2.62 bar BMEP and MBT spark advance [51]

Compared Items	Homogeneous Charge	Stratified Charge
Air/Fuel Ratio (overall)	14.6:1	30:1
MBT Spark Advance (°BTDC)	20	18
Crank Angle of Peak Pressure (°ATDC)		7
MAP (kPa)	42	72
Peak Firing Pressure (bar)	20.5	33.7
Peak Motoring Pressure (bar)	11.6	19.5

without EGR). This is evident in Fig. 6.4-4, which illustrates the effect of air/fuel ratio on the location of peak pressure at 1500 rpm, 2.62 bar BMEP and MBT spark advance. It is clearly shown that at MBT timing the crank angle of peak pressure occurs closer to TDC as the air/fuel ratio is enleaned. This is observed for engines with air-assisted (engine A) or single-fluid (engine L) fuel injection systems, with or without EGR.

A comparison of the incremental burn rate between homogeneous-charge and stratified-charge combustion is presented in the lower portion of Fig. 6.4-3, where the burn time required for the next incremental 1% of mixture mass is plotted as a function of the cumulative percent mass fraction burned. It is shown that a stratified mixture with an air/fuel ratio of 30:1 consumes mass faster than the stoichiometric, homogeneous charge until about

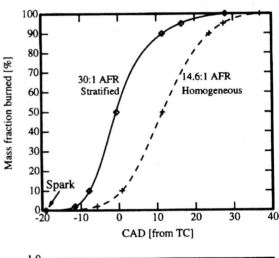

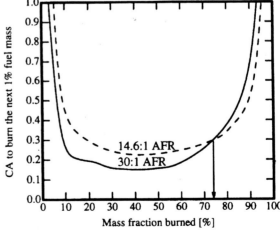

Figure 6.4-3 Comparison of combustion characteristics between homogeneous and stratified combustion; 1500 rpm; 2.62 bar BMEP [51].

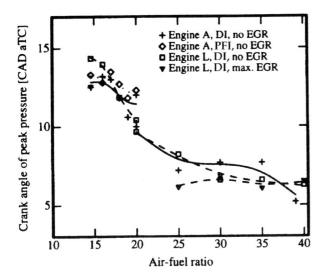

Figure 6.4-4 Effect of air/fuel ratio on the crank angle of peak cylinder pressure for engines with air-assisted (engine A) and single-fluid (engine L) fuel injection systems; 1500 rpm; 2.62 bar BMEP; MBT spark advance [51].

three-quarters of the total mass is burned. At this mass fraction, the stratified-charge combustion rate becomes slower than the stoichiometric-charge combustion, due to the flame front encountering very lean regions in the chamber. As indicated, 50% of the charge mass is burned just prior to TDC [51, 108]. With a homogeneous charge, this would be considered a significantly advanced combustion phasing; however, the reduced burn rate in the late stages of combustion forces the burn to be initiated early for maximum efficiency. If stratified-charge combustion were initiated later, combustion would likely be extinguished before all of the fuel mass was consumed. It would at first seem to be desirable to phase the stratified-charge combustion such that less than 50% of the mass would be burned by TDC; however, retarding the spark timing does not improve the cycle efficiency at all, due to the decrease in heat release during the late portion of the combustion event. This combustion characteristic is confirmed by measurement of engine-out HC emissions as a function of spark timing. As is well established for the PFI engine, HC emissions decrease as spark timing is retarded; however, stratified-charge combustion exhibits exactly the opposite trend. As the spark is retarded for stratified-charge combustion, more time is available for mixing, and some of the stratification is lost. Another possible negative effect is that a portion of fuel may become detached from the main fuel cloud, resulting in the flame having no path to the detached cloud [51].

Flame radiation and observed spectra from flame luminosity measurements for different injection timings are presented in Fig. 6.4-5. The combustion associated with early injection is characterized by flames that are typical of premixed lean or stoichiometric mixtures, and the flame luminescence that is observed is attributed to OH and CH chemiluminescence, as well as to CO-O recombination emission [259]. Luminescence at the longer wavelengths normally associated with soot radiation is generally not observed. For the case of late injection, the major component of flame luminosity consists of continuous blackbody radiation emitted from soot particles formed inside the cylinder, which is typical of a stratified-charge combustion process. Soot radiation decreases abruptly when sufficient air is entrained into the reaction zone. Figure 6.4-6 shows the result of a spectral analysis of flame radiation from both the early and late stages of combustion [15]. For heavy-load operation, soot may be generated for some injection timings as a result of the

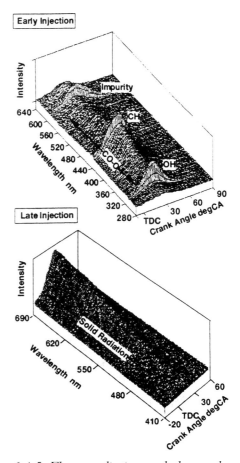

Figure 6.4-5 Flame radiation and observed spectra for early and late injection; 1500 rpm; a fuel injection quantity of 15 mm3/st 4 cylinders; 1.8L [259].

presence of a liquid film on the piston crown. This can occur for injection early in the intake stroke or for a fuel injector having a large spray droplet size and/or high penetration. Another cause of soot generation is an insufficient time for fuel-air mixing, which can occur for high-load injection late in the intake stroke. A number of design parameters such as piston crown geometry, spray cone angle and penetration, injection timing, and in-cylinder air motion must be optimized to minimize soot formation.

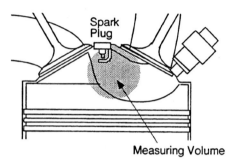

(a) Flame Emission Measuring Volume

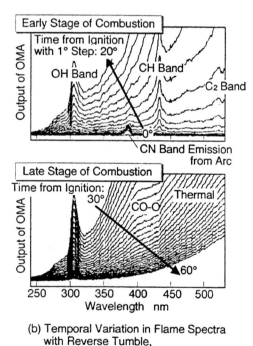

(b) Temporal Variation in Flame Spectra
with Reverse Tumble,
Injection Start Timing: 50°BTDC,
Ignition Timing: 15°BTDC

Figure 6.4-6 Spectral characteristics of flame radiation from the early and late stages of stratified combustion; SOI timing: 50°BTDC; ignition timing: 15°BTDC; 1000 rpm; and air/fuel ratio: 35:1 [264].

6.5 Effects of Engine Operating and Design Parameters on G-DI Combustion

6.5.1 Effects of Injection Timing and Ignition Timing

When an engine operates in stratified-charge mode, the EOI timing is an important parameter, as the EOI marks the time when the last liquid fuel enters the chamber [91]. An important point that was discussed in detail in Chapters 4 and 5 should be recalled: G-DI injectors typically require from 0.35 to 0.68 ms to fully close, thus fuel will continue to enter the chamber for that time interval after the EOI logic command. Combustion diagnostic systems in dynamometer cells normally record SLP and ELP, and typically equilibrate them to SOI and EOI, which they are not. Retarded EOI timing generally results in a more rapid start of combustion, while an advanced EOI tends to shorten the overall combustion duration. The 50% heat release point occurs before TDC when EOI timing is advanced (starting with a baseline of 57°BTDC on compression), and occurs after TDC when EOI timing is retarded [227]. The ISFC is enhanced as EOI timing is retarded; however, HC emissions increase as a result of the degraded mixture quality and an extended combustion duration. It is necessary to carefully phase ignition timing with injection timing in order to achieve accurate control of mixture stratification. By positioning a locally rich mixture around the spark gap, while maintaining a constant overall-lean air/fuel ratio, the COV of IMEP can be reduced by more than 60%. More than 40% of the observed improvement is attributed to the presence of the rich mixture around the spark gap, with the remaining improvement attributed to enhancement of mixing due to the injection event [25].

For stratified-charge operation of a G-DI engine with the reverse-tumble-based combustion system, 40°BTDC on the compression stroke is the most retarded injection timing that can be utilized if the ignition timing is set at 15°BTDC [263]. Charge-stratification is achieved in this engine by the fuel spray cloud being guided after interacting with the piston bowl surface. No significant difference in combustion duration is observed between optimized injection timing and the earliest possible injection timing; however, a marked cycle-to-cycle combustion variation is observed for very advanced injection timing. This is attributed to over-mixing and to a dispersion of the fuel spray plume. The flame propagates more rapidly in the rich region of a stratified charge than in a homogeneous charge, but no significant differences in the flame propagation speed occur as the injection timing is changed. An in-cylinder air flow field that can assist in confining the fuel spray plume inside the piston cavity is effective in achieving a high degree of charge stratification.

The conditions for which stable combustion is achieved for late injection of a G-DI engine with the reverse-tumble-based combustion system are illustrated on the injection and ignition timing map in Fig. 6.5-1 [196]. It is important to create an appropriate interval between EOI and the time of the spark in order to promote fuel evaporation and mixing without creating over-dispersion of the mixture plume. This interval must account for the injector closing delay and the larger mean droplet size of the last fuel droplets to exit the injector. For the tested engine the stable combustion area is the widest at an engine speed of 1500 rpm, and the permissible interval between EOI and spark timing is as wide as 10 to 60 crank angle degrees. With a decrease in engine speed, fuel spray impingement becomes a more important factor in mixture preparation. As a result, the operating area of stable combustion is reduced. When engine speed is increased, the crank-angle increment required for the spray to reach the piston surface becomes smaller, thus EOI must be advanced to maintain the same impingement phasing. The region of stable combustion will be smaller, due to the reduced time available for mixture preparation. This region continues to shrink as engine speed is increased, and in fact, a stable combustion area does *not* exist for this combustion system for an engine speed that exceeds 3500 rpm.

The combustion stability of a tumble-dominated, stratified-charge G-DI engine can be improved when ignition is advanced even beyond the MBT timing [201–203]. However, when ignition is advanced further combustion stability deteriorates very rapidly, which is attributed to insufficient time for fuel evaporation and mixing. HC emissions decrease with ignition advance, whereas fuel economy initially improves before degrading. NOx emissions are also relatively high due to the high local gas temperature. A reduction in the tumble ratio of the in-cylinder flow field has the beneficial effect of shifting the combustion phasing for minimum HC emissions closer to that for minimum BSFC. With ignition timing set for minimum HC emissions, a BSFC improvement of 11% is achievable at the same HC level, whereas with BSFC-optimized ignition timing, a reduction in HC emissions of 34% is obtainable at the

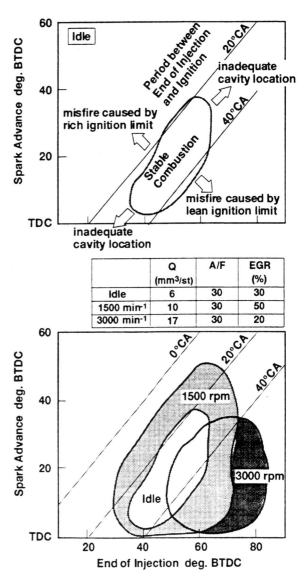

Figure 6.5-1 Map for stable stratified-charge combustion of a G-DI engine with the reverse-tumble-based combustion system [196].

same BSFC. One explanation is that the lower tumble ratio modifies the air/fuel ratio history at the spark gap, thereby producing optimum conditions for combustion at a more retarded timing. The reduction in NOx at all combustion phasings also supports the explanation that the lower tumble ratio modifies the local air/fuel ratio near the spark gap, resulting in a lower peak temperature during combustion.

Injection timing is a key parameter, even when utilizing an air-assisted fuel injection system that introduces its own air/fuel mixture cloud. Figure 6.5-2 shows the COV of gross mean effective pressure (GMEP) as a function of start air injection (SOA) timing at a load of 6.5 bar net mean effective pressure (NMEP). Emissions results for a G-DI engine with air-assisted injection are presented in Figs. 6.5-3 and 6.5-4. As shown in those figures, unthrottled stratified-charge operation is achievable with the air-assisted G-DI system to a substantial engine load. At an overall air/fuel ratio of 23:1, the use of earlier injection timing for more homogeneous-charge combustion yields a larger cycle-by-cycle variation. Retarding the injection timing results in more stable combustion for an overall lean air/fuel mixture due to an enhanced local charge stratification. However, extreme retardation of air injection timing yields a significant increase in CO and smoke emissions, because combustion is initiated in a richer mixture as the charge becomes more stratified. Moreover, retarded air injection timing is associated with elevated cylinder pressures, which diminishes the pressure differential required for fuel atomization. This yields a less-well-atomized fuel spray, as illustrated in Fig. 6.5-5. The fact that measured NOx emissions decrease when injection timing is retarded indicates that the local air/fuel ratio is substantially richer than stoichiometric. The observed decrease in HC emissions with retarded air injection timing may be indicative of a reduction in the amount of fuel in the piston crevices, and an associated reduction in the quenching effect [508].

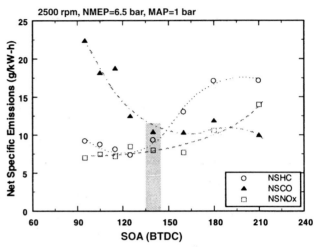

Figure 6.5-3 HC, CO and NOx emissions from an air-assisted DI combustion system at high load without throttling [508].

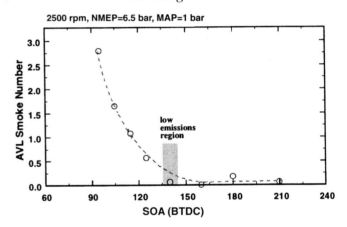

Figure 6.5-4 Soot emissions from an air-assisted DI combustion system at high load without throttling [508].

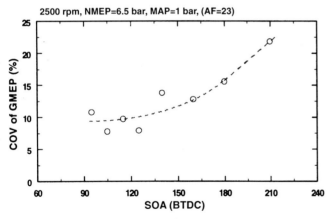

Figure 6.5-2 Combustion stability of air-assisted DI combustion at high load without throttling [508].

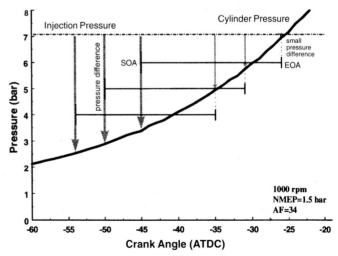

Figure 6.5-5 Variation of effective air-injection pressure differential with air-injection timing for an air-assisted DI combustion system [508].

6.5.2 Effect of Spray Cone Angle

Many investigations have verified that the effective angular extent of the spray, which is normally designated as the spray cone angle, is a very important parameter that very much influences G-DI emissions and performance. The concerns and caveats regarding the interpretation of the cone angle of a spray were addressed in detail in Section 4.6. For a prototype DI engine with a side-mounted injector, a strong correlation of smoke emissions with spray cone angle has been observed that is directly attributable to variations in the fuel film thickness on the piston, as discussed in Section 5.5.1 [190]. Larger spray cone angles can reduce smoke emission dramatically using the same fuel injection timing, as shown in Fig. 6.5-6. When the spray cone angle exceeds 70 degrees for this engine, the effect is less significant. The results from a parametric study of the effect of spray cone angle on emissions with a research DI engine are presented in Fig. 6.5-7 [227]. The optimum spray cone angle for the tested system configuration is 90° for minimum ISFC, 105° for minimum HC emissions, and 75° for minimum NOx emissions. The measured effects of spray cone angle on engine combustion stability, smoke emissions and engine torque from another study are shown in Figs. 6.5-8 and 6.5-9 for a G-DI combustion system with a side-mounted injector. The optimum spray cone angle is defined as one that produces the highest engine torque with minimum smoke emissions and stable combustion [340]. The optimal spray cone angle is strongly dependent on the detailed geometry of the combustion system and a compromise between the part-load combustion stability and the full load performance must normally be made.

n = 2000 rpm; bmep = 2 bar; fuel rail pressure = 60 bar

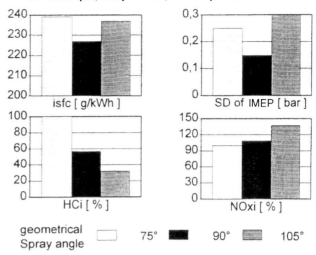

Figure 6.5-7 Effect of spray cone angle on engine performance and exhaust emissions [227].

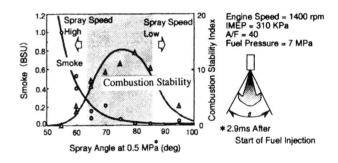

Figure 6.5-8 Effect of spray cone angle on combustion stability and smoke emissions [340].

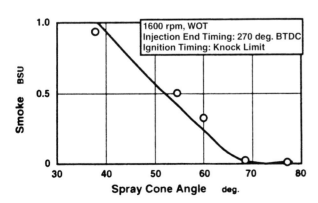

Figure 6.5-6 Effect of spray cone angle on WOT smoke emissions at an engine speed of 1600 rpm [440].

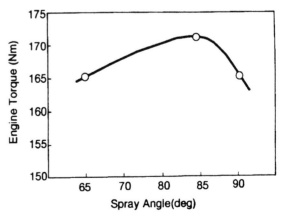

Figure 6.5-9 Effect of spray cone angle on engine torque [340].

6.5.3 Effect of Exhaust Gas Recirculation

The rapid flame development associated with stratified-charge combustion, and the higher O_2 and low CO_2 levels in the exhaust gas, are the reasons for the enhanced EGR tolerance of the G-DI engine. The cyclic variability of combustion becomes excessive at an EGR rate of 20% for the baseline PFI engine, whereas excellent combustion stability is possible for stratified-charge operation in a G-DI engine for an EGR rate of up to 40% [201–203]. For stratified-charge operation, EGR is found to have a slight effect on burn duration, whereas for the PFI engine, the delay and burn durations are significantly increased as the EGR rate is increased. As illustrated in Fig. 6.5-10, EGR does alter combustion phasing and slows the combustion process. G-DI engines without EGR show early combustion phasing with higher peak cylinder pressure, whereas the introduction of 40% EGR delays combustion and improves fuel economy while substantially reducing peak cylinder pressure. At an EGR rate of 40%, G-DI engines exhibit improvements of 3% in BSFC, 81% in NOx and 35% in HC emissions as compared to operation without EGR. The improvement in HC is theorized to be the result of a richer mixture core, resulting in less bulk quenching. The baseline PFI combustion system in comparative tests exhibits poor EGR tolerance, with an associated increase in fuel consumption. As expected, G-DI engines require more EGR for the same NOx reduction when compared to a baseline PFI engine. Another beneficial characteristic of the G-DI engine is improved tolerance to spark retard at higher EGR rates. For a research G-DI engine with a top-entry, bowl-in-piston configuration that is tested at MBT with 40% EGR, the burn rates are similar for both PFI and G-DI engines, although G-DI engines exhibit a mass-burn profile having a rapid initial rate and a slower rate near the end of combustion that is typical of stratified-charge combustion.

It should be noted that the use of a significant percentage of EGR causes a substantial decrease in exhaust temperature due to heat losses in the EGR system and the intake manifold. Figure 6.5-11 shows a typical trend in measured exhaust gas temperatures. For engine operation employing an overall air/fuel ratio of 30:1 with 30% EGR, the decrease in exhaust temperature is about 100°C relative to that for homogeneous, stoichiometric operation with EGR. It must be kept in mind that the temperature of the exhaust gases can directly influence the effectiveness of exhaust aftertreatment. A decrease in engine-out exhaust gas temperature may have serious implications at even light-load conditions where highly-diluted, stratified-charge operation is desirable from a fuel economy perspective [51].

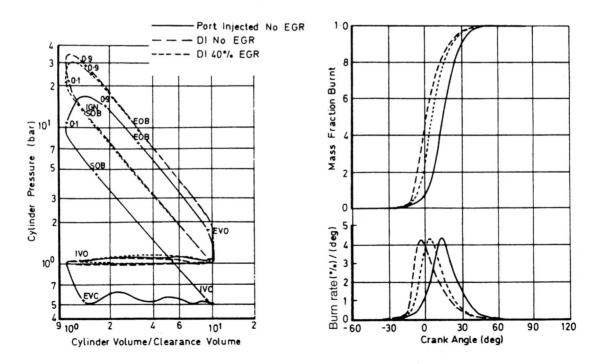

Figure 6.5-10 Effect of EGR rate on the combustion characteristics of PFI and G-DI engines [201].

190

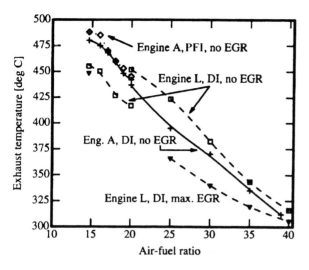

Figure 6.5-11 Effect of EGR on exhaust gas temperature of G-DI engines; engine A having an air-assisted injector; engine L having a single-fluid injector [51].

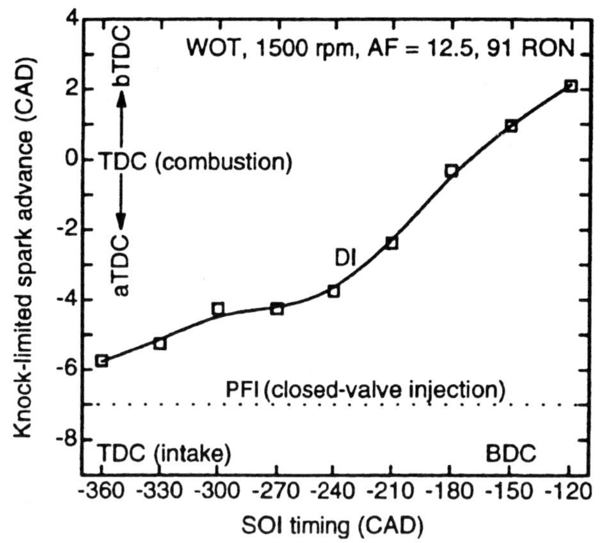

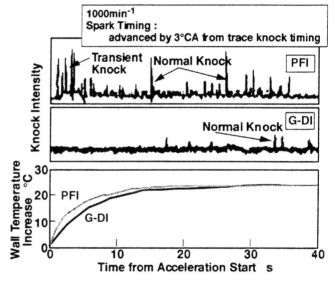

Figure 6.5-12 Comparison of the knock-limited spark advance between G-DI and PFI engines [10].

6.5.4 Knock Resistance Characteristics

Another important advantage of the G-DI engine is enhanced knock resistance compared to the conventional PFI engine, with in-cylinder charge cooling being one of the main contributors. Retarding injection timing can further improve engine knock resistance due to a slight stratification of the charge; however, such timings may not be desirable, as any charge cooling that occurs after intake valve closure does not improve volumetric efficiency, and charge stratification reduces air utilization. As shown in Fig. 6.5-12, for the heavy-load condition, the knock-limited spark advance continues to improve as the injection timing is retarded. Retarding the injection timing not only reduces the extent of wall wetting by the fuel spray, but also reduces charge heating by the walls. This results in a lower mixture temperature at TDC on compression, which translates to the option of a more advanced spark timing [9, 10]. It should be noted that retarding injection timing does reduce the time available for mixing, which usually results in a higher cycle-by-cycle variation in the IMEP. The poor mixing associated with extreme retardation of fuel injection timing for heavy-load operation may also lead to soot emissions.

It should be emphasized that G-DI engines do not exhibit the transient knock that is frequently observed in PFI engines during the first several cycles following vehicle acceleration. Figure 6.5-13 shows a comparison of the measured knock intensity during the acceleration

Figure 6.5-13 Comparison of the knock tendency between PFI and G-DI engines during transient operation [196].

of G-DI and PFI engines [196]. Transient knock is generally caused by the selective transport of a lower-boiling-point component of gasoline that has a lower octane number. In the case of G-DI engines, all of the gasoline components are injected into the cylinder directly and, as a result, transient knock is suppressed.

191

6.5.5 Comparison of Combustion Characteristics Between Air-Assisted and Single-Fluid G-DI Fuel Systems

Alternate fuel injection strategies and injector design characteristics may be utilized to achieve a significantly different in-cylinder fuel distribution, which will provide altered combustion and emissions formation. Two injection strategies that illustrate this precept are those that are based upon the single-fluid and air-assisted injectors, respectively. These two types of injection systems can indeed exhibit different combustion and emissions trends. NOx emissions as a function of air/fuel ratio for both the air-assisted (engine A) and the single-fluid (engine L) systems are shown in Fig. 6.5-14. Engine A exhibits a nearly constant NOx level for stratified-charge operation for air/fuel ratios in the range of 20:1 to 40:1, which is similar to the level achieved for stoichiometric, homogeneous-charge operation. Engine L produces an NOx level that is similar to that of engine A in the air/fuel ratio range from 30:1 to 40:1. Locally rich combustion yields the low NOx level observed for engine L at air/fuel ratios of 20:1 to 25:1, and the addition of EGR reduces NOx emissions significantly for stratified-charge operation.

A comparison of the HC emissions from engines A and L is shown in Fig. 6.5-15. Even for homogeneous-charge operation the HC emissions levels of G-DI

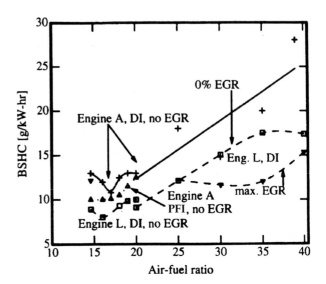

Figure 6.5-15 HC emissions levels for G-DI engines having single-fluid (engine L) and air-assisted (engine A) fuel injection systems; 1500 rpm; 2.62 bar BMEP; MBT [51].

engines are generally higher than those of conventional, state-of-the-art PFI engines. In engine L, the increased compression ratio was partially responsible for the reduced oxidation late in the expansion stroke and the increased HC emissions. For stratified-charge operation the HC emissions from engine A increase monotonically with increasing air/fuel ratio. The improved mixture preparation of the air-assisted injection system in engine A at an air/fuel ratio of 20:1 allows the HC emissions for stratified and homogeneous operations to be identical. However, at a lean-limit air/fuel ratio of 39:1, the enhanced mixing results in the fuel cloud becoming too disperse, leading to overmixing at the periphery of the spray plume, and causing high HC emissions due to flame quenching. The use of EGR reduces the HC emissions for stratified-charge operation, which is in contrast to the trend for homogeneous-charge operation. Although EGR is observed to reduce the exhaust temperature due to heat losses in the EGR system and the intake manifold, it generally provides an increase in air charge temperature, which does provide some benefit in the areas of vaporization and mixing, and thus has a positive effect on brake-specific HC (BSHC) emissions [51].

Figure 6.5-16 shows a comparison of brake-specific CO emissions from engines with air-assisted and single-fluid injection systems. For both engines, CO emissions decrease significantly as the charge is

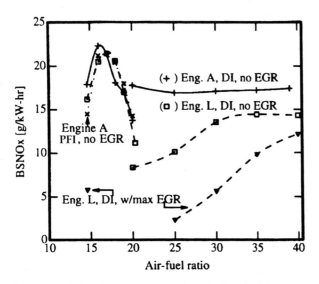

Figure 6.5-14 NOx emission levels for G-DI engines having single-fluid (engine L) and air-assisted (engine A) fuel injection systems; 1500 rpm; 2.62 bar BMEP; MBT [51].

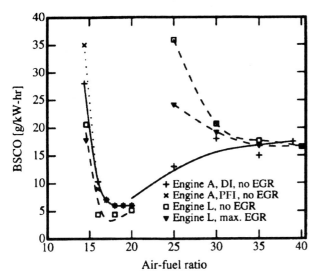

Figure 6.5-16 CO emissions levels for G-DI engines having single-fluid (engine L) and air-assisted (engine A) fuel injection systems; 1500 rpm; 2.62 bar BMEP; MBT [51].

enleaned from stoichiometry. The CO emissions level is nearly constant for homogeneous-charge operation for air/fuel ratios in the range from 16:1 to 20:1, whereas CO emissions increase as engine A is enleaned from 20:1 to 39:1, ultimately attaining the value that is typical for stoichiometric, homogeneous-charge operation. Engine L exhibits a substantial penalty in CO emissions at the transition from homogeneous-charge to stratified-charge operation, which is mainly due to the combustion of very rich mixture. For overall air/fuel ratios in the range from 30:1 to 40:1, the CO emissions are nearly constant, and are at the level that is typical of stoichiometric, homogeneous-charge operation [51]. A comparison of the combustion and emissions characteristics that result from the use of these two diverse fuel injection/combustion systems indicates that some limitations of stratified-charge combustion are inherent, and are relatively independent of the type of fuel system employed.

6.6 Summary

The computer-controlled timing of direct injection, coupled with the integrated use of sub-system models, sensors and actuators, enables a degree of mixture preparation and combustion optimization that cannot be achieved in the conventional PFI engine. The type of combustion can be switched from stratified-charge to homogeneous-charge by altering the injection timing

and fuel pulse width, and split injection may be utilized within a number of control strategies to optimize aspects of combustion, emissions and performance or driveability. This may take the form of reducing catalyst light-off time, achieving a mode transition with less torque fluctuation or enhancing the full-load performance. A wide spectrum of combustion characteristics are encountered with G-DI combustion systems, depending upon the specific system design, control strategy and engine operating conditions. The combustion characteristics for very early injection are fundamentally similar to those of a PFI engine, with a homogeneous, near-stoichiometric mixture formed in the combustion chamber. Combustion phasing with stratified-charge combustion, however, is quite different from that of homogeneous-charge operation. With a stratified mixture the G-DI engine exhibits a significant advantage in both ignition delay and burn duration when compared to a PFI engine of equivalent geometry. The initial flame kernel is developed rapidly in the rich mixture region near the spark gap, but the flame propagation speed is reduced in the outer lean region of the stratified charge. With late-injection combustion, the major components of flame luminosity consist of continuous blackbody radiation that is emitted from soot particles formed inside the cylinder. Soot radiation decreases abruptly at a time late in the combustion process when sufficient air is entrained into the reaction zone. Combustion characteristics are affected by many engine design and operating parameters, with the timing of the injection event being among the most influential. End-of-injection timing has been found to have a significant impact on HC and smoke emissions, as it represents the time when the last liquid fuel enters the cylinder. The interval between the end of fuel injection and the occurrence of the spark determines the maximum time that is available for the last fuel to evaporate and mix with the air. Thus the EOI timing directly and significantly influences the combustion efficiency, the amount of engine-out unburned hydrocarbons and the soot emissions. In the case of stratified-charge combustion, a retardation of the ignition timing at a fixed injection timing provides more time for mixing of the fuel and air. However, it also allows more time for dispersion, and for fuel to detach from the main fuel cloud, leading to an eventual increase in engine-out HC emissions. The rapid development and optimization of G-DI hardware, sensors and control systems will provide new and effective tools with which G-DI combustion can be continuously enhanced.

Chapter 7

Deposit Issues

7.1 Introduction

The subject of deposit formation on the internal surfaces and component parts of PFI engines has been studied extensively over the years. As a result, the severity of deposits in PFI applications has been reduced substantially from the levels of the 1980s using special additives in fuels and lubricants [327].

The working environment of the modern PFI engine is relatively well known, and currently marketed fuels and lubricants have evolved through proven field testing of multi-functional deposit control additives. For example, additives for controlling intake valve deposits (IVD) have been designed to function within the liquid gasoline phase, and to focus the detergent effect by means of spray targeting. The target is almost always the back face of the intake valve. In contrast, deposit control additives for the combustion chamber are designed to disperse rapidly and to have a negligible detrimental effect on combustion chamber deposit (CCD) formation. Lubricants and lubricant additives are also designed to minimize the formation of gums and coke within such areas as intake and exhaust valve guides and combustion chamber surfaces. This continuous-improvement process for PFI engine deposits clearly illustrates that the development of performance-enhancing additives requires a thorough understanding of the complex mechanisms that apply in the operating environment of such engines. Table 7.1-1 summarizes the key factors that may influence engine deposit formation. The relative contribution of each parameter to deposit formation is application-specific [145, 525].

Experience in recent years has shown that many injector deposit formation mechanisms are, in fact, different for PFI versus G-DI engines. For example, PFI injector deposits generally do not form during continuous running, but require a hot-soak period. Injector deposits in G-DI engines, however, do form during continuous running. Therefore, strategies to alleviate injector deposit formation may differ between PFI and G-DI engines. In addition, the negative effects of deposits can be fundamentally different between PFI and G-DI applications. An illustration of this is that the effect of injector deposits on PFI spray quality is known to be less critical than the effect on fuel flow rate reduction. Deposit-induced shifts in G-DI spray geometry, particularly in spray symmetry, which may be perturbed long before the cumulative deposit blockage reduces the flow significantly, can result in a severe degradation in combustion efficiency for stratified-charge operation. The major deposition sites for a generic G-DI engine are summarized in Fig. 7.1-1. The formation mechanisms and measures that may be taken to reduce or alleviate all such deposits are the main topics of this chapter.

7.2 Injector Deposits

Deposit formation is a significant concern with nearly all injector designs for use in G-DI fuel systems, and should be accorded sufficient time and resources in any engine development program. To neglect this important issue is to lengthen development time and decrease the required service interval in the field. The operating environment of the fuel injector is much more harsh for in-cylinder fuel injection than for port injection and, if the system is not carefully optimized, the formation of deposits on the injector tip can be much more rapid than is experienced with PFI systems. Although the operating environment of diesel injectors is even more extreme, deposits tend to form much more gradually in the holes of diesel injection nozzles than on G-DI injector tips. One important factor is the fuel composition, but another major influence is the 50 to 160 MPa injection pressure level. The corresponding pressure of 4 to 13 MPa in the G-DI fuel system is generally not sufficient to mechanically alter the continued deposition of sulfur compounds and carbonaceous material on the injector tip. Thus, the design of the injector tip for inherent resistance to deposit formation is particularly critical for G-DI applications. There are two general types of injector deposits: those that are generated from soot and lubricating oil during normal engine operation, and those that are formed from olefin or aromatic hydrocarbon components of gasoline during either running or cycles of run and hot-soak. For PFI engines, liquid fuel residing on the injector tip

195

Table 7.1-1
Key factors that may influence engine deposit formation

Surface Temperature	Deposit formation mechanisms vary significantly, with surface temperature as the most influential parameter in the presence of a likely deposit precursor.
Operating Conditions	There is a significant difference in deposit formation during continuous operation as compared to hot soak.
Oxygen Concentration	Oxygen concentration affects the type of reaction occurring within the deposits. Regions that are oxygen-depleted tend to form carbonaceous compounds. In comparison oxidation reactions continue in the oxygen-rich regions.
Oil and Fuel Chemistries	Deposit formation mechanisms and deposit compositions vary significantly with engine oil and fuel type.
Combustion Products	Products of combustion or partial oxidation such as blow-by gases and EGR that contain unburned hydrocarbons, particulate and deposit precursors can contribute significantly to deposit formation.
Additives	Deposit-removing and -retarding additives can considerably affect deposit formation chemistry. However, each additive has an effective temperature range beyond which it may become a depositing agent.
Site Material Composition	Deposit formation rate may be enhanced or reduced, depending on the material composition of the site where deposits are formed.
Residence Time	Deposit formation rate is a complex combination of residence time of depositing liquid, surface temperature, fuel and/or oil chemistries and combustion products.

surface during hot soak slowly vaporizes and is then oxidized, leaving a residue that gradually accumulates over repeated hot soak intervals. The hot soak condition necessary for such deposition is a surface temperature greater than 80°C. During normal operation, the temperature of a PFI fuel injector is moderately low due to fuel cooling. Deposit formation during hot soak is a time- and temperature-dependent process during which fuel that is present on the surface of the injector forms a sticky gum, followed by a more solid, lacquer-like deposit. Hence it is beneficial to reduce the liquid fuel left on the surface during hot soak. Reducing fuel drippage is one of the key measures. This soak period is known to be quite important for PFI injector deposit formation, but its impact on G-DI injector deposit formation relative to other operating parameters must be determined by means of additional investigations.

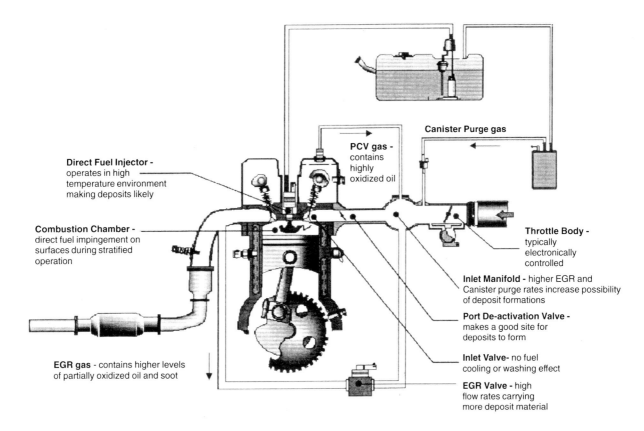

Figure 7.1-1 Key deposition sites within a G-DI engine [145].

7.2.1 Negative Effects of Injector Deposits

The negative effects of G-DI injector deposits are listed in Table 7.2-1. In general, the effects of these deposits are manifested in two distinct areas. The first is a degradation in the spray quality delivered by the injector, and the second is a reduction in the static flow capacity of the injector, with less fuel mass per injection being delivered at the same fuel pulse width. In the case of direct in-cylinder fuel injection, the temperatures and pressures existing within the injector tip may be suitable for polymerization reactions to occur within gasoline. These reactions produce a waxy to hard carbonaceous residue in the internal passages of the injector, which can change the flow calibration. If either the spray characteristics or the injector flow calibration are altered, then the stratification process that has been carefully developed and optimized will be degraded. It may be logically inferred that any deviation from the optimized operating characteristic will have a negative impact on emissions compliance and driveability [202]. For most G-DI injectors the early stage of formation of deposits does not result in a significant reduction in static

flow calibration, but *can* result in substantial changes in spray skew and spray symmetry, and in some changes in droplet-size distribution. The spray geometry, particularly the symmetry, is generally perturbed long before the cumulative deposit blockage reduces the static flow capacity by even 6% from the clean condition. Because the operation of G-DI engines is much more sensitive to spray parameter variations than that of PFI engines, the G-DI engine, in general, exhibits a much greater sensitivity to low levels of injector deposits. For G-DI engine designs in which the spark gap has a critical position relative to the spray periphery for light load, shifts in the spray geometry may be expected to result in a significant degradation of combustion in certain areas of the operating map. Combustion systems that invoke controlled vaporization from fuel-impact surfaces, as is inherent in the wall-guided concept, generally exhibit less sensitivity to changes in spray characteristics, and are thus less affected by injector deposit formation. A comparison of injector deposit tolerance for different combustion systems is provided in Table 7.2-2. It should be emphasized that flow reductions due to deposits do

Table 7.2-1
Negative effects of G-DI injector deposits

Spray Characteristics	• Spray geometry and symmetry degraded significantly • Spray atomization level degraded • Spray penetration characteristics altered • Early stages of deposit formation do not result in significant flow reduction, but can result in significant changes in spray skew, cone angle and spray symmetry and in some changes in droplet size distribution • Droplet size affected only slightly, but the droplet velocity distribution is significantly altered
Injector Flow Characteristics	• A reduction in injector static flow capacity • Spray symmetry is generally perturbed long before the flow rate starts to decrease
Combustion Process	• The stratification process that was carefully developed and optimized will be degraded
Air/Fuel Ratio Control	• Can increase cycle-to-cycle variations in air/fuel ratio

Table 7.2-2
Comparison of injector deposit tolerance for different combustion systems

Centrally Mounted Injector	• Difficult to arrange coolant passages around the injector mounting boss due to the limited space • The injector tip is generally 10°C–15°C hotter than that of the intake-side mounted injector • Generally higher injector deposition rate than for an injector located closer to intake valve seat
Intake-Side-Mounted Injector	• Less difficult to arrange coolant passages around the injector • Tip cooled by induction air flow • Increased distance from the exhaust valve promotes cooler operating temperatures • Generally lower injector deposition rate than for a centrally mounted injector
Combustion System Sensitivity	• Narrow-spacing concept such as spray-guided system is more sensitive to injector deposits • Wide-spacing such as wall-guided and flow-guided systems are less sensitive to injector deposits

not necessarily occur uniformly along a bank of cylinders, thus one very important effect of deposits is to cause cylinder-to-cylinder variations in the air/fuel ratio. All cylinders on the bank will receive the same fuel pulse width command from the EMS in most control systems, but those injectors with more deposits will inject less fuel and will provide a degraded spray geometry.

In comparison with the fuel spray from a new or clean injector, the spray from an injector with significant deposits can exhibit a distorted, non-symmetric spray envelope. Even if the designed spray shape is non-symmetric, as with offset or fan sprays, deposits can alter the designed spray geometry and change the mass distribution of fuel within the combustion chamber. The effect of deposits on spray characteristics is highly dependent on the details of the particular application. For some injectors the distribution of droplet sizes (the droplet-size distribution curve) is affected very little, whereas the droplet velocity distribution is significantly altered. For some injectors both the atomization level and the spray shape are significantly affected, with the spray cross-section changed from near-circular to elliptical.

The effect of injector deposits on the spray characteristics of a 70° swirl-type injector is illustrated in Fig. 7.2-1. The upper left photo shows the spray development for a new injector, which exhibits a toroidal vortex and good symmetry. The upper right photo shows the spray development for the same injector after 28 hours of engine operation on a dynamometer. Although the fuel flow rate is not significantly reduced, the spray symmetry is noticeably degraded. The sac spray is slightly more penetrating, and a prominent spray finger is present. For swirl-type injectors the formation of one or more spray fingers is a very reliable indicator of deposit accumulation. A comparison of the droplet size distributions from these two injectors is plotted in Fig. 7.2-1c. The SMD at various locations in the spray is not greatly affected except within the spray finger, where the SMD is degraded from 16 µm to 21 µm. Internal deposit formation in most G-DI injectors is first manifested as an alteration of spray symmetry, followed by an increase in the SMD and DV90 in a highly localized area of the spray. The sac geometry and penetration rate may also be altered. As deposits build further, additional localized areas of atomization degradation are formed, and the mass of fuel delivered for a given pulse

width exhibits the first small reductions, or flow shift. Except in severe cases, the original flow curve and spray symmetry can be restored to nearly that of the new injector by the use of a proper cleaning cycle using either a commercial injector cleaning solvent or a pump gasoline at high injector rep-rates and fuel pulse width. The original atomization level (SMD) profile can generally be restored to within 1.0 µm to 1.5 µm after flowing about one liter.

A detailed comparison of the spray characteristics delivered by a new injector and the same injector with significant in-service deposits is illustrated in Fig. 7.2-2. PDA evaluation of the spray quality of the injector with deposits reveals significant changes in the distribution profiles for volume flux, droplet size and droplet velocity. The condition exhibiting a 7.5% loss in static flow capacity yields the largest cone angle variation. Spray visualization confirms that the fully developed spray has a maximum penetration distance that is significantly larger for the injector with deposits. Also, the spray from the injector with deposits exhibits spray fingers, and has an altered symmetry. As illustrated in Fig. 7.2-2c, these spray distortions are removed when the injector is filled with fuel containing a Mannich gasoline detergent additive [23].

For most PFI applications, a deposit-induced reduction in static flow capacity of up to 5% can be tolerated. In contrast, only 2.5% to 3.0% can be accepted for the majority of G-DI applications. This is because nearly all deposit-induced flow reductions, both G-DI and PFI, are accompanied by alternations in spray geometry, usually degrading the symmetry if the spray is nominally symmetric. This will have a greater effect on G-DI combustion than on PFI combustion. Very substantial static flow reductions exceeding 10% may also be accompanied by degradations in the spray SMD and DV90, but the limits of combustion stability will likely have been exceeded at a substantially earlier time when the flow reduction is much less. Thus, in many actual G-DI engine applications, deposit-induced injector static flow reductions exceeding 7–10% may not be attainable without first exceeding either emission limits or combustion stability limits. It is worth noting that it is very difficult to generalize on an acceptable flow loss level since this is an issue that is very specific to a particular G-DI engine design and its control system.

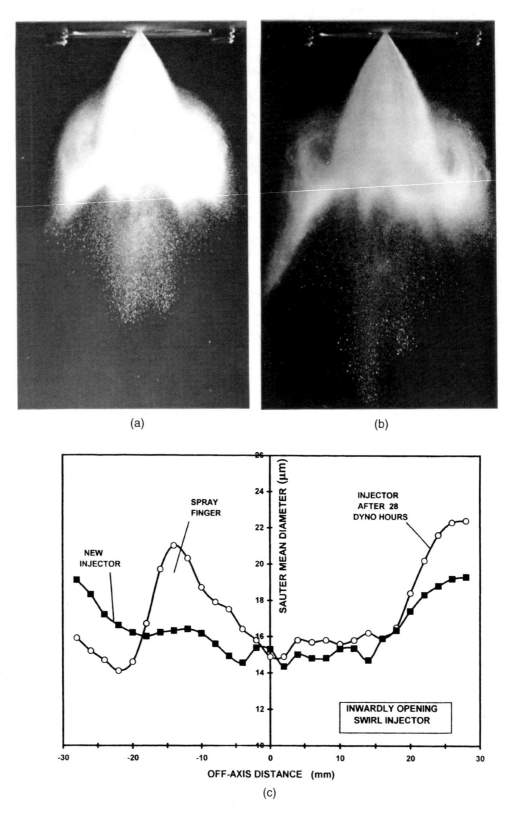

(a)

(b)

(c)

*Figure 7.2-1 Typical manifestation of injector internal deposits on spray characteristics:
(a) spray development from a new swirl injector; (b) spray development from the same
injector with internal deposits; 28 hours of engine operation on dynamometer; (c) measured
mean-drop-size profiles for a G-DI injector; new and after dynamometer testing.*

200

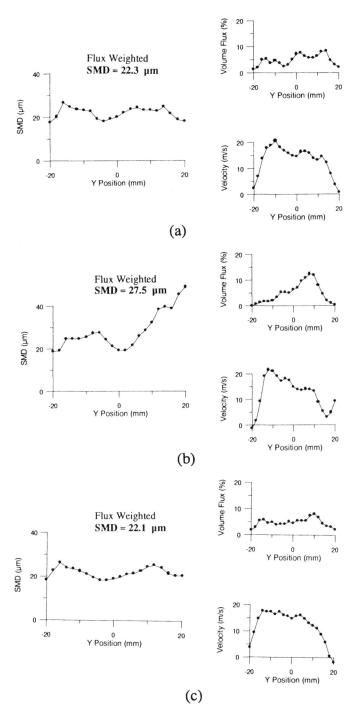

(a)

(b)

(c)

Figure 7.2-2 Spray characteristics of a G-DI injector for the conditions "new" and "with in-service deposits" (fuel: n-heptane; injection pressure: 5 MPa; measurement location: 30 mm from the injector tip): (a) spray characteristics of a new production swirl injector; (b) spray characteristics of the same injector with in-service deposits; (c) spray characteristics of the deposited injector cleaned with the fuel containing a Mannich gasoline detergent additive [23].

7.2.2 Injector Deposit Formation Process and Characteristics

As illustrated in Figs. 7.2-3 and 7.2-4, deposits can form internally, externally, or in both locations. External deposits may form on the downstream face of an isolating director plate, or on the downstream face of a pintle. Internal deposits near the minimum metering area of the fuel flow path or in the swirl-channel exit area cause fuel flow rate reductions due to the decreased flow area, discharge coefficient, or both; whereas external deposits generally result in only a degradation in the spray geometry. The rate at which deposits form on the tip surfaces of a G-DI injector depends upon several factors: the inherent resistance of the injector design to deposit formation, the specific configuration of the combustion chamber, the engine operating condition, and the specific fuel and oil. In a deposit study on an engine dynamometer test of a single-cylinder G-DI configuration, injector deposit effects became evident after 2 hours of operation, with the fuel flow rate for a fixed pulse width continuing to decrease during the initial 8 hours of testing. After 8 hours of engine operation no significant additional decrease in the fuel flow rate was observed. Deposits initially formed at the nozzle orifice exit and progressed into the internal surfaces of the nozzle. The internal injector deposit formed at the position where residual fuel resides after the end of each injection [242, 243].

Figure 7.2-5 shows a cross-section of the effective nozzle orifice for different testing interval hours [21]. It is evident that the amount of injector deposit increases with time, and that the effective flow area decreases. The deposit gradient indicates that the deposit is initially formed at the injector tip, which is directly exposed to

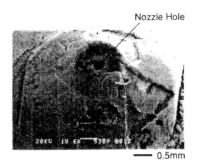

Figure 7.2-3 Photograph of the appearance of a G-DI injector with deposits after 30 hours of engine operation [243].

Nozzle Orifice

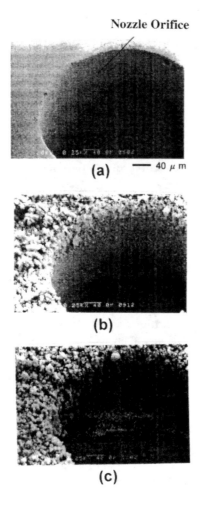

(a) 40 μm

(b)

(c)

Figure 7.2-4 Photographs of G-DI injector nozzles for different dynamometer testing hours: (a) new injector; (b) 4 hours of dynamometer testing; (c) 8 hours of dynamometer testing[243].

the high-temperature combustion gases, and then gradually progresses to the internal surfaces. It may be noted that the deposit layers or "furrows" are orthogonal to the flow direction. This furrow morphology may be related to the velocity vectors of fuel flow inside the nozzle, but further investigation is required to confirm this.

The observed deposits on the critical tip surfaces of G-DI injectors fall into two general classifications. The first class is the thin, brittle-coating type that contains sulfur compounds, and the second is the softer, carbonaceous type. The latter forms fairly rapidly, and is more easily removed with common solvents and cleaners. Within the injector, fuel may be polymerized to form a gum-type deposit. Careful examination of external injector deposits reveal that the deposit layer that accumulates on the tip surface is quite fragile, and that this layer extends slightly into the nozzle exit hole[243]. Examination of deposit compositions by electron probe microanalysis (EPMA) indicates that the major constituents of typical deposits are carbon, oxygen with phosphorous, and calcium in a very low concentration traceable to the engine oil. External injector deposits share many characteristics with combustion chamber deposits; however, there are some differences. The degree of carbonization of nozzle tip deposits is lower than that observed for soot in diesel engines. Furthermore, injector deposits are more fuel-derived, with engine lubricating oil being less important in the formation mechanisms. In comparison, deposits that accumulate inside the nozzle tip cavity strongly adhere, and are more difficult to

Blank   **1 Hour Run**   **2 Hour Run**

Fuel Flow

Figure 7.2-5 Cross sections of effective nozzle orifice for increasing deposit accumulation time [21].

remove. Table 7.2-3 summarizes the results of injector deposit characterization by Fourier transform infrared (FTIR) and scanning electron microscopy (SEM) for the testing fuels listed in Fig. 7.2-6a [21]. Only one of the internal deposit samples was found to contain the three lubricant marker elements of Ca, P and Zn. This indicates that the effect of back flow of combustion gas into the injector nozzle on deposits is not very significant. This is also an indicator that fuel in the nozzle orifice is effective in preventing combustion gas from flowing into the injector. The infrared spectra of these deposits are very similar to those of the testing fuels, further confirming that internal injector deposits are primarily fuel-derived.

The flow loss histories of four injectors over six hours of testing are illustrated in Fig. 7.2-6b for the No. 1 test fuel given in Fig. 7.2-6a [21]. It is interesting to note the periodic nature of the injector deposition process, which involves the accumulation of a critical amount of deposit, which is then partially dislodged by some internal periodic mechanism [21]. This periodic cycling

is found to be quite typical for most injectors and most fuels. It is also apparent that there is a large cylinder-to-cylinder variation in injector flow loss due to deposit buildup, which will likely cause a large cylinder-to-cylinder variation in combustion and emissions.

Knowledge of the inherent resistance of a G-DI injector to deposits is necessary, but not sufficient, for interpreting the observed rate of deposit formation in

Fuel #	T_90 (°C)	Olefins (%)	Sulfur (ppm)
1	160	5	30
2	182	5	30
3	160	20	30
4	182	20	30
5	160	5	150
6	182	5	150
7	160	20	150
8	182	20	150
9	171	12.5	90
10	171	12.5	400
Howell EEE	160	1.2	20

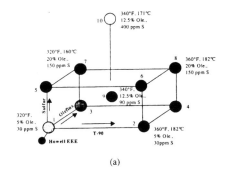

(a)

Table 7.2-3
Characterization of injector deposits by FTIR and SEM [21]

Fuel	Phase	IR Band	P	Ca	Zn
1	External	—	Yes	Yes	Yes
	Internal	Ox[a]	Yes	Yes	Yes
2	External	—	Yes	Yes	Yes
	Internal	Ox/1728/Org[b]	—	—	—
3	External	—	Yes	—	—
	Internal	Ox/Org	—	—	—
4	External	—	—	—	—
	Internal	Ox/Org	—	Yes	—
5	External	—	Yes	Yes	—
	Internal	Ox/1728	—	Yes	—
6	External	—	—	—	—
	Internal	Org	—	Yes	—
7	External	—	Yes	Yes	Yes
	Internal	Ox/Org	Yes	Yes	—
8	External	—	—	—	—
	Internal	Ox/1729/Org	—	—	—
9	External	—	Yes	Yes	Yes
	Internal	Ox	—	Yes	—
10	External	—	Yes	Yes	Yes
	Internal	Org	—	Yes	Yes
EEE[c]	External	—	Yes	Yes	Yes
	Internal	Ox	—	—	—

a) Ox = Oxidation peaks around 1712 and 1735 cm⁻¹
b) Org = Organic Acid peak at 1710 cm⁻¹
c) EEE = Howell EEE gasoline

Fuel Decrease, 2500 RPM, 600 mg/stk, Fuel No. 1

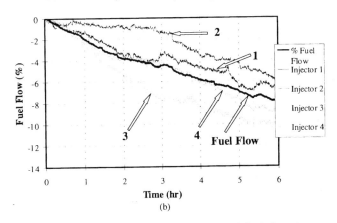

(b)

Figure 7.2-6 Test fuel composition and fuel flow loss histories of four G-DI injectors: (a) physical properties of tested fuels; (b) flow loss histories of four G-DI injectors over 6 hours of dynamometer testing using fuel #1 [21].

a particular application. As with a port fuel injector, the inherent resistance of a G-DI injector to deposit formation would ideally be determined by a standardized bench test using either an oven or a dynamometer in conjunction with a standard fuel, test configuration and test conditions. The lower the level of deposits formed in such a standardized test, the higher the inherent resistance of the particular injector design. Unfortunately, no proven standardized test for deposit resistance has yet been adopted for G-DI injectors, a situation that will have to be quickly resolved if rapid progress in direct-injection gasoline engines is to continue. Such a standardized test must represent true field operating conditions and must be validated by in-service field data.

The inherent resistance, or robustness, of a specific injector design to deposit formation may be either enhanced or degraded by the particular configuration of the injector in the combustion chamber, and by the specific fuel composition that is being used. The injector tip temperature, or more specifically, the mean temperature of the lower portion of the injector body that contains the fuel being injected, is universally regarded as one of the most important parameters in the injector deposit formation process. This tip temperature, however, will be affected by numerous design and operating variables, including the position of the tip relative to the induction air stream, which provides an important source of convective cooling. It is recommended that the tip temperature not only be considered an important variable, but that it be measured and logged whenever feasible during any development program. Whenever possible, micrographs from an SEM device should also be obtained for the orifice area of the injector tip during engine down periods. There is a consensus that, at the current stage of worldwide G-DI development, a significant effort should be directed toward understanding and reducing injector deposits. Table 7.2-4 summarizes the important items that must be considered in minimizing the formation of G-DI injector deposits, with each being discussed in detail in this section.

7.2.3 Formation Mechanism for Internal Injector Deposits

A schematic outlining the steps in internal injector deposit formation is illustrated in Fig. 7.2-7. When an injector tip is subjected to elevated temperatures, thermal decomposition of the fuel inside the nozzle may occur, and deposit precursors may be formed. When the nozzle tip temperature is lower than the threshold

temperature of T90 (the T90 temperature on the distillation curve of the fuel being used), a significant fraction of the residual fuel will remain as liquid, even though some may evaporate before the next injection event occurs. The continued presence of liquid fuel greatly increases the probability that deposit precursors will be washed away during the next injection event [240]. However, when the nozzle tip temperature exceeds the T90 temperature of the tested fuel, most of the residual fuel will vaporize immediately following the end of fuel injection. As a consequence, the deposit precursors are distributed on, and may attach to, the nozzle surface. This makes it more difficult to wash away all the deposit precursors during the next injection event. Eventually deposits may accumulate inside the nozzle orifice and tip cavity. This suggests that, for injector designs that have a small cavity upstream of a delivery orifice or orifices, it is important to retain some of the residual liquid fuel inside the nozzle between the injection events in order to prevent deposit buildup. This will likely degrade the spray quality, however, as this fuel will contribute to the sac or pre-spray. Investigations have determined that maintaining the injector tip temperature below the T90 temperature of the fuel being utilized is a critical factor in minimizing the formation and accumulation of deposits [242, 243]. From this discussion, it is concluded that the temperature that matters is the local temperature of the fuel molecules, which is governed by the upstream fuel temperature, the injector wall temperature and the heat transfer process from the injector to the fuel. The transition for injector deposits to form occurs from a state in which the injector tip is filled with liquid fuel between injection events, to a state in which the liquid fuel inside the injector tip vaporized almost completely between injection events.

7.2.4 Effect of Injector Tip Temperature on Injector Deposit Formation

Both nozzle tip temperature and fuel distillation characteristics can significantly affect the rate of accumulation of injector deposits. Figure 7.2-8 shows the effect of nozzle tip temperature on injector flow rate reduction for five different types of fuels. As shown in Fig. 7.2-8b, the blended fuels A–D have the same T90. Fuel E has a comparatively higher T90. When comparing the blended fuels A and D in Fig. 7.2-8c, it is evident that the measured fuel flow rate is diminished significantly as the nozzle tip temperature exceeds the

Table 7.2-4
Important considerations in minimizing injector deposit formation

Injector Tip Temperature	• As low as possible, but not to exceed 135°C
Injector Installation	• Injector protrusion distance into chamber • Heat path from injector body to engine coolant passages • A more heat-conductive path from the tip to the injector threads or mounting boss and to the cylinder head • More coolant passage area in the vicinity of the injector mounting boss in the cylinder head • Side mounting on intake valve side will be superior to a central mounting regarding deposit formation • Combustion system performance considerations may take precedence over deposition reduction in locating the injector
Injector Design	• Use of a director plate for isolating the needle seat from direct contact with combustion gases • Enhance internal and external surface finish of the nozzle tip • Achieve zero leakage • Avoid needle bouncing on closure • Seat geometry design to minimize deposit effects • Proximity of bulk fuel inside injector to injector tip area • Fuel state in injector orifice after the closure of injector • Special coating or plating of tip surfaces that can delay the onset of deposition • Once deposits are formed, coated and uncoated surfaces may exhibit very little difference in carbon deposition
Environment around Injector Tip	• Air velocity at the injector tip for the engine operating range • Enhance mean air velocity at the injector tip • Avoid fuel droplet recirculation to injector tip • Achieve zero drippage during operation
Operating Cycle	• Identification of the critical operating cycle, including hot-soak intervals (to be verified by future research)
Fuels	• Distillation characteristics—T90 temperature • Fuel composition • Gasoline additives to inhibit cumulative deposit formation

T90 fuel temperature. Even though the blended fuels A and E were tested at the same injector tip temperature, the flow rate reduction with fuel E is much less than that of fuel A, due to the higher T90 point [242, 243]. The correlation of fuel properties with deposit formation, as shown in Fig. 7.2-9, indicates that, in addition to the T90 temperature, the T50 temperature and the end point temperature of the fuel also correlate well with injector deposit formation. It is evident that the history of nozzle tip temperature obtained during G-DI operation is a key parameter in the determination of deposit formation. For a cooler injector tip temperature the amount of deposition is relatively low, and the deposits occur in individual, separate groupings. At higher tip temperatures the deposit morphology changes to a continuous coating across the injector tip and into the orifice cavity [21].

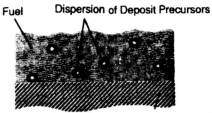

(a) Immediate Situation after Injection

(b) Nozzle Temp. < Fuel 90% Distillation Temp.

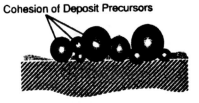

(c) Nozzle Temp. > Fuel 90% Distillation Temp.

Figure 7.2-7 Schematic of injector deposit formation mechanism [243].

The measured injector tip temperature and the associated effect of injector deposits on fuel flow rate reduction under various engine operating conditions are presented in Fig. 7.2-10 [243]. The maximum observed fuel flow rate reduction occurs for operation at 4000 rpm and mid-range load, at which the highest injector tip temperature of near 160°C is measured. The mechanisms of deposit formation inside the nozzle cavity are complex, with many parameters being influential. In general, liquid fuel tends to cool the injector tip and thus helps to retard deposit buildup inside the nozzle. High-flow-rate operation provides an active cooling effect, but it is also associated with high engine loads and combustion temperatures. High-load operation places an additional thermal load on the injector, and results in an elevated injector tip temperature. These two competing processes are always present, and actual measurements have indicated that the maximum injector tip temperature that is obtained for a particular G-DI configuration is for a combination of the highest engine speed and a mid-load fueling rate.

Fuel Pressure	6 MPa
Injection Timing	180°BTDC
Spark Timing	25°BTDC
Engine Speed	1000 rpm

(a)

Test Fuels	Nozzle Tip Temperature (°C)	Air-Fuel Ratio	Smoke (B.S.U.)	T90 (°C)
A	165	12	0	150
B	100	12	0	150
C	154	10	1	150
D	155	15	0	150
E	165	12	0	168

(b)

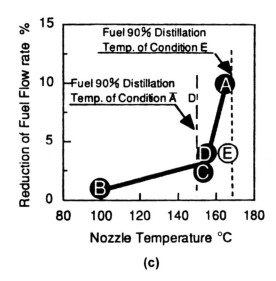

(c)

Figure 7.2-8 Effect of nozzle tip temperature on deposit-induced fuel flow rate reduction [243].

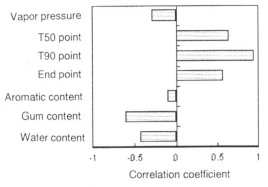

Figure 7.2-9 Correlation of fuel properties with deposit formation [243].

Numerical value in °C shows
the injector tip temperature.

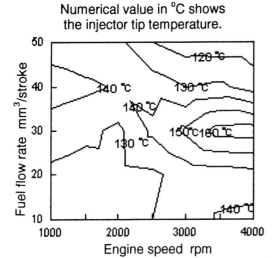

Numerical value in % shows
the fuel flow rate reduction.

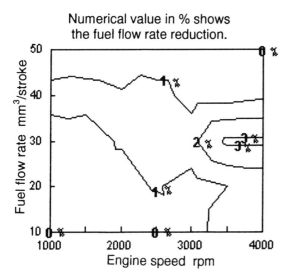

*Figure 7.2-10 Map of measured injector tip temperature
and fuel flow rate reduction by injector deposits [243].*

The specific combustion chamber geometry will influence injector tip temperature, with the protrusion of the tip into the combustion chamber, the conductive path from the injector mounting boss to the coolant passage, and the in-cylinder air velocity history at the tip location being key determinants. Engine load and speed will certainly influence tip temperature, but geometric parameters of the chamber will cause the absolute value to move higher or lower. An injector that is centrally mounted is known to be subject to a higher thermal loading than an injector located beneath the intake port. G-DI combustion systems that use a centrally mounted injector would be expected to yield a 10°C to 15°C higher injector tip

temperature than would systems that derive additional tip cooling from intake air and in which the injector is located far from the exhaust valves. Finite element analysis (FEA) and heat transfer calculations have shown that the highest heat fluxes to a plate in the chamber wall occur at the center of the combustion chamber [203]. For a nearly centrally mounted injector the minimum-temperature location is to be as close as possible to the intake valve seats.

The injector tip surface temperature measured for the Mitsubishi GDI combustion system during continuous, high-speed, full-load operation, followed by a subsequent hot-soak period, is illustrated in Fig. 7.2-11 [196]. This G-DI engine could achieve a running injector tip temperature below 100°C with an injector core temperature of 75°C. During the hot soak period the injector tip and core temperature rise to about 120°C. Extensive measurements of injector tip temperature for the production Mitsubishi GDI engine over a wide range

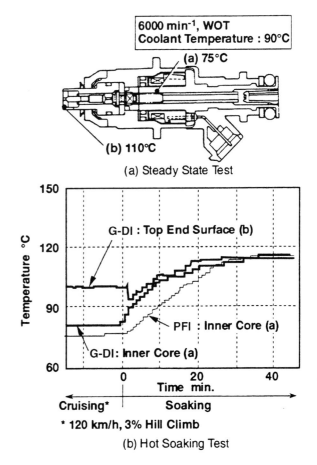

*Figure 7.2-11 Time history of injector tip temperature
for the Mitsubishi GDI engine [196].*

of operating conditions have been conducted, and the results for idle, part-load, full-load maximum temperature and hot soak are summarized in Table 7.2-5 [327]. It is evident from these data that the injector tip temperature is relatively low, and that this moderate temperature is maintained over a wide range of engine operating conditions. The resistance of this injector to deposit formation needs to be evaluated using actual field data to better correlate tip temperature to deposit formation.

Caution must be exercised when interpreting injector tip temperatures reported in the literature. The thermocouple position and response, as well the engine operating parameters, are important factors affecting injector tip temperature. Depending on the details of the injector and thermocouple mounting, the thermocouple could correctly be measuring injector tip metal temperature, but could be measuring the injector tip deposit temperature once deposits start to form on the injector tip. Moreover, the temperature measured at some other injector locations, such as the upper injector body, usually does not directly reflect the thermal condition of the injector tip. Therefore it is important to configure and monitor injector tip temperatures carefully throughout the development program. It may not be appropriate to directly compare injector tip temperatures of two different combustion systems, depending on the thermocouple location and engine operating conditions. Some deposit formation has been noted for combustion systems that use a centrally mounted injector, with a full-load tip temperature of more than 150°C being achieved. The early injector deposit problem for some prototype G-DI engines was alleviated by design alterations that

maintain the injector temperature below 150°C, and by using a special organic material coating on the injector tip [341]. As a design guideline, the injector tip temperature should not exceed 135°C for any extended time period if injector deposit formation is to be avoided.

A study on injector deposit formation inside a tube revealed that the wall temperature is the most significant parameter in determining the carbon deposition rate, with the fuel flow rate ranked second and the fuel inlet temperature ranked third. The variation in rate of deposition is dependent on whether a constant heat flux or a constant wall temperature is provided. The rate of deposition remains approximately at a constant level for wall temperatures below certain threshold values. When the wall temperature exceeds this threshold value, the deposit rate increases sharply. This suggests that the controlling mechanism for deposit formation changes, depending on the wall temperature [298]. This agrees with the results obtained in an investigation of injector deposit formation using a prototype DI engine [242, 243]. The fuel inlet temperature is not as influential as the wall temperature, but it has a noticeable effect on deposit rates when the wall temperature is high. Initially, the deposit rate decreases with increasing fuel temperature. After reaching a minimum value, the rate of deposition begins to increase with a further increase in fuel temperature.

A deposit site on the nozzle surface can act as a heat insulator to diminish the heat transfer rate from the combustion gases. The relationship between nozzle tip temperature and the thickness of the deposit layer on the nozzle surface is shown in Fig. 7.2-12. As illustrated, deposit layer thickness increases with time (dynamometer hours), and the nozzle tip temperature gradually decreases. After 30 hours of testing, however, the tip temperature suddenly increases rapidly. Simultaneously a decrease in deposit layer thickness, namely, a loss of deposits from the nozzle surface, is observed. This indicates that installing a heat insulator on the nozzle surface exposed to combustion gas may be an effective way to lower the injector tip temperature. Figure 7.2-13 shows the predicted effects of heat insulator material and thickness on nozzle tip temperature. An insulator with a lower heat conductivity or a thicker insulator can lower nozzle tip temperature, thus reducing the tendency to form injector deposits [243].

Table 7.2-5

Injector tip temperature for the production Mitsubishi GDI engine for a wide range of operating conditions

Engine Operating Conditions	Measured Injector Tip Temperature
Idle	85°C–95°C
2000 rpm and part load	110°C–115°C
Full throttle-open acceleration (50–80 km/hr.)	110°C–128°C
Maximum temperature	128°C
Hot soak period	No additional increase in nozzle tip temperature

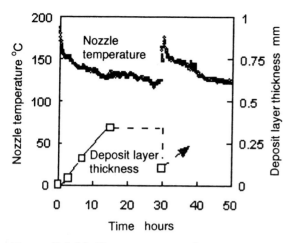

Figure 7.2-12 Variation in nozzle tip temperature with deposit layer thickness [243].

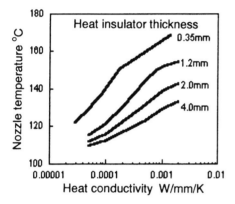

Figure 7.2-13 Predicted effect of insulating layer on injector tip temperature [243].

7.2.5 Effect of Operating Cycle on Injector Deposit Formation

Standard deposit tests for injector deposit resistance have been developed for pintle-type and director-plate-type port fuel injectors, and were proven to correlate with field data from PFI engines [59, 159]. It is considered unlikely, however, that standard PFI deposit tests can be effectively utilized for G-DI injectors. It is well established that the hot soak time and tip temperature history are of critical importance for PFI deposit formation, and that deposit formation rates are very low for continuous operation. The evidence for G-DI applications, however, is that deposits *do* form under continuous operation, thus indicating that the formation mechanism differs from that of the PFI injector. Tip temperature is generally considered

to be an important parameter for both PFI and G-DI injectors, but the contribution of the soak-temperature window and its cycle history are significantly different between PFI and G-DI applications. The interrelationship of the hot soak interval and temperature history is an unknown in the current G-DI literature, making this an important research topic. Without an accepted industry test, the inherent resistance of a particular G-DI design can only be evaluated by an *ad hoc* test within each company.

In order to compare deposit resistance during continuous operation and hot soak, a transient test cycle, as shown in Fig. 7.2-14a, is specified and injector tip temperature is monitored during the test cycle [23]. It should be noted that this is a proposed, *ad hoc* test cycle that has not been verified as providing a correlation with injector deposit formation in G-DI engines in actual field service. The cycle begins with a WOT acceleration to

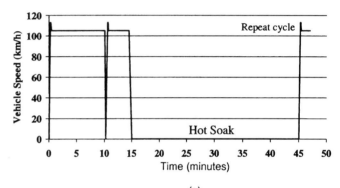

(a)

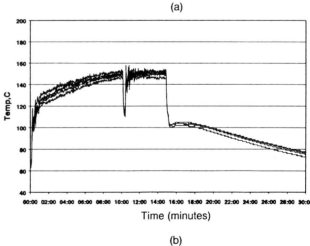

(b)

Figure 7.2-14 Time histories of injector tip temperatures during a transient cycle: (a) transient drive cycle for deposit testing; (b) tip temperature histories of four injectors during a transient drive cycle [23].

113 km/h in 15 seconds and follows a cruise at this speed for 5 seconds before decelerating to 105 km/h in 10 seconds. After cruising at this speed for 9 minutes 30 seconds, the vehicle is decelerated to idle for 15 seconds and is then immediately accelerated at 75% throttle opening to 113 km/h in 15 seconds. After cruising at this speed for 5 seconds, there is a deceleration to 105 km/h in 10 seconds, followed by a cruise at this speed for 4 minutes, and then a deceleration to idle in 15 seconds. At this point the engine is turned off and allowed to hot soak for 30 minutes before the end of the test cycle. The injector tip temperature measured for this test cycle is shown in Fig. 7.2-14b.

Clearly the tip temperatures for the four injectors begin to rise as the vehicle is started, and rise rapidly during acceleration to 113 km/h. The temperatures continue to rise rapidly to a peak near 150°C during the 105 km/h cruise, but fall rapidly to 115°C during the 15 seconds required to decelerate to idle. On acceleration to 113 km/h over 15 seconds at 75% throttle opening, the temperature returns to 150°C before decreasing to 100°C during the final deceleration to idle and stop. By the end of the 30-minute hot soak, the temperature has dropped from 100°C to 75°C. The observed peak tip temperature falls gradually as more of these drive cycles are repeated. By the 17th drive cycle, the peak tip temperature is reduced to 140°C from the 150°C observed during the initial drive cycle. This is attributed to thermal insulation of deposits that have accumulated on the injector tip. It is clear that the injector tip temperature history for G-DI operation is quite different from that of the conventional PFI engine. With the PFI engine, the injector tip temperature reaches 80°C during vehicle acceleration, but rises to 120°C during the hot-soak period because coolant circulation is stopped and air cooling of the injector is no longer available [23]. This illustrates that the G-DI injector tip normally cools during a hot soak, whereas the PFI injector tip normally increases in temperature. This is why the specifics of any hot-soak period in a standard test are ostensibly of more importance for the PFI injector than for the G-DI injector.

Some reports show that lighter load and a lower engine coolant temperature can also increase deposit formation. In addition, the stratified-charge cycle of the G-DI operating map has been found in some engines to produce a greater level of injector deposits than that of the stoichiometric, homogeneous-charge cycle [30]. The mechanisms that control G-DI injector deposit formation are obviously quite complex, and much work remains to be done before they are fully understood.

7.2.6 Effect of Injector Design on Injector Deposit Formation

Injector deposit formation is a very complicated process, which is influenced by many combustion system design, engine operation and fuel parameters. Therefore, it is very difficult to separate each individual impact and generalize the injector deposit formation mechanism. Some contradictory reports exist in the literature. A number of injector design parameters can directly influence the rate of injector deposit formation. The discussion in this section is intended to provide a list of key design parameters that may have a major influence on injector deposit formation. Their final role must be evaluated and determined, based on field testing.

The inherent deposit resistance of an injector may include a director plate for isolating the pintle seat from direct contact with combustion gases, a more heat-conductive path from the tip to the mounting boss, an increased fuel volume closer to the needle seat and special plating or insulation for the tip surfaces. Sharp-edged orifice exits may be beneficial, as they may assist in the mechanical break-up of deposits. Shorter orifice lengths are considered beneficial due to the reduced volume of residual fuel between injectors. Special tip coatings and surface finishes have been found to be effective in enhancing deposit resistance, but this alone may not be sufficient to alleviate the problem. Interior cavity surface finish and surface coating do exhibit an influence on the rate of internal deposit formation. A very fine microfinish has been found to reduce the deposit rate; however, this requires complex and time-consuming fabrication. In contrast, some studies also show a likely increase of deposit formation with a smooth transition from seat angle to exit orifice or internal orifice polish. Some surface coatings have been found to be beneficial in delaying the onset of deposition; however, once a thin deposit layer is formed, the coated and uncoated surfaces exhibit very little difference in deposition. The inherent resistance, or robustness, of a specific injector design to deposit formation may be either enhanced or degraded by the configuration of the injector in the combustion chamber, and by the specific fuel and fuel additive that are being used. It should also be noted that reducing the injector leak rate is another important consideration in minimizing deposit formation, although leakage (very low weeping flow rates) is quite rare among G-DI injectors. The effect of cavitation on injector deposit formation needs to be explored.

Most of the research and development efforts on injector fouling have focused on preventing or minimizing the formation of deposits, but there is another philosophy that can be invoked in parallel, although only by injector manufacturers. An injector can be designed with the important goal of minimizing the influence of deposits on the resulting flow rate and spray. This is known as an inherent tolerance or robustness to the presence of injector deposits. It is desirable, of course, to reduce the basic rate of deposition on the injector during the operation of a G-DI engine. But, bearing in mind the inevitability of some deposition occurring with time, it is also desirable to have an injector flow-path design that exhibits good tolerance to any deposits that *do* form. For example, 100 μg of deposits may form in 100 hours of operation on two different injector designs, which is the same average rate of deposition. However, one injector design may exhibit a 3% flow reduction for this weight of deposits, while the flow in the second is reduced by only 1%. In this regard, some injector designs using swirl have proven to be quite sensitive to small amounts of deposits, particularly in terms of spray-symmetry degradation.

7.2.7 Effect of Fuel Composition on Injector Deposit Formation

Figure 7.2-15 shows the injector flow loss rates as a function of injector tip temperature for the eleven fuels listed in Fig. 7.2-6a [21]. It is clearly shown that increasing injector tip temperature increases the deposit formation rate significantly. Higher olefin levels in the fuel also tend to result in decreased deposit resistance. As shown in Fig. 7.2-9, the correlation between the aromatic content of the fuel and the deposit formation rate is very weak, which is quite different from the correlation observed for PFI engine injector deposits [243]. The sulfur level in the fuel is known to have a non-linear effect, producing a beneficial effect up to 150 ppm, and then reversing its effect for concentrations up to 400 ppm. The non-linear sulfur effect may be related to tip surface catalysis. At low sulfur levels, an increased sulfur content may help to passivate the active sites for deposit formation, whereas a higher level of sulfur begins to contribute to deposits [21]. In general, the sulfur content of the fuel is considered to be influential in the process of deposit formation. Figure 7.2-16

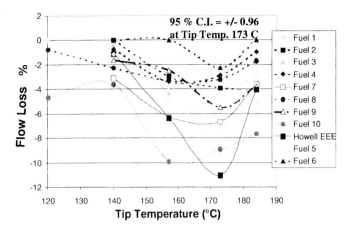

Figure 7.2-15 Effects of fuel properties on injector deposit formation [21].

summarizes the effects of various fuel properties on injector deposit resistance [23].

It should be noted that a reasonable portion of the G-DI deposit problem results from the lack of proven and effective G-DI anti-deposit additives in the fuel supply system and test fuels [29, 470]. As a result, current G-DI development is relying on a compromise of combustion system design to minimize fuel injector deposits by placing the injector in a relatively cooler spot inside the combustion chamber. Fuel supplies are, of course, different in Europe, Japan, and North America, and the higher sulfur content of gasoline in North America could contribute to an elevated level of deposits when that market has G-DI applications. The additives that are currently in the gasoline supply in North America were developed and improved mainly over the time period from 1984 to 1993 to minimize the rates of PFI injector deposit and intake-valve-deposit formation. In the early development of PFI systems, injector

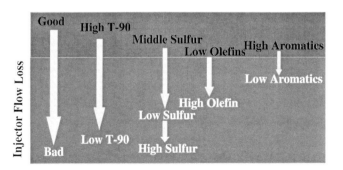

Figure 7.2-16 Injector deposit resistance to various fuel properties [23].

deposits were a very significant field problem that had to be alleviated [59, 516]. The same will have to be done for G-DI systems [28]. It should be pointed out that even if such additives were available for G-DI engines, it would have to be carefully verified that adding them to the fuel supply would not adversely affect the formation of deposits in the 100 million PFI vehicles now in operation.

The results of a study of the effect of fuel additives on injector deposit formation are illustrated in Fig. 7.2-17 [22]. During this study, a six-hour deposit test was conducted using a Howell EEE base fuel (see Fig. 7.2-6a), which yielded an injector flow rate reduction of 9.25%. The fuel was then changed to Howell EEE treated with a Mannich additive for six hours, and a fuel flow rate recovery of 70% was achieved. These test results suggest that for best performance the additive should be used at all times with a concentration at or above the critical threshold level. With additives in the fuel, the deposit formation rate can, in some cases, initially exceed that of the base fuel. This indicates that gasoline detergent additives may be forming a catalytic coating on the nozzle inner surfaces as the fuel is injected. This coating can increase to a finite thickness that yields the initial higher rate of flow loss. A critical thickness is achieved when the rate of additive adhesion to the coating is identical to that of additive desorption from the coating. Test results have confirmed that due to this initial additive layer, the additives have a residual protection effect against injector deposit formation that persists for a finite period of time after the fuel is switched back to no-additive base fuel [22].

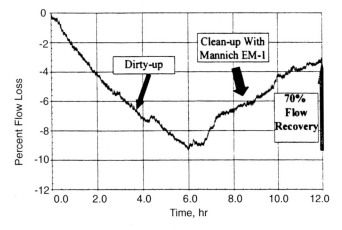

Figure 7.2-17 Effect of gasoline fuel additives on injector deposit formation [22].

7.2.8 Deposit Formation Characteristics of Pulse-Pressurized, Air-Assisted G-DI Injectors

With air-assisted G-DI injectors, the delivery of fuel to an intermediate cavity enables independent control of both the fuel metering and direct injection events. As a result, injector deposits generally do not affect the delivered fuel flow quantity of an air-assisted injector. Air-assisted DI injectors developed for automotive applications employ an external pintle that protrudes into the combustion chamber approximately 8 mm at maximum lift. This pintle projection design is intended to function as an aerodynamic foil to control the shape and stability of the injected air/fuel plume. Thus, the details of the shape of this projection can significantly influence mixture plume geometry and subsequent combustion stability. Deposits that form on the projection may therefore influence spray shape and stability, and could ultimately degrade combustion quality, even though fuel metering remains constant [60]. It should be noted that since injector flow rate is relatively independent of deposit formation, injector flow rate drift is not a significant factor in quantifying the deposit formation process for air-assisted injectors. Smoke, HC emissions and engine combustion stability (COV of IMEP) have been used as metrics to characterize the effect of injector deposits on combustion quality [60]. Hydrocarbon and smoke emissions appear to be a better correlating parameter for deposit formation than COV of IMEP.

As discussed earlier in this chapter, nozzle tip temperature has been identified as a key parameter in the deposit formation process. For the air-assisted injector, at any engine operating point there is a temperature gradient along the axial profile of the pintle. This gradient generally increases with engine speed and load, and there is a heat transfer balance between fuel cooling and combustion gas heating. Figure 7.2-18 shows a typical deposit of an air-assisted injector and the associated temperature distribution across the projection. There is a clear horizontal line demarcating a critical temperature above which fuel-derived carbonaceous deposits do not exist. Above this line only a layer of beaded nodules is present, and below it only the typical fuel-derived carbonaceous deposits are present. The demarcation line for fuel-derived deposits corresponds to a temperature range of 280°C–300°C, which is in general agreement with the reported upper limit for fuel-derived CCD formation. This temperature demarcation line shifts

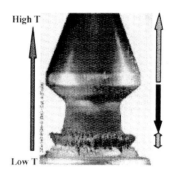

High T

Increasing lubricant content on the injector surface, but very low amount of total deposit in this region.

Increasing deposit mass and decreasing lubricant contribution to deposit mass.

No lubricant contribution to deposit in this region and inside the injector.

Low T

Figure 7.2-18 Typical deposit profile on an air-assisted G-DI injector tip; 15 hours of operation in stratified-charge mode; engine speed: 2500 rpm [60].

toward the injector tip for stratified operation, due to the reduced tip temperature.

Species analysis has revealed that deposits around the air injector gauge line (fuel exit nozzle), namely, the low T region in Fig. 7.2-18, are mainly fuel-derived, with only an insignificant contribution from the lubricant. As shown in Fig. 7.2-19, the major deposit species around this region are carbon and oxygen (stratified-charge operation), with only trace levels of the engine lubricant marker elements calcium (Ca), zinc (Zn), sulfur (S), and phosphorus (P). Components of chromium and iron from the injector surface metal are also identified in the deposits. Air compressor lubricant is not a significant contributor to the injector deposits. However, further downstream along the pintle projection, namely the high T region in Fig. 7.2-18, lubricant marker elements are indeed present in the pintle deposits, and exhibit a trend of increasing concentration toward the pintle tip. Figure 7.2-20 shows the spherical nodules on the high-temperature sections of the pintle surface, which are composed of ash from the combustion of lubricant droplets. However, lubricant contribution to the total quantity of deposit is negligible. The bulk of the deposit formed within the air-assisted injector is located near the fuel exit nozzle, and is almost entirely fuel-derived.

Regarding the mechanism of injector deposit formation, one CCD formation theory postulates that a liquid matrix on the substrate surface is a prerequisite to the initiation of deposit formation. Reactive hydrocarbon species (radicals and polar deposit particles) move from the gas phase into this matrix at a rate that is proportional to the thermophoretic gradient resulting from the temperature difference between the injector surface and the combustion products. With the thermophoretic

mechanism, more material is deposited on the coolest side of the substrate surface, resulting in the deposit profile illustrated in Fig. 7.2-18. This profile is typical of deposit accumulation on the tips of air-assisted DI injectors. The thickness of this deposit is reduced with increasing surface temperature.

As is the case with the single-fluid injector, selected fuel properties are found to have a significant effect on the deposit formation associated with the air-assisted injector. Figure 7.2-21 illustrates a comparison of the measured injector deposit level using a premium unleaded gasoline (F1), and iso-octane (F2). There is practically no deposit formed with iso-octane fuel for either homogeneous-charge or stratified-charge operation. However, significant deposit accumulation is observed with commercial multi-component gasoline fuel. For single-fluid DI injector deposit formation, the T90 temperature of the fuel is found to be the most

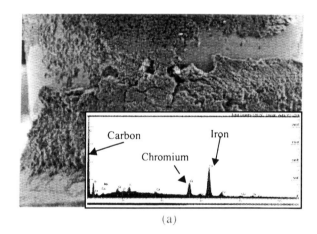

Carbon

Chromium

Iron

(a)

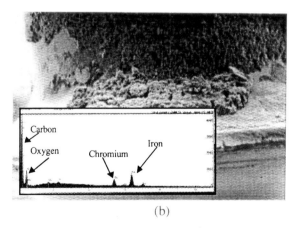

Carbon

Oxygen

Chromium

Iron

(b)

Figure 7.2-19 Species analysis of deposits from an air-assisted G-DI injector for different operating modes: (a) homogeneous-charge operation for 5 hours; (b) stratified-charge operation for 15 hours [60].

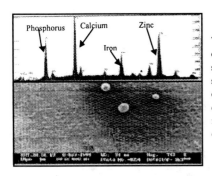

The spherical nodules on the high temperature sections of the injector surface are the ash from combustion of lube droplets. This deposit is inorganic, composed of P, Ca, and Zn.

Figure 7.2-20 Species analysis of deposit nodules on the high temperature region of an air-assisted injector tip [60].

Figure 7.2-21 Comparison of injector deposits from two fuel types for two operating modes; premium unleaded gasoline (F1) and iso-octane (F2): (a) 15 hours of stratified-charge operation at an engine speed of 2500 rpm; (b) 5 hours of homogeneous-charge operation at an engine speed of 4000 rpm [60].

important factor in deposit formation, with a higher T90 temperature yielding less deposits. With air-assisted DI injectors, however, the T90 fuel temperature is found to have only a minor influence on deposit formation, and some experiments suggest that an increased T90 temperature tends to yield slightly *increased* deposit formation rates. Higher levels of aromatic content increase deposit formation for air-assisted DI injectors. Higher levels of olefin content, however, are found to increase deposit formation rates for both air-assisted and single-fluid DI injectors.

When comparing the effect of engine operating mode on injector deposit formation, proportionally more deposits are formed in the stratified-charge operating mode, and are formed closer to the injector tip. This is illustrated in Fig. 7.2-21 for premium unleaded gasoline (F1). The deposits formed under homogeneous-charge operation are characterized by a sharp break in deposit formation in the necked section of the projection, with very little deposit formed at the tip of the pintle. This is consistent with the theory that the local metal temperature is a major determinant of the location of deposit formation zones on the pintle projection. Deposit growth tends to stabilize after a certain period of operation. This effect is consistent with the observed morphology of CCD formation and removal, which is that deposit growth stabilizes when the surface temperature of the deposit increases to the critical limit, and that portions of this deposit eventually dislodge and are regenerated in a periodic cycle.

Investigations of the effect of fuel additives on G-DI injector deposits indicate that additives that effectively control injector coking in single-fluid DI injectors may only be effective in reducing deposits of

air-assisted injectors during stratified-charge operation. In comparison, air-assisted injectors exhibit an increased deposition rate during homogeneous-charge operation when operated with fuels blended with these additives. The reasons for this are not well understood, and additional research is required. This observation, however, does indicate fundamental differences between the deposit mechanisms of single-fluid and air-assisted DI injectors. With single-fluid DI systems, deposits are inhibited from accumulating in the critical injector orifice region by virtue of this region being located in a heavily fuel-wetted zone. This type of environment facilitates the performance of gasoline detergent additives. In comparison, the deposit formation mechanism of the air-assisted injector has many similarities to that proposed for CCD formation, in which fuel additives are postulated to provide little or no benefit, and in some cases can even exacerbate the baseline deposit rate [60].

7.3 Combustion Chamber Deposits

The mechanisms of CCD formation and the associated impact on engine performance and emissions have been extensively investigated for PFI engines. These studies show that chamber deposits accumulate with operating time, and contain fuel, oil and products of combustion. The chemistries of the fuel and oil are known to have a significant influence on the type of deposits and the rate of deposition. Of equal importance is the local surface temperature of the combustion chamber; in fact maintaining the mean combustion chamber temperature above 350°C can minimize or even preclude deposit formation. These deposits are observed to accumulate until a threshold thickness is attained, after which dual processes of deposit formation and removal maintain a mean deposit thickness, with these processes being significantly influenced by the engine speed and load. At higher engine loads using stoichiometric-charge combustion, the levels of in-cylinder gas pressure and temperature are significantly elevated, and conditions are more conducive to the detachment of large CCD deposits. For stratified-charge combustion, however, combustion chamber surfaces are significantly cooler. Although this is favorable for reduced heat losses and increased thermal efficiency, the cooler local surface temperature can exacerbate the formation of chamber deposits.

The role of CCD in altering the effective heat transfer to the combustion chamber wall has been demonstrated in many independent studies over the past two decades. Chamber deposits function as an insulator, and also as a thermal reservoir, storing energy in one cycle and releasing energy to the fresh charge in the next cycle. The CCD also occupy a small combustion chamber volume, thus slightly increasing the compression ratio. Of even more importance to engine-out emissions, the deposits also absorb and release unburned fuel, which is of particular importance in G-DI engines, due to the significant possibility of liquid fuel impinging on the chamber surfaces. The CCD may also absorb and release species that promote detonation and other chemical reactions through catalytic effects. As deposits accumulate on the combustion chamber surfaces, the average heat loss from the combustion chamber decreases. The volumetric efficiency and the heat loss to the coolant decrease; however, the flame propagation rate is observed to increase. As a result, the incoming air is incrementally heated by the deposits. For this reason, the possible benefit obtained from charge cooling for G-DI engines may be reduced significantly. For PFI engines, the weights of stabilized CCD are typically on the order of one to three grams per cylinder, depending upon the driving cycle, fuel, fuel additives, and the size and design of the engine. The presence of chamber deposits normally increases the octane requirement and the NOx emissions, and reduces the volumetric efficiency and maximum power. Such deposits may also increase HC emissions and degrade vehicle acceleration [343, 350, 352].

CCD may be expected to exhibit the same effects in G-DI engines as in PFI engines, potentially leading to an octane requirement increase and increased NOx emissions. However, CCD in G-DI engines adds the increased risk of absorbing fuel as it is directly sprayed at the piston crown in many wall-guided combustion systems. This fuel absorption/desorption process delays fuel vaporization, and may act as an HC source in the same manner as crevices in conventional PFI engines. This could degrade the air/fuel ratio in the vicinity of the spark gap, resulting in combustion instability and increased emissions [29]. For this reason, wall-guided G-DI systems generally exhibit a sensitivity to deposits formed within the piston bowl [385].

It is important to understand the mechanisms of deposit accumulation within the combustion chamber, as CCD may vary significantly with location inside the chamber. Areas such as the piston squish face, piston bowl, cylinder head squish faces, and end gas regions are important. It is important to note that the CCD formation mechanism may also vary markedly with the combustion system. Due to the limited reports available in the literature on this subject, a majority of the observations on CCD in this section were derived from wall-guided combustion systems. A caution is needed when applying these observations to understand the deposit formation issues with spray-guided or flow-guided combustion systems.

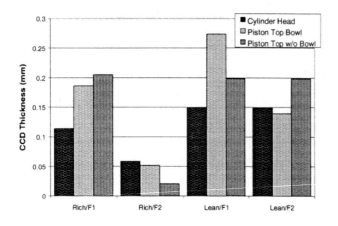

Figure 7.3-1 Effects of fuel type and engine operating mode on G-DI CCD thickness [28].

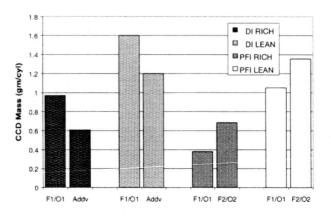

Figure 7.3-2 Effect of fuel and oil types on CCD formation for a range of engine operating conditions [28].

Figure 7.3-1 shows the effects of fuel type and operating mode on G-DI CCD thickness [28]. Here, fuel F1 is an unadditized base fuel having properties that are representative of gasoline for the European market. The same base fuel, but containing deposit control additives, is denoted F2. As illustrated in Fig. 7.3-1, for lean operation, the deposit layer is generally the thickest in the piston bowl, with more deposits formed in the bowl of the piston than on the cylinder head. The deposits that are formed within the piston bowl in the vicinity of the spray-wetted footprint consist of soft, black, porous-type soot. Other areas of the piston generally have a deposit that is much harder [384]. The deposits in the piston bowl are mainly derived from the fuel, whereas the deposits around the periphery of the piston bowl are traceable chiefly to the lubricating oil. Due to the higher temperature of the piston bowl surface, the deposits in this region contain a high percentage of carbonaceous material.

Deposits accumulate on the surfaces of G-DI combustion chambers, with the precursors being the condensed fuel components and partially oxidized hydrocarbons. Lubricating oil is another important contributor to CCD formation and growth. The particular test cycle has a significant impact on the level of CCD produced, with the lean operating cycle generating more deposits, as shown in Fig. 7.3-2. Oil O1 is a multipurpose 5W-30 type crankcase oil representing formulations predominant in the Japanese market. Oil O2 is a 10W-40 partial synthetic crankcase lubricant, which is representative of the type of oil used in the European market. Tests have shown that the mass of CCD is mainly a function of the mileage accumulation cycle

and the type of fuel that is utilized. Experiments also indicate that G-DI engines produce a greater level of CCD than PFI engines under the same engine operating condition [29], with a greater deposition rate obtained when operating in G-DI stratified-charge mode than in homogeneous-charge mode. In fact, the CCD rate in G-DI engines is generally the lowest for high-load, homogeneous-charge operation, due to the higher combustion temperature and higher surface temperature, which hinders CCD formation. Quite high CCD formation rates are observed when the engine is operated at a low load in the stratified-charge mode, an engine durability area to be carefully considered and evaluated [343]. In the homogeneous-charge operating mode the CCD formation rate increases with an increase in the percentage of aromatics, while the effect of olefin percentage is almost negligible. In the stratified-charge mode, the effect of aromatic content is more pronounced, with the CCD that are formed being quite similar to the soot that is formed in diesel combustion. Combustion analyses indicate that the initial burn rate is slowed when the CCD is increased, due to the absorption of fuel. This enleans the mixture in the vicinity of the spark gap and degrades the mixture ignitability [352].

7.4 Intake Valve Deposits

In general, the presence of IVD in PFI engines is associated with increases in exhaust emissions, driveability demerits and degradation in overall engine performance. IVD can affect combustion and emissions through changes to the intake air flow field and air/fuel ratio excursions caused by fuel absorption and desorption. An additional effect of IVD is to increase valve stem friction. In the extreme, such deposits can also result in valve sticking when the clearance between the valve stem and the guide is reduced to zero by the deposit. In addition, valve leakage caused by the deposits is also a concern. Therefore, such deposits must be controlled and minimized for the entire life of the engine.

For PFI engines, the primary cause of IVD is a combination of fuel and lubricating oil from a number of sources including the fuel injector, the positive crankcase ventilation (PCV) and EGR systems and engine oil leakage past the valve seals [145]. All of these effects are engine application specific. The mean surface temperature of the intake valve is the parameter of most significance in determining the deposit formation rate. There is a general consensus in the literature that the mean surface temperature of the intake valve in contact with fuel and/or lubricating oil must be less than 200°C, as the peak deposit formation rate occurs at a mean surface temperature on the order of 230°C. For a mean surface temperature that exceeds 350°C, the deposit formation rate is observed to be significantly reduced. It is worth noting that, for diesel engines, the valve temperature in the vicinity of the tulip region typically exceeds 350°C, with the result being that intake valve deposit formation rate is quite low. For PFI engines, fuel is injected directly onto the tulip region of the valve, which cools that area and also washes the valve surface, thus decreasing the probability of deposit precursor accumulation. G-DI engines have no fuel injected directly onto the surface area of the inlet valve; therefore, the cooling and washing effects are absent. Increases in valve temperature lead to the possibility of increased intake valve deposits, through an elevated surface temperature and the absence of periodic washing by IVD-additive-treated fuel.

IVD in PFI engines have been found to be very dependent on the chemistries of the fuel and oil. Species such as aromatics, olefins and paraffins tend to increase deposit formation. Intake valve deposits in PFI engines are influenced significantly by the engine oil leakage rate through the valve stem seals, with the particular composition of the oil being significant. Oxidized oil exhibits a greater deposit-forming tendency, and oil having high anti-oxidant resistance yields reduced intake valve deposits [145].

The formation mechanisms of PFI intake valve deposits and their effect on engine performance, fuel economy and emissions have been investigated extensively over many decades; however, the knowledge base on G-DI IVD is limited, and it is not at all obvious that the knowledge base for PFI IVD can be directly applied. The lack of fuel residing within the intake port means that detergents designed to clean IVD are not in direct, continuous contact with the intake valve, thus making the G-DI IVD accumulation more difficult to control. In contrast to PFI IVD, which absorb/desorb fuel and create an uncontrolled transient air/fuel ratio delivered to the cylinder, a performance penalty associated with G-DI engines occurs only if the IVD are sufficiently large to either decrease the mass of air inducted or to modify the air flow distribution around the valve periphery. Figure 7.4-1 shows a photo of a significantly deposited G-DI intake valve [29]. Intake valve deposits are wet and fluid in nature, and exhibit poor structural integrity [327]. The observed color of the IVD changes significantly with the particular engine operating condition, with lighter loads yielding a soft, black soot coating and heavy loads resulting in valves with soft, tacky deposits that also cover the lower region of the valve stem [343].

Research data accumulated to date indicate that both the fuel requirements and the lubricant requirements for adequately controlling deposition rates on G-DI combustion chambers, intake valves and exhaust port surfaces are different from those that have evolved for PFI engines [327]. It has been recognized that deposits formed on the intake valves in PFI engines are derived from rather complex reactions involving both the fuel and the lubricant. These include vaporization and partial oxidation of the hydrocarbons and other compounds in these fluids. Although the intake valve is washed with a very small amount of lubricant as compared to fuel, lubricant oxidation can contribute significantly more to PFI deposit formation than the fuel. Current detergent additives in fuel provide cleaning by mobilizing the lubricant and fuel-derived deposits within the liquid phase. In contrast, the intake valves in a G-DI engine are not washed by fuel during each cycle. Hence it has been observed that a fuel additive that was found to be effective in reducing G-DI injector deposits and combustion chamber deposits in G-DI engines has no significant impact on IVD [28]. Note that a parameter to consider when investigating G-DI intake valve deposits is fuel injection timing. If a G-DI engine is operated with early injection timing, such that there is backflow during the valve overlap period, some of the fuel may enter the intake port and may influence either IVD formation or reduction [340]. Further investigation is required to clarify this issue.

Figure 7.4-1 Photo of a deposited G-DI intake valve [28].

Engine lubricating oil has been identified as a key consideration in assessing the amount of IVD formed in G-DI engines [29]. It is understandable that oil can be a contributing factor, as the oil has a direct path to the intake valves through both the valve guides and the PCV system, whereas fuel interacts with the G-DI intake valves in a secondary fashion, such as through backflow. Figure 7.4-2 shows a comparison of intake valve deposits for both G-DI and PFI engines for two types of engine oils and for both lean and rich operating conditions [28]. As a reminder, Fuel F1 is an unadditized base fuel having properties representative of a fuel for the European market, with the same base fuel plus deposit control additives being denoted as F2.

It has been observed that lean-cycle operation produces significantly more deposits than rich-cycle operation for the G-DI engine. In some experimental tests the majority of IVD were found to be accumulated on the intake valves of even-numbered cylinders, and are most pronounced for lean-cycle operations. For the engine in these tests, the PCV system only vented to even-numbered cylinders, and it was determined that 90% of the total IVD came from those cylinders. The odd-numbered cylinders, without PVC, thus contributed only 10% of the total measured IVD mass. This confirms that crankcase oil inducted through the PCV system is a major contributor to IVD in G-DI engines. Tests have also shown that the two intake valves in a four-valve-engine cylinder do not generally experience the same level of deposition, which is attributed to asymmetries in PCV purge flow. It has also been confirmed that increased IVD rates are associated with higher blow-by gas flows [327, 343]. Inductively coupled plasma (ICP) analysis of the deposits shows an elevated fraction of Ca, S, and P on intake

valves that are heavily deposited. These particular elements are present in lubricating oil, and exhibit a positive correlation with IVD, while wear metals such as Fe, Al and Cu are negatively correlated. This further confirms the relationship between G-DI IVD and the lubricating oil.

Oil leakage through the valve stem is yet another source of IVD formation for both G-DI and PFI engines; however, such leakage normally occurs only when there is a vacuum in the intake manifold. G-DI engines are only mildly throttled for reasons of combustion stability control and emissions reduction, which means that the MAP level of a G-DI engine is generally higher than that of a PFI engine. Therefore, it is to be expected that oil leakage through the valve stem, and the associated contributions to IVD, should be less in G-DI engines than in PFI engines. The very aggressive use of EGR in G-DI engines is another contributing factor to IVD formation. It is evident that the deposition rate for a particular intake valve is dependent on nearly all of the factors that affect the valve environment, including valve temperature excursions, PCV flow field, EGR distribution, in-cylinder air flow field, blow-by and valve-stem leakage. This is the reason why such large variations in IVD mass are observed, even within the same engine or cylinder.

A study on the IVD of three different G-DI engines from different vehicle manufacturers revealed that the averaged weight of IVD was nearly equal across the vehicles [350], although large variation was observed in IVD mass among valves from the same vehicle [343]. A

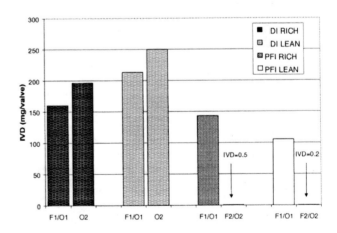

Figure 7.4-2 Effects of fuel and oil types on G-DI intake valve deposits under different operating modes [28].

greater fraction of IVD is present on the intake valves in relatively specific positions in the engine, and this was true for all of the G-DI engines tested. Thermogravimetric analysis of deposits has independently confirmed that IVD is primarily composed of compounds traceable to engine oil. For one G-DI vehicle, the IVD mass for the valves at the rear-facing half of the engine is larger, and the major composition of the IVD is more thermally stable, than for the corresponding IVD on the valves at the forward-facing half of the engine. This is considered to result from a difference in the thermal conditions between the valves at the forward and rear sides of the engine, as the valves on the front side of the engine have an SCV valve installed, with those valves exhibiting less IVD. Deposit level variation for valves in different engines using the same fuel is of the same order of magnitude as the variation resulting from the use of several fuels in one engine. Aftermarket fuel additives designed to reduce injector deposits in the engine were also found to reduce CCD by about 10–50%, depending on the vehicle type, and also reduced the IVD in some vehicles by 8–29%. This IVD reduction is a prime indicator that fuel containing the additives probably contacted the deposit on the intake valves when the engine was operated with fuel injected during the intake stroke [343].

Finally, it should be noted that exhaust valve deposits in G-DI engines are typically an order of magnitude less in mass than intake valve deposits, which is also the case in PFI engines. One major difference between G-DI and PFI exhaust valve deposits is their appearance. Exhaust valve deposits in PFI engines are characteristically gray, whereas such deposits are typically dark and sooty in G-DI engines. This is attributed to the difference in operating temperature between these two engines. PFI engines tend to operate at temperatures considerably above that at which all fuel and lubricant carbon deposits are oxidized. Consequently the only deposits normally present are low levels of inorganic ash components. G-DI engines that operate under stratified-charge conditions, however, yield reduced exhaust valve operating temperatures.

7.5 Deposits on Other G-DI System Components

In addition to the component deposits that have been discussed, deposit formation on other components such as the throttle body, EGR system, port deactivation device, and spark plugs may also occur. Deposit accumulation on these components could significantly degrade vehicle functionality, leading to penalties in engine fuel economy, emissions, performance and vehicle driveability [145]. In extreme cases it is possible for throttle body deposits to accumulate to a point where air flow is significantly reduced at the so-called limp-home-mode position of the plate, which is the mechanically neutral position of the throttle plate. This cannot be compensated for by the engine management system or alleviated by additives. Therefore, it must be prevented by an appropriate design as part of the system development. Throttle body deposits may be minimized by the avoidance of any contact of the throttle body surfaces with potential deposit formation sources such as PCV, canister purge and EGR gases. With G-DI engines, deposit issues concerning the throttle body are typically focused on the throttle plate and the bore in which it rotates, especially with electronic throttle control (ETC). For stratified-charge operation, it is necessary to employ an ETC device to maintain computer control of the necessary torque demanded by the driver or electronic controller such as transmission or traction control.

In general, throttle body deposits only occur during an engine hot soak because the operating temperature is always below 100°C due to air cooling during engine operation. Throttle body deposits for PFI engines can be composed of raw fuel that is circulated back to the throttle body area during backflow from the inlet valve and induction flow pulsing. Raw fuel can also be introduced in proximity to the throttle body during a vapor purge. Another possible source is the PCV gas system that enters the intake system both before and after the throttle body. Such gases contain unburned hydrocarbons and partially oxidized oil fractions, as well as many other chemical species formed during combustion. Those contaminants that are present in PCV gases constitute a significant potential source of throttle body deposits. The large flow rate of EGR gas and the relatively higher temperature in G-DI engines will accelerate the deposition reaction at the throttle body. The lighter fractions of PCV gas are vaporized, leaving a tacky coating on the throttle body surfaces. The solution to this problem is difficult, as it requires the removal of the bulk of the deposit-forming substances from PCV gases before passing them into the intake system. There are PCV valves that incorporate a cyclonic separator to trap a large amount of the oil, fuel and water with dissolved acids that are present in the PCV gas. This, however, must eventually be regenerated and, as a result, places a higher requirement on the engine oil to absorb what would otherwise be consumed and exhausted in the combustion process. With air-assisted G-DI systems, PCV gas may be introduced into the air compressor inlet, and eventually injected into the combustion chamber with the fuel and the air. As a result, the problem with PCV gas in the intake system can be avoided. For the same reason, vapor purging through the air compressor in the air-assisted system can generally reduce the extent of deposit problems within the intake system [145]. It should be noted that careful monitoring is required to ensure that deposits are not formed inside the air compressor or in the air lines.

Also contributing to these deposits are exhaust gas constituents arriving through the EGR system, and circulating to the throttle body. The negative effect of EGR gas on throttle-body deposits is well established for PFI applications, with the design for avoiding PFI throttle-body deposits being to introduce the EGR gas at a point in the intake system such that it cannot be transported to the throttle body in the inlet air flow. For a part throttle opening, a recirculation area exists downstream of the throttle plate, which, if EGR is introduced in the vicinity, can entrain the EGR gas directly upstream and contact the throttle blade, promoting deposit formation. A very large amount of EGR is utilized by G-DI engines for NOx control, thus increasing the potential for contamination of the intake system with partially oxidized fuel and oil, and with the high levels of soot that are present in G-DI exhaust. As a result, such contaminants could lead to a deposit formation rate on the lower throttle body area that exceeds that in PFI engines. EGR gas may be introduced closer to the intake valves through a distribution system, but a compromise with regard to intake valve deposits has to be made.

The EGR system is another area of potential deposit formation, given the increased level of soot and unburned fuel in G-DI exhaust gas. Deposit formation here could lead to reduced EGR gas flow or the sticking of the EGR valve. Maintaining the highest possible temperature in the EGR system may be effective in avoiding excessive deposit formation; however, a conflicting requirement is the need to keep the EGR valve actuator relatively cool. This valve is driven electrically with linear actuation, and the temperature of the valve at the actuator must be kept relatively low to avoid damage. In addition, maintaining the EGR gas at a higher temperature may also limit the maximum mass flow rate of EGR that can be circulated by the system. Pretreatment of exhaust gases before introduction to the intake system

can provide a reduction in deposit formation both within the EGR system and on the throttle body. This may be achieved by obtaining the EGR gases downstream of the TWC, such that the exhaust gas has undergone considerable oxidation. This is possible only if the exhaust back pressure downstream of the TWC is sufficient to provide the amount of EGR needed for NOx control.

Any port deactivation device located in the intake port of the G-DI engine may also be subject to deposit formation, particularly when it is in contact with deposit-forming substances and has a surface temperature capable of enabling such formation. The design of the port deactivation device must be optimized to ensure that it does not become a site for deposit formation resulting from EGR, PCV and/or purged vapor gases.

Finally, the accumulation of carbon deposits on the spark electrode is another important consideration in the G-DI engine development process [296]. A reliable ignition source must be assured for the total range of engine operating conditions over the vehicle lifetime. As illustrated in Fig. 7.5-1, carbon formed on the spark alters the insulator resistance, which can degrade the spark initiation and discharge characteristics. This can significantly alter combustion stability, and can even yield complete misfires. Chemical analyses reveal that the deposits formed on the surface of the plug insulator are chiefly composed of carbon, with calcium and silica also present [437]. Significant development efforts have been directed toward improving the carbon fouling resistance of the spark plug, while preserving a long service life and high ignitability. Figure 7.5-2 illustrates a newly developed spark plug that has been shown to offer advantages over the conventional plug used for G-DI applications. Iridium electrodes have long been identified with higher ignitability and longer life as compared with conventional plugs. A very small electrode has also been shown to be effective for combustion stability, and is achieved by employing a highly wear-resistant iridium-rhodium alloy for the center electrode. The newly developed spark plug has two additional side-group electrodes that will allow discharge to occur under fouled conditions, thus retaining insulation resistance [155].

7.6 Summary

Deposit control is a very important consideration in the design of a G-DI combustion system. Ignoring this important issue can result in increased development time and in significant field service issues. Increased injector

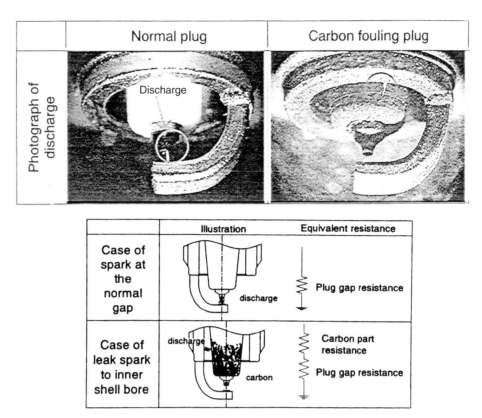

Figure 7.5-1 Comparison of normal and carbon-fouling plugs and their ignition characteristics [437].

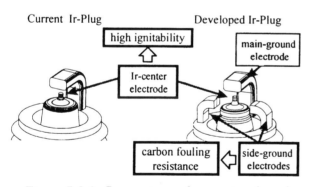

Figure 7.5-2 Comparison of current and newly developed iridium plugs [155].

deposits degrade the spray quality delivered by the fuel injector, and thus diminish the robustness of the combustion system. Among the spray parameters, spray geometry is regarded as among the most important, as it is generally perturbed long before cumulative deposit blockage significantly reduces fuel flow rate. This means that G-DI injector deposits can cause degradations in COV of IMEP, and even misfires, well before significant fuel flow reductions are encountered. This is particularly true for spray-guided and air-guided combustion systems. If deposits do accumulate to the extent that the fuel flow rate is noticeably decreased, the uniform flow rate calibration along a bank of cylinders will yield cylinder-to-cylinder variations in air/fuel ratio.

As contrasted to PFI applications, G-DI injector deposits do indeed form during continuous operation, and can form quite rapidly. The injector tip temperature has been identified and confirmed as a critical operating parameter in the deposit formation process, and maintaining the injector tip temperature below the T90 point of the fuel being utilized is effective in minimizing the injector deposition rate. Correctly positioning the injector in the combustion chamber and shielding the injector tip from direct exposure to hot combustion gases are effective means of promoting injector tip cooling. Injector design and positioning are certainly important in minimizing injector deposit formation, but this may not be adequate for full avoidance.

Fuel additives have been found to be extremely effective for minimizing PFI injector deposits, and research is being conducted on the development of specific additives for G-DI applications. Without effective additives, G-DI combustion system design must be somewhat compromised by placing the injector in a relatively cooler spot inside the combustion chamber. Even if G-DI fuel additives were available, it would have to be carefully verified that adding them to the fuel supply would not adversely affect the formation of injector deposits in the PFI vehicles currently in operation.

The mechanism of injector deposit formation for air-assisted injectors is quite different from that for single-fluid fuel injection systems. Even though fuel metering with the air-assisted fuel injection system is independent of injector deposits, spray characteristics and stability, and ultimately combustion quality, are influenced significantly by the deposits formed on the injector tip projection.

Combustion chamber deposits in wall-guided G-DI engines add the negative factor of absorbing fuel as it is sprayed into the piston cavity. This can occur to a lesser degree for air-guided and spray-guided systems, if any surface containing deposits is impinged. This fuel absorption/desorption process delays fuel vaporization, and may act as an HC source. Such deposits can also alter the air flow field that is critical to mixture preparation for a stratified charge.

Combustion chamber deposits can vary significantly with regard to location inside the cylinder, with more deposits typically being formed in the piston cavity than are formed on the cylinder head. Significant combustion chamber deposit formation is obtained when engines are operated at low load in stratified-charge mode, which makes CCD one of the major challenges in G-DI engine development. The lack of liquid fuel inside the intake port makes the removal of G-DI intake valve deposits more difficult. Engine lubricating oil that is delivered to the head and lower stem area of the intake valve through valve stem leakage and PCV is a significant factor in the amount of intake valve deposits

formed in G-DI engines. As with injector deposits, effective additives must be developed that minimize deposition rates on the G-DI combustion chamber and intake valve, without producing any negative impacts on current PFI engines in the field.

Solutions to G-DI deposit issues should combine prevention, tolerance and removal. An effective design should be developed that can provide an environment limiting deposit formation. This should include thermal and/or mechanical means that limit the presence of depositing chemistries in the area of concern, and that continuously clean and regenerate surfaces to preclude deposit accumulation. Using fuel and/or oil additives can also be very effective in preventing deposit formation, although this must extend to the entire fuel supply in the field. Some deposit formation is inevitable, thus a robust system should be designed so that a certain level of deposit formation can be tolerated without adversely affecting system functionality. Finally a method should be developed to periodically remove deposits either thermally, mechanically or chemically. It is important for engineers involved in G-DI engine durability design and evaluation to consider the range of deposit surfaces and precursors that have been discussed in this chapter.

Much additional research must be conducted before the complex mechanisms of deposit formation, for the numerous components considered, are adequately understood. However, deposit control measures cannot wait until this level of understanding is attained. Key parameters and observed trends that have been detailed should provide a framework to both comprehend the overall potential deposit problem and to develop and incorporate a design package for minimal deposits. If this is achieved for the components that are candidates for minimal deposit formation, a more durable engine will result, with very little degradation in emissions, fuel economy, performance and driveability over the lifetime of the engine.

Chapter 8

Emissions: Formation Mechanisms and Reduction Strategies

8.1 Introduction

The injection of fuel directly into the cylinder has long been regarded as a promising strategy to enhance the thermal efficiency, and hence the resultant fuel economy, of vehicles powered by gasoline engines. Theoretically this fuel economy improvement could be as high as 20% to 25%, depending on the driving cycle; however, only about half of this potential gain has currently been demonstrated in vehicles in the field. This significant shortfall is due to a number of factors, but is largely attributed to system changes that are required to meet stringent emissions standards. Fundamentally, many engine design variables and operating factors that are critical to improving fuel economy, such as throttling reduction, compression ratio increases and an extended load limit for stratified-charge operation, tend to have counteracting impacts on exhaust emissions. Moreover, many emissions-reduction strategies, such as NOx storage catalyst regeneration, post injection to increase the catalyst temperature, and extended stoichiometric operation for fast catalyst light-off during cold start, are detrimental to engine BSFC reduction and the associated vehicle fuel economy.

It is very evident that some fraction of the potential fuel economy gains have to be compromised in order to meet increasingly stringent emissions standards, with the realization that the theoretical thermodynamic potential of the gasoline direct injection engine will likely never be fully achieved [58]. In order to obtain the largest possible fraction of the theoretical benefit, it is crucial to understand the fundamental emissions formation mechanisms associated with the G-DI engine. This will be invaluable in invoking effective control strategies to suppress emissions formation, and in developing aftertreatment systems to remove emissions from the exhaust stream. The fundamental formation mechanisms for unburned hydrocarbons, NOx and particulates, as well as the contemporary technologies and strategies for suppressing pollutant formation and obtaining tailpipe-out reductions, will be discussed in this chapter.

8.2 Hydrocarbon Emissions

8.2.1 Hydrocarbon Emissions During Cold Start and Transient Operations

Direct in-cylinder injection of gasoline completely eliminates the formation of a liquid fuel film on the intake port walls, which has the benefit of eliminating the fuel transport delay and fuel metering errors associated with inducting fuel from a liquid pool in the port of a PFI engine. Injection directly into the cylinder significantly improves metering accuracy and engine response under both cold start and transient conditions. A comparison of the engine performance and emissions levels of a G-DI engine versus a PFI engine under cold start conditions indicates that the G-DI engine exhibits a rapid rise in IMEP following the first injection event, whereas the PFI engine requires about 10 cycles for the engine to attain robust combustion [407]. This is attributed to the formation and growth of a fuel film on the intake valve and port wall of the PFI engine, wherein the fuel mass entering the cylinder on each cycle is not necessarily what is being metered by the injector on that cycle. To compensate for the delay in fuel reaching a steady-oscillatory state for the wall film, significant additional fuel must be added at cold start for the PFI engine. As a result, HC emissions for the PFI engine are typically quite high during these early cycles.

In theory, G-DI engines can be started cold using a stoichiometric or even slightly lean mixture. As a result, the potential exists for both cold-enrichment compensation and acceleration-enrichment compensation to be substantially reduced in the engine calibration, thus reducing total HC emissions. Therefore, one of the significant emissions advantages of the four-stroke G-DI engine is a potential reduction in HC emissions during engine cold start and warm-up; in fact, this has been attained in numerous experimental studies [9, 10, 407, 408, 439]. However, it should be emphasized that this emission benefit has been found to be very much dependent upon the quality of the spray delivered by the fuel system. As discussed in Chapter 3, if the fuel pump delivery pressure, and hence the fuel rail pressure, cannot be generated quickly enough during cold start, the level of fuel atomization in the resultant spray will be significantly poorer than is obtained after full pump priming. This is the case for all current production G-DI engines that use a simple low-pressure pump to prime the main pump. As a result, increased fuel enrichment for cold starting is generally required, which may lead to a liquid pool in the cylinder, as is the case in PFI intake ports. This is also an example of how the realization of the theoretical potential of G-DI is dependent upon the quality of the delivered fuel spray.

Tests with a prototype DI engine have verified that G-DI engines do indeed fire on the second cold-start cycle at a stoichiometric air/fuel ratio. On the other hand, typical PFI engines fail to fire until the seventh cycle, after which unstable combustion events occur [9]. Moreover, G-DI engines exhibit relatively good combustion with little cycle-to-cycle variation following the first combusting cycle. G-DI engines can also be expected to fire on the very first cycle if a slightly greater amount of fuel is injected to account for the fact that there may be no residual gas in the combustion chamber when the first fuel injection occurs. Relatively higher IMEP is obtained for the second cycle of G-DI operation, which is due to this being the first and only firing cycle in which combustion

occurs without residual burned gases. In fact, the cylinder contains residual unburned mixture from the first cycle instead of residual burned mixture. The IMEP produced is therefore greater than that obtained in any later cycle, in spite of the fact that the combustion chamber surfaces are the coldest on the first firing cycle. With the G-DI engine, however, the lack of combustion on the first cycle does result in elevated HC emissions. Fast flame ionization detector (FID) measurements of HC emissions indicate that by the fifth cycle, HC emissions attain a steady-state level, whereas HC level builds steadily in the PFI engine at a stoichiometric air/fuel ratio until the first fire in the seventh cycle. The pattern of high HC emissions continues through as many as 35 cycles as unstable burns continue for the PFI engine.

The improved transient response and significant reduction in HC emissions of a 1.8 liter, 4-cylinder G-DI engine is shown in Fig. 8.2-1 for a cold-starting test that also includes a direct comparison with a standard PFI engine [439, 440]. Four cycles are required for the G-DI engine to attain a stable IMEP when operating at an air/fuel ratio of 14.5:1, whereas the PFI engine requires 12 cycles to reach stable operation at an air/fuel ratio of 13:1. A comparison of the engine emission characteristics of G-DI and PFI engines during cold start was made for a coolant temperature of 23°C, using both early-intake fuel injection (30°ATDC on intake stroke)

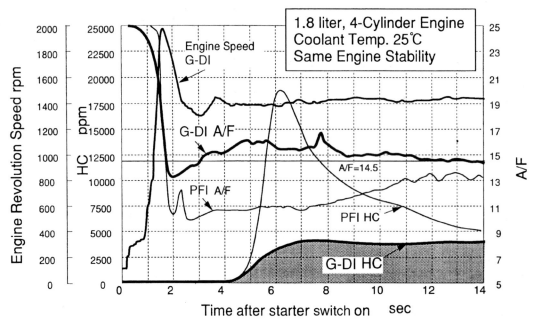

Figure 8.2-1 Comparison of cold-starting air/fuel ratio requirements and emission characteristics between G-DI and PFI engines [440].

and near-BDC injection (170°ATDC on intake stroke) [408]. HC emissions for PFI and G-DI engines for early injection are higher than those obtained for G-DI engines using injection at the start of the compression stroke. Data on the effect of injection timing on transient HC emissions during the first several seconds of a cold start reveal that G-DI HC emissions decrease and reach a minimum at the injection timing of 170°ATDC on intake stroke, and are 60% lower than those obtained for early injection (30°ATDC on intake stroke).

The effect of coolant temperature on part-load HC emissions and fuel consumption was measured for an engine speed of 1400 rpm and a BMEP of 1.5 bar. The data for a coolant temperature of 20°C show that a 50% reduction in steady-state HC emissions is obtained by retarding the injection timing from 30°ATDC to 110°ATDC on the intake stroke. For a constant coolant temperature of 80°C, the minimum BSFC occurs at an injection timing of 30°ATDC on intake. For the tested system, injection timing was optimally set at 110°ATDC on intake during cold operation in order to obtain low HC emissions, whereas a timing of 30°ATDC on intake was determined to be optimum for warm, steady-state operation. This timing schedule provides a good balance between HC emissions and fuel consumption for the given combustion system. The use of an early injection timing of 30°ATDC may, however, result in the loss of a portion of the charge-cooling benefit. Measurements of the effect of the operating air/fuel ratio on cold start performance of G-DI and PFI engines confirm that G-DI engines can operate two full ratios leaner than PFI engines at the same starting cycle number.

A vehicle cold-start test of a prototype G-DI engine using an optimized injection timing and a lean mixture provides verification that the G-DI engine exhibits a clear reduction in both HC and CO [408]. In this engine the injection timing is set at the end of the intake stroke for cranking, and to 110°ATDC on intake following the first firing. The G-DI engine is found to yield a more rapid catalyst warm-up, and the HC emissions level is lower than is obtained with the PFI engine. This is due to the rapid increase in catalyst temperature during a cold start and run, which results from the earlier firing and improved air/fuel ratio control capability afforded by direct injection. For the conventional PFI engine numerous calibration compensations are required to minimize the effect of fuel transport lag caused by the inherent fuel/wall wetting. During speed and load transients the effectiveness of such compensation is limited. As a result, the PFI engine exhibits a large fluctuation in catalyst efficiency during engine transients, indicating a decrease in HC conversion efficiency, and tip-in and tip-out spikes of tailpipe-out HC emissions occur. The more precise air/fuel ratio control in the G-DI engine generally provides an associated improvement in catalyst conversion efficiency, thus resulting in lower HC emissions.

8.2.2 Sources of Hydrocarbon Emissions During Light-Load, Stratified-Charge Operation

The basic sources of HC emissions associated with homogeneous-charge operation of SI engines have been extensively studied over the years, and the important contributing mechanisms are summarized in Table 8.2-1 [66]. In general, the sources of HC emissions from stratified-charge operation have a different priority ranking from those for homogeneous-charge operation. The crevice and oil layer mechanisms are considered to be less important, as there should be only a small amount of fuel vapor within the crevices and close to the cylinder wall. Other mechanisms include overmixing during the significantly longer mixing time, undermixing due to the limited time available for air entrainment and mixing of fuel vapor, and the presence of liquid fuel HC sources that are specific to the stratified-charge G-DI engine. The lower in-cylinder temperature associated with overall lean operation also reduces the contribution of post-flame oxidation relative to that occurring for homogeneous-charge combustion [385]. The key considerations regarding increased HC emissions from stratified-charge G-DI engines operating at light load are listed in Table 8.2-2.

HC emissions for low-load, stratified-charge operation represent a significant difficulty associated with gasoline DISC engines [133]. In the case of gasoline direct injection, the flame is eventually quenched in the extra-lean area at the outer boundary of the stratified-charge region where the mixture is in transition from rich to lean [439]. As a result, a significant amount of unburned HC mass remains in the total charge. The inability of the flame front to propagate from the rich mixture at the spark gap through all of the lean mixture in the outer portion of the combustion chamber is a key factor contributing to the part-load HC emissions in G-DI engines; in fact, substantial research effort is being directed toward the resolution of this problem. Another problem is that there can be overly rich regions near the piston as a result of piston-crown and/or cylinder-wall wetting by the fuel spray.

Table 8.2-1
Basic sources of HC emissions associated with homogeneous-charge operation of SI engines

Flame Quenching	• Flame quenching on the cylinder walls • Flame quenching in the bulk gas due to local mixture leanness and heterogeneity
Crevice Loading	• Unburned mixture in the crevices escaping combustion
Fuel Absorption and Desorption	• Fuel absorbed by lubricating oil during the intake and compression strokes is desorbed during the expansion and exhaust strokes • Absorption and desorption of fuel mass by combustion chamber deposits
Others	• Valve leakage

Table 8.2-2
Key considerations regarding increased HC emissions from stratified-charge G-DI engines operating at light load

Flame Quenching	• Flame extinction occurs within the very lean mixtures near the outer boundary of the stratified charge.
Fuel Wall Wetting	• Poor combustion can occur in overly rich regions near a piston crown or cylinder wall that has been subjected to spray/wall wetting.
Lower Combustion Temperature	• A lower combustion temperature reduces the degree of post-flame oxidation of unburned HC.
Lower Exhaust Gas Temperature	• A lower exhaust gas temperature significantly reduces the rate of HC oxidation that occurs in the exhaust port. • A lower exhaust gas temperature degrades the conversion efficiency of the catalyst system.

Stratified-charge G-DI combustion produces a greatly reduced in-cylinder peak temperature. As a result, the HCs that do not participate in the combustion process have a much lower probability of post-combustion oxidation. The exhaust temperature is also low, making it more difficult for current catalyst systems to operate at the high conversion efficiencies necessary for satisfactory tailpipe emission levels. The low exhaust temperature also precludes the possibility of any significant hydrocarbon oxidation in the exhaust port, such as is observed with PFI engines operating at a comparable load [11].

Modeling of the G-DI engine and aftertreatment system reveals that the exhaust gas during lean-stratified operation is actually cooler than is typically required for catalyst light-off. As the level of engine-out HC emissions increases, an increased catalyst exothermic reaction may occur, which could result in a higher catalyst mid-bed temperature and an enhanced HC conversion efficiency. As a result, tailpipe HC emissions are projected to decrease with increasing HC feedgas levels. Figure 8.2-2 shows the results derived for stratified-charge operation subsequent to varying periods of stoichiometric operation. The results strongly suggest that for stratified-lean operation, some incremental increase in engine-out HC emissions is very desirable, or perhaps even necessary, in order to maintain the catalyst conversion efficiency [168].

Figure 8.2-2 also illustrates the compromise of limiting HC emissions even though it constrains the fuel economy benefit of lean-stratified-charge operation. It may be noticed from this particular example that when engine-out HC emissions for lean-stratified-charge operation are less than 2.5 times those of the base stoichiometric engine, a minimum of 400 seconds of stoichiometric operation after cold start is required in order to meet the EURO III HC emissions standard. During this 400-second period, more than 25% of the emission-unconstrained fuel economy benefit is lost. This assumes that the engine is capable of operating in the lean-stratified-charge mode immediately following a start-up. Since the catalyst has not yet achieved a light-off during a portion of the initial stoichiometric start-up period, all of the HC emissions produced by the engine during this period of time will directly contribute to the tailpipe-out HC emissions. As a result, the cycle-averaged tailpipe HC emissions are very much influenced by the engine-out HC emissions that result from stoichiometric operation.

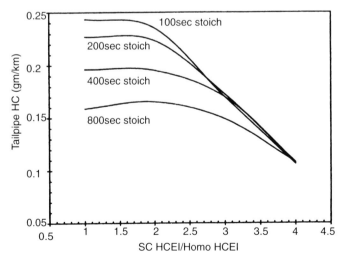

Figure 8.2-2 Tailpipe HC emissions as a function of engine-out HC emissions during stratified-charge operation (SC HCEI: HC emissions index for stratified-charge operation; Homo HCEI: HC emissions index for homogeneous-charge operation) [168].

These results enforce the importance of setting stringent engine-out HC emissions targets for homogeneous stoichiometric operation of the base engine. In fact, the engine-out HC emissions targets for stoichiometric operation may be more important than the HC targets associated with lean-stratified-charge operation [168].

In general, direct gasoline injection yields a slightly elevated level of HC emissions at idle, and can yield a substantial increase at part load. For part-load operation at higher engine speeds, another source of HC emissions is the decrease in the time available for mixture preparation. Without sufficient vaporization time, diffusion burning at the surface of the larger liquid droplets may occur. The use of a higher compression ratio for the G-DI application can result in additional amounts of unburned fuel being forced into crevices. This can result only in an incremental increase in HC emissions for homogeneous-charge operation, since the crevice loading is less critical for a G-DI engine operating with a stratified charge [9, 85, 216, 303].

Some of the very early observations related to HC emissions of DISC engines may still be applied directly to current G-DI engines. The possible reasons for the observed increase in HC emissions for DISC systems at low load were documented for the TCCS combustion system [4, 113]. At low load the mixture becomes locally too lean prior to ignition due to the mixing and diffusion of the injected fuel. As a result, the combustion efficiency of the TCCS system is quite low for many cycles, and the associated HC emissions are excessive. At low load an overall ultra-lean mixture is supplied in this system, which results in overmixing. As a result of the utilization of high-variability, injectors in combination with a small quantity of fuel injected into the combustion chamber, the cyclic variation of the fuel evaporation time becomes a significant factor. HC emissions are traceable mainly to cycles having the lowest combustion efficiency.

Any delay in flame kernel development in the vicinity of the spark gap, or in the propagation rate of the flame through the mixture, is a contributing factor that increases HC emissions, and it is known that engine combustion events with large combustion delays yield a marked increase in HC emissions [113]. This indicates that for stratified-charge combustion of lean mixtures, the resident time of the fuel inside the combustion chamber should be as brief as possible. Thus, the spray characteristics at the end of the fuel injection event are an important factor in determining HC emissions. As has already been discussed in detail in Chapter 4, the last fuel injected as the pintle is closing is not well atomized, and has the least amount of time available to vaporize. In addition, with some pintle designs the last fuel injected may have a trajectory that differs significantly from that of the main spray plume. This degraded atomization and trajectory can contribute to both smoke and an increase in HC emissions.

When operating in the stratified-charge mode, HC emissions increase continuously with an enleanment of the air/fuel ratio [378]. This is attributed to increasingly prevalent pockets of excessively lean mixture and, as a result, an incomplete combustion event. Numerical analysis confirmed that pockets of mixture within the combustible range for overall air/fuel ratios richer than 30:1 are distributed randomly inside the flame [180]. As a result, the flame does not propagate uniformly through these regions, yielding an increase in HC emissions. Thus, elevated HC emissions at low load are an inherent feature of the stratified-charge-combustion concept. To reduce this category of HC emissions, it is considered advantageous to suppress the formation of rich pockets inside the combustion zone. One of the technologies for accomplishing this is the two-stage injection strategy outlined in Section 6.3 [163, 314].

When the fuel spray impinges on the piston, a film of liquid fuel will be formed on the impinged surfaces. Whether this wall film is present or not when the flame arrives is highly dependent on the initial film thickness, the piston temperature and the distillation curve of the fuel. The mean piston temperature for a production G-DI engine, measured at 1 mm below the piston bowl surface, is about 120°C for part load operation. In comparison, the least volatile components in gasoline evaporate at about 180°C. Therefore the wall film evaporates and mixes with air relatively slowly, and some of the high-boiling-point components of the fuel will remain as liquid until the required heat of vaporization is supplied by the combustion event. Due to the limited availability of oxygen above the wall film after the start of combustion, combustion will likely occur as a rich diffusion flame, possibly in the form of a pool fire, with increased levels of engine-out HC, CO and soot emissions to be expected.

After combustion is complete, large-scale motion in the cylinder is of decreased magnitude, and the oxygen concentration in the vicinity of the piston bowl is low. As a result, the probability of oxidizing the bulk of the fuel mass that evaporates after the end of the combustion event is diminished even though the overall air/fuel ratio is lean. For a multicomponent fuel, the lighter fuel fractions evaporate first, leaving the heavier components in the liquid film on the wall, which can increase soot formation and HC emissions even further [385]. When combustion chamber deposits are formed, the amount of residual fuel trapped on the chamber surface following spray impingement is likely to be increased. These deposits will delay fuel vaporization, as the porous deposits adsorb and desorb liquid and gaseous hydrocarbons, and they effectively act as an HC source in the same manner as do crevices in a PFI engine. As a result, the HC emissions characteristics of wall-guided systems are generally sensitive to deposits formed within the piston bowl.

A detailed computational analysis [11] of the HC sources of a stratified-charge G-DI engine using the KIVA code reveals that for significant wall-wetting situations as much as 10% of the injected fuel may remain in liquid form at the time of ignition. This is predicted to occur for combustion systems having both side-mounted and centrally-mounted injectors, with the fuel on the piston crown being the largest source of HC emissions. This impingement and pooling of liquid fuel can result in incomplete combustion of the fuel-rich layer above the piston if it is insufficiently evaporated and mixed before the flame propagates beyond it. The total HC can be as much as 3.4% of the total fuel, with 0.6% resulting from incomplete flame propagation, 2.7% from wall wetting, and 0.1% from the fuel injection system and crevices.

For the late-injection, stratified-charge operating mode, significant fuel vapor is normally present inside a swirl-type spray very soon after the start of the injection, resulting in a "solid-cone" vapor distribution, even within a partially-hollow-cone liquid spray. The mixture is quite rich inside the spray, becoming increasingly leaner in the radial direction, with a thin layer of lean mixture ($\phi < 0.5$) present at the outer periphery of the spray. The amount of fuel in this lean mixture region increases with time and crank angle due to continued mixing. This overly lean region can contain as much as 20% of the total fuel injected by the time when ignition occurs, and, if not completely burned, can be a very significant source of HC emissions. Any unburned lean mixtures outside the primary flame region will contribute to engine-out HC emissions if the in-cylinder temperature is below 1000K during the expansion stroke. Even though the local equivalence ratio may be very high near the piston surface, the oxidation of HCs in this region is slowed significantly by a low temperature. The calculations indicate that the flame can propagate into regions having an equivalence ratio of about 0.3, with oxidation of a mixture having an equivalence ratio of 0.2 being possible for a period of about 1.5 ms if the in-cylinder temperature is at least 1150K. A mixture having an equivalence ratio of less than 0.1 is only partially oxidized, with reaction ceasing after about 3 ms due to a rapid decrease in cylinder pressure and temperature during the expansion stroke.

The deposition of fuel droplets on chamber surfaces is another important HC formation mechanism in G-DI engines. Droplets of fuel from the fuel spray may be transported by the imparted spray momentum or by the mean air flow or turbulence to a combustion chamber boundary such as the cylinder wall, piston or cylinder head, and may contact the surface and form a thin liquid film. A portion of the fuel film formed in this manner can evaporate during the combustion and exhaust strokes, either being oxidized or exiting the chamber as HC emissions [385].

In order to identify the possible sources of HC emissions in the exhaust, analyses of exhaust HC concentration traces have been conducted [385]. Figure 8.2-3 shows an example of both the calculated instantaneous HC mass flow rate and the integrated sum. Blowdown HCs are defined as those emitted during the exhaust blowdown phase, and exhaust-stroke HCs represent the remainder of the HC mass flow. Blowdown HCs are theorized to contain HCs that are traceable to undermixing resulting from the limited time available for air entrainment and mixing, whereas the majority of exhaust-stroke HCs result from overmixing and transport of droplets out of the main combustion zone. Using this representation, the variation in integrated HC as a function of SOI is illustrated in Fig. 8.2-4. It is evident that with an earlier SOI the exhaust-stroke fraction of the total HC emissions increases. This agrees well with the fact that overmixing is a time-dependent process. Total and blowdown HCs also increase, lending credence to the theory that blowdown HCs consist of contributions from all three mechanisms: overmixing, undermixing and liquid fuel effects. If the relative blowdown HCs are taken as a measure of both undermixing and interactions of liquid fuel with the wall, then they should be proportional to the CO emissions, which are shown in Fig. 8.2-5 for three different fuels. The relative blowdown HCs are defined as the fraction of the total HC emissions exhausted before BDC.

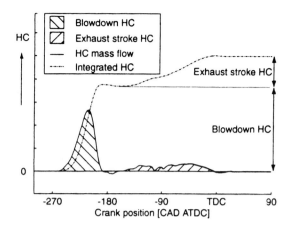

Figure 8.2-3 Definition of HC emissions during different phases of exhaust stroke and the time history of the integrated HC emissions [385].

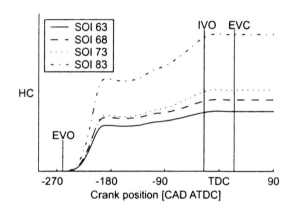

Figure 8.2-4 Time histories of integrated HC emissions for different SOI (°BTDC on compression stroke); 2000 rpm; 3 bar IMEP [385].

Clearly the iso-pentane, which is the most volatile fuel tested, undermixes the least because of the more rapid evaporation and the absence of a persistent liquid wall film. Another measure of undermixing is the mass fraction of fuel that burns late during the combustion event. It has been found that as the available mixture preparation time decreases with a more retarded SOI, both CO and mass fraction HC during blowdown increase, indicating the increased impact of undermixing.

8.2.3 Effects of Operating Parameters on Hydrocarbon Emissions

Effect of Engine Operating Mode. A comparison of HC emissions concentration traces for both homogeneous-charge and stratified-charge operation is shown

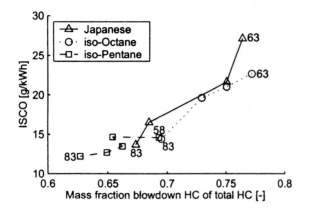

Figure 8.2-5 Indicated specific CO emissions as a function of blowdown HC fraction for different SOI (58, 63, 83°BTDC) and three fuels (Japanese: fuel used in Japanese market) [385].

concentration and the highest degree of oxidation. At the end of the exhaust stroke, HC concentration is observed to increase rapidly. This increase is attributed in homogeneous-charge engines to the roll-up vortex of the cylinder wall boundary layer that contains unburned HC from the piston ring-pack crevice, as well as to fuel desorption from the oil layer. For stratified-charge engines, the observed increase is likely due to a combination of overmixing and quenching. At exhaust valve closing (EVC), the port velocity is low and the HC concentration is observed to decrease slowly until EVO in the subsequent cycle. The reduction in HC concentration is due to the slow oxidation that occurs within the exhaust port. As there is no mean mass flow after EVC, there is subsequently no addition to HC emissions.

Effect of Fuel Injection Timing. In general, the SOI is regarded as the parameter that establishes the total time interval that is available for fuel evaporation and air entrainment, whereas the EOI is related to the time that is available for the last fuel injected. Either of these parameters can thus be descriptive correlating parameters that can significantly influence HC emissions. Figure 8.2-7 shows the effect of SOI on HC concentration [385]. For the case having the earliest injection timing (83°BTDC), the HC concentration trace is the highest, and at intake valve opening (IVO) a small concentration peak is observed. This peak is similar to that associated with the roll-up vortex from a homogeneous-charge SI engine, and is theorized to be due to the transport of fuel-air mixture to the cylinder wall, where it is compressed into the top land crevice. It may also result from the fuel spray penetrating across the bore and impacting the oil layer on the

in Fig. 8.2-6. These are results obtained by averaging the data from 250 consecutive combustion cycles from a single cylinder of a 1.8L production G-DI engine [385]. The measurement was conducted at an engine speed of 2000 rpm and a load of 3 bar IMEP. As is evident, the time history of HC emissions is significantly different for different engine operating modes. The peak that occurs after exhaust valve opening (EVO) for homogeneous-charge operation is most likely due to crevice gases close to the exhaust valve. As more exhaust is expelled, the HC concentration is observed to decrease rapidly as exhaust gas from the center of the cylinder moves into the port. The gas exhausted early in the exhaust stroke (−180° to −90°) has the lowest HC

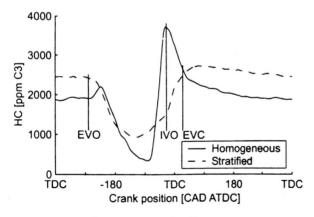

Figure 8.2-6 Comparison of HC concentration traces between homogeneous-charge and stratified-charge operating modes (2000 rpm; 3 bar IMEP) [385].

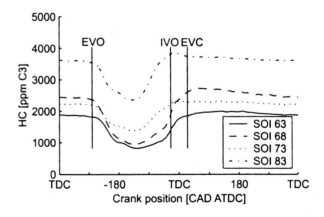

Figure 8.2-7 Time histories of HC concentration trace for a range of start-of-injection (°BTDC) timings (2000 rpm; 3 bar IMEP) [385].

opposing cylinder wall. The HC peak is observed to diminish as mixture preparation time is shortened by utilizing a retarded SOI.

Effect of EGR. HC emissions in G-DI engines are affected significantly by adding EGR, but the mechanism of reduction is complex, involving an increased burning period, a later combustion phasing and an increased post-flame oxidation, with the result being that varying trends are reported in the literature. An increased intake air temperature resulting from the use of hot EGR is beneficial to HC emissions reduction. The heat release of stratified-charge combustion occurs earlier than that of homogeneous-charge combustion; however, the addition of EGR increases the burn duration due to the presence of the diluent, and shifts the heat release to a later timing. But, unlike throttling, adding EGR does not enhance the flame propagation process, as the lean regions are further diluted by the high exhaust gas concentration, which further degrades flame quenching.

In general, HC emissions increase in a manner similar to that in a PFI engine when the EGR percentage is increased at a constant air/fuel ratio [201, 275]. Because there is no need to maintain a constant air/fuel ratio in the stratified-charge G-DI engine, the HC reduction benefits can be obtained by adding EGR through a reduction in air/fuel ratio; however, the specific effect of EGR on HC reduction is combustion-system-dependent. Both HC emissions and BSFC are initially improved as an increasing percentage of EGR is added, with the trend being quite pronounced for light loads. The observed improvement in HC emissions and BSFC are attributed to improvements in fuel-air mixing resulting from the heating effect of EGR [274] and an increase in combustion temperature, which extends the air/fuel ratio limits of flame propagation. As the EGR percentage is further increased, both the BSFC and HC emissions are degraded, as the addition of EGR narrows the flammability limits [11]. The net effect of EGR varies markedly, depending on the details of the combustion system and the engine operating conditions. Therefore, the net effect on HC emissions of using large fractions of EGR must be carefully evaluated for a particular G-DI combustion system design and specific engine operating conditions.

Effect of Throttling. Although the G-DI engine offers the potential for significant enhancements in BSFC by operating at part load without throttling, the use of moderate throttling is found to provide emissions improvements at the expense of some fuel economy [274].

In addition, throttling may be successfully combined with EGR to improve the overall compromise among fuel economy, NOx and HC emissions. The effect of throttling on fuel economy and emissions as a function of EGR has been investigated, and it is found that a 20% reduction in HC emissions can be obtained by the use of light throttling at an EGR rate of 20%, with a fuel economy penalty of only 2.5% [274]. A 50% reduction in HC emissions through the use of throttling can result in a fuel economy penalty of approximately 8%, whereas throttling the engine without any EGR yields the lowest HC emissions at any air/fuel ratio level. Moreover, throttling the engine can be accomplished without an increase in NOx emissions at a fixed EGR rate. If throttling is increased to enrich the mixture, NOx decreases at a fixed EGR rate, but to the detriment of fuel economy. For most G-DI designs, about a 50% reduction in NOx emissions can be achieved with an approximately 2% loss of fuel economy by utilizing EGR without throttling. An 80% reduction in NOx emissions is realized when throttling is used in conjunction with EGR; however, this generally yields an associated 15% shortfall in the achievable fuel economy. Another approach for HC emissions reduction is to limit the degree of stratified-charge operation, and to alternatively operate the engine in the homogeneous-charge, throttled mode. This, however, defeats one of the major incentives for creating a stratified-charge engine, which is to improve fuel economy by significantly reducing the light-load pumping work.

It should be emphasized that the effect of throttling on engine-out HC emissions is dependent on the specific combustion system. In general, throttling the engine will extend the flame further into the quenching zone. This comparatively richer combustion will also elevate the burned gas temperature, which promotes post-flame oxidation of unburned fuel both inside the cylinder and in the exhaust port. As a result, throttling tends to reduce the HC emissions that result from quenching. However, the HC emissions that result from liquid fuel on the piston crown and cylinder wall may actually increase due to the increased spray penetration rate when throttling the engine.

Effect of Fuel Composition. An estimated 65% to 75% of HC emissions from a G-DI engine consist of individual hydrocarbon components that are originally present in the test fuel, as compared to 45% to 70% for a PFI engine, with the HC emissions from the G-DI engine consisting mainly of aromatics and alkanes.

Increasing the mid-range volatility of the fuel is found to reduce G-DI HC emissions by about 7%, as compared to approximately 5% for a PFI engine. Increasing the heavy-end volatility can reduce HC emissions from G-DI engines by about 6%, but yields negligible impact on the HC emissions from a PFI engine. This indicates that G-DI engines are more sensitive to changes in fuel volatility than PFI engines, and that there could be an opportunity to reduce HC emissions by altering the fuel distillation curve [383]. Measurements of the HC emissions from a prototype G-DI engine equipped with an air-assisted fuel injection system indicate that the contribution of individual unburned fuel species to the total HC emissions depends on the volatility of the individual fuel components. As the spark is retarded, the percentage contribution of the high volatility components in the exhaust increases, while that of the low volatility components decrease [217]. Further research investigations in this area may yet reveal interrelationships between fuel composition and G-DI engines emissions that may lead to possible emissions reductions.

Speciation studies of HC emissions from a production G-DI engine reveal that changes in engine-out emissions are accompanied by significant changes in the distribution of exhaust HC species. As the charge stratification of the engine is increased by retarding the EOI, the contribution of olefin partial combustion products (ethylene, propylene and iso-butane) to total HC emissions is increased by a factor of 2.5, whereas the contribution of unburned paraffinic fuel components (iso-pentane and iso-octane) decreases by a factor of 2.0. This increase in the contribution of olefins relative to paraffins in the exhaust contributes to a 25% increase in the specific atmospheric reactivity ratio (g O_3/g HC) of the exhaust HC emissions for forming photochemical smog, as the stratification is increased [216]. This does not necessarily mean that the overall atmospheric impact of the HC emissions is worse, as total HC emissions are most likely decreased. However, it is clear that the total reactivity of the emissions may be influenced by the degree of stratification that is used, as well as by other operating parameters, in a manner that differs significantly from that observed for the PFI engine. Thus, caution is advised when applying "conventional wisdom" that has been accumulated from the study of PFI engines to interpret and extrapolate emissions trends for G-DI engines.

The effects of paraffins, olefins, naphthenes, aromatic compounds, ethers and refinery feedstocks on G-DI emissions have been evaluated in a single-cylinder G-DI engine, and it was ascertained that the combustion of olefins in a stratified-charge G-DI engine yields a shorter combustion duration, a higher IMEP, lower HC emissions and higher NOx emissions as compared to other fuel types [160]. The results for paraffins reveal that the boiling point of the fuel has a direct influence on combustion duration and HC emissions. The results for aromatic compounds suggest that the overly rich region of the air/fuel mixture yields a lower combustion quality and higher smoke emissions. The increased smoke emissions are considered to be partially due to the condensation of aromatic compounds in any overly rich zone.

Effect of Reduced Catalyst Light-Off Time. It is well recognized that the catalyst light-off characteristics have a significant impact on HC emissions during cold start, which comprises a large percentage of the HC emissions that are emitted over the entire test cycle. As discussed in Sections 6.2.4 and 6.3.3, the flexible fuel injection strategy associated with DI gasoline engines provides the opportunity to optimize mixture preparation and combustion processes in order to reduce catalyst light-off time. One approach is to use a post-injection during the expansion stroke to generate heat for enhancing the catalyst light-off process. Another approach is to produce a controlled amount of excess CO and O_2 during the combustion process through a slightly-lean-stratified strategy. The oxidation of CO on the catalyst can significantly reduce catalyst light-off time, and a combination of these two strategies can further enhance the benefits.

8.3 NOx Emissions

8.3.1 NOx Emissions from Stratified-Charge G-DI Engines

As the air/fuel mixture becomes leaner, the conventional lean-burn PFI engine exhibits a maximum NOx formation rate with a mixture that is slightly leaner than stoichiometric, after which it shows continuously decreasing NOx formation rates and engine-out NOx emissions until it reaches the homogeneous, lean combustion limit. This NOx decrease results from a reduction in the maximum reaction zone temperature. However, for G-DI engines that operate with a stratified charge, the temperature within the reaction zone remains high due to the presence of a stoichiometric or slightly rich mixture in the core region of this stratified charge. Therefore, the NOx formation rate is high in these areas, even though the peak cycle thermodynamic temperature is reduced due to the overall lean operation. The aforementioned dependence of engine-out NOx emissions on air/fuel ratio has already been illustrated in Fig. 6.5-14. G-DI engines usually operate with a higher knock-limited compression ratio, which may also elevate the NOx emission levels. Another factor is that the in-cylinder temperature is higher because there is a larger amount of compressed mass in the cylinder than for a comparable load with throttled operation. The net result is that the NOx level of a G-DI engine operating without EGR will generally be quite high, even though some G-DI engines can operate at an overall air/fuel ratio leaner than 50:1. G-DI engines also exhibit significantly elevated NOx emissions at idle as compared to PFI engines. This is the result of locally stoichiometric combustion and the associated high heat release rate at an elevated charge density, as compared to the slower combustion of homogeneous-charge PFI engines at a lower peak temperature level. The higher exhaust residual that is present during idle with throttling is another contributor to lower NOx with a PFI engine. The use of early injection timing for a stoichiometric G-DI engine produces NOx at approximately the same level as that for a PFI engine, whereas retarding the G-DI injection timing results in a significant decrease in peak NOx. However, for a mixture with an air/fuel ratio leaner than 20:1, the NOx level is higher with a retarded timing than is obtained with earlier injection timing.

For a G-DI engine that is designed to operate with lean or even ultra-lean mixtures, it is unfortunately true that a conventional three-way catalyst cannot be used to remove NOx, therefore other techniques for in-cylinder NOx reduction or exhaust aftertreatment must be employed. There is a consensus in the literature that NOx reduction aftertreatment for lean-burn engines is a challenging task. Because part-load, lean operation of G-DI engines is quite frequent, and contributes a large portion of the total NOx emissions depending on the emissions test cycle, the attainment of regulatory levels of tailpipe NOx emissions is one of the major challenges facing DI gasoline engine developers [196].

8.3.2 Exhaust Gas Recirculation for NOx Control

The reintroduction of exhaust gas into the inducted air is widely used for in-cylinder NOx reduction. In this technique, the exhaust gas functions primarily as a diluent during the combustion of the fuel-air mixture. The dilution of the mixture by EGR is a very straightforward method of reducing the peak combustion temperature, thus directly reducing the NOx formation rate. The exhaust gas recirculated during stratified-charge operation of a G-DI engine has a composition relatively close to that of hot air, containing a low percentage of CO_2 and H_2O. Thus, in order to increase the NOx reduction effect, a larger amount of EGR is necessary. As compared to air dilution, however, dilution using exhaust gases decreases the polytropic index of the charge being compressed, due to the presence of CO_2 and H_2O molecules and the associated higher specific heat ratio. Therefore, the effect of EGR dilution on thermal efficiency has been determined to be inferior to that of air dilution. Furthermore, the amount of EGR that can be introduced is limited because it tends to degrade combustion stability. Investigations have confirmed that both the ignition delay and the combustion period are extended with EGR, due to the associated decrease in the laminar flame speed [274, 386]. Consequently, it may be appropriate to use EGR in combination with other techniques [14].

In general, a larger NOx reduction can be realized with EGR in G-DI engines than can be obtained in either PFI engines or diesel engines [262]. For G-DI engines the available fuel-air mixing time is comparatively longer than that of the diesel engine and, as a result, the role of EGR is more effective and NOx emissions can be further reduced. In a conventional PFI engine, NOx reduction using EGR ceases to be effective near the combustion stability limit when the mixture is diluted, resulting in either slow or partial burns. For a G-DI engine using charge stratification, the mixture near the spark gap is ideally either stoichiometric or slightly richer. Furthermore, mixture preparation can be improved due to EGR heating effects. Robust combustion is therefore possible with a much higher level of EGR than is possible with homogeneous combustion. There is however, an associated compromise between a significant NOx reduction and a simultaneous degradation in both HC emissions and fuel consumption due to the degradation of combustion quality.

As a typical example of the effect of EGR usage, Fig. 8.3-1 shows the measured effect of EGR rate on the reduction of NOx from the production Mitsubishi GDI engine. In contrast to the engine-out NOx reduction of a PFI engine operating with a stoichiometric mixture, the NOx reduction for a G-DI engine with EGR can exceed 90% while maintaining an improved BSFC. For this production G-DI combustion system the peak flame temperature is found to be reduced by about 200 K when 30% EGR is employed at light load. The effect of EGR on the lean limit and NOx emissions of the Toyota first-generation D-4 engine is shown in Fig. 8.3-2. Stable combustion is obtained even when the air/fuel ratio is increased to 55:1 without throttling. Without the use of EGR the engine-out NOx emissions for this engine increase as the mixture is enriched from an air/fuel ratio of 55:1, and attain a maximum at an air/fuel ratio of 22:1. NOx emissions continue to decrease as the EGR rate increases until unstable combustion starts to occur when the EGR rate is increased significantly for stratified-charge operation. Stable combustion was maintained for an EGR rate of up to 40% at light load with this engine, which yields a 90% reduction in NOx as compared to engine operation without EGR, and a 35% improvement in fuel economy as compared to the conventional PFI engine.

Providing a G-DI engine with the required amount of EGR is always a design challenge, as a considerably higher EGR mass flow rate must be metered and distrib-

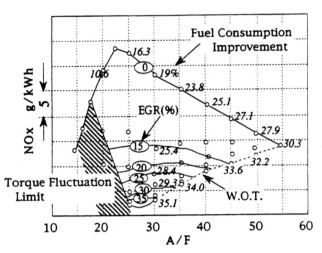

Figure 8.3-2 NOx emissions and fuel economy improvement as a function of air/fuel ratio with various EGR levels for the Toyota first-generation D-4 engine (1200 rpm; 12 mm³/st; 4-cylinder engine; 2.0L) [156].

uted uniformly to individual cylinders at a much lower pressure differential than is typically available with PFI engines [374, 469]. The flow of such a large amount of EGR may require a moderate level of intake vacuum, which will maintain a large portion of the pumping loss that the G-DI engine is supposed to reduce substantially. It is also quite difficult to provide the appropriate amount of EGR during engine transients, such as during G-DI mode transitions, which otherwise may penalize vehicle driveability or NOx emissions.

As an example, a variable-EGR distributing system that may circumvent some of the disadvantages of current EGR systems is illustrated in Fig. 8.3-3 [481]. A distribution plenum in close proximity to the cylinder is designed to be closed by means of a rotary disk valve. The opening of the cylinder feed lines is synchronized with the firing order, with the actual metering occurring within an electrically operated EGR valve. At part load a large cross-section is opened to each individual cylinder, allowing a high EGR mass flow rate while maintaining an even distribution to all cylinders. At full load, or during engine transients, the EGR system close to the cylinder is shut off, improving the EGR dynamics. For the transition into part-load operation, EGR is available close to the cylinder head by opening the distributing plenum. Because only a small section of intake manifold is exposed to hot exhaust gas, the temperature increase of the induction air is less than with a conventional central feed design.

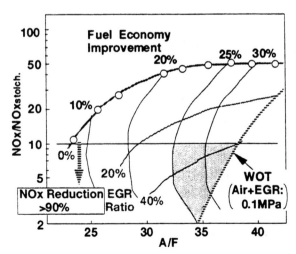

Figure 8.3-1 NOx emissions and fuel economy improvement as a function of air/fuel ratio with various EGR levels for the Mitsubishi GDI engine (2000 rpm; fuel flow rate: 15 mm³/st; 4-cylinder engine; 1.8L) [259].

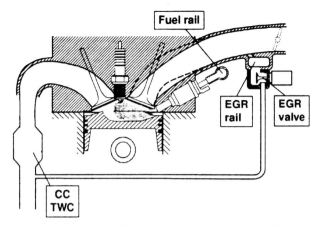

Figure 8.3-3 Schematic of a variable-EGR distributing system [481].

Other strategies such as introducing EGR directly into the cylinder through a separate EGR port or valve have also been considered [259]. However, the location and orientation of the EGR port and valve can influence in-cylinder EGR distribution significantly, and thus fuel stratification, by producing an extra flow of recirculated exhaust gas within the combustion chamber. This impact must be carefully evaluated when implementing this strategy. The sealing of the EGR valve or port and packaging constraints in the vicinity of the cylinder head are also important factors that constrain the application of the direct EGR strategy.

8.3.3 Requirements for Lean-NOx Aftertreatment Systems

Even though EGR is widely employed for reducing NOx emissions from SI engines, there is a general view that the success of G-DI engines is strongly coupled to the development of lean-NOx aftertreatment technologies [187, 188]. This is because EGR cannot reduce NOx emissions over the entire engine speed-load map to meet the scheduled emission limits. For example, within the U.S. FTP emission test cycle the stratified-charge G-DI engine can operate for a significant fraction of the time with a lean-homogeneous mixture at a stable lean limit to maximize the fuel economy gain. At such an operating condition, very large amounts of EGR cannot be used, as the engine combustion stability would be significantly degraded. Also, the important stratified-charge region in the operating map is narrowed considerably when it is restricted to low NOx levels, thus reducing the achievable fuel economy improvement. Therefore, a proven lean-NOx aftertreatment technology would

definitely be a welcome tool in the optimization and extension of stratified-charge operation. Significant research and development efforts are currently being directed toward this important technology [431]. The principal characteristics of several lean-NOx aftertreatment systems are listed and compared in Table 8.3-1. It should be emphasized that any future exhaust aftertreatment systems and associated control strategies must satisfy the following three key requirements in order to be considered for extensive applications:

- Reliable compliance with emissions standards
- Minimal fuel-economy penalty
- No noticeable deterioration in driveability

It is worth noting that for a simple homogeneous, stoichiometric G-DI engine, the aftertreatment system would largely mimic that of a conventional PFI engine, with the possible exception of slightly larger catalyst volumes and increased loadings. This would be necessary to deal with higher engine-out HC emissions resulting from an elevated compression ratio and slightly degraded mixture uniformity as compared to a contemporary PFI engine.

8.3.4 Lean-NOx Catalyst

Lean-NOx catalysts, also called selective reduction catalysts or DeNOx catalysts in some publications, utilize engine-generated hydrocarbons to chemically reduce NO emissions. As illustrated in Fig. 8.3-4, stable conversion capability with a high-sulfur fuel even after lengthy operation is the biggest advantage of this type of catalyst, especially when compared to NOx storage (trap) catalysts, which are very sensitive to the sulfur content of the fuel and will be discussed in detail later in Section 8.3.11. In comparison with the urea-based, selective-catalysis-reduction (SCR) concept where a reducing agent is externally added to the exhaust stream, an advantage of this HC-based lean-NOx catalyst approach is the fact that the hydrocarbon species required for optimal NOx reduction are either emitted by the engine itself or can be supplied by a post-combustion fuel injection event. In this manner the need for an external source of reductant is abrogated.

Depending on the source of hydrocarbons, the system can be designated as either a passive system, which uses hydrocarbons that are naturally emitted by the engine, and that are present in the exhaust; or an active system in which hydrocarbons are actively supplied. For

Table 8.3-1
Comparison of principal characteristics of several lean-NOx aftertreatment technologies

Lean NOx Catalyst	• Precious metal or zeolite catalysts are used to react NOx with unburned HCs in the presence of oxygen to reduce NOx. • Combustion-borne HCs or raw fuel can be used as the reductant. • Maximum conversion efficiency: 30%–50%. • Narrow working temperature window (180°C–300°C for Pt-based system). • It may be necessary to combine with other catalysts to cover the entire engine operating temperature window. • Less complexity for implementation. • High resistance against sulfur contamination. • Precious metal catalyst is only slightly inhibited by water. • High NOx conversion to N_2O (an identified greenhouse gas). • Cannot be used downstream of a close-coupled catalyst. • Sulfate formation at higher temperature.
NOx Storage Catalyst	• Potential for high NOx conversion efficiency (>90%) for lean operation. • Operating temperature window for peak reduction: 200°C–550°C. • No HC required to store NOx. Fuel can be used as a reductant with no additional reductant necessary. • No secondary emissions. • Integrated three-way catalyst functionality. • Applicability for close-coupled catalyst and underfloor NOx storage catalyst for cold start emissions reduction. • Regeneration required by rich operation, leading to complicated system control, high fuel economy penalty, HC slip, and particulate emissions. • Rapidly poisoned by sulfur in the fuel and lubricants, thus limited to low-sulfur gasoline and oil. • Desulfurization required, resulting in possible SO_2 and/or H_2S emissions, and fuel economy penalty. • HC/NOx emissions ratio is critical. Higher HC leads to degraded performance. • CO and HC performance decreases at high NOx performance level. • Aging stability and life cycle durability have to be further improved.
Urea-Based Selective Catalysis Reduction (SCR) System	• Potential for high NOx reduction (>70%). • Operating temperature window: 200°C–550°C. • No fuel economy penalty. • High sulfur resistance. • Additional reductant necessary. • Added-on system and further space requirement for urea tank and injection system. • Metering of urea into the exhaust stream must be accurate to maintain high NOx conversion efficiency while preventing ammonia slip as secondary emissions. • Complexity of urea supply system and optimization of urea injection. • Freezing of urea-water solution in cold climates. • On-board urea storage for a satisfactory refill interval is required. • Urea refill logistics must be resolved.
Non-Thermal Plasma Catalyst	• Potential for high NOx conversion efficiency (>70%) over a wide range of operating temperature (150°C–500°C). • High resistance to sulfur. • HCs in the exhaust system are beneficial. • Secondary emissions is a concern. • Fuel economy penalty, system complexity and robustness are development issues.

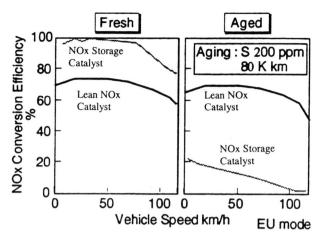

Figure 8.3-4 *Comparison of sulfur poisoning between lean-NOx catalyst and NOx storage catalyst* [342].

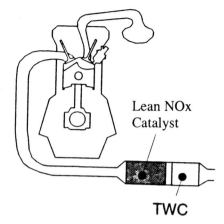

Figure 8.3-5 *Schematic of the system layout for a lean-NOx catalyst* [444].

an active system, the source may be fuel that is injected either into the cylinder during the late stage of the expansion stroke or directly into the exhaust stream. Obviously the injection of fuel into the exhaust stream will require additional fuel delivery hardware, and will complicate the system. Active systems have found application in diesel engines due to the low level of combustion-generated HC emissions from this type of engine. Any fuel provided by such an active system is directly associated with a fuel economy penalty. It is worth noting that the chemical classes and the chain lengths of the particular hydrocarbons in the fuel also play important roles in determining the final NOx conversion efficiency of the lean-NOx catalyst.

A schematic of the system layout for a passive lean-NOx catalyst is illustrated in Fig. 8.3-5[444]. In order to achieve adequate NOx conversion efficiency with this type of exhaust aftertreatment system, a significant surplus of hydrocarbons relative to the nitrogen oxides (high HC/NOx ratio) is necessary. This is because there is a competition for HC between oxygen and nitric oxide that leads to a reduction of NO. The hydrocarbons do not react selectively with the nitrogen oxides, but are oxidized to a large extent by the high oxygen content during lean operation. This leads to a fundamental problem with respect to the use of this type of catalyst, particularly in relation to hydrocarbon conversion during cold start and warm-up through a close-coupled catalyst. If a close-coupled catalyst is utilized upstream of a lean-NOx catalyst to assure good HC conversion during the cold start phase, the HC/NOx ratio at the downstream lean-NOx catalyst is dramatically reduced, thus considerably

lowering NOx conversion efficiency during lean operation at normal engine operating temperature. That is why a close-coupled catalyst cannot be used in combination with this type of catalyst. If, on the other hand, the lean-NOx catalyst is mounted in a close-coupled position, a significant degradation in conversion efficiency resulting from catalyst overheating will be encountered due to its narrow operating temperature window. Therefore, the lean-NOx catalyst has to be installed in the underfloor position upstream of the three-way catalyst in order to avoid heat damage. This is generally not a viable option because of the unacceptably low HC conversion that occurs during the cold-start phase.

The conflicting requirements between converting the hydrocarbons during cold-start phase and converting the NOx during lean-burn phase provide little potential for further development of this type of catalyst aimed at meeting future emissions standards [163]. Figure 8.3-6 contains data on the variation in volumetric NOx conversion efficiency as a function of catalyst inlet temperature for this type of catalyst. The maximum conversion is barely above 45% and it occurs only within a narrow temperature window, which varies with the catalysts utilized, and is in the range of 180°C to 300°C for Pt-based systems and 280°C to 550°C for Ir-based systems [276]. Clearly a narrow operating temperature range with a low conversion efficiency is a significant limitation for this type of catalyst. This will permit only a limited lean operating mode for direct gasoline injection, thus restricting the full realization of the DI fuel economy potential [137]. In addition, this type of catalyst may also generate additional N_2O and NO_2 emissions while reducing NO [201–203].

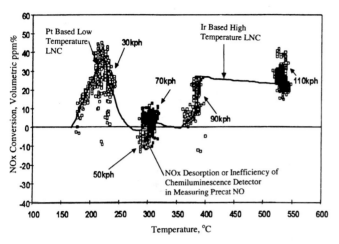

Figure 8.3-6 Volumetric NOx conversion efficiency of a lean-NOx catalyst (LNC) as a function of catalyst inlet temperature [276].

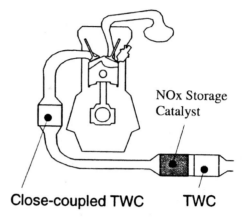

Figure 8.3-7 Schematic of the system layout for an NOx storage catalyst [444].

8.3.5 Operating Principle of NOx Storage Catalysts

NOx storage catalysts, also called NOx traps or NOx adsorbers in some publications, fulfill all of the requirements of contemporary three-way catalysis technologies; however, they incorporate additional components and functionalities that enable the storage of nitrogen oxides under lean operating conditions. A schematic of the system layout for an NOx storage catalyst is illustrated in Fig. 8.3-7 [444]. Figure 8.3-8a illustrates the principle of the NOx storage catalyst for NOx reduction [156]. NOx storage catalysts are essentially characterized by a noble metal component and a basic NOx storage medium. In oxygen-rich (lean) exhaust gas, the noble metal component acts as a catalyst for the conversion of NO and O_2 to NO_2, while the storage medium (M) adsorbs NOx and forms thermally stable nitrates. The elements platinum and rhodium are normally used as the noble metal components, and certain alkaline and alkaline earth oxides (Na_2O, K_2O, Cs_2O, Rb_2O, SrO, BaO) are used as the NOx storage media.

In principle, materials which are suitable as NOx storage materials are those which, by virtue of their basic properties, are capable of forming sufficiently stable nitrates within the temperature range prescribed by G-DI engines. For G-DI engines that utilize an NOx storage catalyst system, the range of the temperature window within which the storage catalyst can effectively adsorb NOx is an important factor in determining the fuel economy potential that can be achieved. As shown in Fig. 8.3-8c, this temperature window strongly depends

on the catalysts used. This will be discussed in detail in Section 8.3.6. It is unfortunate that the materials that can store NOx as nitrates also exhibit a strong tendency to form sulfates, as illustrated in Fig. 8.3-8b. As a result, NOx-storing elements are poisoned by the presence of sulfur in the fuel, and the NOx storing capacity will deteriorate at a rate that is dependent on the sulfur level of the fuel. This will be discussed in detail in Section 8.3.11.

Because NOx storage catalysts are not able to convert the NOx in a lean exhaust, but instead store it when the exhaust gas composition is net-oxidizing, these catalysts have to be purged, or regenerated, periodically in order to maintain a high level of NOx storage capacity. Regeneration is activated by a brief enrichment phase that is programmed into the engine control system. Purging is possible because nitrates are thermodynamically unstable at all relevant temperatures under rich conditions. During the purging process the previously formed nitrates are spontaneously decomposed and the nitrogen oxides are released into the oxygen-lean (rich) exhaust gas where they are converted to elementary nitrogen. This is accomplished by the nitrogen oxides reacting with the excess reducing components of CO, H_2 and HC in a manner similar to that in conventional three-way catalysts. It is evident that the use of an NOx storage catalyst for exhaust gas aftertreatment requires intermittent operation, which places significant demands on the engine management system. In comparison to standard three-way catalysts for PFI engines, lean-NOx storage catalyst technology requires a substantially higher degree of integration into the powertrain system. The engine and drivetrain management systems must satisfy the requirements listed in Table 8.3-2.

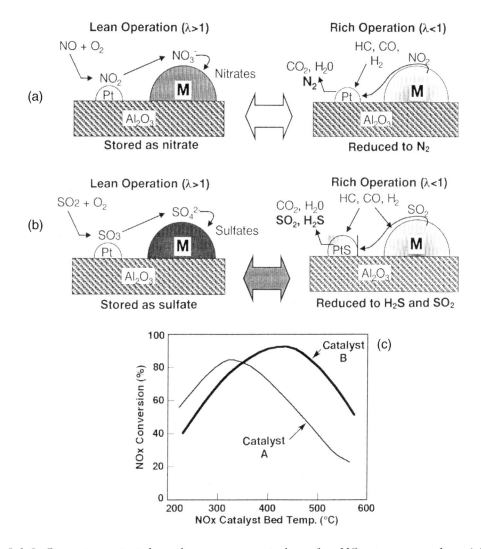

Figure 8.3-8 Operating principle and temperature window of an NOx storage catalyst: (a) NOx storage and reduction processes [156]; (b) sulfur poisoning and desulfurization processes; (c) operating temperature window for two different catalysts [194].

Table 8.3-2
Engine management system requirements for NOx storage catalysts

Determination of Stored NOx Mass	• Depends on a number of key parameters such as engine-out emissions, storage period (duration), exhaust gas mass flow and temperature, and the excess air ratio.
Initiation of Catalyst Purging Events	• Depends on stored NOx mass, exhaust gas temperature and exhaust mass flow. • Determination of the purging (regeneration) duration and the required air/fuel ratio.
Driveability	• Equalization of engine torque outputs during an operating-mode change for purging in order to maintain vehicle driveability.
System Monitoring and Compensation	• Computation of new regeneration parameters based upon system aging or other operating efficiency shifts.

8.3.6 Operating Temperature Window for NOx Storage Catalysts

An effective NOx storage catalyst must adsorb NOx from the lean exhaust stream with a very high efficiency, and must provide a rapid release of NOx and the selective reduction to nitrogen in a rich exhaust stream. Because NO_2 is adsorbed at a significantly higher rate than NO, NOx storage under low temperature operation is limited by the rate of oxidation of NO to NO_2 for the NOx storage catalyst formulation used. This is true even with oxidation of NO to NO_2 also occurring in a close-coupled catalyst. For high-temperature, lean operation, the nitrate species formed during the NOx adsorption process become unstable, and the amount of NOx adsorbed is limited by the formation of nitrates in the adsorbent. These phenomena define a temperature range over which NOx can be stored as a nitrate under excess oxygen conditions. This limited temperature window becomes the key factor in incorporating a lean-NOx storage system, and is a major factor in determining the fuel economy improvement that G-DI technology can provide.

The temperature window for current NOx storage catalysts is in the range of 200°C to 550°C, within which a relatively high conversion efficiency can be maintained. There is also a trade-off for the exhaust system to meet both the low-end and high-end temperature requirements. The low-end temperature fundamentally limits the length of time that a G-DI engine can be operated in stratified-charge mode before it has to be switched back to the stoichiometric, homogeneous mode. This must be done to prevent the exhaust stream temperature from falling below the threshold light-off temperature for three-way catalysis function. The low-end temperature also places a limit on how lean the engine can be operated, as leaner mixtures yield cooler exhaust streams that will eventually cool the catalyst below the light-off threshold. To avoid the decrease in fuel economy that results from catalyst cooling, the exhaust system should be designed to minimize heat losses, which is obviously not desired for high-temperature, lean operation. The high temperature limit of the operating window effectively dictates the maximum sustainable load that may be used for stratified-charge mode or homogeneous-lean operation. At a higher load the exhaust gas temperature will exceed the upper limit of the effective temperature window within which the lean-NOx catalyst is effective, and tailpipe NOx emissions will increase significantly. A key control factor that determines the maximum allowable temperature is the particular formulation of the NOx storage catalyst that is used [194, 329].

The lower limit of the temperature window may not constitute a significant problem in practice, as a close-coupled catalyst is generally installed for removing HC emissions. This provides some temperature increase through the exothermic oxidation of HC and CO. In contrast, the limitation of the maximum operating temperature window is much more critical. It determines the option of operating at higher speeds and loads in lean mode [480]. Therefore thermal management of NOx storage catalysts to maintain the exhaust gas temperature within the optimal range is crucial to the successful implementation of this technology.

Many approaches have been examined to control the exhaust gas temperature within the optimal window for the NOx storage catalyst. For example, an automatic cooling loop controlled by the engine management system has been investigated for providing rapid and precise temperature control of the NOx storage catalyst, but issues of system complexity and packaging have proven to be quite difficult [306]. Attempts have also been made to automatically adjust the exhaust gas temperature by varying the exhaust valve timing, but the system complexity has so far kept this technology from being incorporated into production applications. Another possibility for future development is the incorporation of the variable geometry turbine (VGT) developed for turbocharged DI engines, which could provide a dual benefit. It is well known that a VGT extracts energy from the exhaust gas to operate the compressor, and that the amount of exhaust-stream energy extracted by the turbine varies with turbine inlet vane position. As a consequence, the temperature of the exhaust gas entering the aftertreatment hardware on a turbocharged G-DI engine could be electronically controlled to be within the optimal range. This would permit the maintenance of an acceptable temperature to maximize the conversion efficiency for NOx storage catalysis [57]. Because NOx emissions in this high temperature range are usually associated with a rather limited area within the homogeneous-lean operating map, a recalibration of this load and speed range to stoichiometric, high-EGR operation can substantially reduce NOx emissions through the use of a three-way catalyst [480].

For better exploitation of the working temperature window for NOx storage catalysts and to reduce the high temperature load, two exhaust cooling measures are taken in the Volkswagen DI gasoline system. Figure 8.3-9 shows a schematic representation of this exhaust system [214].

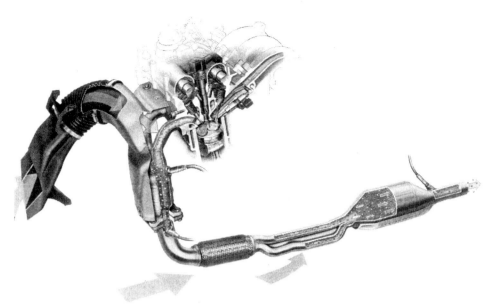

Figure 8.3-9 Exhaust system layout of the Volkswagen FSI engine [214].

Cold air is first directed onto the exhaust-manifold-close-coupled-catalyst module through an air duct that employs the air ram pressure available in the front of the vehicle below the bumper. The second step is to divide the exhaust system between the close-coupled catalyst and the NOx storage catalyst into three separate pipes, which effectively increases the heat rejection to the ambient air by increasing the surface area of the exhaust system. This type of exhaust gas heat exchanger is reported to cause only a minor increase in exhaust gas back pressure.

In order to enable sufficient NOx storing efficiency, it is necessary to prohibit lean operation when the NOx storage catalyst temperature is high. In cases where engine speed and load decrease rapidly, lean operation must be delayed until the NOx storage catalyst temperature is reduced. The required interval is directly related to the thermal inertia of the converter. To maximize the attainable fuel economy benefit for a vehicle, this delay should be minimized. In practice, the heat loss from the NOx storage catalyst brick occurs primarily by means of heat transfer to the exhaust gas, with significantly less heat rejected via convection from the converter shell surface. Therefore, given a fixed starting brick temperature, the time required to attain a new, lower brick temperature during a rapid deceleration is largely dependent upon the inlet gas temperature, which can be enhanced either by a change in location of the converter or by engine calibration. Calibration changes can reduce feedgas temperature, as well as feedgas HC and CO concentration, thus reducing exothermic reactions in a three-way catalyst.

It should be noted that NOx storage catalysts may be subject to elevated temperatures during operation at high engine speeds; in fact, extended operation with an exhaust gas hotter than 750°C will permanently damage the NOx storage capacity [93]. NOx control is achieved through a combination of adsorption, storage, release, and reduction reactions, as well as by maintaining a large, effective surface area for reaction. Storage material must therefore retain sufficient precious metal and adsorber dispersion after being subjected to elevated temperatures for an extended time period. An atmosphere of high oxygen concentration at elevated temperatures (>800°C) could promote the formation of mobile platinum oxides, thus accelerating the rate of sintering of platinum particles, with an associated loss of catalytic effectiveness. Engine operation combining oxygen-rich exhaust gas with high temperatures, such as the condition produced by fuel cut-off during full-load operation of the engine, should be carefully evaluated and restricted if an NOx storage catalyst is present [137].

8.3.7 NOx Storage Catalyst Regeneration Issues

NOx storage catalysts convert and store NOx as nitrates, then release them into the exhaust stream when the stream is oxygen-deficient. To maintain high storing efficiency, NOx storage catalysts must be purged, or regenerated, periodically by rich spikes of exhaust gas. A characteristic of NOx storage catalyst systems is that the rate of NOx adsorption decreases as the NOx filling level increases, thus tailpipe emissions are observed to continuously increase during the storage process. For this reason, the storage medium must be regenerated well before the maximum storage capacity is attained [478]. The required purging frequency is determined by the level of NOx emissions loading, which in turn depends upon the engine-out NOx levels and the catalyst characteristics. Incomplete regeneration of the catalyst will lead to a breakthrough of NOx emissions, whereas unnecessary regeneration of the catalyst will result in an excessive fuel economy penalty. The degradation in fuel economy resulting from NOx storage catalyst regeneration is found to be in the range of 0.5–3.0%, as compared to the fuel economy obtained for continuous lean operation. This fuel economy penalty is larger for both high-speed and high-load lean operation due to the increased frequency of catalyst purging.

A requirement that increases the complexity of applying NOx storage catalyst technology to lean combustion is the generation of a precisely controlled rich spike in the exhaust flow during extended periods of part-load operation. The injection of fuel very late in the expansion stroke is generally regarded as an overall practical method of generating a rich spike for purging lean-NOx storage catalysts in G-DI engines [92–94]. During lean operation, any brief increase in fueling rate can result in a momentary increase in engine torque and power, unless it is properly anticipated and controlled. Optimal strategies must be developed and invoked to allow the engine to be switched from lean to rich operation.

The engine control system must change the air flow rate, fueling rate, spark timing and EGR rate rapidly, but without any noticeable torque fluctuation during the catalyst-purging event. In particular, the catalyst regeneration strategy must be accurately calibrated to prevent HC and CO breakthrough due to over-purging, and must minimize the quantity of fuel used for purging. The use of a three-way catalyst formulation near the exit of the NOx storage catalyst, as illustrated in Fig. 8.3-7, can be effective in catalyzing any excess HC and CO emissions produced during the regeneration process. Figure 8.3-10 shows the impact of the regeneration frequency and the injected fuel mass on the fuel economy penalty for steady-state operation of a production G-DI engine fitted with a prototype NOx storage catalyst [276]. It is clearly shown that more frequent catalyst purging than is required will lead to a significant fuel economy penalty. It is also evident that richer purging at higher frequencies can completely erode the fuel economy benefit of running the engine in stratified-charge mode. As will be discussed in

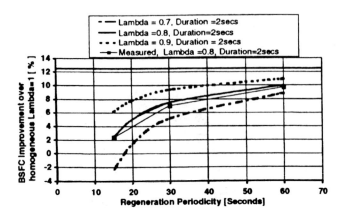

Figure 8.3-10 Effect of regeneration frequency and injected fuel mass on the fuel economy penalty for steady-state operation of a production G-DI engine fitted with a prototype NOx storage catalyst [276].

Section 8.3-12, the use of an NOx sensor can contribute significantly to optimizing the process of regenerating NOx storage catalysts.

During homogeneous-lean operation of a production G-DI engine, the required rich-mixture spike is obtained at intervals of 50 seconds for purging the NOx storage catalyst. Synchronized control of both the fuel pulse width and the spark timing (retardation) is used to maintain a constant engine output torque. For stratified-charge operation this spike is obtained through a complex series of control-system commands that not only provide a brief fuel pulse width and spark retard, but also a simultaneous brief adjustment to the injection timing, throttle opening and positions of the SCV and the EGR valves. The NOx emissions level for this particular engine without either EGR or an NOx storage catalyst is 1.85 g/km for the Japanese 10-15 mode test. With EGR control only, NOx is reduced to 0.60 g/km, which is a 67% reduction. This is then further reduced to 0.10 g/km through the use of a stabilized NOx storage catalyst. Finally, the periodic regeneration of the catalyst through the introduction of the required rich mixture yields a 2.0 % loss of fuel economy potential [156].

Stratified-charge or homogeneous-lean operation of G-DI engines is limited by the NOx emissions and the subsequent regeneration frequency of the NOx storage catalyst. Figure 8.3-11 shows the NOx mass emissions from a 2.0-liter G-DI engine and the associated regeneration interval requirement as a function of engine load [483]. The regeneration interval is obtained assuming an NOx

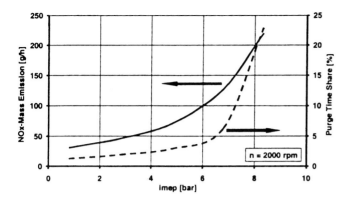

Figure 8.3-11 NOx mass emissions and the associated regeneration interval requirement as a function of engine load for a 2.0 liter G-DI engine [483].

storage capacity of 2 g and a purging duration of 3 seconds. The high-temperature window for the NOx storage catalyst is assumed to be 520°C, and incorporates an allowance for a declining capacity in the upper range of the temperature window. As illustrated in the figure, the required regeneration time shows a steep increase for IMEP levels exceeding 7 bar, which is probably the upper limit of lean operation for this engine.

As vehicle cruise speed increases, lean operation results in elevated levels of engine-out NOx emissions, thus requiring more frequent regeneration of the catalyst. The purging events under these operating conditions can result in significant tailpipe NOx emissions, as opposed to those resulting from purging events at lower vehicle speeds. When the regeneration process begins, the release rate of NOx from the catalyst is initially very high, but rapidly diminishes. It is considered that the mass flow of reductants from the engine at the initiation of the purging event is not sufficient to effectively convert this rapid initial release of NOx from the catalyst at the higher engine speeds, with the result being elevated levels of tailpipe NOx emissions [156, 306, 483].

Sustained vehicle acceleration under stoichiometric engine operation produces a large amount of NOx, primarily due to the unintended release of stored NOx from the storage catalyst. During the transition from lean to stoichiometric operation that is associated with vehicle acceleration, there is a significant decrease in the oxygen level of the exhaust stream, which makes the NOx stored within the catalyst unstable. This results in a release of stored NOx, producing brief periods of elevated NOx tailpipe emissions and reduced NOx

catalyst efficiency. The NOx emissions associated with this transient can be reduced by using an engine control strategy where the engine is operated slightly leaner than stoichiometric with spark retard during acceleration, and purging events are coordinated to occur during periods of vehicle acceleration [43, 156, 306].

8.3.8 System Layout for NOx Storage Catalyst: Close Coupling versus Underfloor

Unlike the HC-based, selective reduction catalyst discussed in Section 8.3.4, NOx storage catalysts do not require any selective reducing agent for the storage phase during lean operation. Thus, as illustrated in Fig. 8.3-7, an exhaust aftertreatment system that is composed of a close-coupled catalyst combined with an underfloor NOx storage catalyst does avoid the design problem described previously for HC-based, lean-NOx catalysts [136]. The close-coupled catalyst provides effective control of engine-out hydrocarbons following start-up, but must also be able to tolerate the very high temperatures that are encountered for high-load operation in homogeneous-charge mode.

The relative location of the underfloor NOx storage catalyst system is dictated by the requirement for maintaining the temperature of the storage brick within the operating window of approximately 200 to 550°C. By positioning the system to keep the storage catalyst temperature from exceeding the upper limit of the window during high engine speeds, stratified-charge operation of the G-DI engine can be realized at a relatively higher load. The light-off of the NOx storage catalyst will, however, be delayed during the cold-start phase, which will eliminate the possibility of stratified-charge operation during the early part of the test cycle. Clearly a compromise is required between the two requirements. By attempting to maximize the potential fuel economy benefit from lean operation during the high-speed part of the cycle, some potential fuel economy gains are sacrificed during the early portion of the test cycle.

Another option is to install the NOx storage catalyst in a close-coupled position. This way the oxidation of NO to NO_2 can take place in the close-coupled catalyst, as well as on the catalytic platinum contacts inside the adsorber. In this manner the NOx storage catalyst can light off much more rapidly than it can in an underfloor (underbody) position. The catalyst can react with a high conversion efficiency shortly after the engine is started, even for low loads and correspondingly low exhaust gas temperatures. However, a major concern with regard to positioning the NOx storage catalyst close to the engine is the thermal durability limitations. In a close-coupled location the storage catalyst would be subject to the same thermal environment as a close-coupled three-way catalyst. To permit close-coupling, the NOx storage catalyst must be aged at a much higher temperature, such as 1000°C.

Relative to the underfloor NOx storage catalyst position, another advantage of close-coupling the NOx storage catalyst is that it can provide extended capabilities in coping with sulfur poisoning through desulfurization. Figure 8.3-12 shows the predicted steady-state NOx storage catalyst temperature as a function of vehicle speed for stoichiometric operation. With a close-coupled NOx storage catalyst, the attainment of the required ~650°C threshold for desulfurization kinetics at vehicle speeds above ~80 km/hr is feasible. Desulfurization at lower vehicle speeds may also be attainable through the application of mild spark retard. In contrast, a threshold vehicle speed in excess of 120 km/hr is required to achieve desulfurization if the NOx storage catalyst is placed in an underfloor location. Moreover, due to the more rapid temperature rise associated with close-coupled NOx storage catalysts (see Fig. 8.3-13), lean engine operation can be initiated earlier in the drive cycle, which can provide significant fuel economy advantages [168]. Clearly, it is very difficult to generalize the system layout for

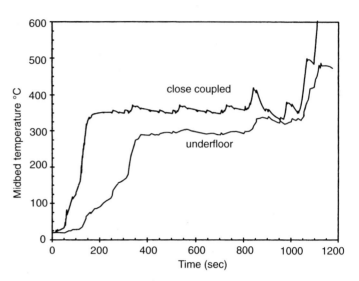

Figure 8.3-13 Comparison of predicted time histories of NOx storage catalyst mid-bed temperature; close-coupled and underfloor catalyst positions [168].

NOx storage catalyst. The system layout illustrated in Fig. 8.3-7 is widely accepted, which certainly requires a special strategy to improve the light-off characteristics of the underfloor catalyst. The optimum position of the storage catalyst strongly depends on the development goal of emissions and fuel economy, test cycle, calibration strategy, and particular exhaust geometry.

8.3.9 Effect of HC/NOx Ratio on NOx Storage Capacity

It is well established that a G-DI engine operating in the stratified-charge mode produces relatively higher HCs as compared to the contemporary PFI engine, and that G-DI aftertreatment systems must be extremely efficient with regard to HC oxidation. Because NO oxidation is the first step in the NOx adsorption process, increasing the HC/NOx ratio will reduce the NOx storage capacity of an adsorber significantly, thus degrading tailpipe NOx emissions, even though higher HC emissions may benefit the catalyst temperature during stratified-charge operation. The fundamental reason is that with an NOx storage catalyst and a close-coupled three-way catalyst, the unburned hydrocarbons are in competition with the nitrogen monoxide for catalyst contacts. This hinders the oxidation of NO to NO_2, which is necessary for subsequent NOx adsorption, and leads to the requirement for an HC/NOx ratio in the engine exhaust being as low as possible. In addition, it is essential to achieve efficient

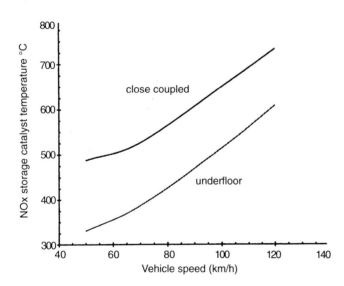

Figure 8.3-12 Comparison of predicted steady-state NOx storage catalyst temperature as a function of vehicle speed; close-coupled and underfloor catalyst positions for stoichiometric operation [168].

HC conversion within the front catalyst in order to achieve high NOx reduction efficiency in the NOx storage catalyst during start-up and stratified-charge operation.

Figure 8.3-14 shows an example of the HC/NOx trade-off for an NOx storage catalyst, with the significant influence of HC/NOx ratio on NOx storage capacity being evident [43]. When the HC/NOx ratio is reduced, NOx storage capacity improves due to the diminished incidence of competing oxidation reactions of HC and NO on the catalytic platinum contacts inside the adsorber. This underscores the significance of achieving low levels of engine-out HC emissions relative to achieving the overall emission-reduction capabilities of G-DI aftertreatment systems. If higher levels of HC emissions are present in the exhaust, the precious metal loading of the catalyst must be increased to improve the HC conversion and NOx storage capacity. In contrast, for HC-based selective reduction catalysts the ratios are inverted, and the functioning of the catalyst is based upon the presence of a sufficiently high hydrocarbon concentration that serves as a reducing agent.

8.3.10 Requirements of NOx Storage Catalysts and Close-Coupled Catalysts

Although the requirement for both low-temperature HC oxidation during start-up and high-temperature durability is identical to that of the close-coupled catalyst commonly used in stoichiometric PFI engines, the close-coupled catalyst that is upstream of the NOx storage catalyst for the system illustrated in Fig. 8.3-7 must also enable effective NO conversion to NO_2 during NOx purging. The conversion of the NO emissions that have not been oxidized in the close-coupled catalyst must occur at the precious metal contacts of the NOx adsorber so that there is an effective storage of all nitrogen oxide as nitrates [43]. During regeneration, some of the reactants generated by rich spikes will be consumed by oxygen that was previously stored within the catalyst system. This requires additional fuel during the regeneration cycle. As a result, a reduced oxygen storage capacity is essential to keep the regeneration events as brief as possible in order to minimize its negative impact on fuel consumption. It is also important for the HC-rich exhaust gas to reach the NOx storage catalyst with as little delay as possible, thus providing a rapid desorption and subsequent reaction of the stored nitrogen oxides. A close-coupled catalyst formulation reportedly has been developed for NOx storage application that has less than 25% of the oxygen storage capability of a conventional three-way catalyst, and can provide the required oxidation capability and thermal durability [306]. The NOx storage catalyst has a dual function, with the first being to act as a stable three-way catalyst with the associated required oxygen storage capability. The second function is to store NOx and convert the HC and CO during lean operation of the engine. Thus, oxygen storage material must be integrated into the NOx storage catalyst in order to provide adequate three-way catalysis activity during periods of stoichiometric operation, or a separate three-way catalyst must be attached to downstream of

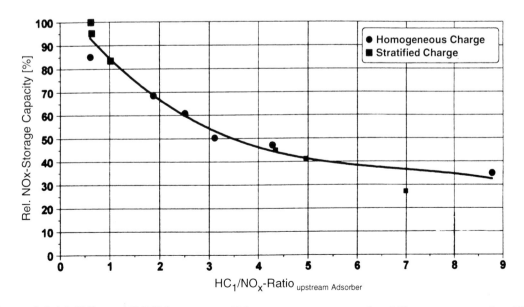

Figure 8.3-14 Effect of HC/NOx ratio on NOx storage capacity of an NOx storage catalyst [43].

the NOx storage catalyst without an oxygen storing capability.

Diagnosing the functionality of the separate close-coupled catalyst is another important issue with regard to G-DI aftertreatment systems. The efficiency of the close-coupled catalyst can be determined by measuring the exothermic reactions, with a model used to calculate catalyst inlet temperature if it is not known. A thermocouple is used to monitor the temperature downstream of the close-coupled catalyst, and the measured temperature is compared to that calculated from a model of the working catalyst. This measured temperature can also be used as feedback for both purging and temperature control of the NOx storage catalyst. For this reason, a thermal model of the system from the close-coupled catalyst to the inlet of the NOx storage catalyst is required [93]. The diagnostic of the NOx storage catalyst will be discussed in detail in Section 8.3.12.

8.3.11 Sulfur Poisoning Issues and Desulfurization of NOx Storage Catalysts

A major disadvantage of NOx storage systems lies in the inherent affinity of the NOx storage materials for sulfur oxides (SOx) released by combustion of fuel. As illustrated in Fig. 8.3-8b, these sulfur oxides interact with the storage components used in the NOx storage catalyst to form extremely stable sulfate compounds. As a result, a steady decrease in NOx storage capacity, and hence a large reduction in the NOx conversion efficiency of the storage catalyst, is observed for sulfur-containing fuel as the total operating time accumulates. This is generally denoted as sulfur poisoning of the catalyst system. As illustrated in Fig. 8.3-4, nearly all of the conversion efficiency of an NOx storage catalyst is lost after being aged using gasoline fuels with a sulfur concentration of 200 ppm for 80000 km.

In general, materials that adsorb NOx effectively also have a high affinity for sulfur species. The resulting sulfate materials in the NOx storage catalyst are thermodynamically more stable than the nitrates, and the sulfur compounds gradually replace the nitrates at the NOx storage sites, eventually leading to complete deactivation of the NOx storage catalyst. Sulfur poisoning of NOx storage catalysts can be circumvented by invoking a desulfurization strategy of operating at high catalyst temperature under reducing exhaust conditions [478], which is schematically illustrated in Fig. 8.3-8b. This is to date possible only with the penalty of increased fuel consumption [136]. Until

sulfur levels in field gasoline are reduced to very low levels, such as 5 ppm or less, desulfurization will continue to be one of the key requirements associated with NOx storage catalyst implementation. The critical questions associated with desulfurization of a particular NOx storage catalyst system are as follows [478]:

- Can the sulfur adsorption of the NOx storage material be completely reversed?
- What compromises are acceptable with respect to additional penalties on fuel consumption and driveability?
- Are there additional problems with respect to H_2S emissions that exceed the smell threshold?

Even though sulfates decompose at elevated exhaust temperatures, the temperatures have to be within the range that can be obtained under realistic driving conditions, and also have to be below the threshold at which massive and irreversible thermal damage to NOx storage catalysts occurs. In general, the oxides of sulfur are released from the storage catalyst above 650°C in an HC-rich environment. One of the key goals in NOx storage catalyst development is to ensure that the sulfur can be completely removed from the storage material by on-board desulfurization. This guarantees that progressive and irreversible sulfur poisoning does not arise as a result of many repeated cycles of partial desulfurization. A further requirement of any desulfurization strategy is that it does not yield emissions of hydrogen sulfide (H_2S), which has a pungent odor and is very objectionable. One study of desulfurization processes indicates that a strategy using an HC-rich exhaust can convert as much as 60% of the sulfur into H_2S, which is not acceptable [136]. Desulfurization strategies have been developed that greatly minimize the production of H_2S by a proscribed modulation of the air/fuel ratio. Air/fuel ratios just slightly richer than stoichiometric are very effective in suppressing the formation of H_2S. In this regard aged NOx storage catalysts are less susceptible to H_2S generation at rich air/fuel mixtures than are non-aged units [137]. Therefore, a periodic oscillation between lean and rich exhaust streams to avoid over-rich exhaust almost completely suppresses any H_2S formation during the desulfurization process. Even very brief periods of rich operation are sufficient to purge at least part of the sulfur from the catalyst so that under selected vehicle operating conditions "natural" or inherent desulfurization occurs. Apart from this inherent process that occurs during medium part-load operation, it must be anticipated that the vehicle will be operated to a great extent within the part load range. Therefore, an active desulfurization strategy is inevitably necessary if sulfur is present in the fuel at anything above trace levels.

The strategy of spark retardation coupled with a rich mixture spike can provide both the elevated exhaust temperature and the reducing exhaust gas condition that is required for desulfurization. This strategy, however, will generally result in a significant fuel economy penalty, depending on the sulfur level of the fuel used. Even with the rich-mixture-spike strategy, desulfurization that is initiated during part-load operation entails an additional fuel consumption penalty. Moreover, there is always a risk that, over the lifetime of the catalyst, the concentration of irreversibly stored sulfur may increase very gradually, leading to a small but continuous decrease in NOx conversion efficiency. Therefore, reducing the sulfur content of the fuel is beneficial when introducing G-DI aftertreatment technology [478].

A study of the performance of a prototype NOx storage catalyst retrofitted to a production G-DI engine has shown that sulfur poisoning of the NOx storage catalyst occurs during urban driving. This occurs during lean operation, with exhaust gas temperatures upstream of the NOx storage catalyst in the range of 250°C to 500°C. For highway operation, a temperature level in the range of 350°C–650°C is measured at the inlet of the NOx storage catalyst. According to the torque requirements, either a lean, stoichiometric, or rich air/fuel mixture is applied. When a lean air/fuel mixture is applied, sulfur poisoning of the NOx storage catalyst

occurs; however, applying a rich air/fuel mixture, along with an inlet temperature above 600°C, results in desulfurization of the NOx storage catalyst, and an associated positive effect on NOx storage efficiency. For highway operation, thermal damage of the NOx storage catalyst does not occur, as the inlet temperature of the NOx storage catalyst does not exceed 650°C. Aggressive driving on the highway can yield an exhaust gas temperature in the range of 650°C to 770°C at the inlet of the NOx storage catalyst, whereas at full load, temperatures that exceed 800°C can be generated. The air/fuel mixture under these conditions is either stoichiometric or rich. Therefore, highway operation can provide optimal conditions for desulfurization of the NOx storage catalyst. However, with these high temperatures and the possible frequent fuel cut-offs, thermal aging of the catalyst becomes a critical issue due to the enhanced formation of mobile platinum oxides. This accelerates the rate of sintering of platinum particles, with an associated loss of catalytic effectiveness [137].

As illustrated in Fig. 8.3-15, the elevated sulfur content of standard gasoline grades in North America and Europe poses a major impediment to the wider application of this promising catalyst technology, as relatively small amounts of adsorbed sulfur can cause a severe suppression of NOx storage activity. The sulfur levels of gasoline in the European and U.S. markets may be seen to be several to ten times higher than that in the Japanese market, presenting a significant challenge to automotive engineers who are designing G-DI hardware that will operate in these markets. In order for G-DI combustion and aftertreatment systems

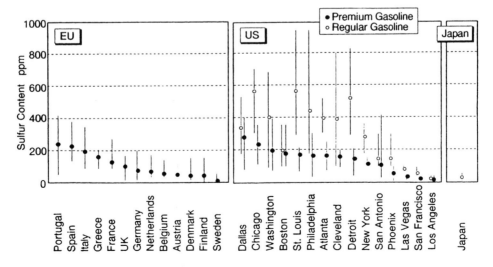

Figure 8.3-15 Worldwide sulfur concentration in gasoline (data published in 1998) [342].

to provide compliance with the stringent NOx emissions regulations in Europe and North America, either the sulfur content of the gasoline supply must be reduced to a trace, or a more efficient NOx storage catalyst that is more resistant to long-term sulfur-poisoning must be developed [341, 431].

The sulfur content of gasoline in the marketplace must be lower than 8 ppm, in order to avoid the poisoning of NOx storage catalysts. Under such a condition, a specific desulfurization strategy is generally not necessary. With a high sulfur level in gasoline and stringent emissions requirements, the fuel economy benefits of G-DI may not be fully achievable. This is because a very high frequency of NOx storage catalyst regeneration would be required [93]. The data in Fig. 8.3-15 were published in 1998. Recently a worldwide effort has been initiated to lower the sulfur level in the fuel. It must be noted, however, that even for the case of 10 ppm to 30 ppm sulfur by weight, the NOx storage material could become saturated with sulfur over time, and the conversion of NOx could degrade to an unacceptable value. As a result, independent of the sulfur concentration level of gasoline that may be achieved in the future, the complete, cyclic purging of sulfur in NOx storage catalysts appears to be a basic requirement as long as the sulfur level in the fuel is above 8 ppm. This is necessary to guarantee an acceptable operational life of the storage catalyst, during which compliance with emissions regulations is maintained [137].

8.3.12 NOx Storage Catalyst Diagnostics

Once NOx storage rates diminish substantially for an NOx storage catalyst, NOx purging must be initiated and then terminated without any breakthrough of the reducing agent. Both control processes must also be conducted with the utmost precision in a dynamic operation mode, and most likely with catalysts that are aged and/or partially poisoned by sulfur. This must be accomplished while maintaining the lowest possible fuel consumption and the best achievable vehicle driveability. In principle the aforementioned control process can be based either on a step-response oxygen sensor downstream from the NOx storage catalyst or on an NOx sensor.

Diagnostic Control Employing an Oxygen Sensor. Diagnostic control of the main NOx storage catalyst can be achieved using a step-response O_2 sensor. Rich purging of the NOx storage catalyst by means of a selected number of lean/rich cycles is continued until such a time as the downstream oxygen sensor detects a

transition from lean to rich, indicating the breakthrough of reductants due to over-purging. From the total purging time it is possible to determine the total NOx and O_2 storage capacity when the lambda value and aspirated air mass are known. For a known catalyst coating, it is possible to acquire a correlation between the available NOx storage capacity and the total measured storage capacity. The correlation is based upon a uniform aging of the NOx and O_2 storage material within the catalytic aftertreatment system. In this manner it is possible to observe any changes of the NOx storage material that result from sulfur poisoning and/or thermal damage. The NOx storage phase can be analyzed with the assistance of a model to determine if the NOx emissions exceed a threshold value. If the threshold is reached earlier than calculated by the model, it can then be concluded that the NOx storage capacity has changed. Through appropriate processing of the signal, this method provides reliable detection of the current storage capacity. In the event of an intolerable decrease in the computed total storage capacity, desulfurization must be initiated. If the overall storage capacity cannot be restored even after several attempts at sulfur regeneration, then irreversible aging must be assumed [93, 261].

One proposed diagnostic control system, illustrated in Fig. 8.3-16, employs three sensors in the exhaust, allowing optimum engine operation while also providing diagnostics of both the close-coupled catalyst (pre-catalyst) and main catalysts. An oxygen sensor incorporating a switching characteristic is installed downstream of the main catalyst to diagnose the NOx and O_2 storage capability, and is used to calibrate the wide-range oxygen sensor that is positioned upstream of the close-coupled catalyst. The required lambda for homogeneous lean operation is controlled by the wide-range oxygen sensor. The torque value for the monitoring operation is determined from both the engine air flow and the oxygen sensor signal for lean operation. The wide-range oxygen sensor has multiple uses, as it is also used as part of the canister purging system to determine the amount of fresh air in the recirculated exhaust gas, and to diagnose the NOx storage capability.

Figure 8.3-17 illustrates the functional principle of model-based NOx storage catalyst control. The temperature of the NOx catalyst is calculated by the model using signals from the exhaust temperature sensor and the engine air flow sensor. NOx mass flow is calculated in another part of the model from the current engine speed, fuel mass and EGR rate. From the NOx

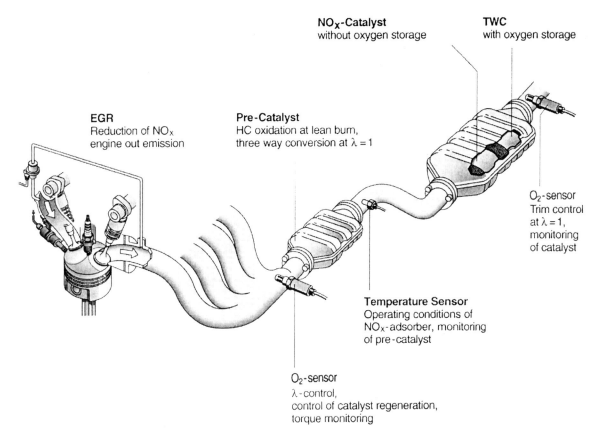

Figure 8.3-16 Proposed diagnostic control system for an NOx storage catalyst using three sensors in the exhaust system [261].

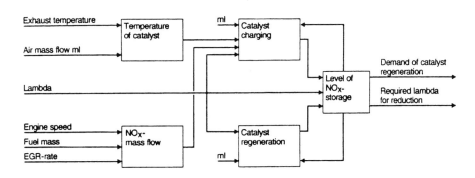

Figure 8.3-17 Functional principle of model-based NOx storage catalyst control [261].

mass flow rate and the fresh air intake flow rate, the load of the storage catalyst is computed. The decrease in NOx load during catalyst regeneration is calculated in the "catalyst regeneration" subfunction. As a result, the actual level of NOx storage and the remaining storage capacity are known. Based upon the lambda signal, either a catalyst-charging or catalyst-discharging

algorithm is used. If the storage capability is below a threshold, regeneration is initiated by a request for a specified lambda. Regeneration can be initiated by switching from stratified-charge mode to homogeneous-charge mode or from homogeneous-charge to stratified-charge mode. Modeling of the storage capacity of the NOx catalyst determines the quality of the catalyst storage control [261]. It is worth noting that when using a lambda sensor for monitoring and controlling the NOx purging process, initiation of NOx regeneration must occur more frequently for a comparatively small amount of stored NOx mass in order to provide a margin of safety for avoiding NOx breakthrough.

Diagnostic Control Employing an NOx Sensor. The NOx storage phase can be directly monitored using an NOx sensor. Irrespective of the mass of nitrogen oxides stored and the engine operating parameters, NOx regeneration can be initiated as required when the NOx saturation point is reached. In each rich-lean cycle, NOx purging is adapted to the storage activity of the catalyst. As a result of conducting purging only as required, unnecessary NOx purging is eliminated. Lower NOx emissions are achievable than that with the oxygen sensor control method. This is because NOx desorption peaks occur correspondingly more rarely at the start of purging, and reducing agent breakthroughs occur more rarely at the end of NOx purging. The lower purging frequency will also lead to a reduced fuel economy penalty.

The considerable merit associated with the use of an NOx sensor is evident when considering catalyst aging and sulfur poisoning problems. To meet future stringent emissions requirement with an extended durability interval, even a small deterioration in NOx storage capacity resulting from sulfur poisoning must be detected. This is required in order to initiate immediate desulfurization, as the deactivation of the NOx storage catalyst is known to increase disproportionately with increasing sulfur loading. Desulfurization may be initiated by a model-based control strategy through detecting the amount of sulfur stored in the NOx storage catalyst. This, however, requires knowledge of the maximum permitted sulfur content in order to guarantee reliable desulfurization. As a result, over-desulfurization may occur when the engine is using a low-sulfur fuel, which will result in an unnecessary penalty in fuel consumption. What is more critical, however, is that the occasional use of an extremely high-sulfur fuel cannot be detected with this procedure, therefore the necessary complete desulfurization cannot be guaranteed. Thus, an NOx sensor is clearly beneficial, as it permits the tailoring of the desulfurization process to the actual NOx storage capacity of the catalyst. In the case of low-sulfur fuel, the desulfurization interval is correspondingly lengthened, while more frequent desulfurization processes will be initiated with high-sulfur

fuel [136]. Because sulfur poisoning detection is normally not within the control of the engine management system, and because of the widely varying sulfur content of commercially available fuels, the use of an NOx sensor offers considerable advantages. The NOx emissions downstream of the storage catalyst can be measured directly, thus desorption can be triggered as required under all operating conditions, without any unnecessary penalty in fuel consumption [43]. NOx sensor and diagnostic systems have been implemented in production DI gasoline vehicles [214].

Compared with the diagnostic approach using an oxygen sensor, monitoring the regeneration of an NOx storage catalyst using an NOx sensor offers significant advantages. In contrast with the NOx sensor, a step-response oxygen sensor does not supply a relevant signal during lean operation. Only complete emptying of the NOx and O_2 storage material leads to sensor transition, and hence a meaningful signal. In addition, the signal from an NOx sensor can be used to detect any sulfur poisoning of the storage catalyst. If the storage capacity is computed as being less than a specified lower limit, the regeneration of the NOx storage catalyst may be initiated.

8.3.13 Urea-Based SCR System

The selective catalytic reduction (SCR) of nitrogen oxides using ammonia (NH_3) or urea ($NH_2)_2CO$ is currently being developed for application in the diesel commercial vehicle sector, with urea as the main agent being considered. Figure 8.3-18 shows a schematic description of a proposed SCR system [132]. This system consists of three different catalysts in series downstream from the urea injection point: a hydrolysis

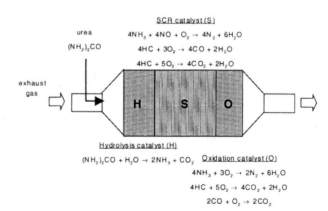

Figure 8.3-18 A schematic description of a proposed urea-based SCR system [132].

catalyst (H), an SCR catalyst (S) and an oxidation catalyst (O). Urea is sprayed evenly into the hydrolysis catalyst, which converts the urea with water selectively to ammonia and carbon dioxide. On the SCR catalyst, the ammonia then reacts with NOx that is present in the exhaust stream to form N_2. The oxidation catalyst works as an active monitor to avoid any breakthrough of ammonia during transient operation.

The application of SCR technology to lean-burn gasoline engines is somewhat promising. However, the vanadium catalysts that are used do exhibit shortcomings with respect to heat stability when they are used in lean-burn, DI gasoline engines. As summarized in Table 8.3-1, a further difficulty in using this technology would be the establishment of a delivery system that would ensure the widespread availability of urea, which would have to be carried in a separate tank on the vehicle. Because this technology is based on the addition of a reducing agent for the exhaust gas, it is also designated as an active SCR process. This is in contrast to a passive process, which uses engine-emitted hydrocarbons as the reducing agent for the exhaust gas. The special advantage of this lean-NOx technology lies in the lower sensitivity to fuel sulfur content. The need to carry an additional reducing agent on board the vehicle is to be considered a significant disadvantage of this approach, and some investigations have been reported regarding the use of a solid reduction agent that only needs to be replenished at the usual maintenance intervals [43]. The details of introducing urea into the exhaust stream under all engine-operating conditions have not been finalized for a production application. In fact, this NOx reduction technique may be best suited for stationary engine applications, for which conversion efficiencies of up to 90% have been measured. For the SCR approach, the NO level must be monitored and measured in order to avoid the breakthrough of either NOx or ammonia, depending on whether the reductant is underdosed or overdosed.

8.3.14 Non-Thermal Plasma Aftertreatment System

Another new technology that is being developed for potential application to G-DI engines is the non-thermal (post-combustion) plasma system, which provides for simultaneous conversion of NOx, HC and CO. Non-thermal plasma can induce a plethora of new chemical reactions through the process of electron excitation, resulting in the abundant production of radicals and excited-state molecules. It has been generally agreed that the important chemistry is not a direct result of ion reactions, but rather of radical chemistry between species resulting from dissociation by electron impact. Present plasma aftertreatment strategies apply a high-voltage to the exhaust stream, resulting in the formation of non-thermal plasma, which is highly reactive, but thermally cool. An energy deposition of 30 J/L (sufficient for typical NOx aftertreatment in automotive application) is only enough to increase exhaust gas temperature by less than 10°C.

The electrons generated in the non-thermal plasma process can effectively influence the chemistry, even in the collision-dominated regimes, whether an engine has just started cold or is already at operating temperature. It has been determined that there is no loss in performance due to fuel contaminants such as sulfur. To date it has also been found that the active electrical field in this system can be developed without the wear of components, that is, with no erosion of electrodes such as occurs with a spark plug.

Recently there have been a large number of investigations into the use of various plasma technologies for pollutant reduction. Various techniques have been proposed for generating plasma, some of which have been used to reduce NOx in a synthetic gas stream. Other studies indicate that in the presence of excess oxygen it is possible to oxidize NO to NO_2 using a plasma device. In certain applications, the non-thermal plasma has been coupled with SCR catalysts in order to enhance overall NOx reduction rates [175].

There are four non-thermal plasma technologies being considered for reducing the NOx content of lean exhaust streams: (1) corona discharge, (2) surface plasma discharges, (3) dielectric barrier discharges and (4) the dielectric packed-bed reactor. However, two issues inhibit the application of plasma discharge aftertreatment systems: low plasma efficiency and an undesirable chemical path. Regarding low plasma efficiency, it is important to note that there are two kinds of efficiencies associated with this plasma technology. One is the electrical conversion efficiency and the other is the chemical processing efficiency. Electrical conversion efficiency refers to the conversion of electrical power to the energy of the electrons in the plasma. Chemical processing efficiency refers to the amount of pollutant removed or decomposed for a given amount of energy deposited into the plasma. The latter is often expressed in terms of the specific energy consumption in units such as grams of NOx per kW-hr. The chemical path is another important factor that must be carefully examined in developing non-thermal plasma after-treatment systems. Oxidation and reduction reaction pathways are both possible for the dissociation of NOx, depending on the system design and the application.

In systems that employ plasma-assisted catalyst reduction, the reactions are primarily oxidizing; NOx does not get reduced to N_2. Instead, HC is partially oxidized, resulting in aldehydes and CO along with various organic species. In addition, NO is converted to NO_2 and other minor species such as CH_3ONO_2 and HNO_3. A suitable catalyst downstream plasma reactor will react the partially oxidized HC and NO_2 to form N_2 with a very small fraction of N_2O. The presence of hydrocarbons has been found to be very critical to the selective partial oxidation of NO to NO_2. Hydrocarbons are beneficial in reducing the plasma energy requirement and in greatly reducing acid formation, and a specific range of HC/NOx ratio is required to achieve and maintain optimal oxidation. In addition, when HC is present in the plasma, the O radicals react preferentially with HC instead of SO_2. As a result, SO_2 passes through the plasma with an extremely low rate of conversion to sulfates, thus leading to a low-sulfur sensitivity for this type of catalyst.

It is evident that the non-thermal-plasma approach to emissions aftertreatment is a unique and interesting alternative (or supplemental) technology. It appears to have the potential to overcome some of the inherent limitations associated with more conventional catalysts, such as sulfur tolerance, reversibility and catalyst light-off. At the current stage of development, gas-phase remediation by discharge technology does not have the necessary energy efficiency and robustness to be a practical replacement for present emissions reduction technologies; however, techniques that combine discharges with surface chemistry may have this potential. Improving the NOx conversion efficiency and system robustness with an extended operating temperature range, while reducing the required power, are the targets for future development of this emerging technology.

8.4 Particulate Emissions

8.4.1 General Characteristics of Particulate Emissions from Engines

In general, particulate matter (PM) is defined as all substances, other than unbound water, that are present in the exhaust gas in the solid (ash, carbon) or liquid phases. Engine particulates basically consist of combustion-generated, solid carbon particles, commonly referred to as soot, that result from agglomeration or cracking, and upon which some organic compounds have been absorbed. The carbon particles become coated with condensed and absorbed organic compounds, including unburned hydrocarbons and oxygenated hydrocarbons. The condensed matter can contain inorganic compounds such as sulfur dioxide, nitrogen dioxide and/or sulfuric acid.

As illustrated in Fig. 8.4-1, the size distribution of an exhaust aerosol from an engine is composed of particles having sizes ranging from several nanometers to several micrometers [140, 244]. Both the peak magnitude(s) and the shape of the size distribution curve can be altered dramatically with changes in engine type and operating condition. The complete size range is most effectively divided into three size regimes that have physical significance. The three regimes, or modes, that are used in the literature to describe particulate size distribution are the nuclei, accumulation and coarse modes.

The first regime, the nuclei mode, is composed of particles with equivalent diameters of less than 50 nm. The particles in this mode are primarily formed during combustion and dilution, usually through both homogeneous and heterogeneous nucleation mechanisms. This mode also normally contains the largest number of particles, thus dominating the number-weighted size distribution, but having an insignificant impact on the mass-weighted size distribution.

The second regime is the accumulation mode, which consists of particles having an equivalent diameter ranging from 50 to 1000 nm. The particles in this size regime are generally formed through the agglomeration of nuclei-mode particles, and may also contain a layer of condensed or absorbed volatile material. The particles in the accumulation mode normally contribute only a moderate amount to the total number-weighted size distribution, but are generally the most significant in the mass-weighted size distribution.

The third and last size regime, denoted as the coarse mode, is composed of particles having an equivalent diameter larger than 1000 nm. These particles are generally not a direct product of the combustion process, but are normally formed from deposits on the valves and chamber walls, which occasionally leave the solid surfaces and enter the exhaust system. The number density of the coarse-mode particles is generally quite low, but can, under some conditions, influence the mass-weighted size distribution.

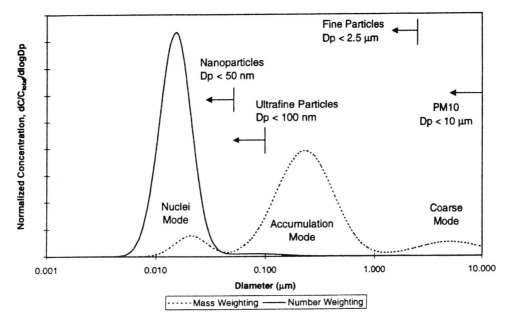

Figure 8.4-1 Typical mass-weighted and number-weighted size distributions of particulate emissions from the engine exhaust gases (Dp: particle diameter) [140].

The significance of each mode is, to some degree, dependent on the specific particulate emissions regulations. The U.S. Environmental Protection Agency (EPA) has regulated PM emissions using mass-weighting, and its PM_{10} designation includes all particulate matters having equivalent diameters of less than 10 μm. The newer EPA "fine particle," or $PM_{2.5}$, standard includes all particles having equivalent diameters of less than 2.5 μm. This indicates the heightened interest in smaller particles that is reflected in the evolution of PM emissions standards. For particles from diesel engines, only those in the size range above 100 nm in diameter have any significant effect on the mass-weighted PM_{10} and $PM_{2.5}$ mean values, and nuclei-mode particles have little or no effect on the mass-weighted distribution, regardless of the number density [244].

8.4.2 Particulate Emissions from PFI Engines

Historically, gasoline engines have been exempt from the requirement to meet the particulate emissions standard for diesel engines. The justification for this has been that gasoline engines produced particulate emissions that were on the order of only 1% of that of diesel engines. This was certainly the case during the development of recent legislation on diesel particulates [18]. Contemporary studies indicate that current gasoline SI engines often emit an increased fraction of nanoparticles

during transient operation, even though steady-state particulate number emissions are generally several orders of magnitude lower than those from modern diesel engines [139, 140]. Particulate number emissions from gasoline PFI engines have also been shown to increase significantly when the engines are operated under high-load, transient and cold-start conditions.

Unlike the PM emissions from diesel engines, the particulate emissions from PFI engines are quite variable [139]. As illustrated in Fig. 8.4-2, a typical baseline concentration of engine-out PM emissions is on the order of 10^5 particles/cm³; however, a "spike" in PM emissions is occasionally observed. These spikes are found to be composed of nearly 100% volatile particles of less than 30 nm in diameter, and a brief spike can exhibit number densities exceeding 100 times that of the baseline concentration. Analyses of particulates from PFI engines reveal that the bulk of the mass is ash, with the second largest fraction being unburned lubricating oil [18]. Carbon emissions are found to be significant only at high load with mixture enrichment, whereas at other operating conditions, carbon comprises less than 10% of the total PM mass. The large ash fraction of gasoline PM emissions includes a large fraction of metal compounds, with calcium and sodium evident for operation at low load without EGR, and copper and magnesium predominant for operation with EGR.

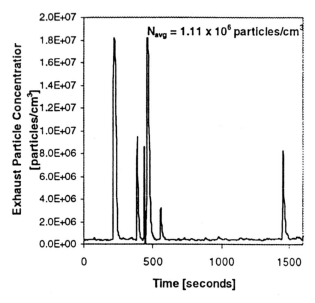

Figure 8.4-2 *Time-resolved exhaust particulate number emissions from a PFI engine at 2500 rpm and a MAP of 55 kPa [139].*

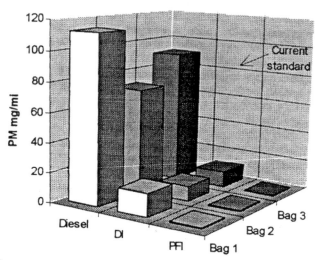

Figure 8.4-3 *Comparison of the particulate emissions from a current PFI engine, a current production G-DI engine and a 1995 European IDI diesel engine operating on the U.S. FTP cycle [299].*

8.4.3 Particulate Emissions from G-DI Gasoline Engines

Research indicates that G-DI engines, as evolving powerplants for automotive applications, emit a larger amount of particulates than do conventional PFI engines, especially during stratified-charge operation. Depending on the degree of combustion system optimization, smoke emissions from prototype G-DI engines could be as high as 1.2 Bosch smoke units (BSU) [183, 190, 340]. A comparison of the particulate emissions from a current PFI engine, a current production G-DI engine and a 1995 European IDI diesel engine operating on the U.S. FTP cycle is illustrated in Fig. 8.4-3 [299]. These data represent mass measurements of particulate matter collected on filter media. It may be seen that the measured level of PM emissions for the G-DI engine is between those of the diesel and PFI engines. The PM emissions from a vehicle powered by a G-DI engine are on the order of 10 mg/mile. In comparison, the PM emissions from a comparable diesel-powered vehicle are on the order of 100 mg/mile. As compared to the 1 to 3 mg/mile PM emissions from a modern PFI engine [18, 299], the PM emissions from current production G-DI engines are relatively high, even though they are well below the current U.S. standard of 80 mg/mile as measured on the FTP test cycle. In interpreting this published comparison, it should be noted that both the PFI and G-DI engines have been

developed, optimized and mass-produced without major specific efforts being directed toward the minimization of particulate emissions.

Measurements of the particulate emissions from a production G-DI engine using a chassis-dynamometer test indicate that the average polydisperse number concentration is on the order of 10^8 particles/cm^3, and that the number-weighted, geometric mean particle diameter is between 68 nm and 88 nm [141]. In contrast, modern PFI engines emit average particulate number concentrations ranging from 10^5 particles/cm^3 at light load to 10^7 particles/cm^3 at high load [139, 140]. Older PFI engines emit particulate emissions with number concentrations in excess of 10^8 particles/cm^3 for conditions corresponding to highway cruise operation. For the operating conditions tested, the number-weighted, geometric mean diameters of the particulate matter emitted from a G-DI engine are larger than are obtained for the PFI engines tested. This relatively large mean particle size tends to increase the particulate mass emissions from the G-DI engine, but could also be indicative of a decrease in the relative fraction of nanoparticles.

As illustrated by the results in Fig. 8.4-4, which are derived from PM emissions measurements on a 1.83L, 4-cylinder, 4-valve, production G-DI engine for a range of operating conditions, the particle number emissions are found to increase by a factor of 10 to 40 when the operating mode is stratified-charge rather than

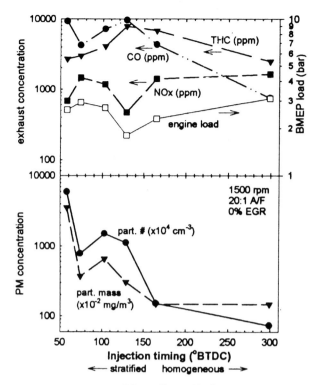

Figure 8.4-4 PM, CO, HC, and NOx emissions as a function of EOI fuel injection timing for a range of engine loads [299].

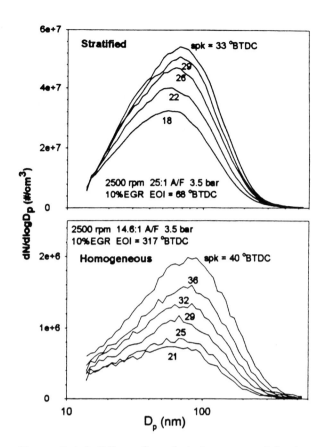

Figure 8.4-5 Effect of spark timing on particle size distribution for homogeneous-charge and stratified-charge operations (spk: spark timing) [299].

homogeneous-charge [299]. Emissions of particulate matter exhibit a strong dependence on the G-DI injection timing, and the particle number and volume concentrations are found to increase markedly as the injection timing is retarded. As is evident from the experimental results, operating a G-DI engine in the stratified-charge mode has a significant effect on the overall particulate emissions from G-DI engines. The PM emissions for this mode are found to vary significantly with small changes in the engine operating point. It is theorized that particulate matter can result from two types of rich combustion: that of a locally rich, gaseous air/fuel mixture and diffusion combustion of incompletely volatilized, liquid fuel droplets. The combustion of liquid fuel droplets is known to be an important source of PM emissions for late injection timings. For homogeneous-charge operation, PM mass emissions remain quite independent of fuel injection timing, but the PM number concentration decreases monotonically as the injection timing is advanced.

Figure 8.4-5 shows the effect of spark timing on PM emissions from a G-DI engine during homogeneous-charge and stratified-charge operations. Advancing the spark timing generally yields an increase in both the particle number concentration and the mean particle size for both homogeneous-charge and stratified-charge operations. The particle size distributions for homogeneous-charge operation have an asymmetric shape, with the peak in the distribution shifting slightly from about 70 nm for a spark timing of 21°BTDC to a value of 85 nm for a spark advance of 40°BTDC. These trends suggest that more nuclei-mode particles are formed as the spark is advanced, possibly due to higher rates of nucleation and coagulation at the higher peak temperatures. Furthermore, the post-flame oxidation rate is slowed due to the decrease in exhaust gas temperature as the spark timing is advanced. All of these factors contribute to the observed increase in peak particle number and mean particle size as the spark is advanced. For homogeneous-charge operation, the trend in particulate emissions with load is found to be quite similar to that measured for a PFI engine.

Investigations of the effect of engine speed and load on size distributions of particulate emissions reveal that an increase in engine speed and load generally leads to an increase in PM emissions; however, this trend is dependent on injection timing when operating in the stratified-charge mode. PM measurements conducted during a chassis-dynamometer test emulating the FTP test cycle are shown in Fig. 8.4-6. It is evident that PM emissions exhibit very substantial fluctuations, with a strong correlation existing between vehicle acceleration and observed increases in PM emissions. These increases are theorized to occur because both the exhaust flow and the PM concentration in the exhaust gas increase with the added load needed to accelerate the vehicle. The observed PM increases during the FTP test cycle are found to be well correlated with changes from homogeneous-charge to stratified-charge operation. These particulates are normally in the size range of 15 nm to 600 nm, with the bulk of the particles being in the range of 50 nm to 100 nm. The number concentration is found to be on the order of 10^6 particles/cm^3 for homogeneous-charge operation, and 10^7 particles/cm^3 for stratified-charge operation.

Measured data for particle sizes and engine-out smoke emissions resulting from lean-stratified-charge operation with different EOI fuel injection timings are shown in Fig. 8.4-7 [276]. The measured smoke emissions as a function of EOI fuel injection timing are plotted in Fig. 8.4-8. The particle size clearly exhibits a distribution having two peaks: one for sub-80 nm particles and one for particles larger than 100 nm. As the injection timing is retarded, the particle size and number represented by the first peak both decrease, whereas the particle size and number represented by the second peak both increase. The particles represented by the first peak are primarily composed of homogeneously nucleated HCs, whereas those of the second peak are composed of carbon with adsorbed HCs. As the fuel injection event occurs later, more carbon is formed, with the increased carbon increasing HC adsorption and reducing nucleation. The overall particle numbers are reduced and the carbon production is increased. It has been observed that there

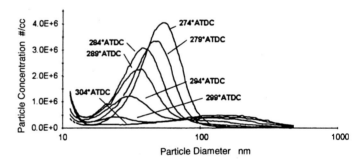

Figure 8.4-7 Effect of EOI fuel injection timing on particle size distribution under stratified-charge operation [276].

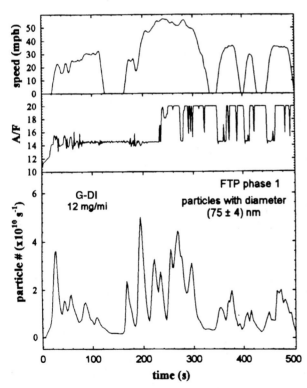

Figure 8.4-6 Transient characteristics of G-DI particulate emissions during the first phase of the U.S. FTP test cycle [299].

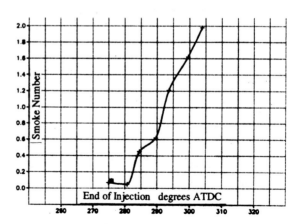

Figure 8.4-8 Smoke emissions as a function of the EOI fuel injection timing [276].

is a positive linear relationship between smoke emissions and the number of particles in the size range of 80 nm to 500 nm, whereas there is a negative linear relationship between smoke emissions and the number of particles in the size range of 10 nm to 80 nm. Also, a negative linear relationship exists between smoke and the number of particles in the size range of 10 nm to 500 nm. This indicates that the number of particles can be the *highest* when smoke levels are very low, and that low levels of smoke emissions *do not necessarily* mean low particle numbers. The fact that there is a relationship between injection timing and particle number concentration may enable injection strategies such as split injection to be used to reduce the particle number concentration. In addition, it is expected that achieving an understanding of the detailed relationship between particle size distribution and smoke would lead to specific injection strategies for reducing visible smoke.

The aforementioned particle/smoke correlation may be partially explained in terms of the chemical composition of the two distribution modes, based upon observations of the combustion processes for a range of injection timings, and upon observations of the formation of particles within a dilution tunnel. Recent studies reveal that at mini-dilution-tunnel dilution rates, hydrocarbons homogeneously nucleate to form nanoparticles in the tunnel if the levels of carbon are low [244]. This nucleation occurs by a mechanism whereby carbon particles absorb or desorb hydrocarbons. Particles having sizes near that represented by the first peak in Fig. 8.4-7 are likely to be predominantly composed of unburned hydrocarbons, which survive to nucleate within the downstream dilution tunnel.

At advanced injection timings, where the mixing of fuel and air is maximized, dispersion of the charge occurs prior to ignition, which allows some hydrocarbon fuel to escape the combustion process. As a result, the number of particles may be high, even though the measured smoke level is low. In comparison, at the most retarded fuel injection timings, less time is available for the fuel and air to mix, which leads to a greater degree of diffusion combustion. As a result, more carbon particles are formed, which can absorb both small particles and hydrocarbon fuel. This, in turn, reduces the number of small particles and hydrocarbons available to nucleate. This explains why a low particle number concentration is observed with extremely retarded injection timing. However, those particles are predominantly large and carbonaceous, and thus contribute significantly

to smoke emissions. Consequently the lowest number of particles should be observed for the condition where most of the nucleated fuel droplets and hydrocarbons are adsorbed by the carbon particles that are generated during combustion. This corresponds to an EOI injection timing of 304°ATDC on the intake stroke, or 56°BTDC on the compression stroke. The optimal injection timing should occur near 286°ATDC on the intake stroke, where the least amount of carbon is subsequently generated during combustion [12].

Analyses of the measured PM emissions from a PFI vehicle with a three-way catalyst, a high-speed direct-injection (HSDI) diesel vehicle equipped with an oxidation catalyst and a production G-DI vehicle equipped with a lean-NOx aftertreatment system show that the particulate compositions of G-DI and HSDI diesels engines are quite similar. The chemical species of particulate emissions from the PFI engine are observed to be dominated by volatile hydrocarbons. It should be noted, however, that the similarity of particulate emissions between G-DI and diesel engines could be partially an artifact of the sampling methodology, which is most applicable to diesel engines. At the temperatures and flow rates under which sampling is conducted, solid matter and low-vapor-pressure hydrocarbons are preferentially collected. As a result, light hydrocarbons, even if captured initially, are likely to vaporize, and may not be included in the collection sample [12].

Studies conducted on the influence of fuel additives indicate that the presence of such additives does not have a strong, consistent influence on particulate emissions from G-DI engines. Polyether amine, which is a common gasoline additive, does seem to reduce particulate number concentrations somewhat as compared to unadditized base fuel. Conversely, polyolefin amine is observed to greatly increase the emissions of particulates with a size larger than 100 nm at vehicle operating conditions of 15 km/hr and 30 km/hr [142].

8.4.4 Comparison of Particulate Emissions Between Single-Fluid and Air-Assisted G-DI Engines

Particulate emissions from a four-valve, single-cylinder G-DI engine with an air-assisted fuel injection system have been investigated, and the observed dependence of particle size on spark timing is presented in Fig. 8.4-9. In this study, the 0.31L test engine incorporated a pentroof chamber modified to accommodate a centrally co-located spark plug and fuel injector combination [300]. The

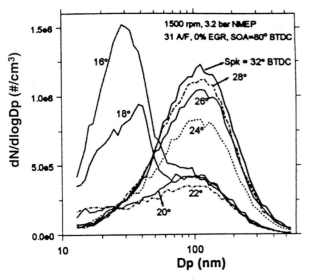

Figure 8.4-9 Effect of spark timing on particle size distribution for stratified-charge operation of a DI engine with an air-assisted fuel injection system (spk: spark timing) [(300)].

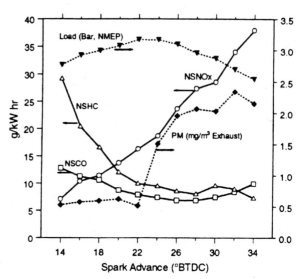

Figure 8.4-10 Effect of spark timing on PM, HC, CO, and NOx emissions for stratified-charge operation of a DI engine with an air-assisted fuel injection system [(300)].

piston crown incorporated a bowl, and a compression ratio of 11.4:1 was employed. The two maxima of the standard bi-modal distribution of particles are evident, with one mode centered near 25 nm and the other in the vicinity of 100 nm. Many, if not most, measurements of particulate matter in G-DI engines exhibit a bi-modal distribution of particle number density versus particle size. The particles in the small size mode of 25 nm are not soot particles, but instead represent aerosols from unburned injector cleaner (a polyether amine) that nucleate as the exhaust gases cool. Due to the small sizes, such aerosols contribute little to either smoke measurement or to particulate mass, whereas particles in the mode larger than 100 nm are typically soot. The detectable particle number emissions increase about fivefold as the spark timing is advanced from 20° to 32°BTDC; however, the change in particle diameter that accompanies this increase in number is almost negligible.

Particle mass emissions are plotted in Fig. 8.4-10, along with other measured exhaust emissions for a DI engine having an air-assisted fuel injection system. As expected from the size distribution shown in Fig. 8.4-9, particulate mass emissions increase by approximately a factor of five as the spark timing is advanced from 14° to 34°BTDC [(300)]. This trend is completely opposite to that for hydrocarbon emissions, which decrease by nearly an equal amount within the same range of spark

advance. CO emissions are comparatively insensitive to spark timing within that testing range, but NOx emissions increase linearly as the spark is advanced. If it is assumed that the increased particulate mass and number emissions result from enhanced nucleation and diminished late-cycle oxidation due to the lower exhaust gas temperatures, the PM trends observed here are most likely not due to temperature changes. In fact, the exhaust gas temperature changes only slightly for the spark-timing range tested. The increase in particulate emissions may be attributed to the reduced mixing time associated with advancing the spark timing while operating in stratified-charge mode.

Figure 8.4-11 shows a comparison of the measured particulate mass emissions as a function of injection timing for G-DI engines that are fitted with both single-fluid and air-assisted fuel systems, with both engines operated at similar conditions [(300)]. It should be noted that the plotted injection timing represents the start of air injection for the air-assisted fuel system and the end of injection for the single-fluid fuel system. Due to the difference in combustion system designs, the results only qualitatively indicate how particulate emissions vary with different fueling systems. It is evident that there are some similarities in particulate emissions between these two fuel injection systems.

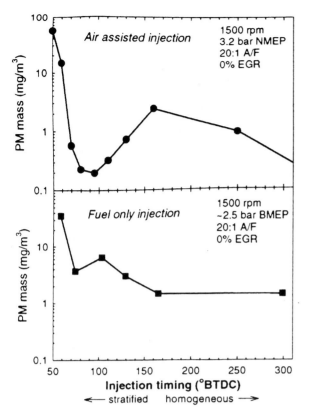

Figure 8.4-11 Comparison of PM emissions from DI engines retrofitted with air-assisted and single-fluid fuel injection systems [300].

For both injection systems, particulate emissions are very sensitive to injection timing and mixing dynamics inside the cylinder, particularly for stratified-charge operation. Particulate emissions at the highest level of stratification approach 50 mg/m³ for both fuel systems. For the early injection mode, however, particulate emissions decrease to a plateau of about 1 mg/m³ for the single-fluid injection system. In contrast, particulate emissions for the air-assisted system are near this level over most of the homogeneous mode, becoming significantly lower with very early injection timing. This difference indicates that the residual level of stratification for the early injection timing remains higher for the single-fluid fuel injection system than for the air-assisted fuel injection system. For the air-assisted fuel system, there is an injection-timing window for stratified-charge operation where particulate emissions are extremely low, and this window does not appear to be present for the single-fluid injection system. This may offer some advantage for air-assisted, stratified-charge G-DI engines in achieving reduced PM emissions levels.

8.5 Summary

The attainment of future regulatory levels of emissions is one of the major challenges facing DI gasoline engine developers. The majority of system design and engine operating variables are interdependent, thus any proposed alteration of a component or strategy must be examined to evaluate the very important compromise between fuel economy and emissions. Undermixing and overmixing of fuel and air are two overall process variables that play important roles in determining the level of engine-out HC emissions. Another overall process variable is the control of the liquid fuel distribution and subsequent vaporization. This means that unintended fuel impingement on combustion chamber surfaces should be avoided.

Both the degree of undermixing or overmixing and the control of the liquid fuel distribution are determined by a number of critical engine design and operating parameters such as the injector spray parameters, spray targeting, injection timing, EGR level, and degree of throttling. These parameters are well established as key tools to be used by engine designers in achieving an optimum system design that has very low engine-out HC and NOx emissions.

A number of alternative technologies such as lean-NOx catalysts, NOx storage catalysts and non-thermal plasma aftertreatment systems are being developed to control tailpipe-out NOx emissions, with lean-NOx catalysts and NOx storage catalysts being used in production applications. There is some doubt as to whether the conversion efficiency of hydrocarbon-based lean-NOx catalysts can be increased to more than 40%; however, it may be possible for light-duty vehicles to employ this type of catalyst for NOx reduction. NOx storage catalysts are promising candidates because of their high conversion efficiency, but the limited operating temperature window, especially the high-end temperature, limits the attainment of the full fuel economy benefit of G-DI engines. In addition, sulfur poisoning and the associated desulfurization process represent major challenges for the wide application of such technology, particularly in markets having gasoline sulfur levels exceeding 10 ppm. Reducing the sulfur concentration of the gasoline supplies of North America and Europe is regarded as the most direct way of expanding the application of NOx storage catalysis; thus it represents a significant enabler of G-DI technology. If this control technology is utilized it is important to develop an effective NOx storage regeneration system and strategy that requires only a minimal fuel economy penalty while maintaining the desired vehicle driveability. Non-thermal plasma systems do offer some unique possibilities for circumventing the common limitations of conventional thermal catalyst systems, but much more research and development must be conducted before such systems can replace thermal catalysts on production G-DI vehicles.

Experimental investigations have revealed that both the mass-based and number-based particulate emissions from G-DI engines lie between those of the diesel and the PFI engines. Particulates emitted during stratified-charge operation are a major concern in the development of gasoline direct-injection engines; hence the reduction of particulate emissions from G-DI engines is an important research area. The specific processes of particulate formation for homogeneous-charge and stratified-charge operation need to be more completely researched, and additional modeling work, as well as experiments under carefully controlled conditions, are required. The understanding that is gained will provide the ability to predict, control and optimize the processes that result in particulate matter, which will lead to significant reductions in PM emissions of G-DI engines.

Even though the engine-out HC emissions of a stratified-charge G-DI engine are more than twice those of a contemporary, homogeneous-charge, PFI engine, attaining both the EURO III and EURO IV tailpipe HC levels should be possible for current stratified-charge systems with moderate further refinement. Stratified-charge engines operating with a lean calibration may be marginally capable of meeting the LEV/ULEV NOx requirements; however, further development of combustion, engine control and aftertreatment systems is definitely required to meet the LEV II emissions standards. This will require the surmounting of many technical challenges.

The future development of G-DI exhaust aftertreatment systems will certainly involve new approaches and concepts, but will be mainly directed toward enhanced system design and integration. This will involve the incorporation of not only new and improved coatings and structures, but enhanced sensors and control-system models. Coating enhancements will address the optimization of faster light-off, enhanced conversion efficiencies, aging characteristics and improved resistance to sulfur poisoning. Reducing the sulfur content in fuel must be regarded as a crucial step in achieving a significantly larger fraction of the theoretical fuel economy benefit of gasoline DI technology.

Chapter 9

Fuel Economy: Potential and Challenges

9.1 Introduction

The areas of significant potential for direct-injection-gasoline, four-stroke, SI engines are quite evident. However, the realization of a viable production engine with significant fuel economy improvement requires the successful implementation of emissions control strategies, hardware and control algorithms in order to achieve certifiable emissions levels and acceptable driveability during load transients [261, 276]. Experience has verified that some fraction of the potential fuel economy gains that are computed from thermodynamic considerations will have to be compromised in order to achieve increasingly stringent emissions standards such as the EURO IV and the U.S. ULEV and SULEV requirements. The key question for OEMs in regard to G-DI engines is whether the final margin in fuel economy for a particular engine system is sufficient to justify the additional hardware and complexities required. Another very important consideration is that, to be meaningful, this fuel economy margin should always be measured relative to the increasingly sophisticated PFI engine, which is, of course, a moving target.

The actual advantage achieved by current production G-DI technology still fails to meet early expectations when compared with that achieved by a best-in-class, stoichiometric + EGR, PFI engine. The main areas for additional optimization are emissions, partial throttling, catalyst purging, further refinement of transient EMS functions, and friction minimization. It is the specific detail of the compromise between BSFC and test-cycle emissions, coupled with the increased system complexity, that is generally the key factor in determining the production feasibility of a particular G-DI engine design. With ever more stringent emissions standards being required around the world, the actual fuel economy benefit from G-DI engines will be emissions-constrained, and will depend to a large degree upon the capability of the aftertreatment system and the fuel quality available in the field [165, 168, 204, 268, 269, 316].

Other options to further improve fuel economy are to combine G-DI technology with other fuel-efficient technologies such as boosting, idle engine shut-off,

continuously variable transmission (CVT) and the hybrid electric vehicle (HEV). Such strategies have not been invoked to a significant degree to date, but will see increasing application in the future. In many instances a G-DI engine combined with one or more of these technologies, will further enhance the advantages associated with each individual strategy. The fuel economy potential and challenges associated with G-DI engines, and the possible strategies for combining G-DI with other technologies will be the topics of this chapter.

9.2 Fuel Economy Potential and Constraints

9.2.1 Factors Contributing to Improved Fuel Economy

One current strategic objective in the automotive application of four-stroke, gasoline engines is a substantial improvement in fuel consumption while meeting the required levels of pollutant emissions and engine durability [359, 360, 389]. The degree of improvement in automotive vehicle fuel economy is an important consideration that could influence the future percentage utilization of SI engines relative to small, high-speed, common-rail, diesel engines [361]. The thermal efficiency of G-DI engines can be enhanced by increasing the compression ratio, and by using an overall lean mixture for load control, thereby reducing the significant losses due to throttling and wall heat transfer. A G-DI engine that uses charge stratification offers the potential for reducing the part-load fuel consumption by 20–25%, depending on the test cycle, when the gas cycle, heat transfer and geometric configuration are optimized [227].

The major factors that contribute to the BSFC advantage of a G-DI engine as compared to that of a conventional PFI engine are summarized in Table 9.2-1. A G-DI engine that can achieve load control without throttling, or with very minimal throttling, can realize significant improvements in fuel economy, resulting mainly from a decrease in pumping work and an increases in the relative magnitude of the expansion stroke work [402]. Indeed, thermodynamic analyses do indicate that the

Table 9.2-1
Major factors contributing to the BSFC advantage of a G-DI engine as compared to that of a conventional PFI engine

Pumping Loss	Reduced pumping losses due to unthrottled part-load operation using overall lean mixtures
In-Cylinder Charge Cooling	Increased cooling of the intake charge due to in-cylinder injection during induction
Compression Ratio	Increased knock-limited compression ratio due to a lower end-gas temperature
Specific Heat Ratio	Increased cycle efficiency due to the incrementally higher specific heat ratio of lean mixtures
Heat Loss	Reduced cylinder wall and combustion chamber heat losses due to stratified-charge combustion
Fuel Cut-Off	Fuel cut-off during vehicle deceleration

reduction of pumping work is the principal factor contributing to the fuel economy improvement of a G-DI system that achieves part-load operation by using an overall-lean, but stratified, mixture, as illustrated in Fig. 1.2-3. Also contributing, but to a smaller degree, are reductions in heat losses and an increase in specific heat ratio [51, 259]. The gas near the cylinder wall is at a lower temperature for part-load, stratified-charge operation, thus there is a smaller, effective temperature differential, with less heat transferred to the wall [448, 449]. As discussed in Section 5.3.1 on charge cooling, early injection can favorably impact the octane number requirement of the engine [9, 10] and, as a result, the knock-limited engine compression ratio for the G-DI engine can generally be increased. As is well known, the beneficial effect of a higher compression ratio on fuel economy is significant. In addition, direct fuel injection into the cylinder permits an effective fuel cut-off during vehicle deceleration, which provides a fuel economy improvement.

Many factors, both positive and negative, can influence the net fuel economy improvement achieved by applying a particular G-DI technology to an engine/vehicle system. However, without knowing the details of the baseline engine and/or vehicle that is used for comparison, it can be difficult to interpret reported fuel economy improvements associated with individual segments of direct-injection technology. Table 9.2-2 [51] lists the negative factors that moderate the fuel economy potential of a stratified-charge G-DI combustion system as compared to a conventional PFI engine.

Figure 9.2-1 shows a computed engine energy balance for 2000 rpm and 0.2 MPa BMEP. In this calculation the engine output power and friction loss are maintained constant as the equivalence ratio is varied. The simulation is for engine operation in the homogeneous-charge mode with excess air ratios in the range of $\lambda = 1.0$ to $\lambda = 1.3$, and in stratified-charge mode with λ in the range of 1.6 to 3.4, a condition produced by reducing the amount of throttling. The exhaust energy and the wall heat transfer for stratified-charge operation are found to be reduced by 10.8% and 12.5%, respectively, even though the total air mass in the cylinder is increased. As a result, the predicted improvement in fuel consumption is 23%. This considerable increase in thermal efficiency cannot be explained by the unthrottling of the thermodynamic gas cycle alone. Figures 9.2-2 and 9.2-3 illustrate the changes in the individual energy balances for the same operating point. The thermodynamic benefits of G-DI at this operating point are seen to result from a reduction in the pumping

Table 9.2-2
Negative factors that moderate the fuel economy potential of a stratified-charge
G-DI engine

Emissions	Emissions constraints that may not allow the engine to operate at the optimum BSFC point
Fuel Consumption	Fuel consumption penalty due to high-pressure pump and/or air compressor for air-assisted injection
Combustion Chamber Design	Negative impact of increased surface-to-volume ratio of a stratified-charge combustion system due to an increased compression ratio coupled with a more complex piston geometry

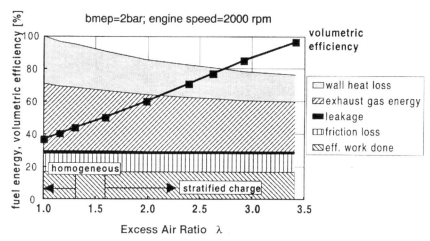

Figure 9.2-1 Estimated energy balance for a G-DI system under various degrees of charge stratification; 2000 rpm; 0.2 MPa BMEP [227].

work by 95%, a reduction in heat work by 42% and a reduction in exhaust energy by 26%. It can also be seen that homogeneous lean operation at $\lambda = 1.3$ leads to an overall thermal efficiency gain of 5%, which can be attributed to a reduction of pumping loss by 30%, wall heat loss by 8% and a reduction in exhaust energy by 6% [227].

Generally the four available operating modes, lean stratified charge, lean homogeneous charge, stoichiometric charge with high EGR, and rich homogeneous charge, have to be utilized in suitable speed and load ranges. Analyses indicate a limitation of the stratified-charge mode to engine speeds at or below 3000 rpm. The upper load limit decreases with increasing engine speed due to a reduced mixture preparation time. Required engine loads that are in excess of the stratified-charge load

	Exhaust gas energy	Wall heat total	Wall heat gas exchange phase	Wall heat high pressure phase
$\lambda = 1$	Basis	Basis	Basis	Basis
$\lambda = 1.3$	-6%	-8%	-19%	-4%
$\lambda = 3.4$	-26%	-42%	-92%	-23%

	High pressure work	Gas exchange work	Total efficiency gain
$\lambda = 1$	Basis	Basis	Basis
$\lambda = 1.3$	-6%	-30%	5%
$\lambda = 3.4$	-20%	-95%	23%

Figure 9.2-2 Change of individual energy categories for a G-DI system under various levels of excess air ratio; 2000 rpm; 0.2 MPa BMEP [227].

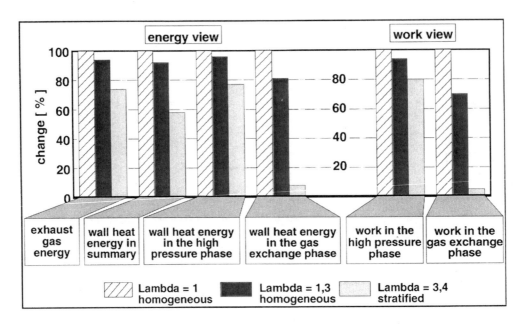

Figure 9.2-3 Comparison of individual energy change for a G-DI system at excess air ratios of 1, 1.3 and 3.4; 2000 rpm; 0.2 MPa BMEP [227].

limit may be achieved by lean, homogeneous-charge operation using a high percentage of EGR. High-EGR, stoichiometric-charge operation is a viable alternative, as three-way catalyst functionality in this range contributes to a significant reduction in emissions [480].

For stoichiometric, homogeneous-charge G-DI operation, a 7–10% improvement in engine BSFC is reportedly achievable without requiring a lean-NOx catalyst. The individual contributions to this lower BSFC include: 1.5–2% attributed to a higher compression ratio; 1.5–2% realized from a higher EGR tolerance; 1–2% due to improved air-fuel control during warm up and transients; and 3–4% achievable by downsizing the engine while maintaining the same torque and power[20]. A comparison of the fuel consumption of a baseline PFI engine with that of a G-DI engine that operates in the early-injection, homogeneous-charge mode reveals that the lean limit achievable using very early injection timing on a G-DI engine is basically the same as that obtained with a PFI engine [407]. However, retarding the fuel injection timing to the later portion of the intake stroke significantly extends the lean combustion limit for the G-DI engine. Engine BSFC is generally improved when lean air/fuel ratios are used, although the effect becomes marginal for air/fuel ratios leaner than 20:1. At an air/fuel ratio of 25:1, BSFC is improved by about 13% over that obtained for a stoichiometric mixture. This air/fuel

ratio resulted in the lowest BSFC for the engine operating condition tested [407].

Even with stratified-charge operation, the fuel economy improvement relative to a baseline PFI engine varies significantly with engine load. The theoretical potential for improving fuel economy by using direct gasoline injection is about 20–25% at part load, with an associated potential reduction of 35% in idle fuel consumption [108, 378, 439, 484]. Figure 9.2-4a shows the measured fuel economy improvement of the Mitsubishi GDI engine for late injection at an engine speed of 2000 rpm [196, 259]. The principal factors contributing to the fuel economy improvement at 2000 rpm and an overall air/fuel ratio of 40:1, but without EGR, are itemized in Fig. 9.2-4b.

Other studies indicate that a conventional PFI lean-burn engine can, when operated at an air/fuel ratio of 25:1, provide an overall fuel consumption improvement of about 10 to 12% when compared to stoichiometric-homogeneous-charge operation without EGR [274]. A stratified-charge G-DI engine operating at an overall air/fuel ratio exceeding 40:1 can provide an overall fuel consumption improvement of up to 25% when compared to the same baseline. The amount of EGR is important in determining the final level of fuel consumption improvement, as the reduction in the rate of heat release with EGR is beneficial not only for NOx suppression but also for fuel economy.

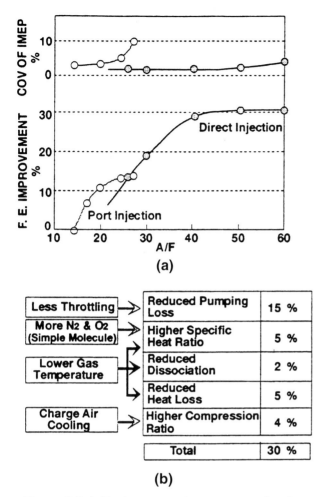

(a)

Less Throttling →	Reduced Pumping Loss	15 %
More N₂ & O₂ (Simple Molecule) →	Higher Specific Heat Ratio	5 %
Lower Gas Temperature →	Reduced Dissociation	2 %
	Reduced Heat Loss	5 %
Charge Air Cooling →	Higher Compression Ratio	4 %
	Total	**30 %**

(b)

Figure 9.2-4 Fuel economy improvement for the Mitsubishi GDI engine: (a) comparison of fuel economy improvement between GDI and PFI engines; (b) itemized list of the principal contributing factors to the fuel economy improvement (2000 rpm; amount of fuel injected: 15mm³/st; air/fuel ratio for G-DI operation: 40:1; compression ratio for PFI engine: 10.5:1; compression ratio for G-DI engine:12:1) [259].

Fuel economy can be improved slightly by adding uncooled EGR at light loads, even though the engine operates unthrottled. This is attributed to the additional enhancement of mixing associated with hot EGR. In addition, the increased charge temperature with a large amount of hot EGR improves mixture formation, which results in reduced HC emissions at moderate EGR rates, leading to an improvement in combustion efficiency. Most important for exhaust gas aftertreatment is the change in exhaust gas temperature that accompanies the reduced intake air flow with EGR [480]. The use of EGR, however, has little effect on engine fuel consumption at intermediate loads, and ultimately degrades the obtainable fuel economy at high loads.

For the best compromise between fuel consumption and emissions characteristics, the system should employ maximum EGR at part load, limiting the overall air/fuel ratio to no leaner than 30:1 [200, 201]. At a constant engine operating condition of 2000 rpm and 0.2 MPa BMEP, an improvement in fuel consumption of 23% is achieved for the unthrottled Mercedes-Benz prototype G-DI combustion system as compared to stoichiometric, homogeneous-charge, throttled operation of the same engine. This is achieved not only through a reduction in throttling loss, but also due to a decrease in peak cycle temperature [227].

9.2.2 Comparison of Fuel Economy Benefits Between Single-Fluid and Air-Assisted G-DI Combustion Systems

Achieving a robust combustion system design that can be operated with a highly-stratified charge while providing an optimal heat release rate is key to obtaining the maximum fuel economy benefit of a given G-DI system. Obviously combustion characteristics and emissions formation processes vary markedly with the specifics of the combustion system. Figure 9.2-5 shows a comparison of the BSFC improvement of two G-DI engines, one retrofitted with an air-assisted injection system (engine A), and one using a single-fluid, high-pressure injection system (engine L). The listed improvements are relative to a stoichiometric DI baseline engine without EGR [51]. Data for both engines were obtained at 1500 rpm and 2.62 bar BMEP. Results for homogeneous-charge operation are tabulated up to an air/fuel ratio of 20:1, and those for stratified-charge operation are listed for the range from 20:1 to 40:1.

It is evident that the lean, homogeneous-charge operation results for both engines are similar, and show a fuel consumption improvement of 9–10% at an air/fuel ratio of 20:1. The fuel consumption benefit for stratified-charge operation without EGR is about 16%. The minimum stratified-charge fuel consumption for engine A (with an air-assisted fuel injection system) is achieved at an air/fuel ratio near 35:1. In comparison, engine L (with a single-fluid fuel injection system) achieves optimum fuel consumption at an air/fuel ratio of 40:1. The level of BSFC improvement for both engines is limited by the combustion stability, which degrades as less throttling is employed. For engine L

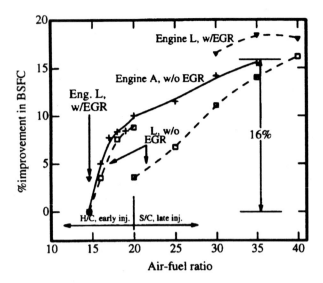

Figure 9.2-5 Fuel economy improvement as a function of air/fuel ratio relative to stoichiometric operation without EGR; engine speed: 1500 rpm, BMEP: 0.262 MPa; MBT spark timing and EOI optimized for the best BSFC (engine A: air-assisted fuel injection system; engine L: single-fluid, high-pressure fuel injection system) [51].

with EGR, the maximum fuel economy benefit is about 18% relative to baseline operation without EGR. However, when comparing stratified-charge operation with homogeneous-charge operation, both with EGR, the fuel consumption improvement due to stratified-charge operation is approximately 15%, which is quite comparable to the 16% achievable without EGR.

As can be seen in Fig. 9.2-5, the fuel economy benefit for engine L at air/fuel ratios of 20:1 to 25:1 is significantly lower than that for engine A, presumably due to the difference in the two combustion systems. The analysis of data obtained from injection-timing hooks indicates that as compared to engine L, engine A exhibits a longer mixing time at air/fuel ratios in the range of 20:1 to 25:1, which operates with a less stratified mixture. This weakly stratified mixture yields a better fuel efficiency than a strongly stratified mixture at the same air/fuel ratio. This permits a smoother transition between combustion modes, and a monotonically improved fuel economy as a function of air/fuel ratio. In this study, the engine using the air-assisted injection system had slightly smaller fuel droplets in combination with a relatively quiescent combustion chamber. This facilitates the fuel cloud remaining intact near the spark gap for an extended period. The strong bulk

motion in engine L, coupled with the lower piston position corresponding to the earlier injection timing, cannot provide a reliable ignitable mixture near the spark gap. This is why engine L exhibits a fuel consumption discontinuity of 5% for the transition from homogeneous-charge to stratified-charge combustion modes at an air/fuel ratio of 20:1.

The factors that contribute to the fuel consumption benefit include a mechanical friction reduction, a pumping work reduction and an indicated thermal efficiency increase. Figure 9.2-6 tabulates these components of the total gain in net fuel consumption and shows that a 9% improvement in BSFC for both engines is

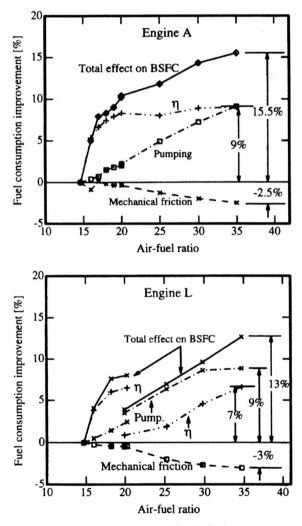

Figure 9.2-6 Components of G-DI fuel consumption improvement; 1500 rpm and 0.262 MPa BMEP without EGR (engine A: air-assisted fuel injection system; engine L: single-fluid, high-pressure fuel injection system) [51].

attributable to a pumping work reduction for operation at an air/fuel ratio of 35:1. Both engines exhibit a penalty in fuel consumption of approximately 2.5–3% due to an increase in mechanical friction, with two sub-factors considered to be responsible for this. One is the higher overall cylinder pressure that is present as the mixture becomes leaner. The second is the faster and earlier combustion for stratified-charge operation. Evidence of the elevated cylinder pressure for stratified-charge operation is a peak pressure that is 1.3 MPa higher than for homogeneous-charge operation. Both of these effects increase the friction of the piston rings and skirt, and also increase the load on the bearings, thus increasing the total mechanical friction.

The key difference exhibited by the two engines is in the magnitude of improvement attributable to changes in indicated thermal efficiency, η, for stratified-charge operation. For homogeneous-charge operation at an air/fuel ratio of 20:1 as shown in Fig. 9.2-6, a 7% improvement in indicated thermal efficiency is observed in engine L. This decreases to 1% for stratified-charge operation at the same air/fuel ratio with an optimum injection timing, and increases to 7% as the air/fuel ratio is enleaned to 35:1. In contrast, the indicated thermal efficiency for engine A improves by about 8% for an air/fuel ratio of 20:1 for both stratified-charge and homogeneous-charge operation. However, further enleanment to an air/fuel ratio of 35:1 only produces an additional 1% improvement in fuel consumption.

Based upon an Otto fuel-air cycle analysis, a 5% improvement in indicated thermal efficiency is expected when the air/fuel ratio is changed from 20:1 to 35:1. However, the observed indicated thermal efficiency with engine A is almost constant for air/fuel ratios ranging from 20:1 to 35:1, thereby indicating about 5% in total penalties for stratified-charge operation. The contributors to this are:

- An earlier combustion phasing than for premixed combustion
- A relatively high concentration of CO and H_2, indicating a lower combustion efficiency
- A local air/fuel ratio near the spark gap at the time of ignition that is richer than the overall air/fuel ratio and closer to stoichiometric

The last item yields a lower specific heat ratio and increased dissociation during the initial burn. The associated energy that is stored is partially recovered as sensible enthalpy later in the expansion stroke. Although

engine L does exhibit a continuous increase in indicated thermal efficiency as the air/fuel ratio is enleaned from 20:1 to 35:1, the values at 20:1 and 25:1 are moderated as a result of the locally rich combustion [51].

9.2.3 Fuel Economy versus Emissions Compromise

Expectations regarding the significant fuel economy benefits of G-DI engines are generally based on the thermodynamic potential of unthrottled operation. The actual fuel economy improvement that can be obtained with G-DI production hardware, however, is constrained by emissions requirements. Future emissions standards, whether in Europe, North America or elsewhere, require an interactive development of the combustion system, engine operating strategies and exhaust gas after-treatment systems. If imposed constraints such as emissions, driveability and manufacturability are satisfied, the actual in-service fuel economy improvement may be significantly less than predicted. It is therefore safe to state that the thermodynamic potential will probably never be achieved with production hardware [482]. For most automotive G-DI applications, some mitigation of the potential fuel economy gain will occur when meeting strict emission standards, and this is considered to be the dominant optimization constraint. The required emissions-related compromises will be dependent upon the specific combustion process, EGR strategy, exhaust aftertreatment adopted, fuels available in the field, and the particular emissions test cycle.

Light Throttling. The potential of G-DI technology for significantly improving vehicle fuel economy results from the ability to operate an engine in the stratified-charge mode without throttling. However, it is difficult to achieve stable, ultra-lean combustion while maintaining an effective catalyst operating temperature during steady engine operation. As a result, extended periods of lean operation may be limited by decreases in the exhaust temperature. It may indeed be necessary to employ some manifold vacuum in order to maintain the exhaust gas temperature above the effective operating temperature of the catalyst for extended periods of lean operation [214]. This constraint contributes to a fuel economy shortfall of about 3–6% in the MVEG test cycle. Some throttling may also be necessary for maintaining acceptable levels of combustion stability at the lowest part loads. A light throttling of the engine has also been identified as effective for HC emissions control; for example, at 20% EGR, a 20% reduction in HC emissions can be achieved by throttling the engine slightly, which results in a 3% penalty in fuel economy [201, 202, 203]. Nearly unthrottled operation may be feasible with some modifications to the combustion system, which would allow combustion development to more readily incorporate future improvements in lean NOx aftertreatment catalysis.

A G-DI engine that is designed to operate in the stratified-charge mode must, of necessity, also operate for conditions such as cold start and high load, which require a homogeneous charge. For such operating conditions it has been found that the use of some throttling is quite advantageous. The application of some throttling has also been shown to be quite effective in achieving smoother transitions between engine operating modes. In addition, the flow of a significant percentage of EGR for essential NOx reduction as well as canister purging may require an intake manifold vacuum. This can be obtained most expeditiously by applying some throttling under the control of the EMS, and may lead to an additional fuel economy penalty of 1%. Also, a conventional vacuum-assisted power brake system also requires some vacuum, which may be optionally provided by throttling rather than by an on-board vacuum pump. A differentiation among the throttling requirements as a function of engine load is illustrated in Fig. 9.2-7 [482]. It should be noted that since over half of the total G-DI fuel economy benefit derives from the reduction of pumping losses [259], the use of the throttling in a G-DI operating system, even though beneficial for emissions and control, should always be carefully evaluated [274].

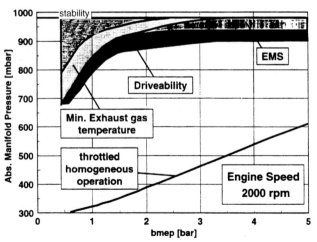

Figure 9.2-7 Throttling requirements of a stratified charge engine as compared to throttled homogeneous-charge operation [482].

Analysis of the enhancement in fuel consumption of a G-DI engine for a range of operating conditions using a two-zone model reveals that an improvement in fuel economy up to 20% can be achieved [180]. This improvement is relative to a current PFI engine using a TWC, with the G-DI engine operating in the stratified-charge mode over the entire operating range. With regard to aftertreatment, however, this may not be practical. The active operating temperature of the catalyst is a very important parameter, and the overall mixture cannot be ultra-lean if the catalyst inlet temperature is to be maintained above a critical threshold value. The analysis also indicates that the operating range of the stratified-charge engine is narrowed when emission constraints are applied, reducing the predicted fuel economy improvement to 10%. This reduction makes it evident that further improvements in catalyst technology will be required in order to take full advantage of the potential benefits of highly stratified combustion.

Because low-load operation with throttling is dominated by pumping losses, a fuel economy improvement results from the ability of a combustion system to operate at higher dilution ratios. Current G-DI combustion systems generally have an optimum dilution ratio of approximately 30:1 to 40:1 for best gross indicated thermal efficiency. However, unthrottled operation at low loads of less than 0.2 MPa BMEP may represent a dilution ratio of 80:1 or even leaner. Therefore an area that requires continued development and refinement is that of part-load operation at very light load. In this area, the goal is to achieve enhanced stratification control without the loss of containment at high dilution ratios. As noted, however, this further enhanced fuel consumption will yield a lower exhaust temperature. The extremely low exhaust temperatures associated with lean, low-throttling operation constitute a significant problem due to the need to maintain the catalyst temperatures and conversion efficiency. It is important to emphasize that increasing the capability of a DI combustion system to operate with even higher dilution ratios to improve fuel economy may have only limited application. The use of this level of dilution ratio may not be a viable approach because of the resulting low exhaust gas temperature [63]. Table 9.2-3 summarizes the key reasons for using light throttling of G-DI engines.

Lean-NOx Aftertreatment System Operating Temperature Window. As discussed in Chapter 8, each exhaust aftertreatment system has an optimum operating temperature window for best conversion efficiency. The

Table 9.2-3
Key reasons for using light throttling of G-DI engines

Combustion Stability	To maintain the combustion stability at very light loads
Exhaust Gas Temperature	To maintain a minimum threshold exhaust gas temperature
Emissions	To achieve operating strategies that reduce HC emissions
Driveability	To maintain driveability at light load
Other Load Operations	To meet the requirements for cold start and high-load operations
EGR	To ensure high rates of EGR
Load Transition	To maintain smooth load transition
EMS Requirement	To meet the engine management system requirements for EGR system and vacuum-assisted power brake system

low-temperature limit of a lean-NOx aftertreatment system defines the highest dilution ratio that the G-DI engine can use if the catalyst temperature and desired conversion efficiency are to be maintained. Operation at air/fuel ratios leaner than this dilution-ratio threshold is not practical even though the combustion system may still be able to tolerate further dilution. In contrast, the high-end temperature limit of the lean-NOx aftertreatment system temperature determines the highest load that stratified-charge or lean, homogeneous-charge operations can support. Lean operation at a load exceeding this threshold will lead to significant NOx emissions. In addition, the highest load that stratified-charge operation will permit is also limited by the level of smoke emissions. Therefore the load ranges for fuel-efficient, stratified-charge or lean, homogeneous-charge operations from which a fuel economy gain is derived are significantly constrained by the imposed thresholds of the lean-NOx aftertreatment system temperature. Thus, improving the performance and expanding the operating range of lean-NOx

aftertreatment systems is one key to exploiting the full fuel economy potential of G-DI technology.

Relationships of NOx Storage Catalyst Regeneration to Fuel Economy. With NOx storage catalysts, a periodic regeneration, or purge, is required. This is accomplished by means of a brief, rich spike, which is associated with a direct fuel economy penalty. This is particularly true for operations at higher engine speeds and loads when the engine is operated in either the lean-stratified or lean-homogeneous mode. For medium, part-load operation, the stratified-charge-operating map area is limited not by the combustion system, but more by the purge requirements of the NOx storage catalyst. With increasing speed and load, the exhaust gas temperature and engine-out NOx emissions increase substantially. For the same region of the engine operating map the effective NOx storage capacity decreases, requiring a more frequent purge, which has a negative impact on fuel economy. At some point the purge frequency becomes a more stringent limiter of lean operation than the combustion system itself. For this reason, lean, homogeneous-charge operation may have to be constrained, switching instead to high-EGR, stoichiometric-charge operation where a TWC can be utilized. Although this switching yields an additional 1.5% fuel economy penalty, engine-out NOx emissions can be reduced by almost 50% for the same testing condition [482]. Even with lean operation only in the lower part of the load range, the fuel economy penalty for purging the NOx storage catalyst is approximately 0.5–1.5%, depending on the emissions target. As discussed in Chapter 8, the optimization of purging strategies and the development and implementation of a reduced oxygen-storage catalyst for close-coupled catalysts are important measures to minimize this fuel economy penalty.

9.2.4 Opportunities for Future Fuel Economy Improvement

Previous sections have emphasized the large discrepancy that has been observed between the expected theoretical fuel economy improvement of 20–25% and the currently achieved value of 10–12%. A detailed analysis of the factors contributing to the shortfall in the fuel economy benefit has been conducted using both the actual 1998 G-DI performance status and the 1996 expectation when G-DI was first launched. This detailed study was undertaken in order to identify the reasons for the large difference between expectation and reality. The individual items and the corresponding

Table 9.2-4
Individual items and their contributions to fuel economy shortfall between theoretical expectation and current practice of G-DI engine

Emissions Constraints	7–10%
Combustion	2–4.5%
Mechanical Friction	2–4%
Engine Management	1.5–2%
Driveability	1–2%
Tolerances	1–1.5%

Table 9.2-5
Factors contributing to the increased friction with unthrottled operation of G-DI engines

- High-pressure fuel pump drive load
- Higher cylinder pressure
- Earlier combustion phasing
- Increased piston weight and less favorable piston mass distribution
- Increased electrical energy consumption
- Increased friction due to a slower engine warm-up

contributions to this 10–15% shortfall in fuel economy improvement are summarized in Table 9.2-4 [482]. It is evident that the major factor limiting the improvement in fuel economy is the emissions compromise that must be incorporated into both design and operation. Although engine friction is a contributor, for G-DI engine operation there is still a potential for further reductions. This is important, as the G-DI engine exhibits a low BSFC gradient with engine load, and is even more sensitive to friction than PFI engines. This also highlights the importance of reduced warm-up times and minimized electrical power consumption, as these contribute to friction load in field operation [483]. The high-pressure fuel pump that is associated with direct fuel injection requires substantially more power to operate than the low-pressure PFI vane pump, thus contributing to the total parasitic load on the engine.

PFI engines generally exhibit a sharp gradient of specific fuel consumption as a function of engine load, thus an increase in engine friction is partially compensated by the increased indicated load. An unthrottled, stratified-charge engine exhibits a much smaller gradient, therefore friction increases are transformed more directly into fuel consumption increases. For example, a 22% improvement in ISFC of a stratified-charge, unthrottled G-DI engine with EGR, when compared to a stoichiometric, homogeneous-charge PFI engine without EGR, only exhibits a 19% improvement in BSFC. This difference between ISFC and BSFC improvements is attributed to the higher cylinder pressure level in the stratified-charge G-DI case, leading to an incremental increase in engine friction. The factors contributing to increased friction with unthrottled operation are summarized in Table 9.2-5. All of the factors listed in Table 9.2-5 must be carefully evaluated and optimized to effectively reduce the mechanical friction and parasitic loads associated with the operation of unthrottled, or lightly throttled, G-DI engines [481].

In general, stratified-charge operation, with its associated higher efficiency, results in slower engine warm-up, with both engine coolant and engine oil temperatures increasing more slowly than occurs during a PFI engine warm-up. As illustrated in Fig. 9.2-8, the oil temperature in some G-DI applications can remain cooler than the target value throughout the entire emissions test cycle. This reduced oil temperature contributes directly to increased friction. The reduced oil temperature together with the high friction sensitivity of the G-DI engine is responsible for 1–2% of the fuel economy shortfall [482]. Some measures such as the addition of a heat exchanger for oil and coolant may be effective in helping to minimize this friction increase. An optimized warm-up strategy is considered to be more effective for G-DI engines than for PFI engines.

As discussed previously, the fuel economy shortfall relative to the theoretical thermodynamic potential can be meaningfully explained as a number of small individual effects. In a similar manner, possible future improvements may be considered to be individual positive increments. The experimental evaluation of these measures is fairly difficult, as the benefit from each may be quite small and, in fact, can be close to the tolerance of the measurement. Thus, many of the values of the individual contributions to future enhancements in fuel economy are derived from modeling analyses. The likely potentials for further fuel economy improvements in both the short term and the long term are listed in Table 9.2-6 [482].

The potential for further fuel economy improvements lies predominantly in emissions-related measures,

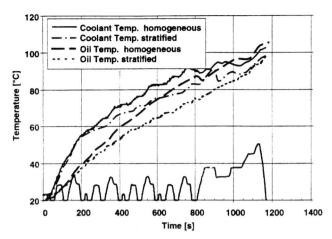

Figure 9.2-8 Comparison of oil and coolant warm-up characteristics between stratified-charge and homogeneous-charge operations [482].

Table 9.2-6
Itemized areas of potential future enhancements to G-DI fuel economy

Area	Short Term	Long Term
Emissions	2.5%	4.0%
Combustion Improvements	0.5%	1.0%
Friction Reduction	1.0%	1.5%
EMS Enhancements	0.5%	0.5%
Driveability Calibration	0.5%	1.0%
Tolerances	0.5%	1.0%

and such enhancements are generally predicated on the successful development and implementation of lean-NOx aftertreatment technology. If the lean-NOx aftertreatment system operating-temperature window can be widened, a lean, homogeneous calibration for medium, part-load operation will be possible, leading to a 1–1.5% fuel economy improvement. The development and inclusion of an optimized close-coupled catalyst with reduced oxygen-storage capacity will enable optimized catalyst purge events for NOx storage catalyst, and will lead to an additional 0.5% improvement in fuel economy. Further, implementing the ability to manage aftertreatment at lower exhaust temperatures will reduce the required throttling and improve fuel economy by another 1–1.5%, of which 0.5–1% can be allocated to the reduction of HC emissions and improved EGR tolerance. Engine friction reductions, both direct

and indirect, can contribute 1–1.5% to improvements in fuel economy through the use of demand-delivery, high-pressure fuel systems; a more efficient, high-pressure fuel pump; a reduced fuel-rail pressure; and faster warm-up of the engine oil and coolant. A more precise EGR metering control system can contribute not only to combustion efficiency improvements, but also to engine-out NOx emissions reduction and a less frequent NOx purge frequency for NOx storage catalyst. This could yield an additional 0.5–1% fuel economy improvement. Considering all of these possible enhancements, an overall further fuel economy improvement in the range of 4–5% is to be regarded as quite reasonable in the short term. In this regard, an important factor for wider applications is the timetable for the availability of low-sulfur fuels or for the development of a sulfur-tolerant lean-NOx catalyst [482].

In the early development of the production G-DI engine, the wall-guided combustion system received significant attention due to the system robustness that can be achieved. However, the slightly-reduced stratification potential with such a system as compared to the spray-guided system reduces the fuel economy potential by at least 1–2% [482]. For HC emissions reduction and combustion stability improvement, stratified-charge combustion is usually tuned to a rather early heat release in the engine cycle instead of the thermodynamic optimum. The fuel economy penalty for this tuning is estimated to be 0.5–1%.

Other areas that can be optimized for further fuel economy improvement are the engine management system and the calibration strategy for improved driveability. An increase in electric power consumption leads to a reduction in fuel economy, particularly at very light load. Some actuators required for G-DI technology such as fuel injectors with a power stage, the EGR valve, the electronic throttle control (ETC), and the SCV all contribute to an elevated power consumption, with the associated fuel economy penalty being as high as 1–1.5% [482]. The vehicle fuel economy improvement that is finally achieved in field operation is strongly dependent upon the fraction of time spent in stratified-charge operation. As illustrated in Figs. 9.2-9 and 9.2-10, the integral energy distributions as a function of excess air ratio, namely different levels of charge stratification, vary significantly with test cycle, vehicle calibration, and system package [482]. It is also important to emphasize that improvement in vehicle fuel economy for a G-DI engine over that for a PFI engine is very dependent on engine/vehicle matching,

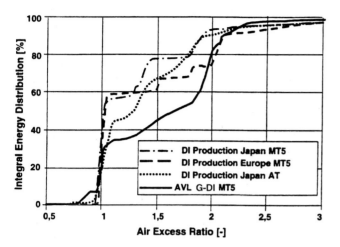

Figure 9.2-9 Integral energy distributions as a function of excess air ratio for a prototype G-DI vehicle over different test cycles [482].

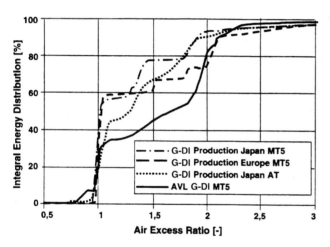

Figure 9.2-10 Integral energy distributions as a function of excess air ratio for different G-DI vehicles over the MVEG test cycle with hot start [482].

driving cycle and G-DI operating-mode transitions [481]. The use of a sophisticated, torque-based, engine control system can provide significant fuel economy benefits. By avoiding direct coupling to the input of the driver, an extended operating time within the stratified charge mode during transient conditions can be realized while maintaining acceptable vehicle driveability [109, 110].

9.3 Gasoline Direct Injection Combined with Other Technologies

A significant fraction of current DI development is based upon the manufacturing consideration of making minimum changes to already existing PFI engines. This can yield an attractive fuel economy improvement, but will not necessarily utilize the full potential of G-DI technology. Clearly, G-DI can be combined with other fuel-efficient technologies such as boosting, idle engine shut-off, CVT and HEV, with a benefit being that G-DI can enhance some of the deficiencies of these technologies. Furthermore, the fundamental advantages of the G-DI engine such as good transient response, superior engine start, precise torque control, in-cylinder charge cooling, flexibility of fueling control, good resistance to knocking, and reduced turbocharger lag time can be employed to significant advantage by combining with other technologies. Thus, an additional benefit of G-DI is that it may be effectively combined with other technologies to achieve distinctly lower fuel consumption. By combining the advantages of G-DI with new drive-train strategies, auxiliary electric power equipment and performance-enhancing technology, potentials can be combined, with fuel economy improvements of up to 25% being achievable, as well as improved transient response and driveability [483].

9.3.1 Boosted G-DI Engine
Boosted G-DI vs. Naturally Aspirated G-DI. The addition of induction-air-pressure boosting in conjunction with engine downsizing is well known as an effective measure for fuel economy improvement. In this case the improvement is achieved by shifting a portion of the load distribution into a more efficient engine map area. However, the achievement of the theoretical potential of downsizing has proven to be difficult in the past in actual field operation, with only a small fraction of the downsized engines able to achieve the full fuel economy potential. The driveability of a vehicle with a downsized, boosted engine is most often inferior to that of a larger-displacement, naturally aspirated PFI engine. The boosting of conventional PFI engines has traditionally been subject to demerits such as reduced low- and mid-range torque due to knocking, increased fuel consumption at high load due to a lower knock-limited compression ratio and turbocharger lag in the initial stage of acceleration when turbocharging is used for boosting. The inherent advantages of the G-DI engine for circumventing some disadvantages of boosted engines are listed in Table 9.3-1.

A critical factor for all boosted engines is the increased octane requirement. Generally speaking, the geometric compression ratio of a boosted engine must

Table 9.3-1
Inherent advantages of the G-DI engine for circumventing some disadvantages of boosted engines

In-Cylinder Charge Cooling	Induction air can be cooled significantly, yielding improved engine anti-knock performance; alternatively, a higher compression ratio may be employed as compared to the baseline PFI engine.
Injection Timing Flexibility	A multi-stage injection strategy may be invoked as an effective means for knock control, enabling the use of the same or higher compression ratio; resulting in more low- and mid-range torque.
Stratified-Charge Operation	Ultra-lean operation may be possible over a wider range of engine speeds because the increased air capacity may extend the region of stratified-charge operation to a higher load.
Reduced Turbocharger Lag	Turbocharger lag time is reduced because the turbine operates at very high speeds even during lean-burn, part-load operation, due to the higher quantity of engine air flow.

be reduced substantially to avoid engine knock, which has a deleterious effect on engine thermal efficiency. G-DI engines, however, can operate at a higher boost level with the same or even higher compression ratio. This is due to the reduced octane requirement resulting from in-cylinder charge cooling and the reduced residence time of the fuel-air mixture. It is worth noting that in-cylinder charge cooling by means of late injection can also be used to improve knock-limited performance, however, this may also increase smoke emissions. With regard to turbocharging technology, another key performance parameter that must be addressed is the characteristic transient response time. With a faster turbine response following a load request, the transient characteristics of a turbocharged G-DI engine are significantly improved over those of a turbocharged PFI engine. Figure 9.3-1 illustrates the difference in turbine acceleration and the resulting torque increase for a step load-change between homogeneous-throttled and stratified-unthrottled operating modes.

The fuel-economy benefits of naturally aspirated G-DI engines are generally maximized at low-speed, low-load conditions due to the large air flow requirements of stratified-charge operation. With boosting, this air flow can be increased, thereby extending the stratified-charge operating regime, and potentially increasing the fuel economy benefit [57]. Some or all of the disadvantage of

a reduced volumetric efficiency resulting from a required swirl or tumble intake port design can also be recovered through boosting. When compared to a naturally aspirated engine with equivalent torque, a boosted G-DI engine also provides advantages in terms of reduced friction losses [254]. An example of an operating map for a boosted G-DI engine is illustrated in Fig. 9.3-2 [427]. The key design considerations associated with a boosted G-DI engine are summarized in Table 9.3-2.

Engine Downsizing through Boosting. It was noted in the previous section that boosting can allow an engine to be downsized for fuel economy improvement while maintaining vehicle performance. On the other hand, downsizing shifts the load distribution of engine operation toward higher specific loads where G-DI provides only a limited fuel economy advantage. Therefore, with downsizing, the importance of stratified-charge, part-load operation to the overall fuel economy is generally reduced. As a result, for the G-DI engine it is very important to provide a very good level of fuel economy at mid-range loads in order to take maximum advantage of boosting. One option available for this portion of the operating map is an expansion of the stratified-charge operational range. Figure 9.3-3 shows a comparison of engine operating maps between a naturally aspirated engine and a downsized, turbocharged, G-DI engine having an equivalent maximum torque.

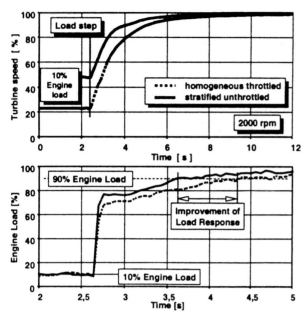

Figure 9.3-1 Comparison of transient response between homogeneous-throttled and stratified-unthrottled operating modes at part load [483].

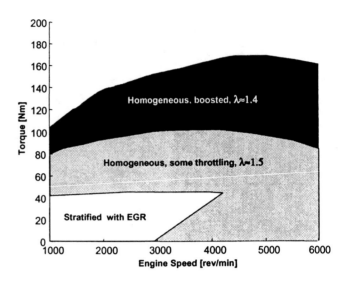

Figure 9.3-2 Example of an operating map for a boosted G-DI engine [427].

Table 9.3-2
Key design considerations for G-DI engine boosting

Stratified-Charge Region	Determine the extent to which the stratified-charge region of the engine operating map can be expanded through boosting
Fuel Economy Benefit	Quantify the fuel economy benefit due to the expansion of the stratified-charge region
EGR Capability	Assess the EGR flow capability throughout the stratified-charge operating region
Maximum Compression Ratio	Evaluate the maximum compression ratio that can be utilized with boosting
Maximum Power	Determine the resulting maximum power increase due to boosting
Load Response	Evaluate the degree of improvement in load response of the engine

Studies on boosted G-DI engines have revealed that the additional induction air does enable the expansion of stratified-charge operation to higher loads and speed, in fact, as high as 3000 rpm and 8 bar IMEP. In order to achieve the lowest ISFC, it is necessary to calibrate the heat release rate of stratified-charge combustion to a thermodynamic optimum. This is generally difficult, as the injection and ignition timings normally have to be relatively advanced to obtain acceptable mixture preparation and HC emissions. For a turbocharged stratified-charge mode, the elevated engine back pressure also degrades the fuel economy, even though the additional

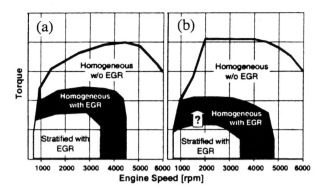

Figure 9.3-3 Comparison of engine operating maps for naturally-aspirated and downsized, turbocharged G-DI engines: (a) naturally-aspirated G-DI engine; (b) downsized, turbocharged G-DI engine [483].

air provides an overall leaner air/fuel ratio. In order to achieve acceptable HC emissions, injection timing for a turbocharged G-DI engine is normally even more advanced than for a naturally aspirated G-DI engine, yielding an earlier crank angle for the 50% mass fraction burned (MFB), increased heat transfer and possibly higher fuel consumption [254].

Boosting a G-DI engine can significantly increase the power available to enhance vehicle performance or provide options to improve fuel economy in a larger vehicle. Analyses of boosted G-DI engines reveal that the fuel economy is not significantly improved by simply extending the stratified-charge region in a mid-size vehicle, as the fraction of time spent in the stratified-charge mode is not increased significantly for any of the cycles. Also, the reduction in pumping work associated with stratified-charge operation is less at higher-speed, higher-load conditions. However, due to the additional flexibility provided by the variable geometry turbine (VGT), constrained fuel economy and NOx emissions may be improved relative to those of a naturally aspirated G-DI system. For example, fuel economy is found to be improved significantly through boosting and charge stratification for a typical mid-size sports utility vehicle with a small displacement engine [57].

An additional method of improving medium-part-load fuel efficiency is through an increased compression ratio for homogeneous operation (stoichiometric, homogeneous-charge operation with EGR or lean, homogeneous-charge operation with or without EGR). This is true as long as the homogeneous-charge operation is unthrottled to the maximum extent possible by

lean operation and/or high EGR dilution. Figure 9.3-4 shows the effect of the geometric compression ratio on the part-load BSFC for throttled homogeneous-charge operation and unthrottled, stratified-charge operation. The throttled, homogeneous-charge operation exhibits a marked increase in efficiency with the use of a higher compression ratio, with 50% MFB occurring at approximately 9°ATDC. In contrast, unthrottled stratified-charge combustion exhibits a 50% MFB at approximately 1°ATDC. This indicates that a very elevated compression ratio of up to 14:1 is near optimum for medium-part-load fuel efficiency.

In contrast, the compression ratio should be *lowered* for low-load, stratified-charge combustion to reduce engine friction, and must be lowered for full-load operation to suppress engine knock. Thus, the G-DI limiting compression ratio can be as high as 15:1 for part-load operation [378]; however, for gasoline currently available in the field, autoignition occurs at high load when operating with a compression ratio higher than 12:1.

In order to address these conflicting requirements, a strategy incorporating a variable compression ratio has been proposed, with the operating map for utilizing this concept being illustrated in Fig. 9.3-5. The map shows that a rather complex optimization could be provided by

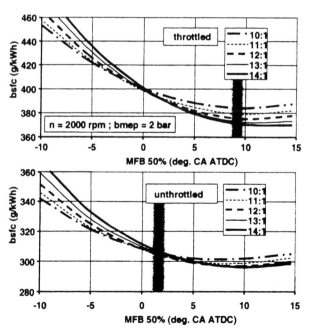

Figure 9.3-4 Effect of geometric compression ratio on part-load BSFC for throttled homogeneous-charge operation and unthrottled stratified-charge operation [483].

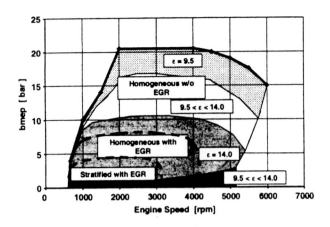

Figure 9.3-5 Operating map for a strategy incorporating a variable compression ratio [483].

the use of a variable compression ratio in a G-DI engine, with the highest compression ratio utilized at part load. An integrated concept combining G-DI turbocharging and a variable effective compression ratio is predicted to provide a potential fuel economy improvement of up to 28%, with the additional advantages of improved transient response and vehicle driveability [483].

9.3.2 Turbocharged versus Supercharged G-DI Engines

Relatively large EGR rates are inherent in stratified-charge G-DI operation, and it is not a simple task to supply the proper amount of EGR required for a boosted G-DI engine over the entire engine operating range. The elevated mean level of manifold pressure does reduce the pressure differential between the exhaust and intake manifolds. When a VGT turbocharger is used, the pressure in the exhaust manifold can be adjusted to allow adequate EGR rates for most operating conditions, although the system must be designed carefully to avoid degradation in turbine performance. When a supercharger is used, however, the pressure in the intake manifold may be elevated to a level that exceeds that in the exhaust manifold. As a result, the introduction of EGR by means of a conventional EGR valve may not be possible, and EGR must be provided by an auxiliary EGR pump or other means such as a variable valve timing (VVT) system. This is normally associated with both increased system complexity and an additional parasitic load. Thus, in EGR system design, the turbocharger offers definite advantages over the supercharger.

As previously discussed in Chapter 8, a VGT is designed to extract energy from the exhaust gas in order to operate the compressor, with the temperature of the turbine exhaust gas being lower than that in the exhaust manifold. The amount of energy extracted varies with the turbine inlet vane position. As a consequence, the temperature of the exhaust gases entering the aftertreatment device can be adjusted to stay within the temperature window of the lean-NOx aftertreatment system, in order to maintain a high operating efficiency. In contrast, the supercharger cannot be used for this purpose [57]. It is worth noting that the turbocharger may be disadvantageous in improving catalyst light-off characteristics. Other means such as exhaust by passing during cold start may have to be considered.

To achieve a sufficiently high air/fuel ratio for stratified-charge operation, the level of boosting must be increased at higher load. The compression work that is associated with supplying additional boosting at part load by the supercharger could result in a decrease in fuel economy as compared to naturally aspirated, stratified-charge operation. In comparison, turbocharging is only associated with high back pressure without the efficiency loss that results from compression work [254]. A turbocharger also offers some advantages with regard to hardware packaging. For all of the above reasons, turbocharging instead of supercharging is considered to be the preferred alternative for boosting a G-DI engine [57, 254, 353, 427, 483].

9.3.3 Combustion and Emissions Characteristics of a Boosted G-DI Engine

Stratified-charge operation can be the preferred mode in a naturally-aspirated engine for loads up to approximately 5.5 bar IMEP. For higher levels of IMEP the fuel economy benefit relative to homogeneous-lean operation diminishes with or without EGR. This IMEP limit is characterized by an excess air ratio of about 1.7 without EGR and 1.45 with EGR. With further enrichment the efficiency of stratified-charge combustion declines, and smoke and CO emissions generally increase. The specific IMEP limit depends upon the volumetric efficiency of the particular G-DI engine. Given these constraints, the operating range of a turbocharged, stratified-charge engine can only be expanded if the compressor is capable of providing sufficient boost pressure at part load.

In this regard, the combustion characteristics of a turbocharged G-DI engine at an engine speed of 2000 rpm are shown in Fig. 9.3-6 for three levels of boost pressure [483]. Both injection and ignition timings are optimized in these data. The smoke limit shifts almost

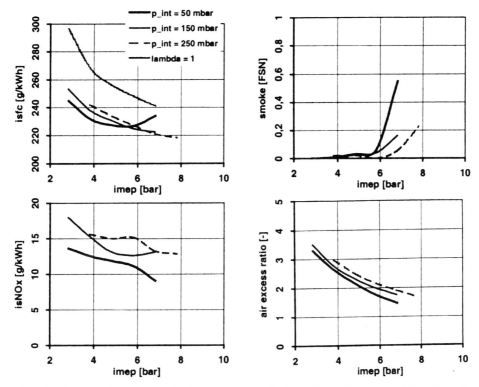

Figure 9.3-6 Combustion and emissions characteristics of a turbocharged G-DI engine for three levels of boost pressure; engine speed: 2000 rpm [483].

linearly to higher IMEP values with increased boost pressure, and the critical limitation of turbocharged G-DI imposed by smoke emissions constraints is elevated by approximately 1.5 bar IMEP. NOx emissions are observed to increase with increased boost pressure due to the higher charge density and flame temperature, whereas at a constant IMEP, HC emissions increase with increased boost pressure due to the relative enleanment. Exhaust gas temperature without EGR is decreased slightly with increased boost pressure. For the entire stratified-charge range of up to 8 bar IMEP, the exhaust gas temperature at 0.25 bar boost pressure remains below 550°C, which is within the temperature window of the typical NOx storage catalyst.

The fuel consumption of turbocharged G-DI engines exhibits some interesting characteristics, with an increase in boost pressure yielding a loss in fuel economy and higher HC emissions at low load due to leaner combustion. At medium part load, stratified-charge operation does achieve improved fuel consumption. Minimum fuel consumption is obtained in a map area where naturally aspirated, stratified-charge combustion provides very little additional fuel economy benefit.

A comparison of the combustion characteristics of a turbocharged G-DI engine at a boost pressure of 0.25 bar is illustrated in Fig. 9.3-7 for the conditions of zero EGR and EGR calibrated for minimum ISFC and minimum NOx emissions. It is evident that with EGR the fuel consumption is lower than without EGR for the load range up to 7 bar IMEP. Also, stratified-charge operation can be extended up to a 7.5 bar IMEP smoke limit. Exhaust gas temperature for operation with EGR is increased by 50°C, but is still below the critical temperature of 550°C for NOx storage catalysts. In summary, the turbocharging option does allow a moderate expansion of the stratified-charge operational range to higher IMEP levels of approximately 7 bar. The limitation is imposed primarily by NOx mass emissions and to a lesser degree by smoke emissions.

9.3.4 Engine Shut-off Strategy During Idle

The strategy of engine shut-off as an alternative to idle has long been recognized as an effective means to reduce vehicle fuel consumption in the field for any engine. In addition, shutting off the engine can also maintain the catalyst temperature above the catalyst

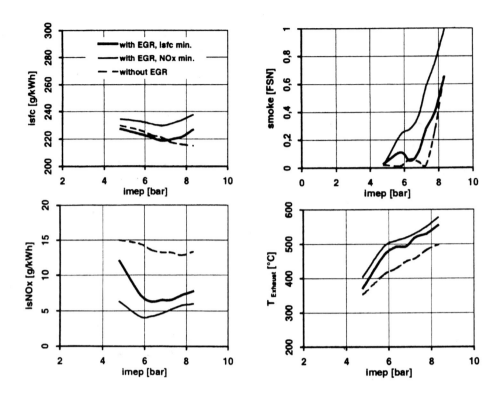

Figure 9.3-7 Combustion and emissions characteristics of a turbocharged G-DI engine at a boost pressure of 0.25 bar; for zero EGR and for EGR calibrated for minimum ISFC and minimum NOx emissions (2000 rpm) [483].

light-off threshold for a longer time period. As shown in Fig. 9.3-8, the catalyst cools much faster when the engine is kept idling following medium to light-load operation than when the engine is shut off [451]. However, the idle shut-off strategy is impractical with PFI engines due to the excessive number of engine cycles required to restart and stabilize combustion. In contrast, the G-DI engine injects fuel directly into the cylinder, and requires significantly fewer cycles to fire and stabilize [215, 451].

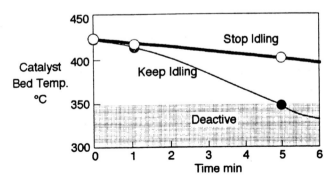

Figure 9.3-8 Comparison of catalyst temperatures between engine idle and engine shut-off [451].

A comparison of the PFI and G-DI starting processes is illustrated in Fig. 9.3-9 [451]. For an optimum starting process in a conventional PFI engine, the cylinder that is in the exhaust stroke during cranking is determined, and fuel is injected into the intake port of that cylinder in order to initiate the quickest start of the engine. However, the fuel-air mixture has to be inducted into the cylinder and compressed before it can be ignited. Therefore, even for an optimum strategy, at least two to three strokes are required before the first combustion event is observed in a PFI engine. For a rapid start in a G-DI engine, however, the cylinder that is in the compression stroke during cranking is determined, and fuel is injected into that cylinder immediately, which requires less than one stroke to get the engine started.

Due to the elevated residual gas temperature that results from wall heat transfer from the cylinder wall during engine shut-off, only a slight compression is sufficient to prepare the mixture for spark ignition near the compression TDC. This strategy may require less than one stroke to achieve the first combustion event for a G-DI engine. As a result, the torque delay is only 1/5 to 1/10 that of a conventional PFI engine during restart. This

278

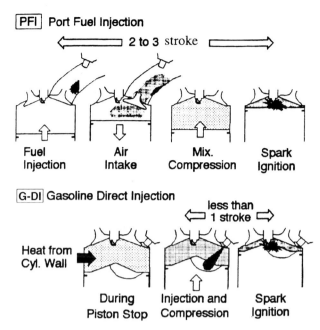

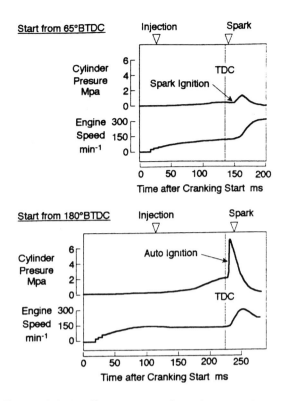

Figure 9.3-9 Comparison of the PFI and G-DI engine cranking and starting processes [451].

Figure 9.3-10 Comparison of combustion characteristics for two initial restart positions of 65°BTDC and 180°BTDC; 1.8L G-DI engine having a compression ratio of 12:1 [451].

potential rapid initiation of the first G-DI combustion event is a key factor in considering the feasibility of incorporating an idle shut-off strategy. If a G-DI engine is employed in a hybrid application, as will be discussed in Section 9.3.6, this rapid restart characteristic represents a very significant advantage that can be readily invoked.

G-DI restart characteristics vary significantly with engine starting position. For an inline, 4-cylinder engine, one piston normally stops at a crank position in the range of 120°–60°BTDC on compression stroke, with 60°–70°BTDC being the most likely position. Figure 9.3-10 presents a comparison of the combustion characteristics for two initial restart positions for a 1.8L G-DI engine having a compression ratio of 12:1. Restart testing was conducted after the engine was shut down for 60 seconds, following light-load, steady-state operation. For a starting position of 65°BTDC, the engine can be readily restarted within 1/6 engine revolution. For a starting position of 180°BTDC, however, autoignition occurs before the spark, resulting in a sharp increase in in-cylinder pressure. From this comparison it is obvious that restart characteristics are strongly dependent on initial piston position, due to the variation in in-cylinder gas temperature. Thermodynamic analyses conducted for an initial cylinder pressure of 0.1 MPa and a polytropic exponent of 1.30 indicate that a

compression temperature of 320°C is attained if compression starts at BDC with an initial temperature of 20°C, as illustrated in Fig. 9.3-11. The same compression temperature is also obtained for starting at 65°BTDC with an initial temperature of 120°C. The initial charge temperature of 120°C is estimated from the chamber wall temperature that occurs when the engine is shut down for a brief period following operation at full warm-up. If compression starts from BDC with the same initial temperature of 120°C, the in-cylinder pressure and temperature at TDC can attain 2.0 MPa and 500°C, respectively. This corresponds to the autoignition condition for regular gasoline with 90 RON [451].

To facilitate a more complete understanding of the restart process, the mixture ignition characteristics during restart are plotted in Fig. 9.3-12 as a function of the amount of fuel injected and the injection timing (EOI) for different piston initial positions. By optimizing the injection timing, a normal spark-ignited combustion is obtainable for equivalence ratios in the range of 0.5 to 2.0. A significantly higher in-cylinder temperature and pressure are reached when the piston initial position is

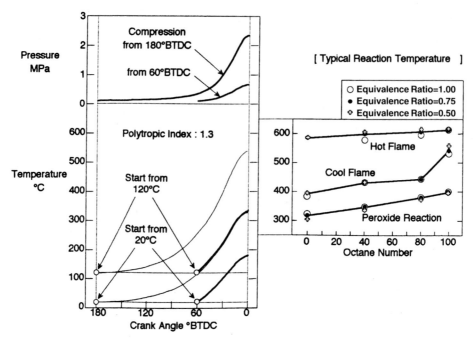

Figure 9.3-11 Thermodynamic analyses of in-cylinder air temperature and pressure histories at various initial restart positions and temperatures; initial in-cylinder pressure: 0.1 MPa; polytropic exponent: 1.30 (451).

close to BDC. When the initial position is near 120°BTDC, autoignition is observed, but only for very retarded fuel injection timing. When the restart position of the piston is at BDC, a wide region of autoignition is observed; however, this autoignition can be avoided for an EOI injection timing near 80°BTDC when a large amount of fuel is injected. It is evident that autoignition on engine restart after idle shut-off can be avoided through an optimal control of fuel injection timing and air/fuel ratio. The

ignition characteristics of mixtures derived from fuels with different octane values at a piston initial position of 180°BTDC are presented in Fig. 9.3-13 (451). The higher the octane number of the gasoline used, the narrower the region of restart autoignition.

It should be noted that, although the analysis was conducted for a piston starting position of BDC, the probability of this occurring is very low. Therefore, G-DI autoignition that is associated with the first firing

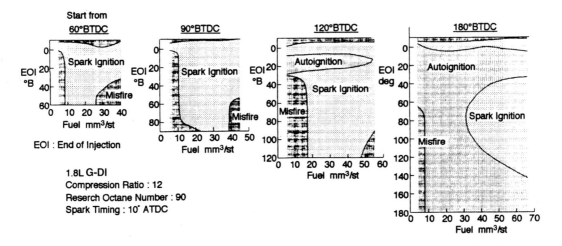

Figure 9.3-12 Restart map for a range of injected fuel amount, EOI injection timing and different piston initial positions (451).

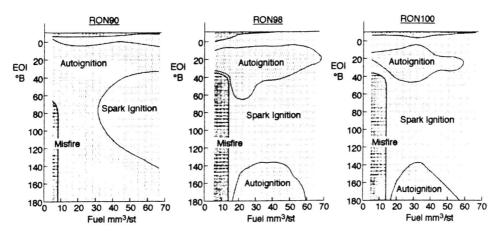

Figure 9.3-13 Restart map for a range of fuels at a piston initial position of 180°BTDC [451].

cylinder at the beginning of the restart is not likely. However, this analysis is important in understanding the combustion characteristics of the second cylinder that restarts from an original position corresponding to the intake stroke. Figure 9.3-14 shows the combustion characteristics of the second cylinder at two restarting piston positions of 260°BTDC and 360°BTDC. At a restart position of 260°BTDC, a large amount of residual gas is trapped in the cylinder, and is heated by free convection during the engine shut-down interval. As a result, a high temperature is attained at the end of the compression stroke. Similar to what is observed for the

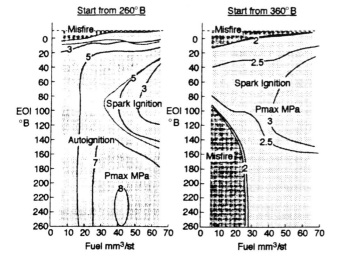

Figure 9.3-14 Restart map of the second cylinder at two restarting piston positions of 260°BTDC and 360°BTDC for a range of injected fuel amount and EOI injection timing [451].

first firing cylinder, a wide region of autoignition is observed when the second cylinder is initially positioned near the middle of the intake stroke. In contrast, when the piston of the second cylinder is at intake TDC, or 360°BTDC, the amount of high-temperature "residual" is much reduced, and the initial cylinder contents will also be diluted and cooled significantly by a large amount of intake air. As a result, the temperature at the end of the compression stroke is below the autoignition temperature. Similar characteristics will also be observable in the third and fourth firing cylinder. The ramification of this is that the possibility of restart autoignition in the second firing cylinder must be carefully considered and avoided for the reasons listed in Table 9.3-3 [451].

In addition to the complex combustion control required, there is also a concern regarding the energy required for restart, which may exceed the energy saved by invoking idle shut-off. Figure 9.3-15 shows a comparison of the measured fuel consumption during restarting of PFI and G-DI engines [451]. For a conventional PFI engine, the fuel consumption for one restart is estimated to correspond to that required to idle the engine for 5.5 seconds, thus the fuel economy will be negatively impacted if the idle shut-off is less than 5.5 seconds. With a G-DI engine, however, the energy consumption for restart can be reduced significantly due to the short cranking, rapid restart and accurate lean-burn torque control. As a result, the fuel consumption for restart corresponds only to that required to idle the engine for 0.4 second. Thus, idle shut-off with a G-DI engine is clearly a proven approach for improving fuel economy. When this strategy is applied to a vehicle retrofitted with a 1.1L G-DI engine, an extra fuel

Table 9.3-3
Reasons to avoid autoignition in the second firing cylinder during restart

Engine Vibration	Engine vibration could be significant due to the constant-volume combustion associated with autoignition. This high-pressure combustion can be controlled by retarding the spark timing.
Engine Deceleration	An early rise in cylinder pressure may lead to a significant decrease in engine speed. In the worse case, a reverse revolution may occur.
Reverse Flow of Burned Gases into Fuel Injector	There is a possibility of a reverse flow of in-cylinder gases into an inwardly opening fuel injector due to the high combustion pressure. A high fuel pressure normally prevents this; however, the fuel rail pressure is expected to be quite low during the initial stages of engine cranking.
Combustion Noise	Excessive and objectionable noise could be generated.

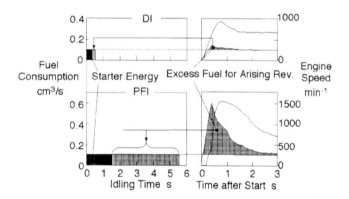

Figure 9.3-15 Comparison of the measured fuel consumption during restarting of PFI and G-DI engines [451].

economy improvement of 10% is achieved in the Japanese 10-15 mode test, in addition to the G-DI benefit itself [451]. The idle shut-off strategy can be invoked in vehicles with a manual transmission without the need for extra hardware. In comparison, a device to rapidly increase the transmission oil pressure may be necessary for G-DI vehicles with automatic or CVT transmissions.

9.3.5 G-DI Engine Combined with CVT
Both G-DI and CVT technologies are considered to have the potential to provide substantial incremental improvements in fuel economy and driveability, and combining

CVT with G-DI can improve some of the individual deficiencies. In mating a CVT to a conventional PFI engine such problems as friction losses in the drive belts, internal losses in the torque converter, vibration in the vehicle body, and non-ideal transmission matching at low-fuel-consumption operating points have traditionally been experienced. The integrated control of the G-DI engine and CVT transmission can circumvent a number of these problems by maximizing the range of superior torque control and low-fuel-consumption inherent in the G-DI engine [251].

Because CVTs use a belt to transmit power, it is necessary to apply force to the pulleys to prevent slippage. However, torque control in conventional PFI engines is difficult, and a high pulley pressure must be applied, resulting in a friction increase. Precise torque control of a G-DI engine enables the application of an appropriate force to the pulleys at all times to achieve better fuel economy. For example, reducing the hydraulic pressure to the pulley system from 1.2 MPa to 0.7 MPa provides a fuel consumption improvement of 5–7% over a wide range of vehicle speed [251]. With a conventional transmission, when high engine output is required, the engine operating point will move to the high torque region. This will force the G-DI engine to operate in homogeneous-charge mode, leading to a less-than-optimum fuel consumption. With a CVT, the engine can operate at a load point with a slight increase in engine torque and a significant increase in engine speed to meet the same demand. As a result, the operating limit for the stratified-charge mode can be extended, thus enhancing fuel economy. At engine speeds of up to 1500 rpm the fuel consumption characteristics of the G-DI engine are maximized, yielding a significant improvement in overall system fuel consumption.

In summary, the G-DI/CVT powertrain has the potential to provide a significantly enhanced matching of points on the engine operating map with vehicle load and speed requirements. It can also provide the advantage of high-precision torque controllability. This integrated control of the G-DI engine and CVT could provide a system that delivers levels of fuel economy and driveability that are significantly enhanced relative to what may be obtained with either G-DI or CVT individually.

9.3.6 G-DI Engine for Hybrid Propulsion System Application

Hybrid propulsion systems are designed to provide high-efficiency and low fuel consumption. However, the multiple drive paths, substantial electric motor/generator units, and high-capacity battery modules make this technology quite complex. Hybrid systems that use PFI engines have traditionally experienced insufficient starting torque and hesitations that are associated with engine restart and stabilization. In addition, when driving on a level road, the limited amount of recoverable kinetic energy means that the engine generator must operate more frequently to

provide the necessary power, leading to a fuel economy penalty. In contrast, utilizing a G-DI engine in a hybrid propulsion system makes it possible to employ a motor/generator unit and battery pack of smaller capacities, thereby providing the advantages summarized in Table 9.3-4. As a result, the smaller-than-usual motor and batteries can be utilized to provide improved vehicle initial acceleration. This is because the G-DI engine requires much less electric power to restart, and starts so rapidly that the engine torque generated on start-up is available in just a few engine cycles [17].

9.4 Summary

The theoretical potential for improving vehicle fuel economy by using a gasoline direct injection engine is well documented. However, the practical realization of such a fuel economy gain is significantly compromised by implementing emissions control strategies. As a result, only about half of the 20–25% theoretical fuel economy improvement has currently been achieved in production engines and vehicles. The key factors contributing to the shortfall in fuel economy benefit include emissions constraints, the use of induction-air throttling, and combustion and control system limitations that constrain engine operation in fuel-efficient modes. The individual components that comprise the overall combustion system must be evaluated carefully with regard to performance optimization. These optimized components must then be incorporated into the design and optimization of the overall G-DI engine system. Possible future improvements in these areas will translate

Table 9.3-4
Key advantages of G-DI engine for hybrid propulsion system application

Superior Restart Characteristics	Superior restart characteristics that require motor torque for only a very short time, after which the positive engine torque can assist in accelerating the vehicle.
Accurate Control of Small Torque	With a G-DI engine, injection of even the smallest quantity of gasoline can result in effective combustion. This makes it possible to more accurately control the torque and torque differential when engine is shut-off or restarted.
Excellent Low-Load Fuel Consumption	The superior low-load fuel consumption of the G-DI engine enables operation with a fuel economy enhancement even when the kinetic energy recovered during deceleration is insufficient.

to individual positive increments of fuel economy. The experimental evaluation of these measures is fairly difficult, as the benefit from each is quite small and, in fact, may be marginally close to the tolerance of the measurement. The integral impact of these items can, however, be quite significant.

Other operating constraints such as any required light throttling, a narrow lean-NOx aftertreatment system temperature window, and the purge requirement for NOx storage catalysts will also significantly degrade the actual fuel economy. Even given the same G-DI engine, the vehicle fuel economy that is finally achieved in the field will be highly dependent upon engine/vehicle matching, drive cycle, and the operating-mode transitions. In order to meet future emissions standards, a coordinated development of the combustion system, engine operating strategies and exhaust aftertreatment systems will be required. Another area of promising future benefit is to combine gasoline direct injection with other technologies such as boosting, idle engine shut-off, CVT, and HEV. Such combinations have the potential to not only enhance G-DI performance, but can also alleviate some basic deficiencies of these technologies. This could permit the realization of a greater fraction of the potential of G-DI technology, such as rapid transient response, superior engine start, precise torque control, in-cylinder charge cooling, flexibility of fueling control, enhanced resistance to knock, and reduced turbocharger lag time. Based upon the consideration of all of these potential enhancement areas, a significant further improvement in fuel economy is to be expected with G-DI technology.

Chapter 10

Production and Prototype Gasoline Direct-Injection Systems

10.1 Early DISC Engine Development

Operating with a single, fixed ignition location imposes very stringent requirements on the mixture preparation process in G-DI engines, as it is very difficult to provide a combustible mixture at a fixed spark gap over the entire engine operating map. This is why most combustion control strategies are primarily directed toward the specifics of preparing and positioning the air-fuel mixture. The generation of a stratified-mixture region of repeatable location, geometry and charge distribution is quite important for achieving the fuel-economy potential of a G-DI combustion system. Conceptually, stable stratification can be directly achieved using a divided chamber in which a subvolume and the main combustion chamber provide well-separated and distinct mixture regions. One of the better-known production applications of this basic technique is the combustion chamber geometry of the Honda compound vortex combustion chamber (CVCC) engine [76]. Although technically not a direct-injection engine, it does represent an early example of a production combustion system that operated with mixture stratification.

The CVCC combustion strategy, illustrated in Fig. 10.1-1, utilizes a secondary mixture preparation system that supplies a fuel-rich mixture through an auxiliary intake valve into a prechamber containing the spark plug. A lean mixture is also supplied to the main combustion chamber through the main fueling system. After combustion is initiated in the prechamber, the rich, burning mixture issues as a jet through an orifice into the main chamber, both entraining and igniting the lean charge in the main chamber. This engine thus invokes the flame-jet ignition technique, which is known to extend the operating limit of conventional SI engines to air/fuel mixtures that are too lean to ignite with a conventional spark discharge. One main disadvantage of flame-jet ignition is that increased wall heat losses and throttling losses between the prechamber and the main combustion chamber reduce the engine combustion efficiency. Additionally, such systems can only be operated at overall lean air/fuel ratios, thus limiting the maximum power output that can be obtained with the specified displacement.

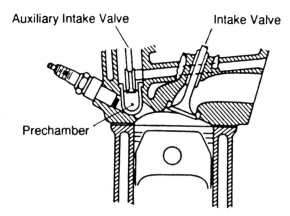

Figure 10.1-1 Schematic of the Honda CVCC combustion system [76].

The concept of injecting gasoline directly into the combustion chamber of an SI engine is certainly not new. It was utilized extensively in radial aircraft engines in the 1940s, and automotive applications were developed nearly five decades ago. In fact, early in 1954 the Benz 300SL engine [391], illustrated in Fig. 10.1-2, utilized a DI system to alleviate some of the performance deficiencies that are directly associated with the use of a carburetor. As one of the first G-DI engines [197], this design used direct injection early in the induction stroke to achieve a homogeneous air-fuel mixture.

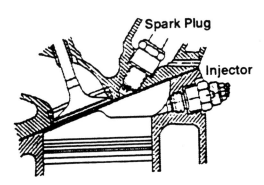

Figure 10.1-2 Schematic of the Benz 300SL G-DI combustion system [391].

During the time frame from 1960 to 1978, numerous DISC systems were proposed to explore the potential of charge stratification. A number of these designs utilized jet-wall interaction and film evaporation to achieve charge stratification by controlling the fuel distribution in an open chamber. A classic example of this type of stratified-charge combustion is the MAN-FM system [307, 453], which is illustrated schematically in Fig. 10.1-3. Extensive testing led to the conclusion that the MAN-FM system exhibits the disadvantages of elevated wall heat losses, as well as substantial HC and soot emissions. Also, as is the case for the CVCC divided chamber, only stratified-charge operation is possible. Even if the elevated compression ratio of the MAN-FM engine is taken into consideration, the engine-specific power is still limited when compared to that for homogeneous, stoichiometric operation of a conventional PFI engine.

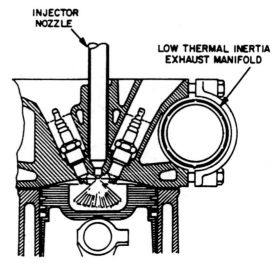

Figure 10.1-4 Schematic of the Ford PROCO combustion system [401].

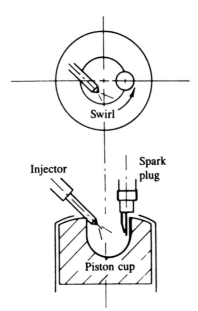

Figure 10.1-3 Schematic of the MAN-FM combustion system [307].

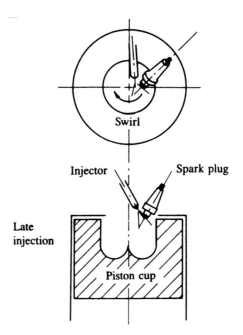

Figure 10.1-5 Schematic of the TCCS combustion system [8].

Other G-DI engines in which there is close proximity between the fuel injector and the spark gap were developed in the period from 1970 to 1979. The well-known systems that place the ignition source directly in the periphery of the fuel spray are the Ford PROCO [401] and the Texaco TCCS [8] systems. The combustion concepts of these two systems are illustrated in Figs. 10.1-4 and 10.1-5. The centrally located injector provides a hollow-cone fuel spray that is either stabilized by an aggressive air swirl rate (PROCO) or provided with a narrow fuel jet injected tangentially into a piston bowl having a substantial air swirl motion (TCCS). In both cases, ignition stability is achieved primarily through the spatial and temporal juxtaposition of injection and ignition. The PROCO system with a swirl-stabilized central mixture plume allows slightly longer ignition delays than does the TCCS system; however, unfavorable mixture ignition conditions occur when crossflow velocities are high. The TCCS system may be regarded as a classic representative of a fully stratified combustion system, whereas the PROCO combustion system was purposely designed with the objective of matching the specific output of the premixed-charge engine using high EGR rates. The PROCO stratified-charge engines employ an injection system with an operating fuel pressure of 2 MPa, with injection occurring early in the compression stroke, while the TCCS engines utilizes a high-pressure, diesel-type injection system.

Although the minimum BSFC values of both of these engines were quite good, the control of HC emissions was extremely difficult for light-load operation. NO_x emissions in both were controlled using EGR rates of up to 50%. The power output of the late-injection TCCS system was also soot-limited. Later studies of stratified-charge engines have shown that high air swirl is desirable for obtaining the required level of fuel-air mixing, and that a high-energy ignition source enhances the multi-fuel capability. In spite of significant measures to reduce autoignition and maximize air utilization, these classic early engines were not able to achieve the specific power output of then-current PFI engines.

Additional G-DI engine concepts from the 1970s that are based upon different geometric combinations and permutations of fuel spray plume and spark gap include the Mitsubishi Combustion Process (MCP) [313], the International Harvester and White Motors system (IH-White) [44], and the Curtiss-Wright Stratified Charge Rotary Combustion system (SCRC) [213]. These systems are shown in Figs. 10.1-6, 10.1-7 and 10.1-8, respectively.

Also proposed was a direct-injection combustion system that can operate in either the gasoline-combustion or diesel-combustion modes by injecting fuel onto a central pedestal in a piston cavity [230, 231]. In this OSKA combustion system, illustrated in Fig. 10.1-9 [230], the fuel is injected by a single-hole nozzle against the flat surface of the pedestal, and achieves fuel atomization by a combination of pneumatic, impingement and sheet-breakup mechanisms. The fuel deflects orthogonally and symmetrically in a disk shape, and forms the air-fuel mixture. As a result of having a comparatively rich mixture present in the vicinity of the pedestal, and of initiating ignition at the center of this mixture, relatively stable combustion is achieved. Because the air-fuel mixture is always formed near the pedestal, there is little fuel present in the squish area. Therefore, it is possible to minimize end-gas detonation, with compression ratios of up to 14.5:1 on gasoline having been successfully utilized. With this

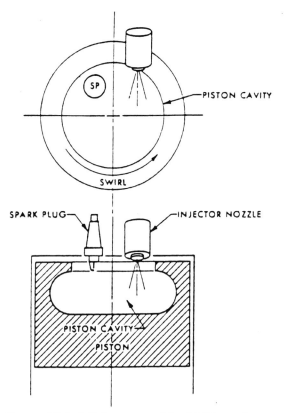

Figure 10.1-6 Schematic of the MCP combustion system [313].

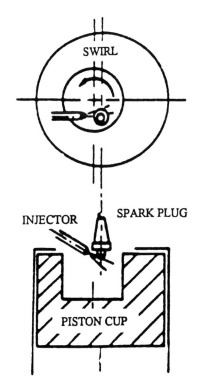

Figure 10.1-7 Schematic of the IH-White combustion system [44].

287

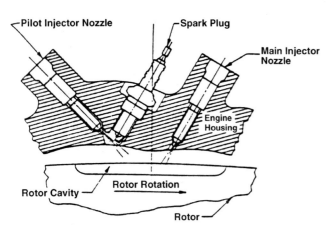

Figure 10.1-8 Schematic of the SCRC combustion system [213].

combustion system, however, the engine-out HC emissions are very high, perhaps due to the relatively poor atomization that results from impingement. This represents a significant development hurdle to be surmounted [229].

Since 1990, a number of innovative mixture-preparation and combustion-control strategies for DI gasoline engines have been proposed and developed by automotive companies, fuel system manufacturers and research institutions, with some of these already having been placed into production in both the Japanese and European markets. The key features of these specific G-DI engines, as well as the related combustion control

strategies developed during this recent time period, were summarized previously in Table 1.3-2. The configurations and operating characteristics of selected contemporary production and prototype systems are discussed in detail in this chapter.

10.2 Mitsubishi Reverse-Tumble-Based, Wall-Guided GDI Combustion System

The Mitsubishi GDI combustion system, designated as GDI, is a pioneering production configuration, and is discussed in detail in a large number of publications and reports, including References 13–16, 33, 34, 37, 70, 71, 79, 122, 196, 197, 221, 259, 263–267, 288, 305, 394–396, 450, 468, 490–492, 495, 497, 499 and 502. A schematic of the combustion system, a combustion chamber cutaway, a photo of the piston, an engine cutaway, and a system layout are shown in Figs. 10.2-1, 2.3-13, 10.2-2, 10.2-3, and 10.2-4. The engine control map, illustrated in Fig. 10.2-5, shows that an early-injection strategy is utilized for engine operation at high loads. In most of this operating regime, the engine uses a stoichiometric mixture, although it operates at a slightly rich condition at full load. For the lowest-load conditions in this operating regime, the engine operates with a homogeneous, lean air-fuel mixture in the range of 20:1 to 25:1, which yields an improvement in the BSFC.

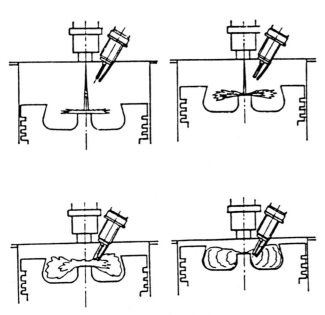

Figure 10.1-9 Schematic of the OSKA combustion system [231].

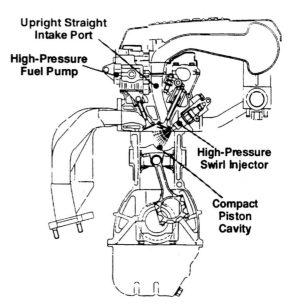

Figure 10.2-1 Schematic of the Mitsubishi wall-guided GDI combustion system employing reverse tumble [259].

288

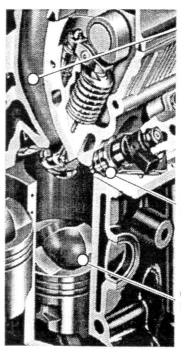

Upright
Straight Port

High-Pressure
Swirl Injector

Compact
Piston Cavity

Figure 10.2-2 Cutaway view of the Mitsubishi GDI combustion system [37, 342].

The first Mitsubishi GDI engine, launched in Japan in 1996, was based on the production 4G93 PFI engine, and incorporated dual-overhead camshafts and four valves per cylinder in an in-line, 4-cylinder configuration. The main differences between the initial G-DI engine and the 4G93 PFI engine are the cylinder head and piston design, and the high-pressure fuel pump and injectors. In the Mitsubishi GDI dual-catalyst system, the first catalyst uses pure iridium, which is effective during lean operation. A normal platinum catalyst is mounted downstream to catalyze the gas stream during periods of non-lean operation.

The fuel economy provided by the Mitsubishi GDI engine exhibits significant improvement when compared to a similar-displacement, conventional PFI engine on the Japanese urban test cycle. Due to the inherent in-cylinder charge cooling and improved intake port design, the volumetric efficiency of the Mitsubishi GDI engine is increased by 5% over the entire engine operating range. The 0-to-100-km/hr vehicle acceleration time is also enhanced by 5%. In combination with an increased compression ratio of 12:1, the total power output is increased by 10% [197], which is verified by the full-load performance data in Fig. 10.2-6.

The principal features of the Mitsubishi 1.8L 4-cylinder GDI engine are summarized in Table 10.2-1. The combustion systems for the Mitsubishi 1.1L I-4, 1.5L I-4, and 3.5L V-6 engines are the same as that of the 1.8L I-4 GDI engine except for the basic specifications of the engines, which are given in Table 10.2-2. The Mitsubishi 1.8L in-line, 4-cylinder, GDI engine was first introduced into the European market in 1997. The major differences between the European version and the Japanese version are listed in Table 10.2-3. The largest GDI engine so far developed by Mitsubishi is a 4.5L, 90-degree, V-8 engine using 32 valves and four camshafts. The piston bowl is slightly shallower than that utilized in the original I-4 and V-6 GDI engines. Two high-pressure, single-plunger, fuel pumps are utilized, one for each cylinder bank.

10.3 Toyota Combustion Systems

10.3.1 Toyota First-Generation, Swirl-Based, Wall-Guided, D-4 Combustion System

The Toyota first-generation direct-injection combustion system, the D-4 system, is described in numerous publications and reports, including References 35, 37, 156,

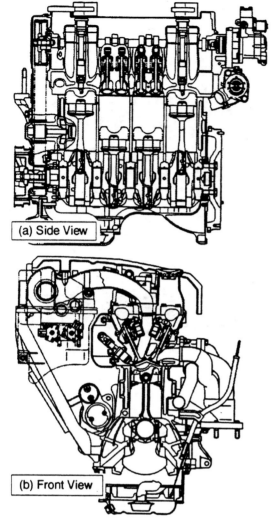

Figure 10.2-3 Cutaway view of the Mitsubishi 1.8L,
I-4, GDI engine [196].

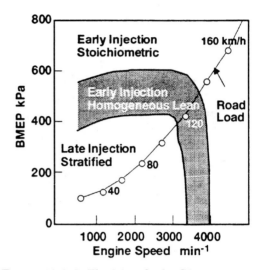

Figure 10.2-5 The Mitsubishi GDI engine control
map [196].

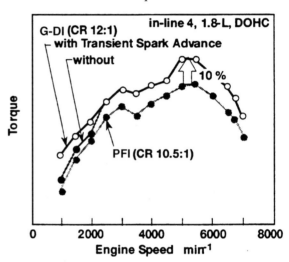

Figure 10.2-6 Full-load performance of the
Mitsubishi 1.8L, I-4, GDI engine [196].

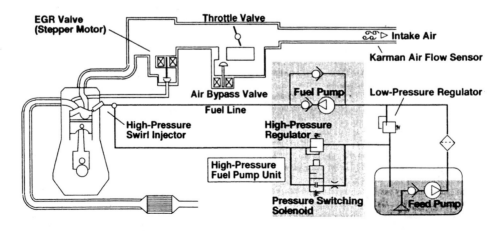

Figure 10.2-4 Schematic of the Mitsubishi GDI engine system [196].

Table 10.2-1
Specifications of the Mitsubishi 1.8L, I-4 GDI engine

Cylinder Configuration	I-4
Displacement (cm^3)	1864
Bore × Stroke (mm)	81 × 89
Compression Ratio	12:1
Chamber Geometry	4-valve, DOHC, pentroof
Stratification Approach	Wall-guided, reverse-tumble-based
Spark Plug Location	Chamber center
Piston Shape	Compact, spherical-segment piston cavity
Intake Port	Upright straight port to generate a reverse tumble
Intake Tumble Ratio	1.8
Injector Location	Under the intake port and between the two intake valves
Injector Type	High-pressure, swirl injector with a tangential slot-type swirler
Fuel Pressure (MPa)	5
Spray Characteristics	Hollow-cone spray
Ignition System	60 mJ in energy and narrow-gap platinum plug
NOx Control	Electric EGR valve actuated by a stepping motor at a maximum rate of 30%; lean-NOx catalyst
Part-Load Operation	Late injection for the stratified-charge mode at an overall air/fuel ratio of up to 40:1
Full-Load Operation	Early injection for the homogeneous-charge mode
Load Transition	Air control through throttle valve with electronic air bypass valve for smooth transition between different operating modes
Fuel Requirement	92 RON
Idle Speed	600 rpm
Power	112 kW JIS net at 6500 rpm
Torque	128 N•m at 5000 rpm

Table 10.2-2
Specifications of three Mitsubishi GDI engines: 1.1L, I-4; 1.5L, I-4; and 3.5L, V-6

Displacement (cm^3)	1094	1468	3496
Cylinder Configuration	I-4	I-4	V-6
Bore × Stroke (mm)	66 × 80	75.5 × 82	93 × 85.8
Compression Ratio	Unavailable	11:1	10.4:1
Spark Plug	Unavailable	Unavailable	A large protrusion into the chamber
Intake System	Unavailable	Unavailable	A variable length induction system for improving low and middle-speed torque
Fuel Requirement	Regular	Regular	Regular (premium recommended)
Power	54 kW at 6000 rpm	Unavailable	180.2 kW at 5500 rpm
Torque	100 N•m at 4000 rpm	Unavailable	343.1 N•m at 2500 rpm
Comment	Combined with idle shut-off system	N/A	N/A

Table 10.2-3
Major differences between the Japanese and European versions of the
Mitsubishi 1.8L, I-4, GDI engine

Compression Ratio	The compression ratio of the European version was increased from 12:1 to 12.5:1.
Spray Targeting	The injector mounting for the European version is more vertical in order to target the spray more toward the center of the piston bowl. This reduces the variation in spray targeting with injection timing, yielding a stable, combustible mixture over a wider operating window.
Calibration Difference	The Japanese GDI engine was tuned to give the best fuel savings at idle speeds, while the European version is tuned to optimize the highway fuel economy.
Mode Shift Point	The crossover point of the European version from lean (40:1) to normal (14.5:1) air/fuel ratio was increased from 2000 rpm to 3000 rpm. The lean range vehicle speed is extended to highway cruise at 120–130 km/hr.
Intake System	The intake manifold was extended from 265 mm to 400 mm to enhance the low-speed torque characteristics for the European version.
Strategy to Improve Low-End Torque	A two-stage injection strategy, denoted as two-stage mixing, was added to the European version to improve low-speed torque and suppress knock. In this strategy part of the fuel is injected during the intake stroke and the remaining portion during compression.
Catalyst Light-Off Strategy	Another two-stage injection strategy, denoted as two-stage combustion, was incorporated into the European version for a more rapid catalyst light-off during cold start. This is done by injecting a fraction of the fuel during the intake stroke for normal combustion, and the remaining portion during the expansion stroke to produce heat for rapidly increasing the catalyst temperature.

212, 225, 232, 233, 302–304, 341, 354, 387, 429, 430, 446, 448, 449, 462, 463, 489, 492, 493, and 496. The principal components and stratification concepts used in the D-4 engine are illustrated in Figs. 10.3-1, 10.3-2 and 2.3-14. As shown in Fig. 2.3-14, this system employs a uniquely shaped cavity in the piston. As illustrated in Fig. 10.3-1, zone (a) of the cavity is the mixture formation area, and is positioned upstream of the spark plug. The wider zone (b) is designed to be the combustion space, and is effective in promoting rapid mixing. The increased width in the swirl flow direction enhances flame propagation after the stratified mixture is ignited. The involute shape (c) is designed to direct vaporized fuel toward the spark plug. The intake system consists of both a helical port and a straight port, which are fully independent. An electronically activated SCV of the butterfly type is located upstream of the straight port, providing a maximum in-cylinder swirl ratio of 2.1. The helical intake

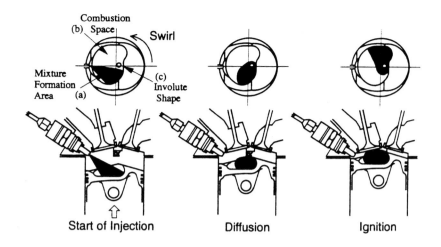

Figure 10.3-1 Schematic of the Toyota first-generation, swirl-based, wall-guided D-4 engine [156].

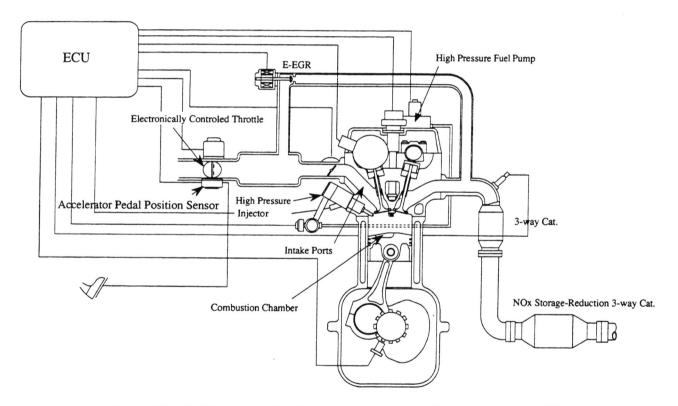

Figure 10.3-2 Schematic of the Toyota first-generation D-4 engine system [156].

port utilizes a variable-valve-timing-intelligent (VVT-i) cam-phasing system on the intake camshaft. These valves are driven by a DC motor so that the desired valve-opening angle can be controlled according to the engine operating conditions.

Figure 10.3-3 shows the SCV operating map. For light-load operation the SCV is closed, which forces the induction air to enter through the helical port, thus creating a swirling flow. Highly atomized fuel is injected into the swirling flow at a time during the latter stages of the compression stroke. The in-cylinder flow field moves the rich mixture to the center of the chamber around the spark plug, while part of the fuel disperses into the air in the combustion chamber, forming a stratified air-fuel

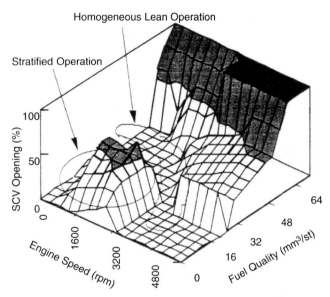

Figure 10.3-3 SCV operating map for the Toyota first-generation D-4 engine [156].

resulting in a homogeneous mixture. The main features of the D-4 system are summarized in Table 10.3-1.

As illustrated previously in Fig. 6.3-1, a special two-stage injection process is used for the transition between light-load and heavy-load operation. This process creates a weakly stratified mixture in order to achieve a smooth transition of torque from ultra-lean operation to either lean or stoichiometric operation. The Toyota first-generation, D-4 engine operates in four operating regimes having distinctly different mixture strategies and/or distributions. The first regime is stratified-charge operation using an air/fuel ratio in the range of 25:1 to 50:1. The vehicle operates in this ultra-lean zone under steady light-load, for road loads of up to 100 km/hr. The second operating regime is a transitional, semi-stratified zone with the air/fuel ratio ranging from 20:1 to 30:1. Within this zone, fuel is injected twice within each engine cycle; once during the intake stroke and once during the compression stroke. The third operating regime utilizes injection during the intake stroke to operate the engine in the homogeneous combustion mode using a lean air/fuel ratio in the range

mixture. For heavy-load operation the SCV is opened, and the intake air is inducted into the cylinder with a lower pressure loss. The fuel is injected during the intake stroke,

Table 10.3-1
Specifications of the Toyota first-generation, 2.0L, D-4 engine

Cylinder Configuration	I-4
Displacement (cm^3)	1998
Bore × Stroke (mm)	86 × 86
Compression Ratio	10:1
Chamber Geometry	4-valve, DOHC, pentroof
Stratification Approach	Wall-guided, swirl-based
Spark Plug Location	Slightly displaced from the chamber center
Piston Shape	Deep bowl in asymmetrical crown; involute-shaped, concave
Intake Port	One helical port with VVT-i system and SCV; one straight port
Injector Location	Under the straight intake port
Injector Type	High-pressure, swirl injector with two-stage injection capability
Fuel Pressure (MPa)	Variable in the range of 8 to 13
Spray Characteristics	Solid-cone spray
NOx Control	Electronically controlled EGR; NOx storage catalyst, and standard three-way catalyst; regenerated once every 50 seconds during lean operation
Operating Modes	Stratified, semi-stratified, lean, and stoichiometric
Part-Load Operation	Late injection for the stratified-charge mode at an overall air/fuel ratio of up to 50:1
Full-Load Operation	Early injection for the homogeneous-charge mode
Load Transition	Two-stage injection and an electronic throttle control system between stratified-charge and homogeneous-charge operation
Fuel Requirement	91 RON
Power	107 kW at 6000 rpm
Torque	196 N•m at 4400 rpm

of 15:1 to 23:1. The fourth regime is denoted as the stoichiometric power zone, and is similar to the third, except that the air/fuel ratio range is from 12:1 to 15:1. The common-rail fuel pressure is variable by design, and is adjusted by the control system over the range 8 to 13 MPa in order to optimize the injection rate and expand the dynamic working range of the injector. An additional fuel injector is employed in an auxiliary throttle body in order to enhance cold startability.

The Toyota first-generation D-4 engine also incorporates an ETC system that diminishes the harshness of torque variations during mode transitions, thus improving driveability. VVT-i system is used to maximize torque in the low to mid-speed ranges, and to maximize power at high speed. This system varies the intake valve opening and closing within a maximum authority of 20° of crank angle. At low engine loads the intake valves are opened earlier, thus increasing the overlap between the intake and exhaust valves. This results in internal EGR, which increases the total effective EGR ratio while requiring less EGR from the external system. As a result, the VVT-i system also contributes to a reduction in NOx emissions. With the VVT-i system and an increase in the compression ratio to 10:1, approximately 10% more torque is obtained from the D-4 engine in the low to mid-rpm ranges than is obtainable with a conventional PFI engine of comparable displacement. At idle, the manifold vacuum at the throttle valve is halved from 67 kPa for a conventional PFI engine to 30 kPa. At a steady vehicle speed of 40 km /hr, the manifold vacuum is only 11 kPa, which is on the same order as the WOT manifold vacuum for a PFI engine.

The first-generation D-4 engine, officially designated the 3S-FSE, shares the basic dimensions of the conventional 3S-FE. It incorporates a dual-overhead-camshaft, four valves per cylinder, in-line, 4-cylinder engine that was first installed in the Japanese Corona Premio compact sedan. Two small catalytic converters, each with 0.5L volume, are positioned immediately downstream from the exhaust manifold. These catalysts have the function of maintaining the exhaust temperature at levels adequate for efficient operation of the 1.3L underfloor catalytic converter. An NOx storage catalyst is also used with the D-4 engine. This catalyst contains rhodium for storage and release, and platinum for cleaning, both on an alumina bed, and efficiently stores NOx during periods of lean-burn operation. The stored NOx is converted at time intervals of 50 seconds during lean operation by utilizing very brief periods of stoichiometric operation. These required cleaning periods are less than one second, and degrade the overall fuel economy by an estimated 2%. This catalytic system reduces the emission levels of oxides of nitrogen by as much as 95% on the Japanese test cycle.

10.3.2 Toyota Second-Generation, Wall-Guided D-4 Combustion System

Toyota marketed a second-generation D-4 combustion system in 1999 [193, 222, 223, 249, 250, 335, 336, 406, 435, 443, 452, 500]. As previously shown in Fig. 2.3-10, this DI gasoline engine employs a well-dispersed, fan-shaped fuel spray of moderate to high penetration that is injected from a slit-type nozzle. The fan-shaped spray is directed into a shell-shaped piston cavity during the second half of the compression stroke to obtain stratified-charge combustion. The fuel spray is targeted at the piston bowl floor surface, which guides the fuel to the spark plug location. Part of the spray impinges, forming a rectangular wetted footprint, with the remainder of the spray plume being redirected toward the spark plug electrode. The velocity vectors of the charge motion induced by the fuel injection event are directed toward the spark gap along the surface of the shell-shaped piston cavity, and develop into a rotating flow in the vertical plane. This results in a spherical mixture cloud in the region surrounding the spark gap [223, 250]. The degree of mixture stratification is controlled by the matching of spray characteristics, piston cavity design and fuel injection timing.

The results of a CFD analysis of the fuel-air mixing process occurring within the combustion chamber near the end of the compression stroke are illustrated in Fig. 10.3-4 [250]. The contours of computed mixture strength and velocity are clearly illustrated, and it is evident that the spray plume first impinges on the bottom of the piston cavity and then progresses toward the spark gap along the piston bowl floor and wall. Although the fan spray entrains less ambient air during the free-space portion of the flight than a spray from a swirl nozzle, the fan spray provides enhanced air entrainment during spray/bowl interactions. During this stage of the spray-wall interaction, the shear flow generates a large eddy at the outer periphery of the spray, promoting a significant degree of air entrainment. As the eddy moves along with the spray, fine fuel droplets and vaporized fuel are entrained to a greater degree than larger droplets, and the spray plume thickness increases gradually. As the injection process proceeds, the spray plume reaches the far-end wall of the piston

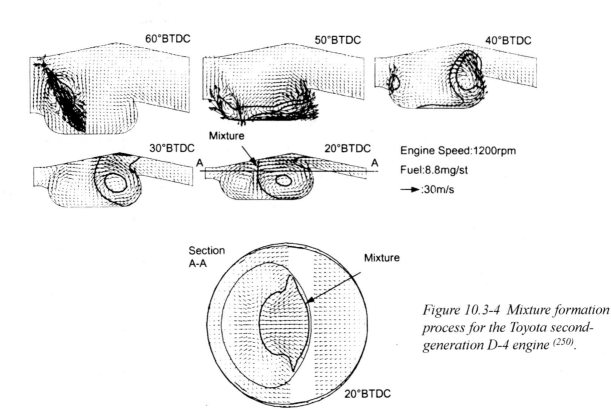

Figure 10.3-4 shown with labels: 60°BTDC, 50°BTDC, 40°BTDC, Mixture, 30°BTDC, 20°BTDC, Engine Speed:1200rpm, Fuel:8.8mg/st, →:30m/s, Section A-A, Mixture, 20°BTDC

Figure 10.3-4 Mixture formation process for the Toyota second-generation D-4 engine [250].

cavity. As the piston approaches TDC on the compression stroke, squish flow from the exhaust valve side of the cylinder begins to influence the fuel distribution.

The exit lip of the piston bowl is designed to direct the spray plume toward the cavity center, thus creating an ellipsoidal mixture plume around the spark gap at the time of the spark discharge for a wide range of engine loads and speeds. Studies indicate that, for a large bowl volume, the flame propagation rate is lower than optimum along the major axis of the cavity during the latter stages of heat release (see Fig. 10.3-5). In order to enhance the heat release rate, particularly during the latter stages of the combustion process, an improved piston bowl cavity having a reduced volume along the major axis of the cavity was developed. As a result, the combustion process was significantly improved, particularly with regard to the COV of IMEP.

A schematic diagram of the enhanced Toyota DI system, also called the New Combustion Process (NCP), is shown in Fig. 10.3-6 [223]. The main features of the new 3.0L D-4 design are summarized in Table 10.3-2. This latest D-4 engine operates in three modes: (1) homogeneous-charge combustion with an air/fuel ratio in the range of 12:1 to 15:1 during cold-temperature starting, high-load, or NOx-reduction operations;

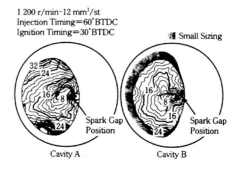

1 200 r/min-12 mm³/st
Injection Timing=60° BTDC
Ignition Timing=30° BTDC
Small Sizing
Spark Gap Position — Cavity A
Spark Gap Position — Cavity B

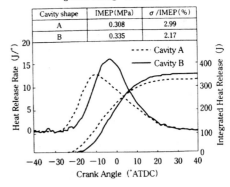

1 200 r/min-12 mm³/st
Injection Timing=60° BTDC
Ignition Timing=30° BTDC

Cavity shape	IMEP(MPa)	σ /IMEP(%)
A	0.308	2.99
B	0.335	2.17

- - - - Cavity A
——— Cavity B

Figure 10.3-5 Effect of two piston bowl geometries on the combustion characteristics of the Toyota second-generation D-4 engine [223].

Table 10.3-2
Specifications of the Toyota second-generation, 3.0L, I-6, D-4 engine

Cylinder Configuration	I-6
Displacement (cm³)	2997
Bore × Stroke (mm)	86 × 86
Compression Ratio	11.3:1
Chamber Geometry	4-valve, DOHC, pentroof
Stratification Approach	Wall-guided, near-quiescent
Spark Plug Location	Chamber center
Piston Shape	Shell-shaped piston cavity; this differs significantly from the involute-shaped cavity of the first-generation D-4 system
Intake Port	A straight intake port with VVT-i; the helical ports and SCV flow control devices of the first-generation D-4 system are not utilized
Air Flow	A weak tumble motion is employed; the tumble ratio is 0.2
Injector Location	Beneath the intake port and between the two intake valves
Injector Type	High-pressure, slit-type injector
Fuel Pressure (MPa)	Variable up to 13
Spray Characteristics	Fan-shaped spray of 80° effective fan angle and 20° offset
NOx Control	Electronically controlled EGR; NOx storage catalyst and standard three-way catalyst
Operating Modes	Stratified, semi-stratified, lean, and stoichiometric
Part-Load Operation	Late injection for stratified-charge mode
Full-Load Operation	Early injection for homogeneous-charge mode
Fuel Requirement	100 RON

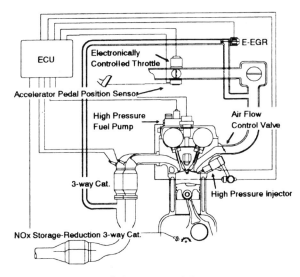

Figure 10.3-6 Schematic of the Toyota second-generation D-4 engine system [223].

(2) weakly-stratified-charge combustion with an air/fuel ratio between 15:1 and 30:1 during medium-load operation; and (3) stratified-charge, lean-to-ultra-lean operation, employing an air/fuel ratio ranging from 17:1 to 50:1 for low-load operation. As compared to the first-generation D-4 combustion system, one of the

major changes is the elimination of the helical port, which enables the latest D-4 combustion system to achieve both excellent homogeneous-charge combustion as well as a wide range of stratified-charge combustion without relying on a variable-flow-control system. The combustion system does not require any particular charge motion such as tumble or swirl, which enables a simplified intake port geometry to be utilized while still enhancing full-load performance. A comparison of torque characteristics for the WOT condition is shown in Fig. 10.3-7. From this figure, the torque improvement of the latest system is evident [223].

Figure 10.3-8 shows the influence of tumble ratio on the Toyota second-generation D-4 combustion chamber, as measured by an impulse meter in continuous flow [223]. The best results for fuel consumption, HC emissions and torque fluctuation are obtained at the lower tumble ratios. Figure 10.3-9 shows a direct comparison of operating regimes for stratified-charge combustion for these two D-4 systems [223]. The shaded area shows the points at which both NOx emissions and torque fluctuation are maintained below an acceptable threshold value. Compared with the first-generation D-4 engine, an extended range of stable stratified-charge combustion is achieved with the latest combustion system at

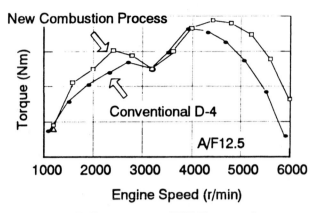

Figure 10.3-7 Comparison of WOT torque characteristics between the Toyota first-generation and second-generation D-4 engines [223].

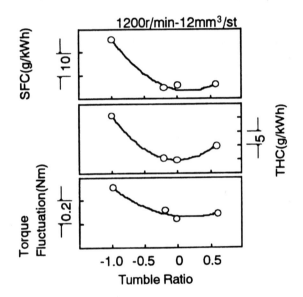

Figure 10.3-8 Effect of tumble ratio on fuel consumption, HC emissions and torque fluctuation for the Toyota second-generation D-4 engine [223].

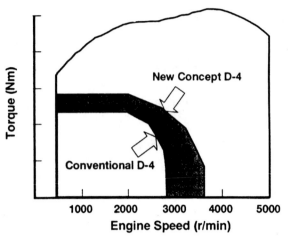

Figure 10.3-9 Comparison of operating regimes of stratified-charge combustion for the Toyota first-generation and second-generation D-4 engines [443].

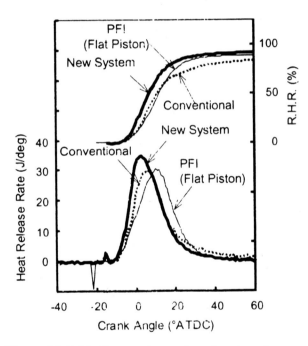

Figure 10.3-10 Comparison of combustion characteristics for the Toyota first-generation D-4 (conventional) and second-generation D-4 (new system), and a baseline PFI engine [250].

both higher load and higher engine speed. The improved fuel distribution characteristics achieved by the fan-shaped, 20°-offset spray enable the avoidance of an overly-rich mixture, even with an increased fuel quantity being delivered at higher loads. In addition, the comparatively higher spray velocity and penetration enables an improved dispersion of the fuel plume, particularly for an increased in-cylinder pressure and higher engine speeds. For purposes of comparison, Fig. 10.3-10 shows the combustion characteristics of the Toyota first-generation D-4 (conventional) and second-generation D-4 (new system), and the baseline PFI engine [250]. The

second-generation D-4 combustion system exhibits the highest burn rate among these three engines.

It is worth noting that a flow-control valve of the butterfly type continues to be utilized in one of the intake ports in the latest D-4 engine, not for assisting in forming charge stratification, but for improving combustion during cold start. This flow-control valve is operated by activating an engine-vacuum-actuated diaphragm, whereas the earlier D-4 engine used an electronically-controlled, electrically-actuated SCV [500]. The new flow-control valve closes during cold, homogeneous-charge combustion operation, which is the operating mode in which fuel is injected during the induction stroke. Turbulence generated by induction air entering from the single intake port improves combustion during this operating mode. The fan-spray fuel injector uses one O-ring and three backup rings to ensure sealing integrity, and is secured by a simple riser clamp. In contrast, the first-generation D-4 fuel injector was encased in a brass receptacle, which was in turn threaded into the cylinder head. Another new engine, the 1AZ-FSE 2.0L, is also retrofitted with the new D-4 combustion system described above. The specifications for this new engine are summarized in Table 10.3-3.

Table 10.3-3
Specifications of the Toyota second-generation 2.0L, I-4, D-4 engine

Cylinder Configuration	I-4
Displacement (cm^3)	1998
Bore × Stroke (mm)	86 × 86.0
Compression Ratio	9.8:1
Chamber Geometry	4-valve, DOHC, pentroof
Intake System	Fitted with VVT-i device, which continuously alters intake valve timing up to 43°
Fuel Requirement	Regular
Power	113 kW at 6000 rpm
Torque	200 N•m at 4000 rpm

The DI gasoline engine that has been newly released into the European market by Toyota uses a stoichiometric-charge combustion concept, with the compression ratio increased to 11.0 from the compression ratio of 9.8 in the Japanese version. It also uses a substantially increased amount of internal EGR, which is supplied by a VVT-i system. Compared to the previous PFI baseline system, this DI engine yields a higher output torque over the entire speed range and improved fuel economy [412]. Similar to the second-generation D-4 engine for the Japanese market, a straight intake port and a fan-spray injector nozzle have been employed in this combustion system. One key difference between the European and the Japanese-market versions is that the European-version, stoichiometric-charge, DI combustion chamber employs a shallow, nearly-flat bowl in the piston, enabling more effective homogeneous-charge mixture formation, whereas the Japanese version employs a shell-shaped piston cavity. Premium fuel with 95 RON is recommended for the European D-4 engine, whereas a regular fuel of 90 RON is recommended for the Japanese-market engine.

10.4 Nissan Swirl-Based, Wall-Guided NEODi Combustion System

The Nissan DI gasoline engine, NEODi (Nissan Ecology Oriented performance and Direct Injection), is designed with a centrally mounted spark plug, and with the injector located underneath the intake port and between the two intake valves [31, 36, 37, 52, 182, 183, 190, 192, 219, 339, 351, 438, 442, 445, 498]. As previously illustrated in Figs. 2.3-11 and 2.3-15, the Nissan combustion system uses a SCV to generate a swirling air flow that aids in forming and maintaining charge stratification. A relatively shallow piston bowl, as compared to that in other production DI gasoline engines, is utilized in this combustion system. The engine can operate in both stratified-charge and homogeneous-charge modes. The engine also operates with stable combustion using an air/fuel ratio leaner than 40:1, resulting in a significant improvement in fuel economy when compared with a baseline PFI engine that operates with a stoichiometric mixture. The principal features of the Nissan prototype DI combustion system are summarized in Table 10.4-1.

The specifications of the latest 3.0L, V-6 production NEODi engine, installed in the Nissan Leopard for the Japanese market, are presented in Table 10.4-2. The major features of the 1.8L, I-4, production DI engine are also included in Table 10.4-2. The piston crown of the production DI engine incorporates a shallow-dish piston bowl that is designed to have a minimal effect on the in-cylinder air flow field. This bowl design is intended to provide efficient combustion during homogeneous-charge operation. This design is contrasted with those employing piston crowns which incorporate deep bowls, and which are well known to degrade high-load operation.

Although it has a minimal effect on the in-cylinder air flow field, the use of the shallow-dish bowl does result in some difficulty in controlling the air/fuel mixture for ultra-lean combustion. An injector using a shaped-spray (casting net) nozzle was developed to provide a

Table 10.4-1
Specifications of the Nissan prototype DI engine

Cylinder Configuration	I-4
Displacement (cm³)	1838
Bore × Stroke (mm)	82.5 × 86.0
Compression Ratio	10.5:1
Chamber Geometry	4-valve, DOHC, pentroof
Stratification Approach	Wall-guided, swirl-based
Spark Plug Location	Chamber center
Piston Shape	Shallow, curved piston crown
Intake Port	Conventional intake port design; reverse tumble obtained by masking the upper side of intake valves
Injector Location	Between the intake valves and beneath the intake port; at an angle of 36° up from the horizontal plane
Injector Type	High-pressure swirl injector
Fuel Pressure (MPa)	10
Spray Characteristics	Hollow-cone fuel spray with a nominal cone angle of 70° and an SMD of 20 μm
Part-Load Operation	Late injection for stratified-charge mode at an overall air/fuel ratio of up to 40:1
Full-Load Operation	Early injection for homogeneous-charge mode

Table 10.4-2
Specifications of Nissan production NEODi engines

Engine Description	Production I-4	Production V-6
Cylinder Configuration	I-4	V-6
Displacement (cm³)	1769	2987
Bore × Stroke (mm)	80 × 88	93 × 73.3
Compression Ratio	10.5:1	11:1
Chamber Geometry	4-valve, DOHC, pentroof	4-valve, DOHC, pentroof
Stratification Approach	Wall-guided, swirl-based	Wall-guided, swirl-based
Spark Plug Location	Chamber center	Chamber center
Piston Shape	Shallow dish-in-piston crown	Shallow dish-in-piston crown
Intake Port	Intake port deactivation by SCV	Intake port deactivation by SCV
Injector Location	Between the intake valves and beneath the intake port	Between the intake valves and beneath the intake port
Injector Type	Shaped-spray (casting net) injector	Shaped-spray (casting net) injector
Fuel Pressure (MPa)	0.3–7.0	7–9
Part-Load Operation	Late injection	Late injection
Full-Load Operation	Early injection	Early injection
Power	Unavailable	171.5 kW JIS at 6400 rpm
Torque	Unavailable	294 N•m at 4000 rpm
Note	Combined with CVT	N/A

good match of the fuel spray geometry to this type of bowl, and is used in this combustion system. The casting-net spray shape has the capability of more uniformly distributing the fuel within a shallow dish in the piston crown.

10.5 Renault Spray-Guided IDE Combustion System

As previously illustrated in Fig. 2.3-6, the Renault spray-guided combustion system, designated as the IDE (Injection Directe Essence) system, is designed with both the spark plug and fuel injector located very near the center of the combustion chamber. It may be observed that a relatively deep piston bowl is used to contain the spray and to minimize fuel impingement on the cylinder wall. The combustion system was developed primarily for the European market, and the specifications are summarized in Table 10.5-1 [47, 138, 399]. This engine is designed to operate only in the homogeneous, stoichiometric-charge mode in order to take full advantage of the three-way catalyst aftertreatment system. An EGR rate of up to 25% is used in the load range of 0–60% to significantly reduce the throttling loss.

Table 10.5-1
Specifications of the Renault spray-guided IDE engine

Cylinder Configuration	I-4
Displacement (cm^3)	2000
Compression Ratio	11.5:1
Chamber Geometry	4-valve, DOHC, pentroof
Stratification Approach	Spray-guided
Spark Plug Location	Near-centrally mounted spark plug
Piston Shape	Deep bowl in piston crown
Air Flow	Tumble
Injector Location	Vertical, near-centrally mounted injector
Injector Type	Single-fluid, high-pressure, swirl injector
Fuel Pressure	A design fuel rail pressure is in the range of 4 to 10 MPa; 8.5 MPa is utilized for most of the operating map
Spray Characteristics	Hollow-cone fuel spray with a cone angle of 40°
EGR	EGR (up to 25%) is used in the load range of 0–60%.
Operating Mode	Homogeneous, stoichiometric operation *only*
Aftertreatment	Standard three-way catalyst system
Power	103 kW at 5500 rpm
Maximum Torque	200 N•m

10.6 Adam Opel Wall-Guided ECOTEC DIRECT Combustion System

The Adam Opel wall-guided DI combustion system, designated as ECOTEC DIRECT, is designed with the spark plug located at the center of the combustion chamber, and with the injector mounted between the intake valves and beneath the intake port. A lean-NOx-storage catalyst is used as an integral part of the aftertreatment system,

with a temperature sensor used to monitor the critical inlet temperature of the NOx storage catalyst. The major features of the ECOTEC DIRECT system are summarized in Table 10.6-1 [393].

Table 10.6-1
Specifications of the Adam Opel ECOTEC DIRECT engine

Cylinder Configuration	I-4
Displacement (cm^3)	2200
Bore × Stroke (mm)	86 × 94.6
Compression Ratio	11.5:1
Chamber Geometry	4-valve, DOHC, pentroof
Stratification Approach	Wall-guided
Spark Plug Location	Chamber center
Piston Shape	Shallow bowl
Injector Type	High-pressure, swirl-type injector
Injector Location	Mounted between the intake valves and beneath the intake port
Fuel Pressure (MPa)	8
Part-Load Operation	Late injection for stratified-charge mode
Full-Load Operation	Early injection for homogeneous-charge mode
Aftertreatment System	TWC + temperature sensor + NOx storage catalyst; two lambda sensors with one before the starter catalyst and the other after the NOx storage catalyst

10.7 Audi Wall-Guided Combustion System

A cutaway view of the Audi 1.2L, 3-cylinder prototype DI gasoline engine is illustrated in Fig. 10.7-1 [45]. This engine incorporates a vertical, straight intake port to induce a reverse-tumble flow for assisting in creating a stratified charge. The combustion system was designed with the injector located beneath the intake port, which should be beneficial in cooling the injector tip. The specifications of the combustion system are summarized in Table 10.7-1 [32, 45].

Figure 10.7-1 Cutaway view of the Audi 1.2L, I-3 DI engine [45].

Table 10.7-1
Specifications of the Audi 1.2L, I-3, DI engine

Cylinder Configuration	I-3
Chamber Geometry	Five valves (three intake valves and two exhaust valves), pentroof combustion chamber
Stratification Approach	Wall-guided, tumble-based
Spark Plug Location	Chamber center
Piston Shape	Spherical compact piston bowl
Intake Port	Vertical straight intake port
Injector Location	Mounted between the intake valves and beneath the intake port
Fuel Pressure (MPa)	10
Part-Load Operation	Late injection for stratified-charge mode
Full-Load Operation	Early injection for homogeneous-charge mode
Power	55 kW at 5500 rpm
Torque	115 N•m at 3000 rpm

10.8 AVL Combustion Systems

10.8.1 AVL Swirl-Based, Wall-Guided Combustion System

The AVL swirl-based, wall-guided, DI combustion system is illustrated schematically in Fig. 10.8-1 [481]. This combustion system is based on the concept of wall-guided mixture transport to a central spark plug, with the extent of spray-wall impingement reduced by using the maximum advancement of injection timing [108–112, 479–481]. A shallow, asymmetric piston bowl design with a radial flow entry is used for directing the swirl flow into the piston bowl [482]. The specifications of this AVL DI system are summarized in Table 10.8-1.

10.8.2 AVL Mixture Injection DMI Combustion System

Gasoline DI combustion systems generally must invoke a compromise between the requirements of minimum mixture formation time and a limiting time interval between injection and combustion if charge stratification is to be maintained. A true separation of these two opposing requirements is generally considered to provide the highest flexibility for a G-DI system. Based upon this consideration, a direct-mixture-injection (DMI)

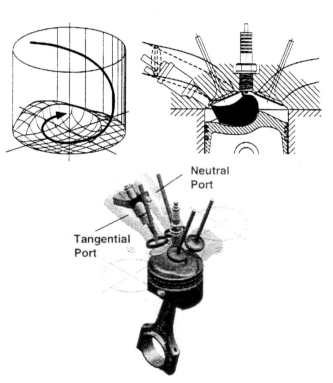

Figure 10.8-1 Schematic of the AVL swirl-based, wall-guided combustion system [481].

Table 10.8-1
Specifications of the AVL, DI, research combustion system

Cylinder Configuration	I-4
Displacement (cm³)	2000
Chamber Geometry	4-valve, DOHC, pentroof
Stratification Approach	Wall-guided, swirl-based
Spark Plug Location	Chamber center
Piston Shape	Shallow, asymmetric piston bowl
Intake Port	One neutral port; one tangential port with a SCV
Injector Location	Mounted between the intake valves and beneath the intake port
Injector	Single-fluid, high-pressure, swirl injector
Part-Load Operation	Late injection for stratified-charge mode
Full-Load Operation	Early injection for homogeneous-charge mode

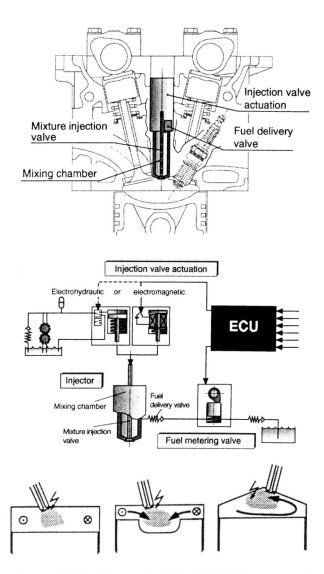

Figure 10.8-2 Schematic of the AVL DMI system and several DMI-based mixture preparation strategies [108].

concept was proposed by AVL to combine the advantages of air-assisted injection and pre-vaporization of fuel, without the need for an external pressurized air supply [108].

The schematics of the DMI system and several DMI-based mixture preparation strategies are illustrated in Fig. 10.8-2 [108]. Timing charts for the DMI concept under different operating conditions are shown in Fig. 10.8-3 [108]. The DMI valve incorporating a standard poppet-valve geometry with electrically-controlled actuation provides for the injection of the mixture and the recharging of the prechamber. A key feature of the system is that the DMI valve recovers the gas pressure required for injection by withdrawing a small amount of the compressed charge from the cylinder in the preceding engine cycle. To avoid combustion in the prechamber, the DMI valve is closed at the time of the spark; however, the mixture can be injected into the main chamber at any time prior to ignition. After the DMI valve is closed, liquid fuel is injected into the prechamber and is vaporized in preparation for injection during the next engine cycle.

The quality of mixture preparation is substantially enhanced by the pre-vaporization of the fuel, and the DMI system effectively decreases the engine crank angle interval required for fuel vaporization as compared to conventional G-DI injection. The processes of fuel

metering and injection into the prechamber can be achieved by means of a constant displacement method using a low fuel pressure. The fuel pressure is required to exceed the maximum prechamber pressure only marginally, as a high degree of fuel atomization from the injection process is not required. A small fuel pressure differential will, however, reduce the rate of fuel flow into the cylinder. Further claims include additional advantages of mixture stratification strategies. For example, a reduced mixture velocity and penetration results from the comparably low pressure difference between the prechamber and the combustion chamber, as well as from the low momentum associated with fuel vapor as compared to fuel droplets.

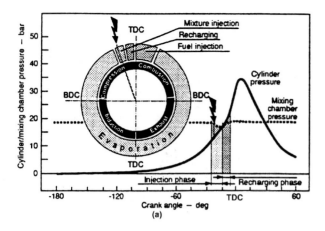

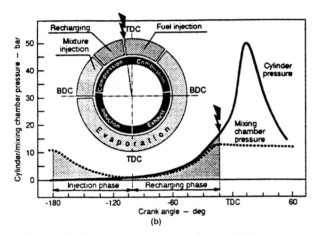

Figure 10.8-3 Timing charts for the DMI concept under different operating conditions: (a) late injection; (b) early injection [108].

It should be emphasized that several issues associated with this DMI system may require further development. The mixture is injected into the main chamber as a consequence of the small pressure difference between the prechamber and the main chamber, thus all of the evaporated fuel may not be injected. This fuel metering error is commonly referred to as the fuel hang-up, and is often associated with injection from intermediate cavities. At high engine speeds or under cold-start conditions, the fuel inside the prechamber may not evaporate completely, and some degree of wetting of the internal cavity wall is likely to occur. These conditions will result in fuel metering errors. Moreover, during engine transients the pressure inside the prechamber, as sampled from the previous cycle, may not be appropriate for the metering of the current cycle. This will very likely yield an engine-out HC spike during engine transient operation.

10.9 FEV Air-Guided Combustion System

In the FEV air-guided DI combustion system, a strong tumble flow is generated by means of a movable control gate [42, 43, 127]. This tumble air flow field is in turn utilized to establish a stratified charge [143, 178], in the manner previously illustrated in Fig. 2.3-24. Figure 10.9-1 contains a schematic of the combustion chamber, whereas the corresponding engine control map is shown in Fig. 10.9-2. The major specifications of the FEV air-guided combustion system are summarized in Table 10.9-1, with the effect of tumble ratio on the combustion characteristics being plotted in Fig. 10.9-3 [127].

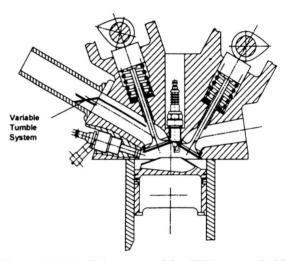

Figure 10.9-1 Schematic of the FEV air-guided DI combustion system [42].

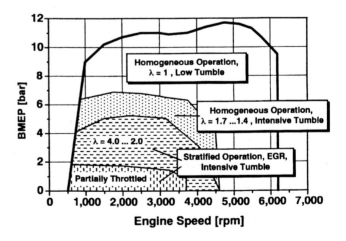

Figure 10.9-2 Proposed engine operating map of the FEV air-guided DI combustion system [143].

It is evident that combustion characteristics are significantly influenced when the tumble ratio is below a certain threshold level. A further increase in the tumble ratio above this threshold does not provide any benefit. Thus, in this application a system, providing a continuously variable tumble ratio is not necessary, and a two-stage tumble system utilizing a simple control gate is adequate. This can significantly simplify system calibration. Stable lean operation may be achieved using this system for loads of up to 6 bar BMEP with the addition of a homogeneous-lean operating mode [43].

Table 10.9-1
Specifications of the FEV air-guided, DI combustion system

Chamber Geometry	4-valve, DOHC, pentroof
Stratification Approach	Air-guided, tumble-based
Spark Plug Location	Chamber center
Piston Shape	Shallow bowl
Air Flow	Variable tumble flow
Injector Location	Side-mounted, beneath the intake port
Part-Load Operation	Late injection for stratified-charge mode
Full-Load Operation	Early injection for homogeneous-charge mode

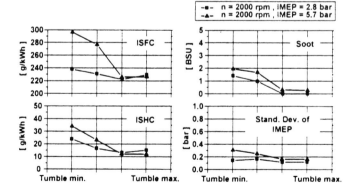

Figure 10.9-3 Effect of tumble ratio on the combustion characteristics of the FEV air-guided DI combustion system [127].

10.10 Fiat Combustion System

The specifications of the Fiat prototype DI engine [19, 20] are summarized in Table 10.10-1. The Fiat DI combustion system was designed with both the injector and spark plug located at the center of the combustion chamber. This 2.0L, I-4 DI engine was developed to operate only in homogeneous-charge mode. A comparison of homogeneous, stoichiometric combustion systems incorporating either centrally mounted or side-mounted injectors revealed that charge homogenization is generally superior for the configuration having a centrally mounted injector. In addition, combustion systems employing a centrally mounted injector generally provide reduced smoke emissions and a higher peak torque [20]. Conversely, the configuration incorporating a side-mounted injector normally yields the better high-speed volumetric efficiency. Results for part-load operation show a slight benefit in HC emissions for the centrally mounted injector configuration, even though the difference is small. The SOI timing for maximizing the volumetric efficiency is not the same as that for the maximum torque due to the limited time for achieving charge homogenization [19]. For the Fiat DI combustion system, maximum torque is obtained at an injection timing that is about 20 to 30 crank angle degrees earlier than that for maximum volumetric efficiency.

Table 10.10-1
Specifications of the Fiat prototype DI combustion system

Cylinder Configuration	I-4
Displacement (cm^3)	1995
Chamber Geometry	4-valve, DOHC, pentroof
Compression Ratio	12:1
Spark Plug Location	Chamber center
Intake Port	Moderate tumble with port deactivation possible
Injector Location	Centrally mounted injector
Operating Mode	Homogeneous-charge mode *only*

10.11 Ford Combustion Systems

10.11.1 Ford Spray-Guided Combustion System

The Ford spray-guided, DI, research engine is designed to operate in homogeneous, stoichiometric-charge mode, and utilizes a centrally located fuel injector and spark plug [9, 10, 506]. A schematic of the combustion chamber has been presented previously in Fig. 2.3-2. The intake ports of the 4-valve, single-cylinder head are optimally designed to yield a higher volumetric efficiency than would normally be expected with the smaller intake valve flow area. The main features of the Ford single-cylinder, research DI engine are summarized in Table 10.11-1. Fuel consumption improvements of up to 5% at part load and 10% at idle with stoichiometric operation are achieved for steady-state operation. Optimum lean operation at part load provides a fuel consumption improvement of up to 12% when compared to a Ford baseline PFI engine, whereas emissions are comparable.

10.11.2 Ford Swirl-Based, Wall-Guided Combustion System

The specifications of the Ford swirl-based, wall-guided, prototype I-3 DI engine are summarized in Table 10.11-1. The combustion system is designed with the spark plug mounted at the center of the combustion chamber and the fuel injector located on the intake side. A lean-NOx storage catalyst is incorporated as part of the exhaust aftertreatment system. To minimize emissions a catalyst warm-up strategy with a stoichiometric air/fuel ratio, retarded ignition and increased idle speed is used for the first 65 seconds of the new European driving cycle (NEDC). Electronic throttle control allows this warm-up strategy to be used during all engine operating conditions (idle, cruise and acceleration) without impacting vehicle driveability. To enhance combustion stability during engine warm-up, a swirl-control valve is used in the intake port. Coil-on-plug ignition with multiple-strike discharge is used to prevent combustion misfires. Although lean operation is feasible early in the NEDC, the engine is operated at stoichiometric with MBT spark timing from 65 seconds to 150 seconds to achieve optimal emissions and fuel efficiency. Lean operation earlier in the cycle inhibits engine warm-up and degrades fuel efficiency due to the increased mechanical friction associated with cold engine oil. This period of operation also allows time for thermal stabilization of the substrate in the lean-NOx storage catalyst, which improves the storing efficiency during lean operation.

Table 10.11-1
Specifications of Ford research and prototype DI engines

Engine Description	Research Engine	Prototype
Cylinder Configuration	Single cylinder	I-3
Displacement (cm^3)	575	1125
Bore × Stroke (mm)	90.2 × 90.0	79.0 × 76.5
Compression Ratio	11.5:1	11.5:1
Chamber Geometry	4-valve, DOHC, pentroof	4-valve, DOHC, pentroof
Stratification Approach	Spray-guided	Wall-guided, swirl-based
Spark Plug Location	Near-centrally mounted spark plug	Near-centrally mounted spark plug
Piston Shape	Flat crown	Central piston bowl
Intake Port and Flow	No net swirl, but small tumble component	Split ports retrofitted with SCV
Injector Location	Vertical, near-centrally mounted injector	Mounted at the intake side
Injector Type	High-pressure, swirl injector	High-pressure, swirl injector
Fuel Pressure (MPa)	5.0	12.0
Spray Characteristics	Unavailable	Hollow-cone spray with a cone angle of 35°
Ignition System	Unavailable	Coil-on-plug ignition with multiple-strike discharge; high energy; 3 electrode full surface plug
Aftertreatment System	N/A	Three-way catalyst (0.6L) and lean NOx storage catalyst (1.6L)
Operating Modes	Early injection for homogeneous-charge mode *only*	Early injection for homogeneous-charge mode and late injection for stratified-charge mode
Fuel Requirement	91 RON	95 RON

The engine control system employs an algorithm to estimate both the amount of NOx produced by the engine and the capacity of the lean-NOx storage catalyst to store NOx at any given operating condition. When the engine control system determines that the lean-NOx storage catalyst has attained its maximum NOx storage limit, a NOx purge event is scheduled. An exhaust oxygen sensor is used for feedback control of the purge event duration, resulting in minimal HC and CO tailpipe emissions. To reduce the time required for purging the lean-NOx storage catalyst, a low oxygen-storage catalyst formulation was specified for the close-coupled three-way catalyst. This minimizes the unintended conversion of carbon monoxide in the three-way catalyst during the purge event.

A fuel cut-off strategy is used during deceleration events. Fuel cut-off provides an oxygen-rich environment in the exhaust system, which helps to stabilize and retain the stored NOx in the lean-NOx storage catalyst. This yields very low tailpipe NOx emissions during deceleration, even though the apparent catalyst NOx conversion efficiency is lower. At a load point of 1500 rpm/2.62 BMEP, an 18.5% fuel economy improvement is obtained for the Ford swirl-based, wall-guided system. This is attributed to a 9% contribution due to pumping loss reduction, a 12% contribution resulting from a higher thermal efficiency, and a −2.5% contribution resulting from increased engine friction. The contributions listed were measured with a calibration optimized to achieve a reasonable balance of fuel economy, emissions, driveability and engine operating stability.

10.12 Honda Spray-Guided Combustion System

The Honda spray-guided DI combustion system is designed with both the spark plug and fuel injector located near the center of the combustion chamber. The inclined spark plug is mounted in close proximity to the fuel injector. For this engine, which was developed for a hybrid-vehicle application, an in-cylinder swirl flow is generated by the VTEC valvetrain mechanism. The schematic of this 1.0L 3-cylinder prototype G-DI engine is shown in Fig. 10.12-1 [494], with the specifications of the combustion system summarized in Table 10.12-1. In the design application a motor/generator is attached to the engine and serves as an additional source of power when accelerating. This configuration provides overall vehicle performance exceeding that provided by a 1.5L conventional PFI engine.

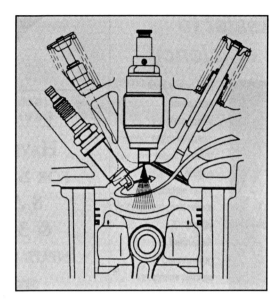

Figure 10.12-1 Schematic of the Honda DI combustion system [494].

Table 10.12-1
Specifications of the Honda spray-guided DI engine

Cylinder Configuration	I-3
Displacement (cm^3)	1000
Chamber Geometry	4-valve, DOHC, pentroof
Stratification Approach	Spray-guided, swirl-based
Spark Plug Location	Inclined spark plug in close proximity to the injector
Piston Shape	Shallow bowl
Air Flow	Swirl
Injector Location	Centrally mounted injector
Part-Load Operation	Late injection for stratified-charge mode
Full-Load Operation	Early injection for homogeneous-charge mode

10.13 Isuzu Combustion System

A single-cylinder, pentroof, 4-valve, PFI engine is the base system from which the Isuzu DI system was developed [407]. The principal features of this system are summarized in Table 10.13-1. This prototype research engine has the provision for incorporating either a centrally mounted or an intake-side-mounted fuel injector. A comparison of the engine performance variation with these alternative injector locations has been conducted, and it was determined that the side-mounted location

Table 10.13-1
Specifications of the Isuzu DI combustion system

Cylinder Configuration	Single cylinder and V-6
Displacement (cm³)	528 per cylinder
Bore × Stroke (mm)	93.4 × 77.0
Compression Ratio	10.7:1
Chamber Geometry	4-valve, DOHC, pentroof
Spark Plug Location	Near-centrally mounted
Piston Shape	Flat
Intake Port	Standard with a tumble ratio of 0.63 and a swirl ratio of 0.0
Injector Location	Vertical, near-centrally mounted injector
Injector Type	Single-fluid, high-pressure, swirl injector
Fuel Pressure (MPa)	5
Spray Cone Angle	55°
Operating Mode	Early injection for homogeneous-charge mode *only*

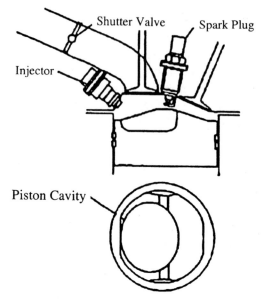

Figure 10.14-1 Schematic of the Mazda swirl-based, wall-guided, DI combustion system [503].

yields an increase in HC emissions and fuel consumption as compared to the baseline PFI engine. This is considered to be the result of fuel impingement on the cylinder liner near the exhaust valves, with some of the fuel being absorbed in the oil film. For the centrally mounted injector location, emissions and engine performance are quite similar to those of the baseline PFI engine for operating conditions in which the fuel is injected during the early portion of the intake stroke. Based upon this experimental evaluation, the centrally mounted injector location was selected for the Isuzu V-6 prototype DI gasoline engine [408].

10.14 Mazda Swirl-Based, Wall-Guided Combustion System

The Mazda gasoline direct-injection combustion system employs a moderate to high air swirl ratio in a wall-guided configuration. The details of this combustion system are provided in Fig. 10.14-1 [503]. The specifications of the Mazda swirl-based, wall-guided DI combustion system are summarized in Table 10.14-1. In the pentroof combustion chamber, the intake valves have an incrementally smaller inclination angle to provide more space for mounting the injector on the intake side of the chamber. An independent intake port with a flow control (shutter) valve is used to generate in-cylinder air swirl during the intake stroke.

Table 10.14-1
Specifications of the Mazda swirl-based, wall-guided DI engine

Cylinder Configuration	I-4
Displacement (cm³)	1992
Bore × Stroke (mm)	83 × 92
Compression Ratio	11:1
Chamber Geometry	4-valve, DOHC, pentroof combustion chamber with a smaller intake-side valve inclination angle
Stratification Approach	Wall-guided, swirl-based
Spark Plug Location	Centrally mounted
Piston Shape	Shallow bowl
Intake Port	Independent intake port with shutter valve for swirl generation
Injector Location	Intake-side-mounted, with an inclination angle of 36°
Injector Type	Single-fluid, high-pressure, swirl injector
Fuel Pressure (MPa)	7
Spray Characteristics	Hollow-cone spray with a cone angle of 60° and an SMD of 20 µm at 7 MPa
Part-Load Operation	Late injection for stratified-charge mode
Full-Load Operation	Early injection for homogeneous-charge mode

10.15 Mercedes-Benz Spray-Guided Combustion System

The Mercedes-Benz spray-guided DI combustion system utilizes a vertical, centrally mounted, fuel injector. A schematic of the combustion chamber and the control map are presented in Fig. 10.15-1 [227]. Dynamometer tests of an engine using this combustion system for a range of fuel rail pressures from 4 to 12 MPa indicate that the fuel consumption, HC emissions and COV of IMEP are minimized at 8 MPa; however, the NOx emissions levels are not optimum for this fuel rail pressure. The combustion duration is the shortest at a fuel rail pressure of 8 MPa, with the corresponding 50% heat release point occurring before TDC on the compression stroke. In addition to the fuel economy improvement provided by this spray-guided system, engine-out NOx emissions are reduced by approximately 35% as compared to a conventional PFI engine; however, the engine-out HC emissions are significantly elevated. For the load range of 0.2 to 0.5 MPa IMEP, the minimum ISFC is obtained at an engine speed of 2000 rpm. The fuel consumption increases at lower or higher engine speeds as a result of degradation of mixture homogeneity. The specifications of the Mercedes-Benz single-cylinder DI engine are summarized in Table 10.15-1.

Table 10.15-1
Specifications of the Mercedes-Benz single-cylinder, research DI engine

Cylinder Configuration	Single cylinder
Displacement (cm³)	538.5
Bore × Stroke (mm)	89 × 86.6
Compression Ratio	10.5:1
Chamber Geometry	4-valve, pentroof; major portion of the combustion chamber is formed by a bowl in the piston
Stratification Approach	Spray-guided
Spark Plug Location	Located between the intake valves, in the immediate vicinity of the injector tip
Piston Shape	Bowl-in-piston centered on cylinder axis
Injector Location	Vertical, centrally mounted
Fuel Pressure (MPa)	Tested in the range of 4 to 12
Spray Characteristics	75°, 90°, 105° cone angle
Part-Load Operation	Late injection for stratified-charge mode
High-Load Operation	Early injection for homogeneous-charge mode

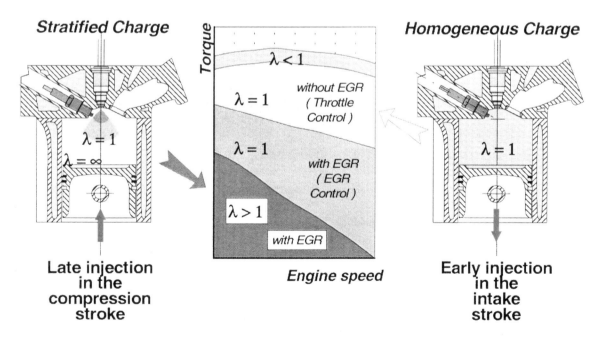

Figure 10.15-1 Schematic of the Mercedes-Benz spray-guided, DI combustion system [227].

10.16 Orbital Combustion System Employing Pulse-Pressurized, Air-Assisted Fuel Injection System

The Orbital air-assisted fuel injection system has been applied to both two-stroke engines and automotive 4-stroke DI engines [54, 176, 177, 255, 337, 338, 426, 508]. This injection system is of the pulse-pressurized, air-assisted (PPAA) type that was discussed in detail in Chapters 3 and 4. The inherent qualities of the air-assisted fuel system in combination with careful matching of the spray to the combustion chamber geometry permits the attainment of a high degree of charge stratification with late injection timings, with stable combustion achievable over a wide range of engine operating conditions. The schematic of the Orbital DI system using a pulse-pressurized, air-assisted fuel injection system is illustrated in Fig. 10.16-1 [176]. The specifications of the Orbital spray-guided DI engine are summarized in Table 10.16-1.

10.17 PSA Reverse-Tumble-Based, Wall-Guided HPi Combustion System

The PSA DI gasoline engine, called the HPi, has a reverse-tumble-based, wall-guided combustion system [126, 135, 363]. Figure 10.17-1 schematically illustrates this gasoline direct-injection combustion system [363]. The piston incorporates a spherical-segment compact cavity to direct the mixture plume to maintain charge stratification, which is assisted by a reverse-tumble charge motion that is generated by straight, vertical intake ports.

Table 10.16-1
Specifications of the Orbital spray-guided DI engine

Cylinder Configuration	I-4
Displacement (cm³)	2000
Bore × Stroke (mm)	80.6 × 88
Compression Ratio	10.4:1
Chamber Geometry	4-valve, DOHC, pentroof
Stratification Approach	Spray-guided, tumble-based
Spark Plug Location	Near-centrally mounted
Air Flow	Low tumble and zero swirl
Injector Location	Near-centrally mounted
Injector Type	Pulse-pressurized, air-assisted (PPAA)
Spray Characteristics	Narrow (25°–30°) solid-cone spray plume; 17 μm SMD
Fuel Pressure (MPa)	Fuel pressure: 0.72; air pressure: 0.65
Part-Load Operation	Late injection for stratified-charge mode
Full-Load Operation	Early injection for homogeneous-charge mode

A lean NOx storage catalyst is employed to reduce the tailpipe NOx emissions during lean operation of the engine. The stratified-charge mode may be employed for engine speeds of up to 3500 rpm for urban driving conditions or at more moderate speeds for steady highway driving. At high speeds the engine operates in stoichiometric-homogeneous-charge mode. The specifications of the PSA DI combustion system are summarized in Table 10.17-1.

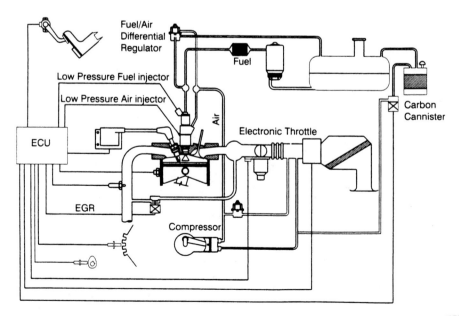

Figure 10.16-1 Schematic of the Orbital spray-guided, DI combustion system [177].

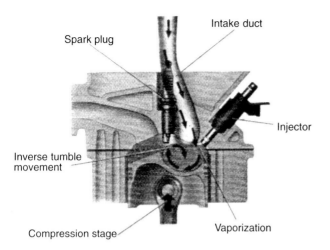

Figure 10.17-1 Schematic of the PSA reverse-tumble-based, wall-guided, HPi combustion system [363].

Table 10.17-1
Specifications of the PSA reverse-tumble-based, wall-guided HPi engine

Cylinder Configuration	I-4
Displacement (cm³)	1998
Compression Ratio	11.4:1
Chamber Geometry	4-valve, DOHC, pentroof
Stratification Approach	Wall-guided, reverse-tumble-based
Spark Plug Location	Chamber center
Piston Shape	A spherical compact piston cavity
Intake Port	Vertical straight intake ports to generate a reverse tumble
Injector Location	Located underneath the intake port and between the two intake valves
Injector Type	High-pressure, swirl injector
Spray Cone Angle	70°
Fuel Pressure (MPa)	Variable; 3 to 10 with 7 MPa for idle, 10 MPa for full load and 3 MPa for load transition
NOx Control	NOx storage catalyst
Part-Load Operation	Late injection for stratified-charge mode
Full-Load Operation	Early injection for homogeneous-charge mode
Power	103 kW at 6000 rpm
Torque	170 N•m at 2000 rpm

10.18 Ricardo Tumble-Based, Wall-Guided Combustion System

The Ricardo tumble-based, wall-guided, DI combustion system incorporating a top-entry-port head was initially used to investigate the combustion and charge motion requirements of a DI gasoline engine that operates on the strategy of early injection to achieve a homogeneous mixture [273]. The original project goal was to achieve improvements in cold start emissions and transient response of a DI engine system that could utilize a TWC, and to document the performance parameters of a DI gasoline engine for homogeneous, stoichiometric-charge combustion. The resulting Ricardo DI gasoline engine produces HC emissions at levels that are nearly equivalent to those from the PFI baseline engine for part-load operation using a stoichiometric-charge mixture. The NOx emission levels are also identical, indicating a similar air/fuel ratio at the spark gap for these two engines. This DI gasoline engine system does exhibit an exceptional EGR tolerance when compared with the baseline PFI engine. This is attributed to the increased turbulence intensity that results from the direct injection of fuel into the cylinder. The DI engine is clearly superior in cold engine transient response and cold-start HC emissions, and the fuel economy for stoichiometric operation is 6% better than for a comparable PFI engine.

A later version of the Ricardo DI combustion system incorporates a top-entry-port head in combination with a curved piston, as depicted in Fig. 10.18-1. This is

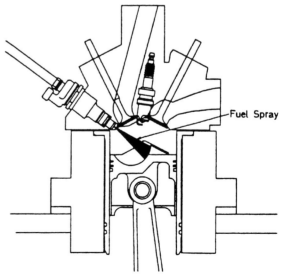

Figure 10.18-1 Schematic of the Ricardo tumble-based, wall-guided DI combustion system [201].

designed to operate using both early and late injection modes [199–202]. In the stratified-charge operating mode, the minimum values of HC emissions and COV of IMEP are obtained using a narrow cone fuel spray having a relatively low injection rate for such a top-entry configuration. The combustion characteristics of this system are found to be more sensitive to the fuel injection rate than to the spray cone angle. For a specified fuel rail pressure, the use of an injector with a reduced fuel injection rate (a lower static flow capacity) generally yields a lower spray penetration rate. This, in turn, can reduce the percentage of spray that impinges on the piston crown and cylinder walls, which is a primary source of HC emissions. A limitation of this is that the injector static flow capacity must accommodate operation at rated power, which requires an acceptable injection pulse width at maximum speed and load.

From the engine speed-load map shown in Fig. 10.18-2, it may be seen that this DI engine is designed to operate in the homogeneous-charge mode above 50% rated load. Above 70% load, the early injection mode, with either a stoichiometric or a slightly rich mixture, is used. Stratified-charge operation is thus restricted to combinations of load and speed where fuel consumption is critical. The engine runs fully unthrottled from 70% to 20% of full-load, and with light throttling below 20% load to control HC emissions. The typical features of this DI combustion system are summarized in Table 10.18-1.

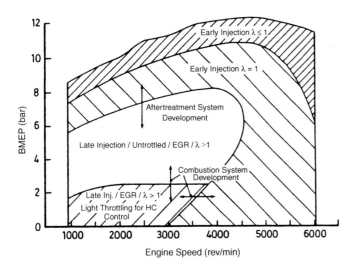

Figure 10.18-2 Proposed operating map of the Ricardo tumble-based, wall-guided DI engine [201].

Table 10.18-1
Specifications of the Ricardo tumble-based, wall-guided, single-cylinder, DI engine

Cylinder Configuration	Single cylinder
Displacement (cm³)	325
Bore × Stroke (mm)	74 × 75.5
Compression Ratio	12.7:1
Chamber Geometry	4-valve, pentroof
Stratification Approach	Wall-guided, reverse-tumble-based
Spark Plug Location	Chamber center
Piston Shape	Curved piston crown
Intake Port	A top-entry port to generate a reverse tumble
Injector Location	Mounted between the two intake valves and beneath the intake port
Part-Load Operation	Late injection for stratified-charge mode
Full-Load Operation	Early injection for homogeneous-charge mode

10.19 Saab Spray-Guided SCC Combustion System

The Saab Combustion Control (SCC) system is based on a combination of DI gasoline, variable valve timing (VVT) and variable spark gap [54, 353, 379]. Figure 10.19-1 illustrates the system layout of the SCC concept. The associated specifications are summarized in Table 10.19-1. A stoichiometric air-fuel mixture is employed in order to permit the use of a conventional TWC. The pulse-pressurized, air-assisted injector and spark plug are integrated into one unit known as the spark plug injector (SPI), and the fuel is injected into the cylinder by means of compressed air. Immediately before the fuel is ignited, a brief air blast is introduced to create turbulence in the cylinder, which assists combustion and shortens the combustion duration. VVT is applied to both the intake and exhaust valves, with a result that up to 70% of the cylinder contents consists of exhaust gases, depending on the engine operating conditions.

The spark plug gap is varied in the range of 1 mm to 4 mm. The spark is struck from a central electrode in the SPI either to a fixed ground electrode at a distance of 4 mm or to a ground electrode on the piston. The variable spark gap together with a very elevated spark firing

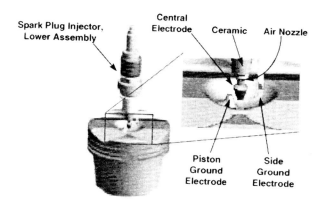

Figure 10.19-1 Schematic of the Saab spray-guided, SCC combustion system [353].

Table 10.19-1
Specifications of the Saab spray-guided, SCC combustion system

Stratification Approach	Spray-guided
Spark Plug and Injector	Integrated into one close-spacing unit
Injector Location	Central
Spark Plug Gap	Variable; 1.0 mm to 4 mm
Intake and Exhaust Systems	VVT is applied to both intake and exhaust valve actuation
Fuel System	Pulse-pressurized, air-assisted (PPAA) fuel injection system
Operation Mode	Homogeneous, stoichiometric operation *only*
Aftertreatment	Three-way catalyst

10.20 Subaru Spray-Guided Combustion System

The Subaru spray-guided DI combustion system is designed with a high-pressure fuel injector located at the center of the combustion chamber, and with the fuel injected vertically into the cylinder. An inclined spark plug is located between the intake valves on the intake side of the engine. With this configuration, the fuel spray does not impinge onto the electrode of the spark plug [505]. The specifications of the Subaru DI engine incorporating this concept are summarized in Table 10.20-1.

Table 10.20-1
Specifications of the Subaru spray-guided, single-cylinder DI engine

Cylinder Configuration	Single cylinder
Displacement (cm^3)	554
Bore × Stroke (mm)	97 × 75
Compression Ratio	9.7:1
Chamber Geometry	4-valve, pentroof
Stratification Approach	Spray-guided
Spark Plug Location	Located between the intake valves
Piston Shape	Curved piston crown
Intake Port	Conventional straight tumble port at a tumble ratio of 0.7
Injector Location	Centrally mounted injector
Injector Type	Single-fluid, high-pressure, swirl injector
Fuel Pressure (MPa)	7

10.21 Volkswagen Tumble-Based, Wall-Guided FSI Combustion System

As illustrated in Fig. 2.3-12, the Volkswagen tumble-based, wall-guided DI combustion system, designated as Fuel Stratified Injection (FSI), is designed with the spark plug mounted at the center of the chamber, and with the fuel injector installed between the intake valves and beneath the intake port. A variable-tumble-generation device is mounted inside the intake port. The major features of this DI engine are summarized in Table 10.21-1 [49, 136, 214, 256, 400]. As previously illustrated in Fig. 8.3-9, a close-coupled catalyst is combined with a compact

energy (80 mJ) is essential for igniting an air-fuel mixture that is so highly diluted by exhaust gases. At light load, the spark arcs from the central electrode in the SPI to a fixed ground electrode with an effective gap of 4 mm. At high load, the spark occurs somewhat later, and the gas density in the combustion chamber is then too high for the spark to bridge a gap of 4 mm. A pin on the piston is then used instead as the ground electrode. The spark will be struck to the electrode on the piston as soon as the voltage becomes sufficient to arc across the decreasing gap.

manifold in order to enhance the light-off characteristics. The position of the wide-range oxygen sensor upstream of the close-coupled catalyst was selected so that the gases flowing from each exhaust port are equally monitored. Downstream of the close-coupled catalyst is a temperature sensor that monitors the exhaust gas and catalyst temperatures for precise use as input to a catalyst model within the control system.

The NOx storage catalyst is located in an underfloor position similar to that used in a normal TWC design. An NOx sensor is positioned downstream of the NOx storage catalyst to monitor the catalyst and provide feedback to control and optimize the exhaust purging event. As has already been described in Section 8.3.6, two exhaust cooling measures of cold air and exhaust gas separation are incorporated into this system. This better exploits the working temperature window for the NOx storage catalyst and reduces the high temperature load. This type of exhaust gas heat exchanger is reported to yield only a minor increase in the exhaust back pressure.

Table 10.21-1
Specifications of the Volkswagen tumble-based, wall-guided FSI engine

Cylinder Configuration	I-4
Displacement (cm³)	1390
Compression Ratio	12:1
Chamber Geometry	4-valve, pentroof
Stratification Approach	Wall-guided, tumble-based
Spark Plug Location	Chamber center
Piston Shape	Two shallow bowls in piston
Intake Port	Variable tumble charge motion
Injector Location	Mounted between the intake valves and beneath the intake port
Injector Type	Single-fluid, high-pressure, swirl injector
Part-Load Operation	Late injection for stratified-charge mode
Full-Load Operation	Early injection for homogeneous-charge mode
Aftertreatment	TWC + temperature sensor + NOx storage catalyst + NOx sensor
Power	77 kW JIS at 6200 rpm
Torque	130 N•m at 4250 rpm

10.22 Summary

The direct-injection gasoline systems and key control strategies that have been recently developed and reported in the literature have been discussed in detail. It is quite evident from the worldwide technical literature that significant incremental gains in fuel economy, engine performance and emissions parameters are indeed indicated in research and prototype engines and, to some degree, have been achieved in production vehicles. The specific technical issues and compromises related to BSFC, HC, NOx, particulate matter and fuel sprays in G-DI combustion systems have been addressed and discussed. This clearly accentuates the numerous practical considerations that will have to be addressed if the G-DI engine is to realize its full potential and become a major automotive powerplant in the future. These key issues are summarized in Table 10.22-1.

In order to specifically identify the particular G-DI configuration that delivers the most advantages, much difficult work remains to be done. The field experiences of a number of production DI systems are being accumulated, and will be analyzed and discussed by automotive engineers. The delivered data on emissions indices and brake specific fuel consumption are being critically evaluated for each mode of the operating map, and are being compared to those obtained with alternative DI configurations and strategies. In addition, the relative merits of spray-guided, wall-guided and air-guided strategies using both tumble and swirl are being determined.

The field of G-DI fuel injectors, pumps and regulators is one of rapid development, with new developments occurring quarterly, rendering older systems, strategies and limitations obsolete, and providing new options for combustion system development. This new hardware with enhanced capabilities and an extended portfolio of available characteristics will permit a more complete evaluation of the relative merits of each system concept. Even when the merits of each of these production and prototype G-DI engines are fairly well established, much research and development will still be required in the area of system optimization in order to develop production configurations. It is safe to say that a DI combustion chamber configuration that is an enhancement of one of the systems discussed in this chapter will incorporate the new hardware and control technology, and will indeed emerge as the primary G-DI system of choice.

Table 10.22-1
Practical considerations associated with the design and development of a G-DI engine

Fuel Economy	Can a sufficient enhancement in operating BSFC be achieved in a stratified-charge G-DI to offset the increased system complexity as compared to a contemporary PFI engine?
System Complexity and Reliability	Can a production-feasible compromise be obtained between system complexity and overall system reliability, considering that the required G-DI hardware could incorporate multi-stage injection, variable swirl and tumble-control hardware and variable fuel pressure?
Emissions	Can the U.S., European, and Japanese emission standards that are applicable in the near-term time frame be achieved and maintained for the required durability intervals?
Deposits	Can deposit formation on various components be minimized for the wide range of fuel quality and composition in the field such that a reasonable service interval is achieved?
Driveability	Can control-system strategies and algorithms be developed and implemented such that sufficiently smooth transitions from stratified-charge, late-injection operation to mid-range to homogeneous-charge, early-injection operation are obtained, thus yielding driveability levels that are comparable to current sequential-injection PFI systems?
Classification	Can the advantages of stratified-charge operation outweigh the increased complexity and development time as compared to a homogeneous-chargy-only G-DI?

References

1. Abraham, J., "Entrainment characteristics of sprays for diesel and DISI applications," SAE Technical Paper No. 981934 (1998).

2. Ader, B. et al., "Simulation of mixture preparation," *Proceedings of the 3rd International FIRE User Meeting*, June 16–17, 1997 (1997).

3. Alain, F. et al., "In-cylinder flow investigation in a gasoline direct injection four valve engine: bowl shape piston effects on swirl and tumble motions," 1998 FISITA Technical Paper No. F98T049 (1998).

4. Alger, T. et al., "Fuel-spray dynamics and fuel vapor concentration near the spark plug in a direct-injected 4-valve SI engine," SAE Technical Paper No.1999-01-0497 (1999).

5. Alger, T. et al., "Effects of swirl and tumble on in-cylinder fuel distribution in a central injected DISI engine," SAE Technical Paper No. 2000-01-0533 (2000a).

6. Alger, T. et al., "The effects of in-cylinder flow fields and injection timing on time-resolved hydrocarbon emissions in a 4-valve, DISI engine," SAE Technical Paper No. 2000-01-1905 (2000b).

7. Allen, J. et al., "Comparison of the spray characteristics of alternative GDI fuel injection systems under atmospheric and elevated pressure operation," JSAE Technical Paper No. 9935086 (1999).

8. Alperstein, M. et al., "Texaco's stratified charge engine—multifuel, efficient, clean, and practical," SAE Technical Paper No. 740563 (1974).

9. Anderson, R. et al., "A new direct injection spark ignition (DISI) combustion system for low emissions," FISITA-96 Technical Paper No. P0201 (1996a).

10. Anderson, R. et al., "Understanding the thermodynamics of direct injection spark ignition (DISI) combustion systems: an analytical and experimental investigation," SAE Technical Paper No. 962018 (1996b).

11. Anderson, R. et al., "Challenges of stratified charge combustion," *Direkteinspritung im Ottomotor*, Haus Der Technik E. V., 45117 Essen, March 12–13, 1997 (1997).

12. Andersson, J. et al., "Particle and sulfur species as key issues in gasoline direct injection exhaust," JSAE Technical Paper No. 9935842 (1999).

13. Ando, H., "Combustion control technologies for direct-injection gasoline engines," *Proceedings of the 73rd JSME Annual Meeting* (V) (in Japanese), No. WS 11-(4), pp. 319–320 (1996a).

14. Ando, H., "Combustion control technologies for gasoline engines," *IMechE. Seminar on Lean Burn Combustion Engines*, S433, December 3–4, 1996 (1996b).

15. Ando, H., "Mitsubishi GDI engine strategies to meet the European requirements," *Proceedings of AVL Conference on Engine and Environment*, Vol. No. 2, pp.55–77 (1997).

16. Ando, H. et al., "Combustion control for Mitsubishi GDI engine," *Proceedings of the 2nd International Workshop on Advanced Spray Combustion*, Nov. 24–26, 1998, Hiroshima, Japan, Paper No. IWASC9820, pp. 225–235 (1998a).

17. Ando, H., "Key words for the future vehicle powertrain," *SAE-Japan Automobile Technology* (in Japanese), Vol. 54, No. 7, Paper No. 20004337, pp. 7–9 (2000).

18. Andrews, G. et al., "The composition of spark ignition engine steady state particulate emissions," SAE Technical Paper No. 1999-01-1143 (1999).

19. Andriesse, D. et al., "Experimental investigation on fuel injection systems for gasoline DI engines," *Direkteinspritung im Ottomotor*, Haus Der Technik E. V., 45117 Essen, March 12–13, 1997 (1997a).

20. Andriesse, D. et al., "Assessment of stoichiometric GDI engine technology," *Proceedings of AVL Engine and Environment Conference*, pp. 93–109 (1997b).

21. Aradi, A. et al., "The effect of fuel composition and engine operating parameters on injector deposits in a high-pressure direct injection gasoline (DIG) research engine," SAE Technical Paper No. 1999-01-3690 (1999).

22. Aradi, A. et al., "A study of fuel additives for direct injection gasoline (DIG) injector deposit control," SAE Technical Paper No. 2000-01-2020 (2000a).

23. Aradi, A. et al., "The effect of fuel composition, engine operating parameters and additive content on injector deposits in a high-pressure direct injection gasoline (DIG) research engine," *Proceedings of Aachen Colloquium—Automobile and Engine Technology*, pp. 187–211 (2000b).

24. Araneo, L. et al., "Effects of fuel temperature and ambient pressure on a GDI swirled injector spray," SAE Technical Paper No. 2000-01-1901 (2000).

25. Arcoumanis, C. et al., "Optimizing local charge stratification in a lean-burn spark ignition engine," *IMechE*, Vol. 211, Part D, pp. 145–154 (1997).

26. Arcoumanis, C. et al., "Modeling of pressure-swirl atomizer for gasoline direct-injection engines," SAE Technical Paper No.1999-01-0500 (1999).

27. Arcoumanis C. et al., "Pressure-swirl atomizers for DISI engines: further modeling and experiments," SAE Technical Paper No. 2000-01-1044 (2000).

28. Arters, D. et al., "A comparison of gasoline direct injection and port fuel injection vehicles; Part I—fuel system deposits and vehicle performance," SAE Technical Paper No. 1999-01-1498 (1999a).

29. Arters, D. et al., "A comparison of gasoline direct injection and port fuel injection vehicles; Part II—lubricant oil performance and engine wear," SAE Technical Paper No. 1999-01-1499 (1999b).

30. Arters, D. et al., "The effect on vehicle performance of injector deposits in a direct injection gasoline engine," SAE Technical Paper No. 2000-01-2021 (2000).

31. Ashizawa, T. et al., "Development of a new in-line 4-cylinder direct-injection gasoline engine," *Proceedings of the JSAE Fall Convention* (in Japanese), No. 71-98, Paper No. 9838237, pp. 5–8 (1998).

32. AUDI, *Audi Frankfurt Autoshow*, September 1997 (1997).

33. Auer, G., "Mitsubishi re-engineers GDI engine for Europe," *Automotive News Europe*, May 12, 1997.

34. AUTOMOTIVE ENGINEER, "Healthy future of GDI predicted by Mitsubishi," *Automotive Engineer*, No. 12, p. 6 (1997a).

35. AUTOMOTIVE ENGINEER, "Toyota's D4 direct injection gasoline engine," *Automotive Engineer*, No. 12, pp. 60–61 (1997b).

36. AUTOMOTIVE ENGINEER, "Nissan develops new direct injection engines," *Automotive Engineer*, Vol. 22, No. 9, p. 6 (1997c).

37. AUTOMOTIVE ENGINEERING, "Getting more direct," *Automotive Engineering*, No. 12, pp. 81–85 (1997).

38. Baby, X. et al., "Investigation of the in-cylinder tumble motion in a multi-valve engine: effect of the piston shape," SAE Technical Paper No. 971643 (1997).

39. Bae, C. et al., "Fuel-spray characteristics of high pressure gasoline direct injection in flowing fields," *Proceedings of the 4th JSME-KSME Thermal Engineering Conference*, October 1–6, 2000 (2000).

40. Balles, E. et al., "Fuel injection characteristics and combustion behavior of a direct-injection stratified-charge engine," SAE Technical Paper No. 841379 (1984).

41. Baranescu, G., "Some characteristics of spark assisted direct injection engine," SAE Technical Paper No. 830589 (1983).

42. Baumgarten, H. et al., "Vehicle application of a 4-cylinder tumble DISI engine," SAE Technical Paper No. 2001-01-0735 (2001).

43. Baumgarten, H. et al., "Development of a charge motion controlled combustion system for DI SI-engines and its vehicle application for EU-4 emission regulations," SAE Technical Paper No. 2000-01-0257 (2000).

44. Bechtold, R., "Performance, emissions, and fuel consumption of the White L-163-S stratified-charge engine using various fuels," SAE Technical Paper No. 780641 (1978).

45. Birch, S., "Advances at Audi," *Automotive Engineering*, No. 11, p. 30 (1997).

46. Birch, S., "Gasoline direct-injection developments," *Automotive Engineering*, p. 88, No. 2 (1998).

47. Birch, S., "Direct gasoline injection from Renault," *Automotive Engineering*, 7, pp. 28–29, (1999).

48. Bladon, S., "Carisma GDI vs. Rover 420 DI," *Diesel Car & 4x4*, No.1, pp. 36–42 (1998).

49. Block, B. et al., "Luminosity and laser-induced incandenscence investigations on a DI gasoline engine," SAE Technical Paper No. 2000-01-2903 (2000).

50. Boulouchos, K., "Strategies for future combustion systems—homogeneous or stratified charge?," SAE Technical Paper No.2000-01-0650 (2000).

51. Brehob, D. et al., "Stratified-charge engine fuel economy and emission characteristics," SAE Technical Paper No.982704 (1998).

52. Brooks, B., "Nissan to direct-inject VQ engines," *Ward's Engine and Vehicle Technology Update*, August 15, 1997 (1997).

53. Brooks, B., "DI gasoline engine problems outlined," *Ward's Engine and Vehicle Technology Update*, February 1, 1998, p. 2 (1998).

54. Brooks, B., "Saab shows advanced DGI combustion system," *Ward's Engine and Vehicle Technology Update*, October 15, 2000, pp. 1–2 (2000).

55. Buchheim, R. et al., "Ecological and economical aspects of future passenger car powertrains," FISITA Technical Paper No. P1404 (1996).

56. Buchholz, K., "Chrysler updates two-stroke engine progress," *Automotive Engineering*, No. 1, p. 84 (1997).

57. Buckland, J. et al., "Technology assessment of boosted direct injection stratified charge gasoline engines," SAE Technical Paper No. 2000-01-0249 (2000).

58. Burk, P. et al., "Future aftertreatment strategies for gasoline lean burn engines," *Proceedings of AVL Engine and Environment Conference*, pp. 219–231 (1997).

59. Caracciolo, F. et al., "An engine dynamometer test for evaluating port fuel injector plugging," SAE Technical Paper No. 872111 (1987).

60. Carlisle, H. et al., "The effect of fuel composition and additive content on injector deposits and performance of an air-assisted direct injection spark ignition (DISI) research engine," SAE Technical Paper No. 2001-01-2030 (2001).

61. Casarella, M., "Emission formation mechanisms in a two-stroke direct-injection engine," SAE Technical Paper No. 982697 (1998).

62. Castagne, M. et al., "Advanced tools for analysis of gasoline direct injection engines," SAE Technical Paper No. 2000-01-1903 (2000).

63. Cathcart, G. et al., "Fundamental characteristics of an air-assisted direct injection combustion system as applied to 4 stroke automotive gasoline engines," SAE Technical Paper No. 2000-01-0256 (2000).

64. Chaouche, A. et al., "NSDI-3: a small bore GDI engine," SAE Technical Paper No.1999-01-0172 (1999).

65. Chehroudi, B., "Gasoline direct injection (GDI)," *Powertrain International*, No. 2, pp. 6–7 (1999).

66. Cheng, W. et al., "An overview of hydrocarbon emissions mechanisms in spark-ignition engines," SAE Technical Paper No. 9832708 (1993).

67. Chinn, J. et al., "Computational analysis of swirl atomizer internal flow," *Proceedings of ICLASS-97* (1997).

68. Cho, N. et al., "Effect of in-cylinder air motion on fuel spray characteristics in a gasoline direct injection engines," SAE Technical Paper No.1999-01-0177 (1999).

69. Choi, K. et al., "A research on fuel spray and air flow fields for spark-ignited direct injection engine using laser image technology," SAE Technical Paper No.1999-01-0503 (1999).

70. Cole, R. et al., "Exhaust emissions of a vehicle with a gasoline direct-injection engine," SAE Technical Paper No. 982605 (1998).

71. Cole, R. et al., "Gaseous and particulate emissions from a vehicle with a spark-ignition direct-injection engine," SAE Technical Paper No.1999-01-1282 (1999).

72. Cousin, J. et al., "Transient flows in high pressure swirl injectors," SAE Technical Paper No. 980499 (1998).

73. Daisho, Y. et al., "A fundamental study on charge stratification," *Proceedings of COMODIA-85*, pp. 423–432 (1985).

74. Das, S. et al., "A study of air-assisted fuel injection into a cylinder," SAE Technical Paper No. 941876 (1994).

75. Das, S. et al., "A new approach for linking experimental data to spray modeling for an outwardly opening direct injection gasoline (DI-G) injector," *Proceedings of the 18th ICLASS*, Pasadena, USA (2000).

76. Date, T. et al., "Research and development of the Honda CVCC engine," SAE Technical Paper No. 740605 (1974).

77. Davy, M. et al., "Effects of injection timing on liquid-phase fuel distributions in a centrally-injected four-valve direct-injection spark-ignition engine," SAE Technical Paper No. 982699 (1998).

78. Davy, M. et al., "Effects of fuel composition on mixture formation in a firing direct-injection spark-ignition (DISI) engine: an experimental study using Mie-scattering and planar laser-induced fluorescence (PLIF) techniques," SAE Technical Paper No. 2000-01-1904 (2000).

79. Demmler, A., "Smallest GDI engine," *Automotive Engineering*, No. 3, p. 40 (1999).

80. Deschamps, B. et al., "Combined catalytic hot wires probe and fuel-air-ratio-laser induced-exciplex fluorescence air/fuel ratio measurements at the spark location prior to ignition in a stratified GDI engine," SAE Technical Paper No. 1999-01-3536 (1999).

81. Diwakar, R. et al., "Liquid and vapor fuel distributions from an air-assist injector—an experimental and computational study," SAE Technical Paper No. 920422 (1992).

82. Dodge, L., "Fuel preparation requirements for direct-injected spark ignition engines," *Proceedings of ILASS-America*, pp. 120–124 (1996a).

83. Dodge, L., "Fuel preparation requirements for direct-injected spark ignition engines," SAE Technical Paper No. 962015 (1996b).

84. Douaud, A., "Tomorrow's efficient and clean engines and fuels," FISITA Technical Paper No. K0006 (1996).

85. Drake, M. et al., "Crevice flow and combustion visualization in a direct-injection spark-ignition engine using laser imaging techniques," SAE Technical Paper No. 952454 (1995).

86. Duclos, J. et al., "3D modeling of intake, injection and combustion in a DI-SI engine under homogeneous and stratified operating conditions," *Proceedings of the 4th International Symposium COMODIA 98*, pp. 335–340 (1998).

87. Duggal, V. et al., "Review of multi-fuel engine concepts and numerical modeling of in-cylinder flow processes in direct injection engines," SAE Technical Paper No. 840005 (1984).

88. Durest, P. et al., "The air assisted direct injection ELEVATE automotive engine combustion system," SAE Technical Paper No. 2000-01-1899 (2000).

89. Duret, P., "The fields of application of IAPAC compressed air assisted DI fuel injection," *Modern Injection Systems for Direct Injection in Spark-Plug and Diesel Engines, ESSEN-HAUS DER TECHNIK*, September 23–24, 1997 (1997).

90. Duret, P. et al., "A new two-stroke engine with compressed-air assisted fuel injection for high efficiency low emissions applications," SAE Technical Paper No. 881076 (1988).

91. Edward, R. et al., "Influence of fuel injection timing over the performances of a direct injection spark ignition engine," SAE Technical Paper No. 1999-01-0174 (1999).

92. Eichlseder, H. et al., "Gasoline direct injection—chances and risks considering future emission scenarios," *Aachen Colloquium—Automobile and Engine Technology*, pp. 749–772 (1999).

93. Eichlseder, H. et al., "Gasoline direct injection—a promising engine concept for future demands," SAE Technical Paper No. 2000-01-0248 (2000a).

94. Eichlseder, H. et al., "Challenges to fulfill EU IV emission standards with gasoline-DI," *International Wiener Motor Symposium*, Vol. 1, pp. 105–134 (2000b).

95. Ekenberg, M. et al., "Fuel distribution in an air-assisted direct injected spark ignition engine with central injection and spark plug measured with laser induced fluorescence," SAE Technical Paper No. 2000-01-1898 (2000).

96. El-Emam, S.H. et al., "Performance of a stratified charge spark-ignition engine with an injection of different fuels," *Proceedings of the 4th International Symposium COMODIA 98*, pp. 329–334 (1998).

97. Ellzey, J. et al., "Simulation of stratified charge combustion," SAE Technical Paper No. 981454 (1998).

98. Emerson, J. et al., "Structure of sprays from fuel injectors part III : the Ford air-assisted fuel injector," SAE Technical Paper No. 900478 (1990).

99. Enright, B. et al., "A critical review of spark ignited diesel combustion," SAE Technical Paper No. 881317 (1988).

100. Evers, L., "Characterization of the transient spray from a high pressure swirl injector," SAE Technical Paper No. 940188 (1994).

101. Fan, L. et al., "Comparison of computed spray in a direct-injection spark-ignited engine with planar images," SAE Technical Paper No. 972883 (1997).

102. Fan, L. et al., "Intake flow simulation and comparison with PTV measurements," SAE Technical Paper No. 1999-01-0176 (1999).

103. Fansler, T. et al., "Swirl, squish and turbulence in stratified-charge engines: laser-velocimetry measurements and implications for combustion," SAE Technical Paper No. 870371 (1987).

104. Fansler, T. et al., "Fuel distribution in a firing direct-injection spark-ignition engine using laser-induced fluorescence imaging," SAE Technical Paper No. 950110 (1995).

105. Farrell, P. et al., "Intake air velocity measurements for a motored direct injection spark ignited engine," SAE Technical Paper No.1999-01-0499 (1999).

106. Faure, M. et al., "Application of LDA and PIV techniques to the validation of a CFD Model of a direct injection gasoline engine," SAE Technical Paper No. 982705 (1998).

107. Felton, P., "Laser diagnostics for direct-injection gasoline engines," *IMechE. Seminar of Lean Burn Combustion Engines*, S433, December 3–4, 1996 (1996).

108. Fraidl, G. et al., "Gasoline direct injection: actual trends and future strategies for injection and combustion systems," SAE Technical Paper No. 960465 (1996).

109. Fraidl, G. et al., "Gasoline direct injection - an integrated systems approach," *Proceedings of AVL Engine and Environment Conference*, pp. 255–278 (1997a).

110. Fraidl, G. et al., "Straight to the point," *Engine Technology International*, November 1997, pp. 30–34 (1997b).

111. Fraidl, G. et al., "Gasoline DI engines: the complete system approach by interaction of advanced development tool," SAE Technical Paper No. 980492 (1998).

112. Fraidl, G. et al., "Gasoline engine concepts related to specific vehicle classes," *International Wiener Motor Symposium*, Vol. 1, pp. 85–104 (2000).

113. Frank, R. et al., "Combustion characterization in a direct-injection stratified-charge engine and implications on hydrocarbon emissions," SAE Technical Paper No. 892058 (1989).

114. Frank, R. et al., "The importance of injection system characteristics on hydrocarbon emissions from a direct-injection stratified-charge engine," SAE Technical Paper No. 900609 (1990a).

115. Frank, R. et al., "The effect of fuel characteristics on combustion in a spark-ignited direct-injection engine," SAE Technical Paper No. 902063 (1990b).

116. Frank, R. et al., "The effect of piston temperature on hydrocarbon emissions from a spark-ignited direct-injection engine," SAE Technical Paper No. 910558 (1991).

117. Fry, M. et al., "Direct injection of gasoline—practical considerations," SAE Technical Paper No.1999-01-0171 (1999).

118. Fujieda M. et al., "Influence of the spray pattern on combustion characteristics of the direct injection SI engine," *Proceedings of ILASS-Japan* (in Japanese), pp. 173–177 (1995).

119. Fujikawa, T. et al., "Quantitative 2-D fuel distribution measurements in a direct-injection gasoline engine using laser-induced fluorescence technique," *Proceedings of the 4th International Symposium COMODIA 98*, pp. 317–322 (1998).

120. Fujimoto, H. et al., "Study on the oil dilution of a DI gasoline engine—1st report: dilution on the cylinder wall," *Proceedings of the JSAE Spring Convention* (in Japanese), Paper No. 9934014 (1999).

121. Fukui, H. et al., "The effect of octane number on stratified charge combustion of direct injection stratified charge engine," JSAE Technical Paper (in Japanese), No. 9934186 (1999).

122. Fukushima, C., "Mitsubishi clears California NOx requirement with new version of GDI engine, plans U.S. launch," *The Japan Automotive Digest*, Vol. III, No. 37 (1997).

123. Furuno, S. et al., "The effects of inclination angle of swirl axis on turbulence characteristics in a 4-valve lean burn engine with SCV," SAE Technical Paper No. 902139 (1990).

124. Gajdeczko, B. et al., "Application of two-color particle image velocimetry to a firing production direct-injection stratified-charge engine," SAE Technical Paper No.1999-01-1111 (1999).

125. Gastaldi, P. et al., "Development of new methods for investigating gasoline direct injection engines," *Aachen Colloquium—Automobile and Engine Technology*, pp. 775–791 (1999).

126. Gavine, A., "Do or DI?" *Engine Technology International*, No. 3, pp. 48–49 (2000).

127. Geiger, J. et al., "Direct injection engines—combustion and design," SAE Technical Paper No.1999-01-0170 (1999).

128. Georjon, T. et al., "Characteristics of mixture formation and combustion in a spray-guided concept gasoline direct injection engine: an experimental and numerical approach," SAE Technical Paper No. 2000-01-0534 (2000).

129. Ghandhi, J. et al., "Investigation of the fuel distribution in a two-stroke engine with an air-assisted injector," SAE Technical Paper No.940394 (1994).

130. Ghandhi, J. et al., "Fuel distribution effects on the combustion of a direct-injection stratified-charge engine," SAE Technical Paper No. 950460 (1995).

131. Ghandhi, J. et al., "Mixture preparation effects on ignition and combustion in a direct-injection spark-ignition engine," SAE Technical Paper No. 962013 (1996).

132. Gieshoff, J., "Improved SCR systems for heavy duty applications," SAE Technical Paper No. 2000-01-0189 (2000).

133. Giovanetti, A. et al., "Analysis of hydrocarbon emission mechanisms in a direct injection spark-ignition engine," SAE Technical Paper No. 830587 (1983).

134. Glaspie, C. et al., "Application of design and development techniques for direct injection spark ignition engines," SAE Technical Paper No.1999-01-0506 (1999).

135. Glover, M., "PSA shows: direction for injection," *Automotive Engineer*, July/August 2000, pp. 76–77 (2000).

136. Gluck, K. et al., "Cleaning the exhaust gas of Volkswagen FSI engines," MTZ Worldwide, No. 6, pp. 19–23 (2000).

137. Gobel, U. et al., "Durability aspects of NOx storage catalysts for direct injection gasoline vehicles," *Proceedings of Direkteinspritzung im Ottomotor II*, pp. 427–456 (1999).

138. Goppelt, G., "Der neue ottomotor mit direkteinspritzung von Renault," *MTZ*, No. 9, pp. 530–533 (1999).

139. Graskow, B. et al., "Characterization of exhaust particulate emissions from a spark ignition engine," SAE Technical Paper No. 980528 (1998).

140. Graskow, B. et al., "Particle emissions from two PFI spark ignition engines," SAE Technical Paper No.1999-01-1144 (1999a).

141. Graskow, B. et al., "Particle emissions from a DI spark ignition engine," SAE Technical Paper No.1999-01-1145 (1999b).

142. Graskow, B. et al., "Influence of fuel additives and dilution conditions on the formation and emission of exhaust particulate matter from a direct injection spark ignition engine," SAE Technical Paper No. 2001-01-2018 (2001).

143. Grigo, M. et al., "Charge motion controlled combustion system for direct injection SI engines," *Proceedings of Global Powertrain Conference on Advanced Engine Design and Performance*, pp. 66–75 (1998).

144. Grimaldi, C. et al., "Analysis method for the spray characteristics of a GDI system with high pressure modulation," SAE Technical Paper No. 2000-01-1043 (2000).

145. Guthrie, P. et al., "A review of fuel, intake and combustion system deposit issues relevant to 4-stroke gasoline direct fuel injection engines," SAE Technical Paper No. 2001-01-1202 (2001).

146. Habchi, C. et al., "Multidimensional modeling of gasoline spray impingement and liquid film heat transfer and boiling on heated surfaces," *Proceedings of the 18th ICLASS*, Pasadena, USA (2000).

147. Hall, D. et al., "Measurement of the number and size distribution of particles emitted from a gasoline direct injection vehicle," SAE Technical Paper No. 1999-01-3530 (1999).

148. Hall, M. et al., "In-cylinder flow and fuel transport in a 4-valve GDI engine: diagnostics and measurements," *Proceedings of Direkteinspritzung im Ottomotor*, pp. 132–146 (1998).

149. Han, Z. et al., "Modeling the effects of intake flow structures on fuel/air mixing in a direct-injected spark ignition engine," SAE Technical Paper No. 961192 (1996).

150. Han, Z. et al., "Effects of injection timing on air/fuel mixing in a direct-injection spark-ignition engine," SAE Technical Paper No. 970625 (1997a).

151. Han, Z. et al., "Multi-dimensional modeling of spray atomization and air/fuel mixing in a direct-injection spark-ignition engine characteristics," SAE Technical Paper No. 970884 (1997b).

152. Han, Z. et al., "Multidimentional modeling of DI gasoline engine sprays," *Proceedings of ILASS-America*, pp. 75–79 (1997c).

153. Han, Z. et al., "Internal structure of vaporizing pressure-swirl fuel sprays," *Proceedings of ICLASS-97*, pp. 474–481 (1997d).

154. Han, Z. et al., "Modeling atomization processes of pressure-swirl hollow-cone fuel sprays," *Atomization and Sprays*, Vol. 7, pp. 663–684 (1997e).

155. Hanashi, K. et al., "Development of new concept iridium plug," SAE Technical Paper No. 2001-01-1201 (2001).

156. Harada, J. et al., "Development of a direct injection gasoline engine," SAE Technical Paper No. 970540 (1997).

157. Harrington, D., "Interactions of direct injection fuel sprays with in-cylinder air motions," *SAE Transactions*, Vol. 93, (1984a).

158. Harrington, D., "Analysis of spray penetration and velocity dissipation for non-steady fuel injection," *ASME Technical Paper* 04-DGP-13 (1984b).

159. Harrington, D. et al., "Deposit-induced fuel flow reduction in multiport fuel injectors," SAE Technical Paper No. 892123 (1989).

160. Hashimoto, K. et al., "Effects of fuel properties on the combustion and emission of direct-injection gasoline engine," SAE Technical Paper No. 2000-01-0253 (2000).

161. Haslett, R. et al., "Stratified charge engines," SAE Technical Paper No. 760755 (1976).

162. Hatakeyama, S. et al., "A study of lean burn of a 4 stroke gasoline engine by the aid of low pressure air assisted in-cylinder injection (part II)," SAE Technical Paper No.1999-01-3689 (1999).

163. Hattori, H. et al., "Fundamental study on DISC engine with two-stage fuel injection," *JSME International J.*, Series B., Vol. 38, No. 1, pp. 129–135 (1995).

164. Heisler H., *Advanced engine technology*, SAE (1995).

165. Heitland, H. et al., "Can the best fuel economy of today's engines still be improved?," SAE Technical Paper No. 981912 (1998).

166. Henriot, S. et al., "NSDI-3, a small bore GDI engine," SAE Technical Paper No. 1999-01-0172 (1999).

167. Hentschel, W. et al., "Investigation of spray formation of DI gasoline hollow-cone injectors inside a pressure chamber and a glass ring engine by multiple optical techniques," SAE Technical Paper No.1999-01-3660 (1999).

168. Hepburn, J. et al., "Engine and aftertreatment modeling for gasoline direct injection," *SAE Technical Paper* No. 982596 (1998).

169. Heywood, J., *Internal combustion engine fundamentals*, McGraw-Hill, (1988).

170. Hiroyasu, H., "Experimental and theoretical studies on the structure of fuel sprays in diesel engines," *Proceedings of ICLASS-91*, Keynote Lecture, pp. 17–32 (1991).

171. Hochgreb, S. et al., "The effect of fuel volatility on early spray development from high-pressure swirl injectors," *Proceedings of Direkteinspritzung im Ottomotor*, pp. 107–116 (1998).

172. Hoffman, J. et al., "Spray photographs and pre liminary spray mass flux distribution measurements of a pulsed pressure atomizer," *Proceedings of ILASS-America*, pp. 288–291 (1996).

173. Hoffman, J. et al., "Comparison between air-assisted and single-fluid pressure atomizers for direct-injection SI engines via spatial and temporal mass flux measurements," SAE Technical Paper No. 970630 (1997).

174. Hoffman, J. et al., "Mass-related properties of various atomizers for direct-injection SI engines," SAE Technical Paper No. 980500 (1998).

175. Hoard, J. et al., "Plasma-catalysis for diesel exhaust treatment: current state of the art," *SAE Tchnical Paper*, No. 2001-01-0185 (2001).

176. Houston, R. et al., "Application of Orbital's low pressure, air-assisted fuel system to automotive direct injection 4-stroke engines," *Modern Injection Systems for Direct Injection in Spark-Plug and Diesel Engines, ESSEN-HAUS DER TECHNIK*, September 23–24, 1997 (1997).

177. Houston, R. et al., "Combustion and emissions characteristics of Orbital's combustion process applied to a multicylinder automotive direct injected 4 stroke engine," SAE Technical Paper No. 980153 (1998).

178. Hupperich, P. et al., "Direct injection gasoline engines—combustion and design," SAE Technical Paper No.1999-01-0170 (1999).

179. Iida, Y., "The current status and future trend of DISC engines," *Preprint of JSME Seminar* (in Japanese), No. 920-48, pp. 72–76 (1992).

180. Iiyama, A. et al., "Current status and future perspective of DISC engine," *Proceedings of JSAE* (in Japanese), No. 9431030, pp. 23–29 (1994).

181. Iiyama, A., "Direct injection gasoline engines," *Technical Seminar at Wayne State University*, February 26, 1996 (1996).

182. Iiyama, A. et al., "Attainment of high power with low fuel consumption and exhaust emissions in a direct-injection gasoline engine," *1998 FISITA Technical Paper* No. F98T048 (1998a).

183. Iiyama, A. et al., "Realization of high power and low fuel consumption with low exhaust emissions in a direct-injection gasoline engine," *Proceedings of GPC '98, Advanced Engine Design & Performance*, pp. 76–88 (1998b).

184. Ikeda, Y. et al., "Spray formation of air-assist injection for two-stroke engine," SAE Technical Paper No. 950271 (1995).

185. Ikeda, Y. et al., "Size-classified droplet dynamics and its slip velocity variation of air-assist injector spray," SAE Technical Paper No. 970632 (1997a).

186. Ikeda, Y. et al., "Cycle-resolved PDA measurement of size-classified spray structure of air-assist injector," SAE Technical Paper No. 970631 (1997b).

187. Ikeda, Y. et al., "Development of NOx storage-reduction 3-way catalyst for D-4 engines," *Proceedings of the 1998 JSAE Fall Meeting* (in Japanese), No. 101-98, Paper No. 9839597, pp. 9–12 (1998).

188. Ikeda, Y. et al., "Development of NOx storage-reduction 3-way catalyst for D-4 engines," SAE Technical Paper No. 1999-01-1279 (1999).

189. Inagaki, H. et al., "Influence of cylinder wet in a direct injection gasoline engine on piston ring lubrication," *Proceedings of JSAE* (in Japanese), No. 20005449 (2000).

190. Iriya, Y. et al., "Engine performance and the effects of fuel spray characteristics on direct injection S.I. engines," *Proceedings of JSAE Fall Convention* (in Japanese), Paper No. 9638031 (1996).

191. Ismailov, M. et al., "Laser-based techniques employed on gasoline swirl injector," *Proceedings of the 4th International Symposium COMODIA 98*, pp. 499–504 (1998).

192. Ito, Y. et al., "Study on improvement of torque response by direct injection gasoline engine and its application," *Proceedings of the JSAE Fall Convention*, No. 71-98, Paper No. 9838246, pp. 9–12 (1998).

193. Ito, Y. et al., "A new concept of direct injection gasoline engine—Part 6: development of a 3.0 liter in-line-6-cylinder engine," JSAE Technical Paper (in Japanese), No. 20005190 (2000).

194. Iwachido, K. et al., "Development of the NOx adsorber catalyst for use with high-temperature condition," SAE Technical Paper No. 2001-01-1298 (2001).

195. Iwakiri, Y. et al., "Effectiveness and issues of various measurement techniques used in evaluating spray characteristics in a direct-injection gasoline engine," JSAE Technical Paper No. 9935095 (1999).

196. Iwamoto, Y. et al., "Development of gasoline direct injection engine," SAE Technical Paper No. 970541 (1997a).

197. Iwamoto, Y. et al., "Development of gasoline direct injection engine," *Proceedings of JSAE Spring Convention* (in Japanese), No. 971, pp. 297–300 (1997b).

198. Iwata, M. *et al.*, "The spray characteristics and engine performance of EFI injector," *Proceedings of the Technical Conference of JSAE* (in Japanese), No. 861, pp. 29–32 (1986).

199. Jackson, N. et al., "A direct injection stratified charge gasoline combustion system for future European passenger cars," *IMechE. Seminar on Lean Burn Combustion Engines*, S433, December 3–4, 1996 (1996a).

200. Jackson, N. et al., "Gasoline combustion and gas exchange technologies for the 3 litre/100 km car—competition for the diesel engine," *International Symposium on Powertrain Technologies for a 3-Litre Car*, pp. 45–56 (1996b).

201. Jackson, N. et al., "Stratified and homogeneous charge operation for the direct injection gasoline engine–high power with low fuel consumption and emissions," SAE Technical Paper No. 970543 (1997a).

202. Jackson, N. et al., "A gasoline direct injection (GDI) powered vehicle concept with 3 litre/100 km fuel economy and EC stage 4 emission capability," *Proceedings of the EAEC 6th European Congress*, Italy, July 2–4, 1997 (1997b).

203. Jackson, N. et al., "Research and development of advanced direct injection gasoline engines," *Proceedings of the 18th Vienna Motor Symposium*, April 24–25, 1997 (1997c).

204. Jackson, N. et al., "A gasoline direct injection (GDI) powered vehicle concept with 3 litre/100 km fuel economy and EC stage 4 emission capability," *Paper presented at the EAEC 6th European Congress "Lightweight & Small Cars—the Answer to Future Needs,"* July 2–4, 1997, Italy (1997d).

205. Jang, C. et al., "Spray characteristics of an intermittent air-assisted fuel injector," *Proceedings of ICLASS'97*, Vol. 1, pp. 553–560 (1997).

206. Jang, C. et al., "Performance of prototype high pressure swirl injector nozzles for gasoline direct injection," SAE Technical Paper No. 1999-01-3654 (1999).

207. Jeong, K. et al., "Initial flame development under fuel stratified conditions," SAE Technical Paper No. 981429 (1998).

208. Jewett, D., "Direct injection boosts mileage, maintains muscle," *Automotive News*, March, 1997, p. 21 (1997a).

209. Jewett, D., "Engines run leaner and burn fuel more completely," *Automotive News*, March, 1997, p. 21 (1997b).

210. Joh, M. et al., "Numerical prediction of stratified charge distribution in a gasoline direct-injection engine—parametric studies," SAE Technical Paper No.1999-01-0178 (1999).

211. Johnson, D. et al., "Electronic direct fuel injection system applied to an 1100cc two-stroke personal watercraft engine," SAE Technical Paper No. 980756 (1998).

212. Johnson, B. et al., "Effects of fuel parameters on FTP emissions of a 1998 Toyota with a direct injection spark ignition engine," SAE Technical Paper No. 2000-01-1907 (2000).

213. Jones, C., "A progress report on Curtiss-Wright's rotary stratified charge engine development," SAE Technical Paper No. 740126 (1974).

214. Jost, K. et al., "Fuel-stratified injection from VW," *Automotive Engineering*, No. 1, pp. 63–65 (2001).

215. Kaihara, K. et al., "Development of automatic idling stop and start system utilizing a direct-injection gasoline engine," *SAE-Japan Automobile Technology* (in Japanese), No. 20004395, Vol. 54, No. 9, pp. 49–55 (2000).

216. Kaiser, E. et al., "Engine out emissions from a direct-injection spark-ignition (DISI) engine," SAE Technical Paper No. 1999-01-1529 (1999).

217. Kaiser, E. et al., "Exhaust emissions from a direct-injected spark-ignition (DISI) engine equipped with an air-forced fuel injector," SAE Technical Paper No. 2000-01-0254 (2000).

218. Kakuhou, A. et al., "LIF visualization of in-cylinder mixture formation in a direct-injection SI engine," *Proceedings of the 4th International Symposium COMODIA 98*, pp. 305–310 (1998).

219. Kakuhou, A., "Characteristics of mixture formation in a direct-injection SI engine with optimized in-cylinder swirl air motion," SAE Technical Paper No. 1999-01-0505 (1999).

220. Kalghatgi, G., "Deposits in gasoline engines— a literature review," SAE Technical Paper No. 902105 (1990).

221. Kamura, H. et al., "Development of in-cylinder gasoline direct injection engine," *Automobile Technology* (in Japanese), Vol. 50, No. 12, pp. 90–95 (1996).

222. Kanda, M. et al., "A new concept of direct injection SI gasoline engine—Part 2: combustion method and the application," JSAE Technical Paper (in Japanese), No. 9939604 (1999).

223. Kanda, M. et al., "Application of a new combustion concept to direct injection gasoline engine," SAE Technical Paper No. 2000-01-0531 (2000).

224. Kaneko, Y., *Studies of the gasoline direct injection engine*, Sankaido (in Japanese), 2000.

225. Kano, M. et al., "Analysis of mixture formation of direct injection gasoline engine," *Proceedings of JSAE Fall Convention* (in Japanese), No. 976, pp. 9–12 (1997).

226. Kano, M. et al., "Analysis of mixture formation of direct injection gasoline engine," SAE Technical Paper No. 980157 (1998).

227. Karl, G. et al., "Analysis of a direct injected gasoline engine," SAE Technical Paper No. 970624 (1997).

228. Kataoka, M. et al., "Measurement of fuel distribution in cavity of DI-SI engine by using of LIF," JSAE Technical Paper (in Japanese), No. 9939640 (1999).

229. Kato, S., "DISC engine technologies," *JSME Seminar* (in Japanese), No. 890-65, pp. 71–81 (1989).

230. Kato, S. et al., "New mixture formation technology of direct fuel injection stratified combustion SI engine (OSKA)," SAE Technical Paper No. 871689 (1987).

231. Kato, S. et al., "Direct fuel injection stratified charge engine by impingement of fuel jet (OSKA)—performance and combustion characteristics," SAE Technical Paper No. 900608 (1990).

232. Kato, S. et al., "Piston temperature measuring technology using electromagnetic induction (measurement of gasoline direct injection engine)," *Proceedings of the 16th Internal Combustion Engine Symposium* (in Japanese), pp. 325–330 (2000).

233. Kawamura, K. et al., "Development of instrument for measurement of air-fuel ratio in vicinity of spark plug—applications to DI gasoline engine," *Proceedings of the 14th Japan Internal Combustion Engine Symposium* (in Japanese), pp. 133–138 (1997).

234. Kech, J. et al., "Analyses of the combustion process in a direct injection gasoline engine," *Proceedings of the 4th International Symposium COMODIA 98*, pp. 287–292 (1998).

235. Keller, P. et al., "CFD analysis of the effect of injection parameters on combustion performance for a stoichiometric direct injection spark ignition engine," *International Wiener Motor Symposium*, Vol. 1, pp. 253–275 (2000).

236. Kenny, R. et al., "Application of direct air-assisted fuel injection to a SI cross-scavenged two-stroke engine," SAE Technical Paper No.932396 (1993).

237. Kenney, T., "Partitioning emissions tasks across engine and aftertreatment systems," SAE Technical Paper No. 1999-01-3475 (1999).

238. Kihara, Y. et al., "Numerical analysis for mixture formation of a fuel-direct injection engine," JSAE Technical Paper (in Japanese), No. 9939631 (1999).

239. Kim, C. et al., "Aldehyde and unburned fuel emission measurements from a methanol-fueled Texaco stratified charge engine," SAE Technical Paper No. 852120 (1985).

240. Kim, C. et al., "Deposit formation on a metal surface in oxidized gasolines," SAE Technical Paper No. 872112 (1987).

241. Kim, K. et al., "Spray characteristics of an air-assisted fuel injector for two-stroke direct-injection gasoline engines," *Atomization and Sprays*, Vol. 4, pp. 501–521 (1994).

242. Kinoshita, M. et al., "Study of nozzle deposit formation mechanism for direct injection gasoline engines," *Proceedings of JSAE Fall Convention* (in Japanese), No. 976, pp. 21–24 (1997).

243. Kinoshita, M. et al., "A method for suppressing formation of deposits on fuel injector for direct injection gasoline engine," SAE Technical Paper No.1999-01-3656 (1999).

244. Kittleson, D., "Engines and nanoparticles: a review," *J. Aerosol Sci.*, Vol.29, No.5/6, pp. 575–588 (1998).

245. Klenk, R. et al., "Investigations on a GDI-engine with airguided combustion system," *Aachen Colloquium—Automobile and Engine Technology*, pp. 817–830 (1999).

246. Koga, N. et al., "An experimental study on fuel behavior during cold start period of a direct injection spark ignition engine," SAE Technical Paper No. 2001-01-0969 (2001).

247. Koike, M. et al., "Vaporization of high pressure gasoline spray," *Proceedings of JSAE Spring Convention* (in Japanese), No. 971, pp. 325–328 (1997).

248. Koike, M. et al., "Influences of fuel vaporization on mixture preparation of a DI gasoline engine, *Proceedings of the 1998 JSAE Spring Convention* (in Japanese), No. 982, Paper No.9832378, pp. 103–106 (1998).

249. Koike, M. et al., "A new concept of direct injection SI gasoline engine—Part 1: mixture preparation method," JSAE Technical Paper (in Japanese), No. 9939596 (1999).

250. Koike, M. et al., "Research and development of a new direct injection gasoline engine," SAE Technical Paper No. 2000-01-0530 (2000).

251. Kondo, K. et al., "Development of new power train consisting of direct injection engine and CVT," *J. of Automobile Technology* (in Japanese), Vol. 54, No. 9, pp. 18–23 (2000).

252. Kono, S. et al., "A study of spray direction against swirl in D.I. engines," *Proceedings of COMODIA-90*, pp. 269–274 (1990).

253. Kou, Y. and Itohga, H., *Gasoline direct injection*, Sankaido (in Japanese), 2000.

254. Kramer, F. et al., "Effect of compression ratio on the combustion of a pressure charged gasoline direct injection engine," SAE Technical Paper No. 2000-01-0250 (2000).

255. Krebs, S. et al., "A cooperative approach to air-assisted direct gasoline injection," *Proceedings of GPC'98, Advanced Engine Design & Performance*, pp. 89–102 (1998).

256. Krebs, R. et al., "FSI–gasoline direct injection engine for the Volkswagen Lupo," *International Wiener Motor Symposium*, Vol. 1, pp. 180–205 (2000).

257. Kubo, M. et al., "Technique for analyzing swirl injectors of direct-injection gasoline engines," SAE Technical Paper No. 2001-01-0964 (2001).

258. Kuder, J. et al., "Optimizing parameters for gasoline direct injection engines," *MTZ Worldwide*, No. 6, pp. 8–11 (2000).

259. Kume, T. et al., "Combustion control technologies for direct injection SI engine," SAE Technical Paper No. 960600 (1996).

260. Kusell, M. et al., "Motronic MED7 for gasoline direct injection: system architecture and diagnosis," *Aachen Colloquium—Automobile and Engine Technology*, pp. 831–851 (1999a).

261. Kusell, M. et al., "Motronic MED7 for gasoline direct injection engines: control strategies and calibration procedures," SAE Technical Paper No.1999-01-1284 (1999b).

262. Kuwahara, K. et al., "A study of combustion characteristics in a direct injection gasoline engine by high-speed spectroscopic measurement," *Proceedings of the Internal Combustion Engine Symposium—Japan* (in Japanese), pp. 145–150 (1996).

263. Kuwahara, K. et al., "Mixture preparation and flame propagation in gasoline direct-injection engine," *Proceedings of the 14th Japan Internal Combustion Engine Symposium* (in Japanese), pp. 115–120 (1997).

264. Kuwahara, K. et al., "Mixing control strategy for engine performance improvement in a gasoline direct injection engine," SAE Technical Paper No. 980158 (1998a).

265. Kuwahara, K. et al., "Control of mixing and combustion for Mitsubishi GDI engine," *Proceedings of the 1998 JSAE Spring Convention*, No. 984, Paper No. 9833791, pp. 35–38 (1998b).

266. Kuwahara, K. et al., "Two-stage combustion for quick catalyst warm-up in gasoline direct injection engine," *Proceedings of the 4th International Symposium COMODIA 98*, pp. 293–298 (1998c).

267. Kuwahara, K. et al., "Intake-port design for Mitsubishi GDI engine to realize distinctive in-cylinder flow and high charge coefficient," SAE Technical Paper No. 2000-01-2018 (2000).

268. Kwon, Y. et al., "The effect of fuel sulfur content on the exhaust emissions from a lean burn gasoline direct injection vehicle marketed in Europe," SAE Technical Paper No.1999-01-3585 (1999).

269. Kwon, Y. et al., "Emissions response of a European specification direct-injection gasoline vehicle to a fuels matrix incorporating independent variations in both compositional and distillation parameters," SAE Technical Paper No.1999-01-3663 (1999).

270. Lacher, S. et al., "In-cylinder mixing rate measurements," SAE Technical Paper No.1999-01-1110 (1999).

271. Laforgia, D. et al., "Structure of sprays from fuel injectors—part II, the Ford DFI-3 fuel injector," SAE Technical Paper No. 890313 (1989).

272. Lai, M. et al., "Characteristics of direct injection gasoline spray wall impingement at elevated temperature conditions," *Proceedings of the 12th Annual Conference on Liquid Atomization and Spray Systems*, May 1999 (1999).

273. Lake, T. et al., "Preliminary investigation of solenoid activated in-cylinder injection in stoichiometric S.I. engine," SAE Technical Paper No. 940483 (1994).

274. Lake, T. et al., "Simulation and development experience of a stratified charge gasoline direct injection engine," SAE Technical Paper No. 962014 (1996).

275. Lake, T. et al., "Comparison of direct injection gasoline combustion systems," SAE Technical Paper No. 980154 (1998).

276. Lake, T. et al., "Development of the control and aftertreatment system for a Euro IV G-Di vehicle," SAE Technical Paper No.1999-01-1281 (1999a).

277. Lake, T. et al., "Development of the control and aftertreatment system for a GDI engine," *MTZ Worldwide*, No. 12, pp. 2–5 (1999b).

278. Ledoyen, S. et al., "Experimental investigation on the characteristics and on the reproducibility of the flow issuing from a high-pressure direct-injection nozzle," SAE Technical Paper No.1999-01-3655 (1999).

279. Leduc, P. et al., "Gasoline direct injection: a suitable standard for the (very) near future," *Proceedings of Direkteinspritzung im Ottomotor*, pp. 38–51 (1998).

280. Lee, C. et al., "Initial comparisons of computed and measured hollow-cone sprays in an engine," SAE Technical Paper No. 940398 (1994).

281. Lee, C. et al., "Experimental measurements of the thickness of a deposited film from spray/wall interaction," *Proceedings of ILASS-America Annual Conference,* Dearborn, Michigan (2001).

282. Lee, S. et al., "A comparison of fuel distribution and combustion during engine cold start for direct and port fuel injection systems," SAE Technical Paper No. 1999-01-1490 (1999).

283. Lee, S. et al., "Engine cold-start testing," *Automotive Engineering*, No. 3 (2000a).

284. Lee, S. et al., "Effects of swirl and tumble on mixture preparation during cold start of a gasoline direct-injection engine," SAE Technical Paper No. 2000-01-1900 (2000b).

285. Lefebvre, A., *Atomization and sprays*, Hemisphere Publishing Corporation (1989).

286. Lenz, U. et al., "Air-fuel ratio control for direct injecting combustion engines using neutral networks," SAE Technical Paper No. 981060 (1998).

287. Lewis, J.M., "UPS multifuel stratified charge engine development program—field test," SAE Technical Paper No. 860067 (1986).

288. Lewis A., Mitsubishi GDI, *Diesel Car & 4x4*, No. 12, pp. 20–21 (1997).

289. Li, G. et al., "Modeling fuel preparation and stratified combustion in a gasoline direct injection engine," SAE Technical Paper No.1999-01-0175 (1999).

290. Li, J. et al., "Preliminary investigation of a diffusing oriented spray stratified combustion system for DI gasoline engines," SAE Technical Paper No. 980150 (1998).

291. Li, J. et al., "Further experiments on the effects of in-cylinder wall wetting on HC emissions from direct injection gasoline engines," SAE Technical Paper No. 1999-01-3661 (1999).

292. Li, J. et al., "Flow simulation of a direct-injection gasoline diaphragm fuel pump with structural interactions," SAE Technical Paper No. 2000-01-1047 (2000).

293. Li, S. et al., "Spray characterization of high pressure gasoline fuel injectors with swirl and non-swirl nozzles," SAE Technical Paper No. 981935 (1998).

294. Lindgren, R. et al., "Modeling gasoline spray-wall interaction," SAE Technical Paper No. 2000-01-2808 (2000).

295. Lippert, A. et al., "Modeling of multicomponent fuels using continuous distributions with application to droplet evaporation and sprays," SAE Technical Paper No. 972882 (1997).

296. Lykowski, J., "Spark plug technology for gasoline direct injection GDI engines," SAE Technical Paper No. 980497 (1998).

297. MacInnes, J. et al., "Computation of the Spray from an Air-Assisted Fuel Injector," SAE Technical Paper No. 902079 (1990).

298. Mao, C., "Investigation of carbon formation inside fuel injector systems," *Proceedings of ILASS-America '98*, pp. 344–348 (1998).

299. Maricq, M. et al., "Particulate emissions from a direct-injection spark-ignition (DISI) engine," SAE Technical Paper No. 1999-01-1530 (1999).

300. Maricq, M. et al., "Sooting tendencies in an air-forced direct injection spark-ignition (DISI) engine," SAE Technical Paper No. 2000-01-0255 (2000).

301. Martin, J., "Fuel/air mixture preparation in SI engines," *Proceedings of 1995 KSEA International Technical Conference*, PART I: Automotive Technology, pp. 155–161 (1995).

302. Matsushita, S. et al., "Mixture formation process and combustion process of direct injection S.I. engine," *Proceedings of JSAE* (in Japanese), No. 965, Oct. 10, 1996, pp. 101–104 (1996).

303. Matthews, R. et al., "Effects of load on emissions and NOx trap/catalyst efficiency for a direct injection spark ignition engine," SAE Technical Paper No. 1999-01-1528 (1999).

304. Matthews, R. et al., "Effect of fuel parameters on emissions from a direct injection spark ignition engine during constant speed, variable load tests," SAE Technical Paper No. 2000-01-1909 (2000).

305. Mccann, K., "MMC ready with first DI gasoline engine," *WARD's Engine and Vehicle Technology Update*, Vol. 21, No. 11, June 1, 1995, pp. 1–2 (1995).

306. Menne, R. et al., "Meeting future emission standards with the new Ford direct injection gasoline engine," *International Wiener Motor Symposium*, Vol. 1, pp. 206–230 (2000).

307. Meurer, S. et al., "Development and operational results of the MAN FM combustion system," SAE Technical Paper No. 690255 (1969).

308. Meyer, J. et al., "Spray visualization of air-assisted fuel injection nozzles for direct injection SI-engines," SAE Technical Paper No. 970623 (1997a).

309. Meyer, J. et al., "Study and visualization of the fuel distribution in a stratified spark ignition engine with EGR using laser-induced fluorescence," SAE Technical Paper No. 970868 (1997b).

310. Miok, J. et al., "Numerical prediction of charge distribution in a lean burn direct-injection spark-ignition engine," SAE Technical Paper No. 970626 (1997).

311. Mitchell, E. et al., "A stratified charge multifuel military engine—a progress report," SAE Technical Paper No. 720051 (1972).

312. Miyajima, A. et al., "A study on fuel spray pattern control of fuel injector of gasoline direct injection engines," SAE Technical Paper No. 2000-01-1045 (2000).

313. Miyake, M., "Developing a new stratified-charge combustion system with fuel injection for reducing exhaust emissions in small farm and industrial engines," SAE Technical Paper No. 720196 (1972).

314. Miyamoto, N. et al., "Combustion and emissions in a new concept DI stratified charge engine with two-stage fuel injection," SAE Technical Paper No. 940675 (1994).

315. Miyamoto, T. et al., "Structure of sprays from an air-assist hollow-cone injector," SAE Technical Paper No. 960771 (1996).

316. Morello, L. et al., "Global approach to the fuel economy improvement in passenger cars," *International Symposium on Powertrain Technologies for a 3-Litre-Car*, pp. 33–43 (1996).

317. Mori, K. et al., "Numerical analysis of gas-exchange process in a two cycle gasoline direct injection engine," JSAE Technical Paper (in Japanese), No. 9939613 (1999).

318. Moriyoshi, Y. et al., "Proposition of a stratified charge system by using in-cylinder gas motion," SAE Technical Paper No. 962425 (1996).

319. Moriyoshi, Y. et al., "Control of flow field during compression stroke by that of intake flow," *Proceedings of the Japan 13th Internal Combustion Engine Symposium* (in Japanese), pp. 299–304 (1997a).

320. Moriyoshi, Y. et al., "Combustion characteristics of a direct injection gasoline engine with enhanced gas motion," *Proceedings of JSAE Spring Convention* (in Japanese), No. 971, pp. 341–344 (1997b).

321. Moriyoshi, Y. et al., "Analysis of flame propagation phenomenon in simplified stratified charge conditions," *Proceedings of JSAE Fall Convention* (in Japanese), No. 976, pp. 17–20 (1997c).

322. Moriyoshi, Y. et al., "Combustion analysis of a direct injection gasoline engine from theoretical and experimental viewpoints," *Proceedings of the 14th Japan Internal Combustion Engine Symposium* (in Japanese), pp. 127–132 (1997d).

323. Moriyoshi, Y. et al., "Evaluation of a concept for DI combustion using enhanced gas motion," SAE Technical Paper No. 980152 (1998a).

324. Moriyoshi, Y. et al., "Analysis of stratified charge combustion in an idealized chamber," *Proceedings of the 1998 JSAE Spring Convention*, No. 984, Paper No. 9831018, pp. 31–34 (1998b).

325. Moriyoshi, Y. et al., "Combustion control of gasoline DI engine using enhanced gas motion," *Proceedings of the 4th International Symposium COMODIA 98*, pp. 299–304 (1998c).

326. Moriyoshi, Y. et al., "Analysis of mixture formation process with a swirl-type injector," SAE Technical Paper No. 2000-01-2057 (2000).

327. Morris, S.W., "The evaluation of performance enhancing fluids and the development of measurement and evaluation techniques in the Mitsubishi GDI engine," SAE Technical Paper No. 1999-01-1496 (1999).

328. Moser, W. et al., "Gasoline direct injection - a new challenge for future engine management systems," *MTZ Worldwide*, Vol. 58, No. 9, pp. 1–5 (1997).

329. Muller, W. et al., "Durability aspects of NOx absorption catalysts for direct injection gasoline vehicles with regard to European application," SAE Technical Paper No.1999-01-1285 (1999).

330. Mundorff, F. et al., "Direct injection—development trends for gasoline and diesel engines," *Proceedings of JSAE Spring Convention*, No. 971, pp. 301–304 (1997).

331. Nagashima, M. et al., "Combustion analysis of a direct injection stratified charge gasoline engine," JSAE Technical Paper (in Japanese), No. 9939622 (1999).

332. Naitoh, K. et al., "Synthesized spheroid particle (SSP) method for calculating spray phenomena in direct-injection SI engines," SAE Technical Paper No. 962017 (1996).

333. Naitoh, K. et al., "Numerical simulation of the fuel mixing process in a direct-injection gasoline engine," *Proceedings of the 14th Japan Internal Combustion Engine Symposium* (in Japanese), pp. 139–143 (1997).

334. Naitoh, K. et al., "Numerical simulation of the fuel mixing process in a direct-injection gasoline engine," SAE Technical Paper No. 981440 (1998).

335. Nakanishi, K. et al., "Application of a new combustion concept to direct injection gasoline engine," *International Wiener Motor Symposium*, Vol. 1, pp. 59–84 (2000).

336. Nakashima, T. et al., "A new concept of direct injection gasoline engine—Part 4: a study of stratified charge combustion characteristics by radical luminescence measurement," *JSAE Technical Paper* (in Japanese), No. 20005141 (2000).

337. Newmann, R., "Being direct," *Engine Technology International*, November 1977, pp. 66–70 (1997).

338. Newmann, R. et al., "Air to the DI engine," *Engine Technology International*, pp. 32–37 (1999).

339. Noda, T. et al., "Effects of fuel and air mixing on WOT output in direct injection gasoline engine," *Proceedings of JSAE Fall Convention* (in Japanese), No. 976, pp. 1–4 (1997).

340. Nogi, T. et al., "Stability improvement of direct fuel injection engine under lean combustion operation," SAE Technical Paper No. 982703 (1998).

341. Nohira, H. et al., "Development of Toyota's direct injection gasoline engine," *Proceedings of AVL Engine and Environment Conference*, pp. 239–249 (1997).

342. Noma, K. et al., "Optimized gasoline direct injection engine for the European market," *SAE Technical Paper*, No. 980150 (1998).

343. Ohkubo, H. et al., "Deposit formation and control in direct injection spark ignition engines," JSAE Technical Paper (in Japanese), No. 20005030 (2000).

344. Ohm, I. et al., "Initial flame development under fuel stratified conditions," SAE Technical Paper No. 981429 (1998).

345. Ohm, I. et al., "Fuel stratification process in the cylinder of an axially stratified engine," SAE Technical Paper No. 2000-01-2842 (2000a).

346. Ohm, I. et al., "Mechanism of axial stratification and its effect in an SI engine," SAE Technical Paper No. 2000-01-2843 (2000b).

347. Ohsuga, M. et al., "Mixture preparation for direct-injection SI engines," SAE Technical Paper No. 970542 (1997).

348. Ohyama, Y. et al., "Effects of fuel/air mixture preparation on fuel consumption and exhaust emission in a spark ignition engine," IMechE Paper, No. 925023, C389/232, pp. 59–64 (1992).

349. Ohyama, Y. et al., "Mixture formation in gasoline direct injection engine," *Proceedings of Direkteinspritzung im Ottomotor*, pp. 79–106 (1998).

350. Okabe, N. et al., "Study on deposits of direct injection gasoline engine," JSAE Technical Paper (in Japanese), No. 9934177 (1999).

351. Okada, Y. et al., "Development of high-pressure fueling system for a direct-injection gasoline engine," SAE Technical Paper No. 981458 (1998).

352. Okamoto, K. et al., "Study on combustion chamber deposits from direct injection gasoline engine," JSAE Technical Paper (in Japanese), No. 20005111 (2000).

353. Olofsson, E. et al., "A high dilution stoichiometric combustion concept using a wide variable spark gap and in-cylinder air injection in order to meet future CO_2 requirements and worldwide emissions regulations," SAE Technical Paper No. 2001-01-0246 (2001).

354. Ortmann, R. et al., "Methods and analysis of fuel injeciton, mixture preparation and charge stratification in different direct injected SI engines," SAE Technical Paper No. 2001-01-0970 (2001).

355. Park, J. et al., "Characteristics of direct injection gasoline spray wall impingement at elevated temperature conditions," SAE Technical Paper No. 1999-01-3662 (1999).

356. Park, Y.K. et al., "Spray characteristics of a gasoline direct swirl injector," *Proceedings of the 18th ICLASS*, Pasadena, USA (2000).

357. Parrish, S. et al., "Transient spray characteristics of a direct-injection spark-ignited fuel injector," SAE Technical Paper No. 970629 (1997).

358. Parrish, S. et al., "Intake flow effects on fuel sprays for direct-injection spark-ignited engines," *Proceedings of the 4th International Symposium COMODIA 98*, pp. 311–316 (1998).

359. Piccone, A. et al., "Strategies for fuel economy improvement of gasoline powertrains," *International Symposium on Powertrain Technologies for a 3-Litre-Car*, pp. 77–93 (1996).

360. Piock, W. et al., "Future gasoline engine concepts based on direct injection technology," *Proceedings of the 1998 JSAE Spring Convention*, No.984, Paper No.9831009, pp. 27–30 (1998).

361. Pischinger, F. et al., "Future trends in automotive engine technology," *FISITA Technical Paper* No. P1303 (1996).

362. Plackmann, J. et al., "The effects of mixture stratification on combustion in a constant-volume combustion vessel," SAE Technical Paper No. 980159 (1998).

363. Ponticel P., "High-pressure fuel injection by PSA," *Automotive Engineering*, No. 11, pp. 71–72 (2000).

364. Pontoppidan, M. et al., "Direct fuel-injection—a study of injector requirements for different mixture preparation concepts," SAE Technical Paper No. 970628 (1997).

365. Pontoppidan, M. et al., "Improvements of GDI-injector optimization tools for enhanced SI-engine combustion chamber layout," SAE Technical Paper No. 980494 (1998).

366. Pontoppidan, M. et al., "Experimental and numerical approach to injection and ignition optimization of lean GDI-combustion behavior," SAE Technical Paper No.1999-01-0173 (1999).

367. Pontoppidan, M. et al., "Enhanced mixture preparation approach for lean stratified SI-combustion by a combined use of GDI and electronically controlled valve-timing," SAE Technical Paper No. 2000-01-0532 (2000).

368. Preussner, C. et al., "Gasoline direct injection, a new challenge for future gasoline control systems—part 2: injector and mixture formation," *MTZ*, Vol. 58, No.10 (1997).

369. Preussner, C. et al., "GDI: interaction between mixture preparation, combustion system and injector performance," SAE Technical Paper No. 980498 (1998).

370. Ranini, A. et al., "Turbocharging a gasoline direct injection engine," SAE Technical Paper No.2001-01-0736 (2001).

371. Ren, W.M. et al., "Geometrical effects on flow characteristics of a gasoline high pressure swirl injector," SAE Technical Paper No. 971641 (1997).

372. Ren, W. et al., "Computation of the hollow cone spray from a pressure-swirl injector," *Proceedings of ILASS-America '98*, pp. 115–119 (1998a).

373. Ren, W. et al., "Computations of hollow-cone sprays from a pressure swirl injector," *SAE Technical Paper* No.982610 (1998b).

374. Reuter, W. et al., "Innovative EGR systems for direct-injection gasoline engines," *Aachen Colloquium—Automobile and Engine Technology*, pp. 793–814 (1999).

375. Richter, M. et al., "Investigation of the fuel distribution and the in-cylinder flow field in a stratified charge engine using laser techniques and comparison with CFD-modeling," SAE Technical Paper No. 1999-01-3540 (1999).

376. Risi, A. et al., "A study of H_2, CH_4, C_2H_6 mixing and combustion in a direct-injection stratified-charge engine," SAE Technical Paper No. 971710 (1997).

377. Robeck, C. et al., "Simulation of stratified charge combustion," SAE Technical Paper No. 981454 (1998).

378. Ronald, B. et al., "Direct fuel injection—a necessary step of development of the SI engine," *FISITA Technical Paper* No. P1613 (1996).

379. SAAB, Paris motor show, September 2000 (2000).

380. Sagawa, T. et al., "Study on the oil dilution of a DI gasoline engine—2nd report: dilution in the oil pan," *Proceedings of the JSAE Spring Convention* (in Japanese), Paper No. 9934023 (1999).

381. Saito, A. et al., "Improvement of fuel atomization electronic fuel injector by air flow," *Proceedings of ICLASS-88*, pp. 263–270 (1988).

382. Salters, D. et al., "Fuel spray characterization within an optically accessed gasoline direct injection engine using a CCD imaging system, SAE Technical Paper No. 961149 (1996).

383. Sandquist, H. et al., "Influence of fuel volatility on emissions and combustion in a direct injection spark ignition engine," SAE Technical Paper No. 982701 (1998).

384. Sandquist, H. et al., "Comparison of homogeneous and stratified charge operation in a direct injection spark ignition engine," JSAE Technical Paper No. 9935509 (1999).

385. Sandquist, H. et al., "Sources of hydrocarbon emissions from a direct injection stratified charge spark ignition engine," SAE Technical Paper No. 2000-01-1906 (2000).

386. Sasaki, S. et al., "Effects of EGR on direct injection gasoline combustion," *Proceedings of JSAE Spring Convention* (in Japanese), No. 971, pp. 333–336 (1997).

387. Sawada, O., "Automotive gasoline direct injection," *JSME Seminar* (in Japanese), No. 97-88, pp. 57–64 (1997).

388. Schapertons, H. et al., "VW's gasoline direct injection (GDI) research engine," SAE Technical Paper No. 910054 (1991).

389. Schdidt, G. et al., "Comparison of direct injection petrol and diesel engines with regard to fuel efficiency," 1998 FISITA Technical Paper No. F98T053 (1998).

390. Schechter, M. et al., "Air-forced fuel injection system for 2-stroke D.I. gasoline engine," SAE Technical Paper No. 910664 (1991).

391. Scherenberg, H., "Ruckbliik uber 25 Jahre Benzin-Einspritzung in Deutchland," *MTZ*, 16, (1955).

392. Schmidt, D. et al., "Pressure-swirl atomization in the near field," SAE Technical Paper No. 1999-01-0496 (1999).

393. Scholten, I. et al., "2.2 L ECOTEC DIRECT from Opel—gasoline direct injection," *International Wiener Motor Symposium*, Vol. 1, pp. 231–252 (2000).

394. Schreffler, R., "MMC, Japanese work to expand gasoline DI tech," *Ward's Engine and Vehicle Technology Update*, February 1, 1998, pp. 2–3 (1998a).

395. Schreffler, R., "Mitsubishi says GDI is LEV-ready," *Ward's Engine and Vehicle Technology Update*, February 15, 1998, p. 2 (1998b).

396. Schreffler, R., "Mitsubishi seeks help in expanding GDI range," *Ward's Engine and Vehicle Technology Update*, March 1, 1999, p. 5 (1999).

397. Schutte, M. et al., "Spatially resolved air-fuel ratio and residual gas measurements by spontaneous Raman scattering in a firing direct injection gasoline engine," SAE Technical Paper No. 2000-01-1795 (2000).

398. Scott, D., "Euro gasoline engines follow DI trend," *Ward's Engine and Vehicle Technology Update*, Vol. 24, No. 7, pp. 1–2 (1998).

399. Scott, D., "Renault first Euro automaker with gasoline direct injection," *WARD's Auto World*, 5, p. 138 (1999a).

400. Scott, D., "VW's first DGI engine handles lean-burn," *WARD's Auto World*, October 1, 1999, p. 1 (1999b).

401. Scussei, A. et al., "The Ford PROCO engine update," SAE Technical Paper No. 780699 (1978).

402. Seiffert, U., "The automobile in the next century," FISITA Technical Paper No. K0011 (1996).

403. Selim, M. et al., "Application of CFD to the matching of in-cylinder fuel injection and air motion in a four stroke gasoline engine," SAE Technical Paper No. 971601 (1997).

404. Sendyka, B., "A description of the shape of an air-fuel mixture and determination of the injection advance angles related to the spark discharges in a gasoline direct-injection engine," SAE Technical Paper No. 980496 (1998).

405. Shelby, M. et al., "Early spray development in gasoline direct-injected spark ignition engines," SAE Technical Paper No. 980160 (1998).

406. Shimizu, R. et al., "A new concept of direct injection gasoline engine—the third report: visualization of fuel mixture distribution using LIF technique," JSAE Technical Paper (in Japanese), No. 20005250 (2000).

407. Shimotani, K. et al., "Characteristics of gasoline in-cylinder direct injection engine," *Proceedings of the Internal Combustion Engine Symposium—Japan* (in Japanese), pp. 289–294 (1995).

408. Shimotani, K. et al., "Characteristics of exhaust emission on gasoline in-cylinder direct injection engine," *Proceedings of the Internal Combustion Engine Symposium—Japan* (in Japanese), pp. 115–120 (1996).

409. Sick, V. et al., "Experimental Investigation of Droplet Size Distributions in a Fan Spray," *Proceedings of ILASS-America 2001 Annual Conference* (2001).

410. Simko, A. et al., "Exhaust emissions control by the Ford Programmed combustion process—PROCO," SAE Technical Paper No. 720052 (1972).

411. Solomon, A., Anderson, R., Najt, P., and Zhao, F. (editors), "Direct fuel injection for gasoline engines," *Progress in Technology Series (PT-80)*, SAE, 2000.

412. Sonoda, Y. et al., "Development of a new direct injection gasoline engine system for European market," *Proceedings of Aachen Colloquium—Automobile and Engine Technology*, pp. 873–885 (2000).

413. Spicher, U. et al., "Gasoline direct injection (GDI) engines—development potentialities," SAE Technical Paper No. 1999-01-2938 (1999).

414. Spiegel, L. et al., "Mixture formation and combustion in a spark ignition engine with direct fuel injection," SAE Technical Paper No. 920521 (1992).

415. Stan, C. et al., "Fluid dynamic modeling of gasoline direct injection for compact combustion chambers," SAE Technical Paper No. 980755 (1998).

416. Stan, C. et al., "Car hybrid propulsion strategy using an ultra-light GDI two stroke engines," SAE Technical Paper No. 1999-01-2940 (1999a).

417. Stan, C. et al., "Concept for modeling and optimization of the mixture formation using gasoline direct injection in compact high speed engines," SAE Technical Paper No. 1999-01-2935 (1999b).

418. Stan, C. et al., "Development, modeling and engine adaptation of a gasoline direct injection system for scooter engines," SAE Technical Paper No. 1999-01-3313 (1999c).

419. Stan, C. et al., "Concept of interactive development of a GDI system with high-pressure modulation," SAE Technical Paper No. 2000-01-1042 (2000a).

420. Stan, C. et al., "Direct injection of variable gasoline/methanol mixtures: a future potential of SI engines," SAE Technical Paper No. 2000-01-2904 (2000b).

421. Stan, C. (editor), *Direct injection systems for spark ignition and compression ignition engines*, SAE, 2000.

422. Stanglmaier, R. et al., "Fuel-spray/charge motion interaction within the cylinder of a direct-injection, 4-valve, SI engine," SAE Technical Paper No. 980155 (1998).

423. Stanglmaier, R. et al., "The effect of in-cylinder wall wetting location on the HC emissions from SI engines," SAE Technical Paper No.1999-01-0502 (1999).

424. Stevens, E. et al., "Piston wetting in an optical DISI engine: fuel films, pool fires, and soot generation," SAE Technical Paper No. 2001-01-1203 (2001).

425. Stocker, H. et al., "Gasoline direct injection and engine management—challenge and implementation," *Proceedings of AVL Engine and Environment Conference*, pp. 111–133 (1997).

426. Stocker, H. et al., "Application of air-assisted direct injection to automotive 4-stroke engines—the 'total system approach'," *Aachen Colloquium Automobile & Engine Technology*, October 5–7, 1998, pp. 711–729 (1998).

427. Stokes, J. et al., "A gasoline engine concept with improved fuel economy—the lean boost system," SAE Technical Paper No. 2000-01-2902 (2000).

428. Stone, R., *Introduction to internal combustion engines*, SAE, 2000.

429. Stovell, C. et al., "Emissions and fuel economy of 1998 Toyota with a direct injection spark ignition engine," SAE Technical Paper No. 1999-01-1527 (1999).

430. Stovell, C. et al., "Effect of fuel parameters on speciated hydrocarbon emissions from a direct injection spark ignition engine," SAE Technical Paper No. 2000-01-1908 (2000).

431. Strehlau, W. et al., "Lean NOx catalysis for gasoline fueled European cars," *Automotive Engineering*, No. 2, pp. 133–135 (1997).

432. Stutzenberger, H. et al., "Gasoline direct injection for SI engines—development status and outlook, VDI," *The 17th International Vienna Engine Symposium*, April 25–26, 1997 (1997).

433. Su, J. et al., "Towards quantitative characterization of transient fuel sprays using planar laser induced fluorescence imaging," *Proceedings of ILASS-America'98*, pp. 106–110 (1998).

434. Sugimoto, T. et al., "Toyota air-mix type two-hole injector for 4-valve engines," SAE Technical Paper No. 912351 (1991).

435. Sugimoto, T. et al., "A new concept of direct injection gasoline engine—Part 5: development of a slit nozzle injector," JSAE Technical Paper (in Japanese), No. 20005191 (2000).

436. Sugiyama, M. et al., "Oil dilution reduction study with direct injection S.I. engine," *Proceedings of JSAE Spring Convention* (in Japanese), No. 972, pp. 173–176 (1997).

437. Suzuki, T. et al., "Combustion monitoring by use of the spark plug for DI engine," SAE Technical Paper No. 2001-01-0994 (2001).

438. Takagi, Y., "The role of mixture formation in improving fuel economy and reducing emissions of automotive S.I. engines," FISITA Technical Paper No. P0109 (1996a).

439. Takagi, Y., "Combustion characteristics and re search topics of in-cylinder direct-injection gasoline engines," *Proceedings of the 73rd JSME Annual Meeting* (V) (in Japanese), No. WS 11-(3), pp. 317–318 (1996b).

440. Takagi, Y. et al., "Simultaneous attainment of low fuel consumption, high power output and low exhaust emissions in direct injection S.I. engines," SAE Technical Paper No. 980149 (1998).

441. Takagi, Y., "Simultaneous attainment of improved fuel consumption, output power and emissions with direct injection SI engines," *Japan Automobile Technology* (in Japanese), Vol. 52, No.1, No. 9830460, pp. 64–71 (1998a).

442. Takagi, Y., "Challenges to overcome limitations in S.I. engines by featuring high pressure direct injection," *Proceedings of the 2nd International Workshop on Advanced Spray Combustion*, Nov. 24–26, 1998, Hiroshima, Japan, Paper No. IWASC9819, pp. 214–224 (1998b).

443. Takeda, K. et al., "Slit nozzle injector for a new concept of direct injection SI gasoline engine," SAE Technical Paper No. 2000-01-1902 (2000).

444. Tatschl R. et al., "PDF modeling of stratified-charge SI engine combustion," SAE Technical Paper No. 981464 (1998a).

445. Tatsuta, H. et al., "Mixture formation and combustion performance in a new direct-injection SI V-6 engine," SAE Technical Paper No. 981435 (1998).

446. Tomoda, T. et al., "Development of direct injection gasoline engine—study of stratified mixture formation," SAE Technical Paper No. 970539 (1997).

447. Topfer, G. et al., "Optical investigation of knocking location on S.I.-engines with direct injection," SAE Technical Paper No. 2000-01-0252 (2000).

448. TOYOTA, "Direct-injection 4-stroke gasoline engine," *TOYOTA Press Information '96*, August, 1996 (1996).

449. TOYOTA, "Toyota's D-4 direct-injection gasoline engine," *Frankfurt Autoshow*, September 1997 (1997).

450. Treece J., "Mitsubishi develops cleaner-burning engine—gasoline direct-injection unit slated for U.S. in '99," *Automotive News*, February 23, 1998 (1998).

451. Ueda, K. et al., "Idling stop system coupled with quick start features of gasoline direct injection," SAE Technical Paper No. 2001-01-0545 (2001).

452. Ueda, S. et al., "Development of a new injector in gasoline direct injection systems," SAE Technical Paper No. 2000-01-1046 (2000).

453. Urlaub, A. et al., "High-speed, multifuel engine: L9204 FMV," SAE Technical Paper No. 740122 (1974).

454. VanDerWege, B. et al., "The effect of fuel volatility on sprays from high-pressure swirl injectors," *Paper for the 27th Symposium (International) on Combustion*, No. 2E08, University of Colorado at Boulder, August 2–7, 1998 (1998a).

455. VanDerWege, B. et al., "The effect of fuel volatility on sprays from high-pressure swirl injectors," *Proceedings of the 4th International Symposium COMODIA 98*, pp. 505–510 (1998b).

456. VanDerWege, B. et al., "Effects of fuel volatility and operating conditions on fuel sprays in DISI engines: (1) imaging investigation," SAE Technical Paper No. 2000-01-0535 (2000a).

457. VanDerWege, B. et al., "Effects of fuel volatility and operating conditions on fuel sprays in DISI engines: (2) PDPA investigation," SAE Technical Paper No. 2000-01-0535 (2000b).

458. VanDerWege, B., "The effects of fuel volatility and operating conditions on sprays from pressure-swirl fuel injectors," PhD Thesis, MIT (1999).

459. Van Nieuwstadt, M. et al., "Heat release regressions for GDI engines," SAE Technical Paper No. 2000-01-0956 (2000).

460. Varble, D. et al., "Design, modeling and development of a unique gasoline direct injection fuel system," *ISATA 98VR028*, Dusseldorf, Germany (1998a).

461. Varble, D. et al., "Development of a unique out wardly-opening direct injection gasoline injector for stratified-charge combustion," *Aachen Colloquium Automobile & Engine Technology*, October 5–7, 1998 (1998b).

462. Visnic, B., "Toyota moving cautiously with D-4 engine," *Ward's Engine and Vehicle Technology Update*, June 1, 1997 (1997).

463. Wagner, V. et al., "Fuel distribution and mixture formation inside a direct injection SI engine in vestigated by 2D Mie and LIEF techniques," SAE Technical Paper No.1999-01-3659 (1999).

464. Wang, J. et al., "Application of low-pressure air-assisted fuel injection system on two-stroke motorcycle," SAE Technical Paper No.911253 (1991).

465. Wang, W. et al., "Spray measurement and visualization of gasoline injection for 2-stroke engine," SAE Technical Paper No. 930496 (1993).

466. Warburton, A., "GDI head-to-head interview," *Automotive Engineer*, No. 3, pp. 18–19 (1998).

467. Warburton, A. et al., "GD-I: tools and techniques," *Engine Technology International*, pp. 39–42 (1999).

468. WARD, "Mitsubishi to go all-out with GDI engines," *Ward's Engine and Vehicle Technology Update*, August 19, 1997.

469. WARD, "EEGR valve may help reduce NOx," *Ward's Engine and Vehicle Technology Update*, February 15, 1998, p. 8 (1998).

470. Warnecke, W. et al., "Requirements for automotive fluids for gasoline direct injection engines," *International Wiener Motor Symposium*, Vol. 1, pp. 162–179 (2000).

471. Weimar, H. et al., "Optical investigations on a Mitsubishi GDI-engine in the driving mode," SAE Technical Paper No.1999-01-0504 (1999).

472. Wensing, M. et al., "Spray formation of high pressure gasoline injectors investigated by two-dimensional Mie and LIEF techniques," SAE Technical Paper No.1999-01-0498 (1999).

473. Whitaker, P. et al., "Comparison of top-entry and side-entry direct injection gasoline combustion systems," *International Conference "Combustion Engines and Hybrid Vehicles," IMechE*, London, April 28–30, 1998 (1998).

474. Wicker, R. et al., "SIDI fuel spray structure investigation using flow visualization and digital particle image velocimetry," SAE Technical Paper No. 1999-01-3535 (1999).

475. Wigley, G. et al., "Droplet velocity and size fields in the near nozzle region of a dual fluid gasoline direct injector," *Proceedings of the 18th ICLASS*, Pasadena, USA (2000).

476. Willand, J. et al., "The knocking syndrome—its cure and its potential," SAE Technical Paper No. 982483 (1998).

477. Williams, P. et al., "Effects of injection timing on the exhaust emissions of a centrally-injected four-valve direct-injection spark-ignition engine," SAE Technical Paper No. 982700 (1998).

478. Winkler, K. et al., "The development of an emission aftertreatment system for gasoline direct injection passenger cars," 1998 FISITA Technical Paper No. F98T218 (1998).

479. Wirth, M. et al., "Actual trends and future strategies for gasoline direct injection," *IMechE. Seminar of Lean Burn Combustion Engines*, S433, December 3–4, 1996 (1996).

480. Wirth, M. et al., "Direct gasoline injection engine concepts for future emission regulations," *Proceedings of GPC '98, Advanced Engine Design & Performance*, pp. 103–112 (1998a).

481. Wirth, M. et al., "Gasoline DI engines: the complete system approach by interaction of advanced development tools," SAE Technical Paper No. 980492 (1998b).

482. Wirth, M. et al., "Gasoline DI engines in Europe: achievements and future concepts for fuel economy and emissions," *ARO/ERC Engine Symposium*, pp. 227–241 (1999).

483. Wirth, M. et al., "Turbocharging the DI gasoline engine," SAE Technical Paper No. 2000-01-0251 (2000).

484. Wojik, K. et al., "Engine and vehicle concepts for low consumption and low-emission passenger cars," FISITA Technical Paper No. P1302 (1996).

485. Wood, C., "Unthrottled open-chamber stratified charge engines," SAE Technical Paper No. 780341 (1978).

486. Worth, D. et al., "A new approach to meet future European emissions standards with Orbital direct injection gasoline engine," SAE Technical Paper No. 2000-01-2913 (2000).

487. Xu, M. et al., "CFD-aided development of spray for an outwardly opening direct injection gasoline engine," SAE Technical Paper No. 980493 (1998a).

488. Xu, M. et al., "Recent Advances in direct injection gasoline injector technology and fuel preparation strategy," *Proceedings of the 2nd International Workshop on Advanced Spray Combustion*, Nov. 24–26, 1998, Hiroshima, Japan, Paper No. IWASC9818, pp. 201–213 (1998b).

489. Yamada, T., "Trends of S.I. engine technologies in Japan," FISITA-96 Technical Paper No. P0204 (1996).

490. Yamaguchi, J., "Mitsubishi DI gasoline engine prototype," *Automotive Engineering*, pp. 25–29, September 1995 (1995).

491. Yamaguchi, J., "Mitsubishi Galant sedan and Legnum wagon," *Automotive Engineering*, pp. 26–29, November, 1996 (1996a).

492. Yamaguchi, J., "Toyota readies direct-injection gasoline engine for production," *Automotive Engineering*, pp. 74–76, November, 1996 (1996b).

493. Yamaguchi, J., "Direct-injection gasoline engine for Toyota," *Automotive Engineering*, pp. 29–31, No. 5 (1997a).

494. Yamaguchi, J., "Honda integrated motor assist to attain the world's top fuel efficiency," *Automotive Engineering*, No. 12, pp. 49–50 (1997b).

495. Yamaguchi, J., "Mitsubishi extends gasoline direct-injection to V6," *Automotive Engineering*, August 1997, pp. 77–81 (1997c).

496. Yamaguchi, J., "Direct-injection gasoline engine for Toyota," *Automotive Engineering*, May 1997, pp. 29–31 (1997d).

497. Yamaguchi, J., "Mitsubishi extends gasoline direct-injection to V6," *Automotive Engineering*, No. 8, pp. 77–81 (1997e).

498. Yamaguchi, J., "Nissan direct-injection gasoline V6," *Automotive Engineering*, No. 1, pp. 91–93 (1998a).

499. Yamaguchi, J., "Mitsubishi prototype GDI V8," *Automotive Engineering*, No. 1, pp. 93–94 (1998b).

500. Yamaguchi, J., "Opa—Toyota's new age vehicle," *Automotive Engineering*, No. 9, pp. 23–34 (2000).

501. Yamamoto, S. et al., "Mixing control and combustion in gasoline direct injection engines for reducing cold-start emissions," SAE Technical Paper No. 2001-01-0550 (2001).

502. Yamamoto, S. et al., "Analysis of the characteristics of the spray for GDI engine," *Proceedings of JSAE Spring Convention* (in Japanese), No. 971, pp. 329–332 (1997).

503. Yamashita, H. et al., "Mixture formation of direct gasoline injection engine—in cylinder gas sampling using fast response ionization detector," *Proceedings of JASE Fall Convention* (in Japanese), No. 976, pp. 5–8 (1997).

504. Yamauchi, T. et al., "Computation of the hollow-cone sprays from high-pressure swirl injector from a gasoline direct-injection SI engine," SAE Technical Paper No. 962016 (1996).

505. Yamauchi, T. et al., "Numerical analysis of stratified mixture formation in direct injection gasoline engines," *Proceedings of Direkteinspritzung im Ottomotor*, pp. 166–185 (1998).

506. Yang, J. et al., "Use of split fuel injection to increase full-load torque output of a direct-injection SI engine," SAE Technical Paper No. 980495 (1998a).

507. Yang, J. et al., "Simulation of the effect of wakes behind fuel droplets on fuel vapor diffusion in direct-injection SI engines," *Proceedings of the 4th International Symposium COMODIA 98*, pp. 323–328 (1998b).

508. Yang, J. et al., "Study of a stratified-charge DISI engine with an air-forced fuel injection system," SAE Technical Paper No. 2000-01-2901 (2000).

509. Yasuoka, M. et al., "A study of a torque control algorithm for direct injection gasoline engine," *Proceedings of the 14th Japan Internal Combustion Engine Symposium* (in Japanese), pp. 121–125 (1997).

510. Yoo, J. et al., "Visualization of direct-injection gasoline spray structure inside a motoring engine," *Proceedings of ILASS-America '98*, pp. 101–105 (1998a).

511. Yoo, J. et al., "Visualization of direct-injection gasoline spray and wall-impingement inside a motoring engine," SAE Technical Paper No. 982702 (1998b).

512. Zeng, Y. et al., "Modeling of spray vaporization and air-fuel mixing in gasoline direct-injection engines," SAE Technical Paper No. 2000-01-0537 (2000).

513. Zhang, H. et al., "Integration of the smart NOx-sensor in the exhaust line of a gasoline high pressure direct injection system," *International Wiener Motor Symposium*, Vol. 2, pp. 288–310 (2000).

514. Zhao, F. et al., "Quantitative imaging of the fuel concentration in a SI engine with laser Rayleigh scattering," SAE Technical Paper No.932641 (1993).

515. Zhao, F. et al., "PLIF measurements of the cyclic variation of mixture concentration in a SI engine," SAE Technical Paper No. 940988 (1994).

516. Zhao, F. et al., "The spray characteristics of automotive port fuel injection—a critical review," SAE Technical Paper No. 950506 (1995a).

517. Zhao, F. et al., "The spray characteristics of dual-stream port fuel injectors for applications to 4-valve gasoline engines," SAE Technical Paper No. 952487 (1995b).

518. Zhao, F. et al., "Spray characteristics of direct-injection gasoline engines," *Proceedings of ILASS-America*, pp. 150–154 (1996a).

519. Zhao, F. et al., "Spray dynamics of high pressure fuel injectors for DI gasoline engines," SAE Technical Paper No. 961925 (1996b).

520. Zhao, F. et al., "The spray structure of air-shrouded dual-stream port fuel injectors with different air mixing mechanisms," *Proceedings of the 1996 Spring Technical Conference of the ASME Internal Combustion Engine Division*, ICE-Vol. 26-2, pp. 21–29 (1996c).

521. Zhao, F. et al., "A review of mixture preparation and combustion control strategies for spark-ignited direct-injection gasoline engines," SAE Technical Paper No. 970627 (1997a).

522. Zhao, F. et al., "In-cylinder spray/wall interactions of a gasoline direct-injection engine," *Proceedings of ILASS-America* (1997b).

523. Zhao, F. et al., "Characteristics of gasoline direct-injection sprays," *Proceedings of ILASS-America* (1997c).

524. Zhao, F. et al., "Characterization of direct-injection gasoline sprays under different ambient and fuel injection conditions," *Proceedings of ICLASS-97* (1997d).

525. Zhao, F. et al., "Injector deposit issues with gasoline direct injection engines," *Proceedings of the 2001 ILASS-Americas Annual Conference* (2001).

Index

Abbreviations are used to indicate figures (*f*) and tables (*t*).